AF352939

Digital Signal, Image and Video Processing for Emerging Multimedia Technology

Digital Signal, Image and Video Processing for Emerging Multimedia Technology

Editor

Byung-Gyu Kim

MDPI • Basel • Beijing • Wuhan • Barcelona • Belgrade • Manchester • Tokyo • Cluj • Tianjin

Editor
Byung-Gyu Kim
Sookmyung Women's University
Korea

Editorial Office
MDPI
St. Alban-Anlage 66
4052 Basel, Switzerland

This is a reprint of articles from the Special Issue published online in the open access journal *Electronics* (ISSN 2079-9292) (available at: https://www.mdpi.com/journal/electronics/special_issues/dsem).

For citation purposes, cite each article independently as indicated on the article page online and as indicated below:

LastName, A.A.; LastName, B.B.; LastName, C.C. Article Title. *Journal Name* **Year**, *Volume Number*, Page Range.

ISBN 978-3-03943-857-0 (Hbk)
ISBN 978-3-03943-858-7 (PDF)

Contents

About the Editor

Byung-Gyu Kim has received his BS degree from Pusan National University, Korea, in 1996 and an MS degree from Korea Advanced Institute of Science and Technology (KAIST) in 1998. In 2004, he received his PhD degree from the Department of Electrical Engineering and Computer Science of Korea Advanced Institute of Science and Technology (KAIST). In March 2004, he joined the real-time multimedia research team at the Electronics and Telecommunications Research Institute (ETRI), Korea, where he was a senior researcher. At ETRI, he developed many real-time video signal processing algorithms and patents and received the Best Paper Award in 2007. From February 2009 to February 2016, he was an associate professor in the Division of Computer Science and Engineering at SunMoon University, Korea. In March 2016, he joined the Department of Information Technology (IT) Engineering at Sookmyung Women's University, Korea, where he is currently a professor. Since 2007, he has served as an associate editor of *Circuits, Systems and Signal Processing* (Springer); *The Journal of Supercomputing* (Springer); *The Journal of Real-Time Image Processing* (Springer); and *International Journal of Image Processing and Visual Communication* (IJIPVC). Since March 2018, he has served as the Editor-in-Chief of The *Journal of Multimedia Information System* and associate editor of IEEE Access. In 2019, he was appointed as associate editor and topic editor of *Heliyon Computer Science* (Elsevier), *Journal of Image Science and Technology* (IS &T), *Electronics* (MDPI), *Applied Sciences* (MDPI) and *Sensors* (MDPI). He also has served on the organizing committee of CSIP 2011, as a co-organizer of CICCAT2016/2017 and as a program committee member of many international conferences. He has received the Special Merit Award for Outstanding Paper from the IEEE Consumer Electronics Society at IEEE ICCE 2012, the Certification Appreciation Award from the SPIE Optical Engineering in 2013 and the Best Academic Award from the CIS in 2014. He has been honored as an IEEE senior member in 2015. He is serving as a professional reviewer for many academic journals, including those of IEEE, ACM, Elsevier, Springer, Oxford, SPIE, IET and MDPI. He has published over 250 international journal and conference papers and patents in his field. His research interests include image and video signal processing for content-based image coding, video coding techniques, 3D video signal processing, deep/reinforcement learning algorithms, embedded multimedia systems and intelligent information systems for image signal processing. He is a senior member of IEEE and a professional member of ACM and IEICE.

Preface to "Digital Signal, Image and Video Processing for Emerging Multimedia Technology"

Recent developments in image/video-based deep learning technology have enabled new services in the field of multimedia and recognition technology. The technologies underlying the development of emerging services are based on essential signal and image processing algorithms. In addition, the recent realistic media services, mixed reality, augmented reality and virtual reality media services also require very high definition media creation, personalization and transmission technologies, and this demand continues to grow. To accommodate these needs, international standardization organizations and industries are studying various digital signal and image processing technologies to provide a variety of new or future media services. In this issue, we present a collection of quality papers concerning advanced signal/image processing and video data processing, including deep learning approaches. This book comprises 23 peer-reviewed articles covering a review of the development of deep-learning-based approaches and presenting original research on learning mechanisms and video signal processing. This book also covers topics including data security and protection and advanced digital signal/image processing. This volume will be a good collection for designers and engineers in both academia and industry that would like to develop an understanding of emerging technology in digital signal, image and video processing, and students will also find this book to be a useful reference.

Byung-Gyu Kim
Editor

Article

Multiple Feature Dependency Detection for Deep Learning Technology—Smart Pet Surveillance System Implementation

Ming-Fong Tsai [1,*], Pei-Ching Lin [1], Zi-Hao Huang [1] and Cheng-Hsun Lin [2]

[1]   Department of Electronic Engineering, National United University, Miaoli 36063, Taiwan; a851103123@gmail.com (P.-C.L.); main668888@gmail.com (Z.-H.H.)
[2]   Electronic and Optoelectronic System Research Laboratories, Industrial Technology Research Institute, Hsinchu 31040, Taiwan; akiolin@itri.org.tw
*   Correspondence: mingfongtsai@gmail.com

Received: 29 June 2020; Accepted: 25 August 2020; Published: 27 August 2020

Abstract: Image identification, machine learning and deep learning technologies have been applied in various fields. However, the application of image identification currently focuses on object detection and identification in order to determine a single momentary picture. This paper not only proposes multiple feature dependency detection to identify key parts of pets (mouth and tail) but also combines the meaning of the pet's bark (growl and cry) to identify the pet's mood and state. Therefore, it is necessary to consider changes of pet hair and ages. To this end, we add an automatic optimization identification module subsystem to respond to changes of pet hair and ages in real time. After successfully identifying images of featured parts each time, our system captures images of the identified featured parts and stores them as effective samples for subsequent training and improving the identification ability of the system. When the identification result is transmitted to the owner each time, the owner can get the current mood and state of the pet in real time. According to the experimental results, our system can use a faster R-CNN model to improve 27.47%, 68.17% and 26.23% accuracy of traditional image identification in the mood of happy, angry and sad respectively.

Keywords: multiple feature; dependency detection; deep learning; surveillance system

---

1. Introduction

In modern society, the population of pets such as cats and dogs is increasing. However, when owners are at work, pets at home will inevitably be alone, and owners might be worried about the safety of pets. Hence, this paper proposes a smart pet surveillance system to automatically identify the pet's mood and state and initiatively send identification results to the owner. In this way, even if owners are busy at work, the pet status will be sent through the smart pet surveillance system. It can quickly grasp the current situation of pets so that owners can work with peace of mind. However, traditional image identification cannot effectively identify the pet's mood and state from a single image or instantaneous state. The pet displays its mood and states through actions of certain barks or continuous movements of several different key parts on its body. Hence, the multiple feature dependency detection algorithm proposed in this paper can be used on most object detection models. The multiple feature dependency detection algorithm can correctly identify the mood and state of the pet when the object detection has sufficient accuracy. Tensorflow architecture of deep learning image identification technology and its Faster-RCNN network architecture extracts a conv-feature map of input images through convolutional layers [1–6]. Then the region propose network (RPN) will process the extracted convolution feature maps and provide a large number of ROIs (region of interest means regions that may contain feature points). It lets ROIhead (responsible for processing the ROIs proposed by RPN) determine whether there is a feature target in ROIs and correct the position and coordinates of ROIs. Finally, it records specific labels of features (the set featured part). The most important of which is ROIhead, which is responsible for determining whether ROIs contains feature targets and modifying the coordinates and position of ROIs directly affects the accuracy of identification [7–15]. In view of the influence of network architecture design, it is important to improve the identification ability of the identification model, directly change the network architecture and improve the number and quality of input training samples. An effective training sample enables RPN to provide better ROIs so that ROIhead can more accurately detect the target featured parts to be identified to improve the level of identification confidence. The identification system of the smart pet surveillance system proposed in this paper uses deep learning image identification technology, and its pretrained model uses the Faster-RCNN neural network using the COCO (Common Object in Context) data set. In order to identify multi-point of pet features, the proposed system collected training samples of multiple relevant features and trained the identification model. By identifying multiple featured parts of pets and analyzing the continuous changes and relative relationships between featured parts, pet's mood and state are analyzed and determined.

Faster R-CNN can effectively identify more subtle features and then identify the key parts of pets, but it cannot identify the meaning behind images and objects, such as mood and status. Even if the training data of the faster R-CNN can directly label the mood and state, the mood and state cannot be successfully identified, because it cannot be identified only through a single image. Hence, the proposed method in this paper is based on faster R-CNN and extracts the identification results of the key parts of the pet to judge the pet's mood and state. Moreover, the multiple features of the proposed method can be the key parts of the pet itself in the image or the sounds made by the pet. The proposed method uses the KNN (K nearest neighbor) algorithm to establish the speech identification model. The characteristic frequency bands of sound waves emitted by pets in different moods and states are needed to extract and identify and eliminate the noises. The proposed method uses MFCC (Mel-scale frequency cepstral coefficients) to suppress noise in sound waves and extract the sound wave characteristic frequency bands of pets in different moods and states. Since the identification objects are pets of different breeds, colors and sizes, the relative position of the featured part will be different. Even if the breed and color are the same, uncertain factors such as age and hair size will affect the identification ability of the system. In order to enable the identification system to respond to the above-mentioned factors, the system uses the identification system to successfully identify the image of featured parts and recreate a new training sample based on the identification results. By inputting new and customized training samples into the identification system for training, the identification model's ability to identify featured parts of specific pets can be improved. The related literature will be discussed in Section 2, and the system architecture will be explained in Section 3, which will be divided into four small chapters for detailed explanation. In Section 4, we analyze the data recorded by the system during the actual identification process. In Section 5, we present the conclusions of this paper.

2. Literature Review

In recent years, there have been many related literatures and applications for image identification, but most of them focused on face identification. Literatures related to animal identification are relatively rare. Related literature [16] focused on wild animal protection, using the image identification system to identify endangered wild animals and protect them. This paper uses convolutional neural networks to train images of endangered wild animals and common animals. The identification system can effectively identify endangered wild animals. However, this paper only identifies the appearance of animals. Limited by its system and training samples, the system can only identify specific creatures, not endangered animals. However, although the system proposed in our paper can only identify common pet species at home, the system can not only identify their appearance but also judge the pet's mood and state by identifying their subtle behaviors. Related literature [17] identified a monkey face by the identification system trained by a convolutional neural network. Monkeys can communicate through facial expression. It plays an important role in their communication. This paper uses the image identification system to capture the movement of the monkey's facial muscles and its facial expression to determine what the monkey wants to express. This paper uses the movement state of the monkey's facial muscles and changes in the five senses to determine what the monkey wants to express. In the exchange of information, although the face sends a lot of messages, gestures of other body parts can also convey more specific or obvious meaning. The training sample of this paper is sufficient, but it only identifies monkeys' faces. If it includes other physical behavior performance, it should be better at identifying what the monkey wants to express. The system proposed in our paper not only identifies faces but also identifies continuous behaviors of other featured parts including information limb movement can convey. Related literature [18] used transfer learning to design a method for dog identification from shallow to deep. In this paper, different features of dogs such as eyes, noses and ears are identified. Different breeds of dogs are included for identification. Related literature [19] proposes using surveillance systems to monitor the location of pets based on the faster R-CNN identification model. However, our proposed system will identify the mood and state of the pet. Related literature [20] observes the difference in horse behavior before and after surgery, inferring different pain levels based on the condition of the surgical wound and observes the horse's behavioral performance at different levels of pain. This paper only observes the horse's displacement in space. It is concluded that the horse will reduce the displacement when it is in pain. However, our proposed method identifies the dependence of the continuous behavior of different key parts of the pet, instead of identifying the entire pet. Moreover, the system proposed in our paper needs to identify common home pets, using transfer learning to improve the accuracy of identification. It not only identifies the pet's breed, body shape and hair condition to confirm the owner's pet but also the pet's behaviors of featured parts to ensure the pet's mood and state can be correctly identified. Due to individual preferences, the selection of pets is different. In training data collection of the identification system in this paper, the open data set is used and the public pet videos are divided into screenshots and other image data. Therefore, as long as the information of the input training data can be clearly defined, the identification function of the identification system can be improved.

3. Materials and Methods

The smart pet surveillance system overview is shown in Figure 1. In order to capture pet videos, the system uses the loop recording subsystem and webcam to capture real-time videos of pets at any time. The loop recording subsystem will continuously capture images and output them as short videos. The recording is not interrupted and the latest short video can be sent to the data processing subsystem. To use image identification and the KNN audio identification model, we need to perform preliminary processing to extract an image and audio signal from the short video. The identification network architecture is shown in Figure 2. This paper was based on the faster R-CNN identification network and customized the key parts of the pet to be identified. Moreover, we used the KNN audio identification model for audio identification. The system automatically processed the images and audio for faster R-CNN and KNN audio identification model to identify the key parts of the pet and the pet's barking. The feature extraction subsystem was used to extract information about the key parts of the pets in these images and the mood contained in the audio. It also generated the training samples for automatic optimization module for supplements when performing automatic training according to the results of the identification of the faster R-CNN identification network. The multiple feature dependency detection algorithm extracted the information from the list of the feature of the pet generated by the feature extraction subsystem. According to the information of these features, we determined the pet's mood and state. Sending the identified pet's message to the owner through the message transmission subsystem so that the owner can understand the pet's latest mood and status. In order to synchronize the identification ability of the identification system with the change of pet's appearance, this system proposed an automatic optimization identification module subsystem that uses the latest system identification results as samples and sends them to the module for training. It effectively synchronizes the ability of identification model with the changes in the pet's appearance. Functions of the smart pet surveillance system are shown in Figure 3. The proposed system includes an image identification system, which is responsible for identifying the key parts of the pet and the sound of the pet. Multiple feature dependency detection extracts the identified information to determine the pet's mood and status and transmits the information to the owner. Automatic system optimization responds to changes in the pet's appearance.

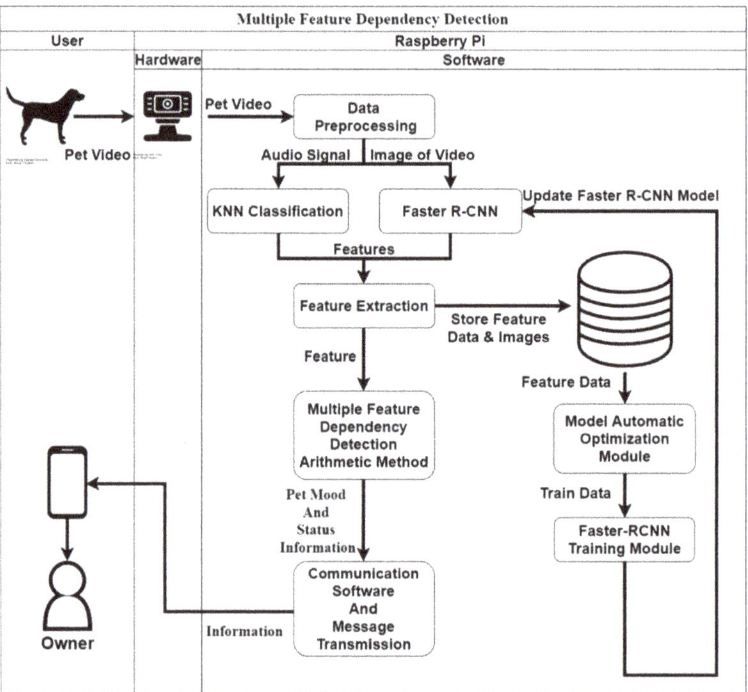

Figure 1. Smart pet surveillance system overview.

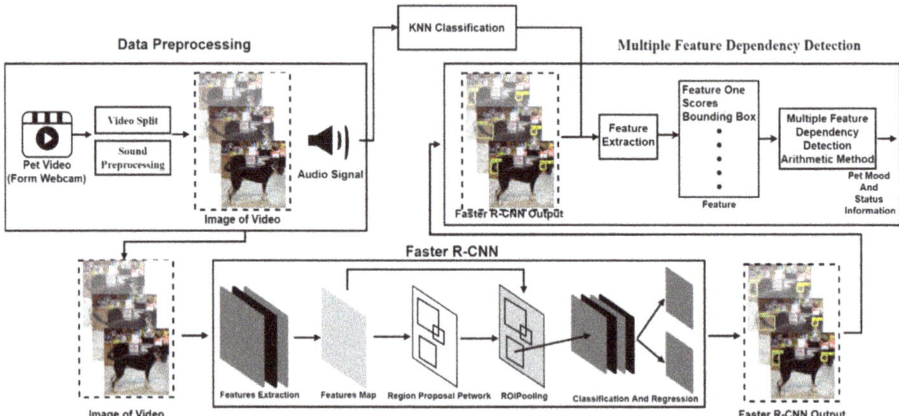

Figure 2. Identification network architecture.

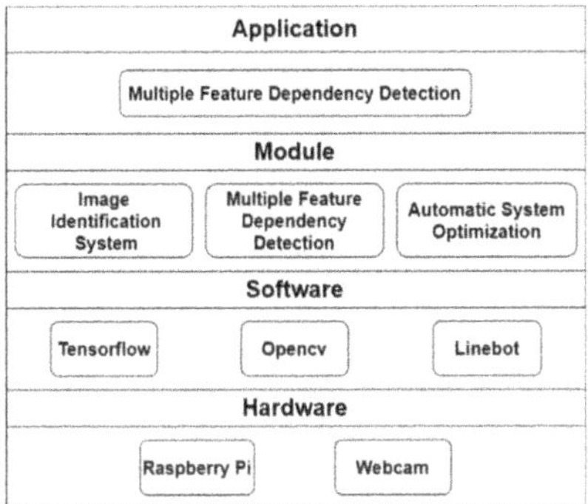

Figure 3. Functions of the smart pet surveillance system.

3.1. Pet Camera

The smart pet surveillance system proposes to continuously capture real-time images through pet cameras. Since the pet camera and the identification system were implemented in the same operation system, when videos captured by the pet camera were saved in the storage space designated by the identification system, the identification system could use these videos. For convenience of the subsequent image identification system, captured images were saved in short videos when recorded by pet cameras. Therefore, the recorded images need to be continuously stored as shown in Algorithm 1. The loop recording processing recorded the video according to the video format and stored the video, which had a duration setting by time in the video path location.

Algorithm 1 Loop_Recording_Processing

Require: threading
Require: video_store_path *video_path*
Require: count variable *time_sup* = 0
1. **def** timer():
2.   *time_sup* += 1
3. **main**():
4.   **threading**.timer()
5.   set video format and information: *v_format*
6.   **while**(1):
7.   set video duration: *time*
8.   **if** *time_sup* **==** *time*:
9.   store the video to *video_path*
10.   *time_sup* = 0
11.   **else:**
12.   recording the video according to *v_format*
13.   **return 0**

3.2. Identification System

As shown in Algorithm 2, the identification subsystem of the smart pet surveillance system was implemented by object identification. By detecting and identifying the key part of the pet, it is able to identify the different pet's mood and state by changing different key parts of the pet. When the identification subsystem is activated, it will automatically read the short videos recorded and stored by the pet camera. Owing to the identification method that uses faster R-CNN and the KNN audio identification model, the video needs to extract image and audio files, then these are sent to the respectively identification system for identification. After the video is divided into images, it will have different results according to different sampling parameters. When the sampling parameters are set too low, time interval between two images will be too long, resulting in few identifiable images to effectively identify the mood and state of pets. When the sampling parameters are set too high, there will be too many segmented images, resulting in excessive load of the system and greatly increasing identification time. Therefore, in this paper, videos were cut to five to six images per second for identification. After the data processing subsystem red the video, it separated and stored the image and feature of the audio according to the sampling parameters set by the system. The system processed through MFCC and extracted the feature parts in the audio. When the convert process ended, the system threw the images extracted from the video and feature of the audio into the faster R-CNN identification system and KNN audio identification system for identification.

Algorithm 2 Data_Preprocessing

Require: Loop_Recording_Processing_output *video_path*
Require: image_save_path *i_path*
Require: audio save_path a_*path*
1.   **def** video_to_img(*video_path, i_path*):
2.   images obtained from the video: *i_image*
3.   i_image store to i_path
4.   **def** video_to_wav(*video_path, v_path*):
5.   *audio* = audio from the video
6.   *mfcc_audio* = audio use mfcc for feature extraction
7.   mfcc_*audio* store to a_*path*
8.   **main**():
9.   video_to_img(*video_path, i_path*)
10.   video_to_wav(*video_path, a_path*)
11.   **return** 0

As shown in Algorithm 3, the feature extraction subsystem will extract all the key parts of pets from the identification results of faster R-CNN information in the feature list. Additionally, it will then extract the sounds identification result from the KNN audio identification model. The identification results of the faster R-CNN and KNN audio identification system were used to generate feature lists with feature categories for subsequent processing. Identifying the mood and state condition of pets requires observation of the continuous changes of their characteristics. This is a natural phenomenon of time continuity and object dependence, so it is impossible to determine the pet's mood and state via a single identification result. It is necessary to continuously identify that the featured parts meet specific conditions to determine that the pet conforms to a specific mood or state. It is also necessary to filter information in identification results and exclude images in specific situation where no featured parts are detected to improve the identification effectiveness of the system. The multiple feature dependency detection algorithm is shown in Algorithm 4. The multiple feature dependency detection algorithm contains a number of emotions, which was identified through continuous image and sound features. The multiple feature dependency detection algorithm obtains a feature list via feature extraction and extracts features needed to identify emotions from the feature list according to different emotion requirements. The proposed algorithm determines whether pets are in a certain emotion state according to the nature of the continuity of emotions. When there are insufficient features to make judgments, the proposed algorithm defines pets as in the normal state. For example, after the identification system receives images, it automatically identifies whether there are specified featured parts in the image, such as the dog's mouth keeps opening and tail shaking. After capturing multiple specific features, the system will record the above information. Then, it judges if the above-mentioned image is a continuous one. Therefore, discontinuous images without time continuity cannot represent the mood and state of pets. The proposed system that obtains the above-required information can identify the pet's mood and state.

Algorithm 3 Feature_Extraction

Require: Faster R-CNN output list *faster_output_list*
Require: KNN audio model output *knn_output*
Require: path of all unprocessed training data *data_path*
1. a dict of feature information: *feature_dict*
2. **for** *feature_info* **in** *faster_output_list*:
3.   *feature_dict* append feature information from *feature_info*
4. *feature_dict* append *knn_output*
5.   store feature data & image to *data_path*
6. **return** *feature_dict*

Algorithm 4 Multiple_Feature_Dependency_Detection_Arithmetic_Method

---

Require: Feature_Extraction output *feature_dict*
1. **def** mood(*feature_dict*):
2.   from *feature_dict* get information to judge pet mood: *data*
3.   **if** *data* is enough to judge mood:
4.   **return** "mood"
5.   **else**:
6.   **return** "normal"
7. **main**():
8.   list of mood: *mood_list*
9.   list of pet status: *pet_status*
10.   **for** mood **in** *mood_list*:
11.   *pet_status* = mood(*feature_dict*)
12.   **return** *pet_status*

3.3. Communication Software and Message Transmission

The system uses the multiple feature dependency detection subsystem to transform images and audio into information, which includes the mood and state of pets. It uses communication software with a high penetration rate allowing information identified by the system to be directly transmitted to the application on the owner's smart phone. As shown in Algorithm 5, after setting the information required by the communication software, this subsystem will set different messages according to different emotions identified in the multiple feature dependency detection algorithm and send to users' communication software so that users obtain the latest information about their pets.

Algorithm 5 Message Transmission

Require: pet_status *status*
1.   owner_id: *id*
2.   dict of message: *dict_msg*
3.   **for** msg_mood **in** dict_msg.keys():
4.   **if** msg_mood == *status:*
5.   send pet's status message to the *id*

3.4. Automatic System Optimization

The smart pet surveillance system proposes to identify the pet's mood and state by identifying the dependencies of multiple featured parts of pets. The above-mentioned specific featured parts will vary with age of the pet and seasons. The model used by the identification subsystem must be updated with time. Multi-featured parts dependency identification model in this subsystem needs to be continuously optimized. The identification model synchronizes with the changes of pets so that the system can get the most accurate identification results. Since the image identification system uses deep learning architecture, sufficient training samples are required in order to send data to the identification model for training. The acquisition, labeling and training of the above-mentioned samples are functions of this subsystem. Core functions are shown in Algorithm 6. This subsystem uses images used in previous identification and identification results to extract featured parts of original images. In addition to the image itself, the training samples also need to label the featured parts contained in this image. Figure 4 shows an image that has been identified, this picture contains two features that the system needs to determine the pet's mood and state, namely the opened mouth and tail, and thus, it can be extracted based on the above results to form a new training sample. The system will mark it with the relevant information obtained during identification, the image and its contained information are completely saved, and the subsequent generated identification model is more complete.

Algorithm 6 Model_Automatic_Optimization_Module

Require: path of all unprocessed training data from Feature_Extraction: *data_path*
Require: path of all processed training data *t_path*
1. **def Retraining_Data_Generation_Module**(*data_path*):
2.   get data and images stored for training: *data_list*
3.   **for** data **in** *data_list***:**
4.   *image* = cut the original image according to x, y
5.   labeling *image* base on *x, y* coordinates and store to *t_path*
6. **main**():
7.   Retraining_Data_Generation_Module(*data_path*)
8.   Faster R-CNN Training Module(*t_path*)

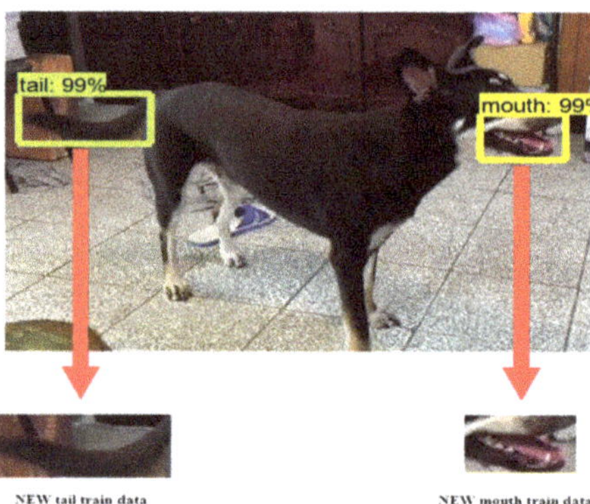

Figure 4. Retraining data samples.

4. Results and Discussion

4.1. System Execution Result

In order to evaluate the practicability and reliability of the proposed system, this paper used an actual pet video and invited testers to verify whether the model correctly identified the mood and state of pets. The environment of the proposed system used Tensorflow-gpu-1.14.0, and the pretraining model used the Faster-RCNN-Inception-V2. The system implementation is shown in Figure 5. The proposed system used training samples to retrain the model for capturing specific featured parts. The training samples included the Stanford Dogs Dataset [21], which contains at least

40 breeds and more than 500 dogs. As shown in Figure 6, there were various pets in training samples to enhance identification ability of faster R-CNN, ssd mobilenet and yolo identification models and avoid over-fitting. The total training images were 2761 and training step was 50,000. In addition to the established Tensorflow-gpu-1.14.0 environment, Raspberry Pi was used by the pet camera with the main system, all subsystems and related modules as shown in Figure 7. As shown in Algorithm 7, the main system will automatically execute the related subsystems for image recording, file conversion, image identification, mood and status determination and other related subsystems.

Algorithm 7 main

Require: threading
Require: Loop_Recording_Processing
Require: Data_Preprocessing
Require: Faster R-CNN
Require: KNN audio model
Require: Feature_Extraction
Require: Multiple_Feature_Dependency_Detection_Arithmetic_Method
Require: Message Transmission
Require: Model_Automatic_Optimization_Module
1. **def** recording:
2.   **threading**.Loop_Recording_Processing()
3. **def** Model_Automatic_Optimization_Module:
4.   **threading**.Model_Automatic_Optimization_Module()
5. **main**():
6.   recording()
7.   **while**(1):
8.   Data_Preprocessing()
9.   *feature_dict* = Feature_Extraction(Faster R-CNN(), KNN audio model())
10.   *pet_status* = Multiple_Feature_Dependency_Detection_Arithmetic_Method(*feature_dict*)
11.   Message Transmission(*pet_status*)

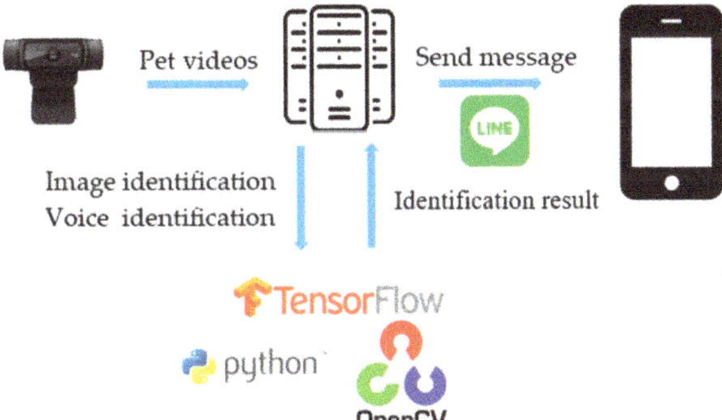

Figure 5. System implementation.

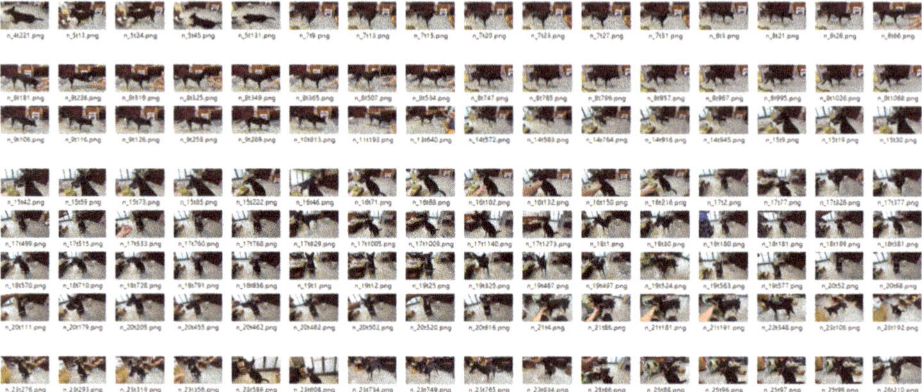

Figure 6. Training samples.

Figure 7. Pet camera.

A loop recording subsystem was set in the server to control the webcam and the Tensorflow environment was used for identification and training. The server also stored all videos captured by the webcam. The training sample file generated by the system program was used for subsequent automatic optimization for the identification model. The system continuously captured pet images through the front-end webcam and then saved the image to the path specified by the identification environment. After the image of the pet was segmented and processed, an image to be identified was obtained. The above-mentioned image was identified and the identified featured parts on the image were recorded. In order to verify the effectiveness of mood and state identification of the smart pet surveillance system, we selected to judge happy and angry mood of dogs. In order to reduce the computational burden of the identification system, this paper set the confidence level as 90%. When the confidence level of the object detection was lower than 90%, subsequent algorithms would directly give up this identification result. After the images and audio images extracted from the short movie were identified by faster R-CNN and the features extracted using the feature extraction subsystem, the pet mood and status identified by the multiple feature dependency detection algorithm would send the message to the owner's smartphone through the message transmission subsystem. After the information was filtered and the state was identified, the information would be transmitted to users through the Internet as shown in Figure 8.

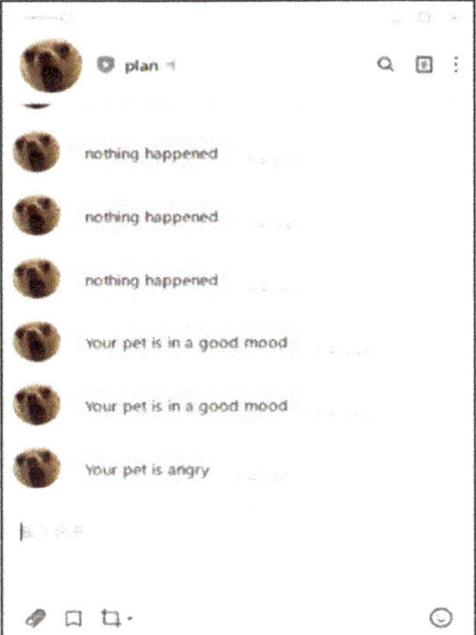

Figure 8. Communication software.

4.2. Types of Mood and Ways of Judging

The happy characteristics were a continuous mouth open and swinging tail. In order to confirm whether there was a dog in the picture, we needed to mark and identify the dog. To identify the dog's mood and state it as happy, the two marked and identified key parts of the dog were the dog's mouth and the dog's tail. The dog's mouth needed to open continuously and the tail needs to swing in a significant angle. The multiple feature dependency detection algorithm needed to extract the classes of the dog, dog's mouth and dog's tail from the feature list as the basis for determining happy emotions. The system first captures the position of the dog in the image and then identifies whether the dog's mouth is open. When the dog's mouth is open, the system will capture the coordinate value of the dog's tail. Then, we stored the image number and coordinate value to determine whether the number of the image in the list was continuous and the change in the coordinate value of the tail. The identify results of the multiple feature dependency detection algorithm are shown in Figure 9. The multiple feature dependency detection algorithm can effectively use the key parts of the pet to identify the pet's mood and state. The different barking changes made by pets can directly express different moods and states. When we cannot identify the mood and state of the pet through the image alone, it will temporarily determine that the pet's mood is normal, and then perform mood detection with image features and sound features. The mood can be identified as normal, angry and sad when images and sounds are used as features for identification. Since sound is added as a feature, the image will be identified whether there is a change in the state of the dog's mouth to determine that the barking is made by the dog. The multiple feature dependency detection algorithm will extract the dog and the dog's mouth open and close state from the feature list. The class identification based on the above three images was used to determine whether the barking sound was made by the dog in the screen. Moreover, the mood represented behind the barking identified using the KNN audio identification model must be extracted. When the dog barks, the dog's mouth will open and close alternately. As shown in Figure 10, the system first captured the position of the dog in the image, and then identified the state of the dog's mouth, and stored the state of the dog's mouth in the list according to the image number. When the state of the dog's mouth in the list continuously opens and closes, the algorithm will determine the mood and state of the pet based on the audio identification result. Since the pet displays its mood and states through actions of certain barks or continuous movements of several different key parts on its body, we designed the rules of multiple feature dependency detection for the pet's mood and states in Table 1.

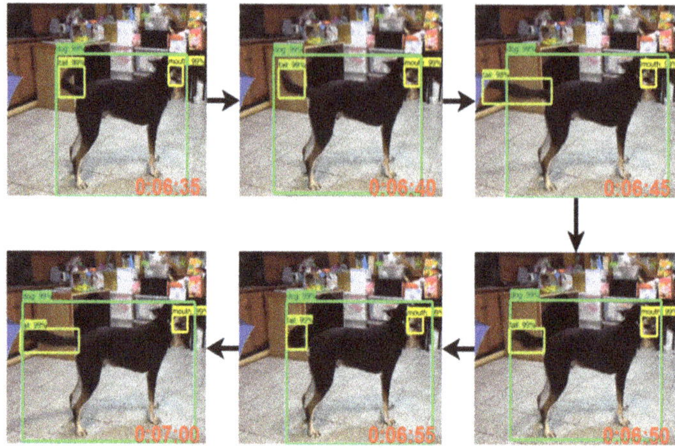

Figure 9. Identify mood and state with images.

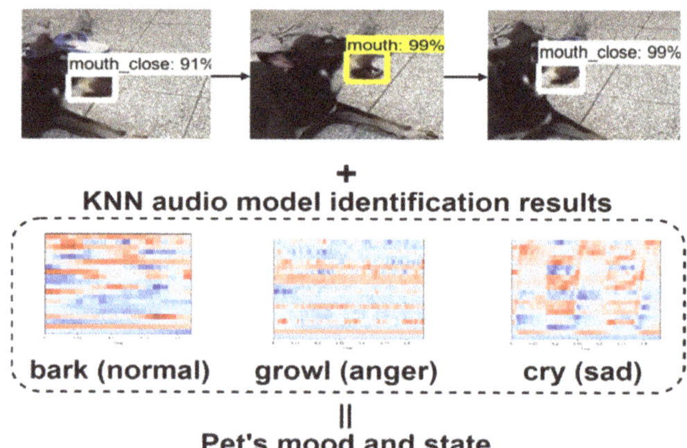

Figure 10. Continuous image combined with sound identification results.

Table 1. Rules of multiple feature dependency detection.

Mood \ Feature	Image	Voice
Happy	mouth keeps opening and tail swing continuously	Not needs
Angry	mouth keeps opening and closing	Growl
Sad	mouth keeps closing	Crying
Normal	other actions cannot be identified as any of the above moods	

4.3. The Accuracy of the Model for Identifying Features

As shown in Figure 11, we tested the identification accuracy of each model in identifying the key parts of the pet and combined the identification results of key parts with multiple feature dependency detection algorithm to determine the pet's mood and state results. Faster R-CNN had the higher identification accuracy for key parts of pets, under an average accuracy of 80.95%, than ssd mobilenet v2 at least 55% identification accuracy. Since the proposed method was based on the accuracy of identified key parts of the pet, the accuracy of faster R-CNN obtained 61.1% accuracy of happy category. The reason for low accuracy of the happy state was that one of two features could not be identified by the faster R-CNN in the continuous image. Theoretically, the identification accuracy of the pet's feature parts improving that multiple feature dependency detection algorithm will more accurately identify the pet's mood and state, such as yolo v3. Hence, the proposed method of this paper could choose fast R-CNN or yolo v3 to identify key parts of pets. This paper tested the identification accuracy of directly using emotions as labels. A faster R-CNN model could reach 33.63% accuracy in identifying the emotions, but the yolo v3 model could reach 55.75% accuracy in identifying the emotions.

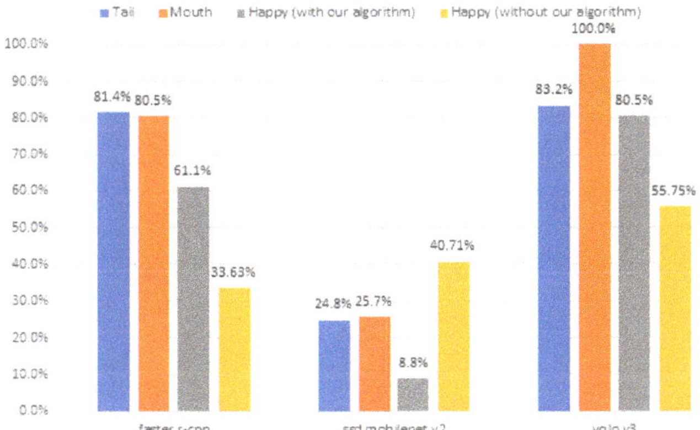

Figure 11. Identification of the accuracy for determining the happy state.

This paper tested the accuracy of the KNN audio identification model where the K value (k nearest neighbors) as 3 was used to identify the mood represented by the dog's voice. Moreover, we tested the impact of the different number of features extracted from each training data on the accuracy of the KNN audio identification model. There were 111 audio data. The time of each audio data was one second, including 37 barking (representing normal), 37 growling (representing anger) and 37 crying (representing sadness) [22]. The number of features extracted from each audio data for the KNN audio identification model training was 20, 30 and 40, respectively. We used the leave-one-out cross validation to test the accuracy of the KNN audio model. As shown in Table 2, we tested the impact of MFCC preprocessing on the accuracy of the KNN audio identification model and the accuracy of each audio training data extracting different amounts of data as features. The average accuracy was 60.06% when the audio training data was not preprocessed with MFCC. The average accuracy of the KNN audio identification model generated by training data was 80.48% when the audio training samples using MFCC for preprocessing were used. In Figure 12, we tested not only the identification accuracy of each model in directly identifying the mood of the pet but also the identification accuracy of our method to determine the effectiveness of our system. Without our method, a yolo v3 model averagely reached 19.78% accuracy. Conversely, our system averagely reached 55.81% accuracy in comparison to a yolo v3 model. Without our method, a faster R-CNN model averagely reached 22.85% accuracy. Conversely, our system averagely reached 70.05% accuracy in comparison to a faster R-CNN model. Our system could identify the mood and state of the pet more accurately when the accuracy of image and sound identification is improved.

Table 2. The accuracy of the K nearest neighbor (KNN) audio identification model.

Feature Number	20 (without/with) MFCC			30 (without/with) MFCC			40 (without/with) MFCC		
	Barking	Growl	Crying	Barking	Growl	Crying	Barking	Growl	Crying
Bark	33/34	8/1	9/14	35/35	6/1	6/13	32/35	5/1	3/13
Growl	2/0	19/33	16/2	1/0	18/33	17/2	3/0	21/33	18/2
Cry	2/3	10/3	12/21	1/2	13/3	14/22	2/2	11/3	16/22
Average	57.66%/79.28%			60.36%/81.08%			62.16%/81.08%		
Accuracy	60.06%/80.48%								

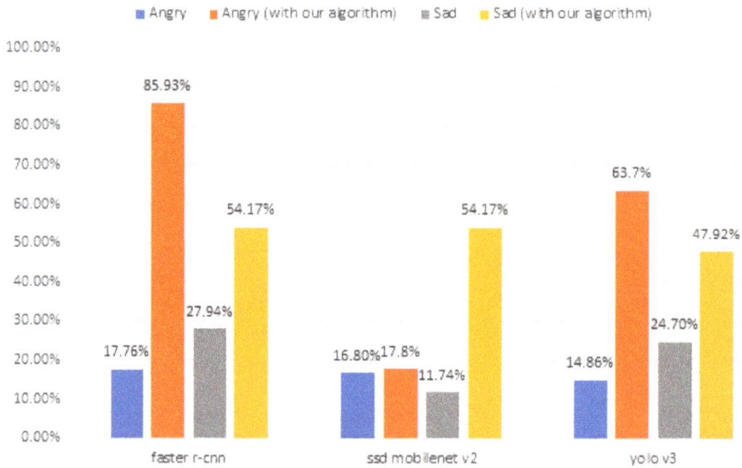

Figure 12. Identification of the accuracy for determining angry and sad states.

5. Conclusions

This paper proposed multiple featured parts and a time-continuous state of featured parts to identify the pet's mood and state. Additionally, the feature was not only the image, but the sound could also be one of the features. Extending image identification from the identification of a single image had time continuity and multiple object dependencies, which greatly increased the scalability of image identification technology applications. The proposed system used images of specific featured parts that were successfully identified and sent feedback to the system's automatic optimization subsystem to achieve model identification capabilities that kept up with the times. The proposed image identification module, KNN audio identification model and the multiple feature dependency detection algorithm were checked at various levels to make sure core functions of the smart pet surveillance system were feasible. Therefore, as long as the mood and state can be defined by featured parts of pets, the system can immediately identify other moods or states. With the automatic optimization subsystem included in the system, it is able to improve the identification ability. The smart pet surveillance system proves that the image identification with time continuity can be applied in a wider range than the traditional single image identification.

Author Contributions: M.-F.T. developed the main idea, supervision, writing—original draft and writing–review & editing; P.-C.L. and Z.-H.H. developed the writing—original draft, and C.-H.L. developed project administration. All authors have read and agreed to the published version of the manuscript.

Funding: This research was funded by Industrial Technology Research Institute, Taiwan and the APC was funded by National United University, Taiwan.

Conflicts of Interest: The authors declare no conflict of interest.

References

1. Girshick, R.; Donahue, J.; Darrell, T.; Malik, J. Region-based convolutional networks for accurate object detection and segmentation. *IEEE Trans. Pattern Anal. Mach. Intell.* **2016**, *38*, 142–158. [CrossRef] [PubMed]
2. Girshick, R. fast R-CNN. In Proceedings of the 2015 IEEE International Conference on Computer Vision (ICCV), Santiago, Chile, 7–13 December 2015; pp. 1440–1448.
3. Ren, S.; He, K.; Girshick, R.; Sun, J. Faster R-CNN: Towards real-time object detection with region proposal networks. *IEEE Trans. Pattern Anal. Mach. Intell.* **2017**, *39*, 1137–1149. [CrossRef] [PubMed]
4. Hao, X.; Yang, G.; Ye, Q.; Lin, D. Rare animal image recognition based on convolutional neural networks. In Proceedings of the 2019 12th International Congress on Image and Signal Processing, BioMedical Engineering and Informatics (CISP-BMEI), Suzhou, China, 19–21 October 2019; pp. 1–5.
5. Wu, M.; Chen, L. Image recognition based on deep learning. In Proceedings of the 2015 Chinese Automation Congress (CAC), Wuhan, China, 27–29 November 2015; pp. 542–546.
6. Ruan, F.; Zhang, X.; Zhu, D.; Xu, Z.; Wan, S.; Qi, L. Deep learning for real-time image steganalysis: A survey. *J. Real Time Image Process.* **2020**, *17*, 149–160. [CrossRef]
7. Nishani, E.; Cico, B. Computer vision approaches based on deep learning and neural networks: Deep neural networks for video analysis of human pose estimation. In Proceedings of the IEEE Mediterranean Conference on Embedded Computing, Bar, Montenegro, 11–15 June 2017; pp. 1–4.
8. Mukai, N.; Zhang, Y.; Chang, Y. Pet face detection. In Proceedings of the 2018 Nicograph International (NicoInt), Tainan, Taiwan, 28–29 June 2018; pp. 52–57.
9. Kumar, S.; Singh, S.K. Biometric recognition for pet animal. *J. Softw. Eng. Appl.* **2014**, *7*, 470–482. [CrossRef]
10. Lin, C.; Lin, Y.; Chang, C.; Chen, C.; Tsai, M. The design of automatic bird data capture systems. In Proceedings of the IEEE International Conference on Consumer Electronics, Taichung, Taiwan, 19–21 May 2018; pp. 1–2.
11. Jothi, G.; Inbarani, H.; Azar, A.; Koubaa, A.; Kamal, N.; Fouad, K. improved dominance soft set based decision rules with pruning for leukemia image classification. *Electronic* **2020**, *9*, 794–822. [CrossRef]
12. Mao, Q.; Sun, H.; Liu, Y.; Jia, R. Mini-YOLOv3: Real-time object detector for embedded applications. *IEEE Access* **2019**, *7*, 133529–133538. [CrossRef]
13. WON, J.; LEE, D.; LEE, K.; LIN, C. An improved YOLOv3-based neural network for de-identification technology. In Proceedings of the IEEE International Technical Conference on Circuits/Systems, Computers and Communications, JeJu, Korea, 23–26 June 2019; pp. 1–2.
14. Kong, W.; Hong, J.; Jia, M.; Yao, J.; Cong, W.; Hu, H.; Zhang, H. YOLOv3-DPFIN: A dual-path feature fusion neural network for robust real-time sonar target detection. *IEEE Sens. J.* **2020**, *20*, 3745–3756. [CrossRef]
15. Li, S.; Tao, F.; Shi, T.; Kuang, J. Improvement of YOLOv3 network based on ROI. In Proceedings of the IEEE Advanced Information Technology, Electronic and Automation Control Conference, Chengdu, China, 20–22 December 2019; pp. 2590–2595.
16. Arruda, M.; Spadon, G.; Rodrigues, J.; Gonçalves, W.; Machado, B. Recognition of endangered pantanal animal species using deep learning methods. In Proceedings of the IEEE International Joint Conference on Neural Networks, Rio de Janeiro, Brazil, 8–13 July 2018; pp. 1–8.
17. Blumrosen, G.; Hawellek, D.; Pesaran, B. Towards automated recognition of facial expressions in animal models. In Proceedings of the IEEE International Conference on Computer Vision Workshops, Venice, Italy, 22–29 October 2017; pp. 2810–2819.
18. Tu, X.; Lai, K.; Yanushkevich, S. Transfer learning on convolutional neural networks for dog identification. In Proceedings of the IEEE International Conference on Software Engineering and Service Science, Beijing, China, 23–25 November 2018; pp. 357–360.
19. Hammam, A.; Soliman, M.; Hassanein, A. DeepPet: A pet animal tracking system in internet of things using deep neural networks. In Proceedings of the IEEE International Conference on Computer Engineering and Systems, Beijing, China, 18–20 August 2018; pp. 38–43.
20. Reulke, R.; Rues, D.; Deckers, N.; Barnewitz, D.; Wieckert, A.; Kienapfel, K. Analysis of motion patterns for pain estimation of horses. In Proceedings of the IEEE International Conference on Advanced Video and Signal Based Surveillance, Genova, Italy, 2–4 September 2018; pp. 1–6.

21. Khosla, A.; Jayadevaprakash, N.; Yao, B.; Li, F. Stanford Dogs Dataset. Available online: http://vision.stanford.edu/aditya86/ImageNetDogs/ (accessed on 29 November 2019).
22. Google, AudioSet. Available online: https://research.google.com/audioset/ontology/dog_1.html (accessed on 11 August 2020).

Article

Detecting Objects from Space: An Evaluation of Deep-Learning Modern Approaches

Khang Nguyen [1,2,3,*], **Nhut T. Huynh** [2,3], **Phat C. Nguyen** [2,3], **Khanh-Duy Nguyen** [1,3], **Nguyen D. Vo** [1,3] **and Tam V. Nguyen** [4]

1   Multimedia Communications Laboratory, University of Information Technology, Ho Chi Minh City 700000, Vietnam; khanhnd@uit.edu.vn (K.-D.N.); nguyenvd@uit.edu.vn (N.D.V.)
2   Department of Software Engineering, University of Information Technology, Ho Chi Minh City 700000, Vietnam; 15520589@gm.uit.edu.vn (N.T.H.); 15520601@gm.uit.edu.vn (P.C.N.)
3   Vietnam National University, Ho Chi Minh City 700000, Vietnam
4   Department of Computer Science, University of Dayton, Dayton, OH 45469, USA; tamnguyen@udayton.edu
*   Correspondence: khangnttm@uit.edu.vn

Received: 13 February 2020; Accepted: 26 March 2020; Published: 30 March 2020

Abstract: Unmanned aircraft systems or drones enable us to record or capture many scenes from the bird's-eye view and they have been fast deployed to a wide range of practical domains, i.e., agriculture, aerial photography, fast delivery and surveillance. Object detection task is one of the core steps in understanding videos collected from the drones. However, this task is very challenging due to the unconstrained viewpoints and low resolution of captured videos. While deep-learning modern object detectors have recently achieved great success in general benchmarks, i.e., PASCAL-VOC and MS-COCO, the robustness of these detectors on aerial images captured by drones is not well studied. In this paper, we present an evaluation of state-of-the-art deep-learning detectors including Faster R-CNN (Faster Regional CNN), RFCN (Region-based Fully Convolutional Networks), SNIPER (Scale Normalization for Image Pyramids with Efficient Resampling), Single-Shot Detector (SSD), YOLO (You Only Look Once), RetinaNet, and CenterNet for the object detection in videos captured by drones. We conduct experiments on VisDrone2019 dataset which contains 96 videos with 39,988 annotated frames and provide insights into efficient object detectors for aerial images.

Keywords: object detection; VisDrone2019; aerial imagery; Faster R-CNN; SSD; RFCN; YOLOv3; RetinaNet; SNIPER; CenterNet

---

1. Introduction

Object detection is a fundamental yet difficult task in image processing and computer vision research. It has been an important research topic for decades. Its development in the past two decades can be regarded as an epitome of computer vision history [1]. Since it plays a principal role in understanding and absorbing the contexts of images, therefore, object detection is considered to be a prerequisite measure that offers the computer to detect various objects. Giving a testing image, object detection could localize the coordinates of the objects and assign the corresponding labels to the objects in terms of the object category, i.e., human, dog, or cat. The coordinates of a detected object represent the object's bounding box [2,3]. Object detection has many applications in robot vision, autonomous driving, human-computer interaction, intelligent video surveillance. The deep-learning technology has brought significant breakthroughs in recent years. In particular, these techniques have produced remarkable development for object detection. Object detection can detect a specific instance, i.e., Obama's face, Eiffel Tower, Golden Gate Bridge; or objects of specific categories, i.e., humans, cars, bicycles. Historically, object detection has mainly directed on the detection of a single category, for example, person class [4]. In recent years, the research community has started moving towards other categories than the well-known categories like person, cat, or dog. Here are some common challenges that object detectors face on aerial images: viewpoints, illuminations, scale variations, perspectives, intra-class variations, low resolutions, and occlusions. For example, the main challenges in pedestrian detection come from crowded scenes with heavy overlaps, occlusion, and low-resolution images.

Generic object detection has received significant attention. There are many competition benchmarks, i.e., PASCAL-VOC [5,6], ImageNet Large Scale Visual Recognition Challenge (ILSVRC) [2], MS-COCO [7], and VisDrone-DET [8]. There are several notable studies on specific object detection like face detection [9], pedestrian detection [10] and vehicle detection [11]. Recently, the research community has focused on deep learning and its applications towards the object recognition/detection tasks. In the past few years, Convolutional Neural Networks (CNNs) [12,13] have brought breakthroughs in speech, audio, image, and video processing. CNNs have driven notable progress in visual recognition and object detection. Many successful CNN architectures, e.g., OverFeat [14], R-CNN [15], Fast R-CNN [16], Faster R-CNN [17], SSD [18], RFCN [19], YOLO [20], YOLOv2 [21], Faster R-CNN [17], RetinaNet [22], YOLOv3 [23], and SNIPER [24] have performed well on the task of object detection. For example, SNIPER, RetinaNet and YOLOv3 are the top models for object detection on MS-COCO dataset [7] with mAP 46.1%, 40.8%, 33.0%, respectively.

The object detection has now been widely used in many practical scenarios [1]. Its use cases ranging from protecting personal security to boosting productivity in the workplace. While the challenges of normal viewpoints have been considered to be the prevalence, in recent years, there has been increasing interest in flying drones and their applications in healthcare, video surveillance, search-and-rescue, and agriculture. The drones are now common devices that enable us to record or capture many scenes as the bird's-eye view. Visual object detection is an essential component in the drone application. However, object detection is a very challenging task since video sequences or images captured by drones vary significantly in terms of scales, perspectives, and weather conditions. Aerial images are often noisy and blurred due to the drone motion. The ratio of object size to the video resolution is also small. Therefore, in this paper, we investigate the performance of state-of-the-art object detectors in the aerial images. Please note that this paper is the extension of our earlier version which is the best paper in MITA 2019 conference [25]. Our contributions are three-fold.

- To the best of our knowledge, we are among the first ones who investigate the impact of different deep-learning object detection methods on the given problem.
- Second, we double the number of benchmarking methods compared to the MITA paper [25].
- Last but not least, we also increase the number of benchmarking classes in VisDrone dataset. In particular, we evaluate full 10 classes in this paper (vs. 2 human classes, namely *pedestrian* and *people*, in the MITA paper).

The remainder of the paper is organized as follows. In Section 2, the related works are presented. Section 3 and Section 4 present the benchmarked methods and the experimental results, respectively. Finally, Section 5 concludes our work.

2. Related Work

2.1. CNN Models

CNN-based architectures have been backbones in many detection frameworks. Popular architectures include AlexNet [12], ZFNet [26] VGGNet [27], GoogLeNet [28], ResNet [29] and DenseNet [30]. We briefly introduce these models as follows.

Known as a pioneering work, AlexNet [12] consists of eight layers: five convolutional (conv1, conv2, conv3, conv4 and conv5) layers and three fully connected (fc6, fc7 and fc8) layers. The fc8 is a SoftMax classifier. The convolutional layers are connected directly: conv3, conv4 and conv5. The convolutional layers are connected via an Overlapping Max-Pooling layer: conv1–conv2, conv2–conv3, conv5–fc6. AlexNet used 11×11, 5×5 and 3×3 kernels. Later, Reference [26] proposed ZFNet by modifying AlexNet. ZFNet uses 7×7 kernels. The small kernels retain more information than the big kernels. By proposing a deeper network, VGGNet [27] outperforms AlexNet in the image classification task. Similarly, GoogLeNet [28] won the ILSVRC2014 competition by increasing the number of layers. In particular, it includes 22 layers: 21 convolutional layers and a fully connected layer.

The number of layers is increasing in CNN architectures. The CNNs require a lot of computational resources. There are several problems: gradient vanishing, exploding, and degrading. Degradation occurs when we add more layers into deep networks, the accuracy becomes saturated and then decrease quickly. To overcome this problem, ResNet [29] introduces many residual blocks. In a residual block, each layer is fed directly to the layers about 2–3 hops away using skip-connections.

DenseNet [30] was proposed by Huang et al. in 2017. It includes many dense blocks. A dense block consists of composite layers which are densely connected together. The input of one layer is the output of all previous layers, so input information is shared.

2.2. Object Detection Methods

Object detection methods are mainly divided into one-stage frameworks and two-stage frameworks. Two-stage frameworks are more accurate than one-stage frameworks, but one-stage frameworks usually achieve real-time detection. The two-stage approach includes two steps: the first stage creates region proposals, the second stage classifies region proposals. The one-stage approach predicts object regions and object classes at the same time.

CNN-based Two-Stage frameworks: Two-stage frameworks mainly include R-CNN [15], SPP-Net [31], Fast R-CNN [16], Faster R-CNN [17], RFCN [19], Mask R-CNN [32], and SNIPER [24].

R-CNN [15] is a method for detecting objects based on the ImageNet pre-trained model. R-CNN uses Selective Search algorithm to generate region proposals. Then, these regions are warped and fed into the pre-trained model to extract high-level features. Finally, several SVM classifiers are trained based on these features to identify object classes. Fast R-CNN [16] was introduced to solve some R-CNN's limitations, i.e., the computational speed. Fast R-CNN feeds the whole image into ConvNet to create convolutional feature map instead of 2000 regions as R-CNN.

Faster R-CNN [17] proposes a Region Proposal Network (RPN) to detect region proposals instead of Selective Search, which is used in R-CNN and Fast R-CNN. Faster R-CNN is $10\times$ faster than Fast R-CNN, and $250\times$ faster than R-CNN in reference time. RFCN introduces the positive sensitive score map, which improves speed but remains accurate compared to Faster R-CNN. Mask R-CNN extends Faster R-CNN with instance segmentation and introduces Align Pooling.

CNN-based one-stage frameworks: The most common examples of one-stage object frameworks are YOLO, YOLOv2, YOLOv3 [23], SSD [18], and RetinaNet. You Only Look Once (YOLO [20]) is one of the first approaches to build a one-stage detector. Unlike R-CNN family, YOLO does not use a region proposal component. Instead, it learns to regress bounding-box coordinates and class probabilities directly from image pixels. This significantly boosts the speed of the detecting process. Single-Shot MultiBox Detector (SSD [18]) is also a one-stage detector which also aims at high speed object detection. However, unlike YOLO, SSD adopts a multi-scale approach. It then adds many convolutional layers decreasing in size sequentially. This can be regarded as a pyramid representation of an image, where earlier levels contain feature maps that are useful to detect small objects and deeper levels are expected to detect larger objects. Each of these layers has a set of predefined anchor boxes (also known as default boxes or prior boxes) for every cell. The model will learn and predict the offsets corresponding to correct anchor boxes. The approach has made a successful attempt on creating an efficient detector for objects in various sizes while maintaining a low inference time.

3. Benchmarked State-of-the-Art Object Detection Methods

In this section, we provide further details of the benchmarked object detection methods. Figure 1 visualizes the methods in the chronological order. Meanwhile, Figure 2 depicts the framework structures of different state-of-the-art object detectors.

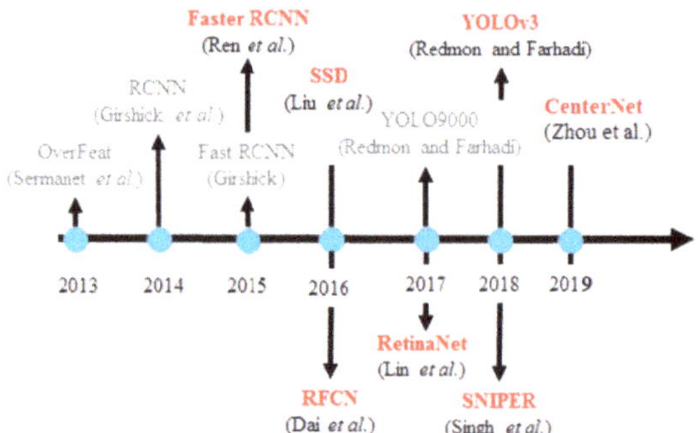

Figure 1. Timeline of state-of-the-art object detection methods. The benchmarked methods are marked in red and boldfaced font.

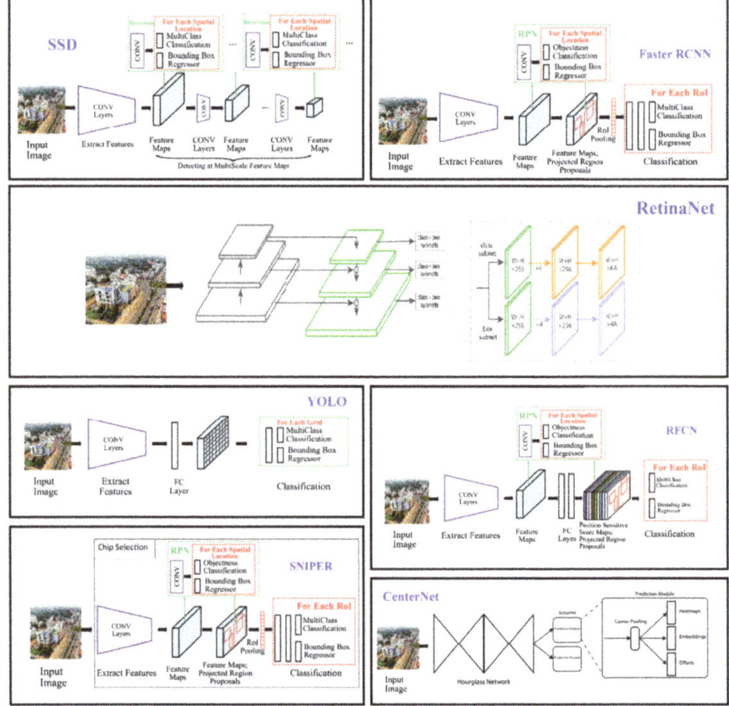

Figure 2. The benchmarked methods adopted in this paper: Faster R-CNN, SSD, RFCN, RetinaNet, SNIPER, YOLOv3, and CenterNet.

3.1. Faster R-CNN

Faster R-CNN [17] is an extension of the R-CNN [15] and Fast R-CNN [16] methods for object detection. R-CNN requires a forward pass of the CNN for around 2000 region proposals (ROI) for every single image. Later, Fast R-CNN was able to solve the problem of R-CNN by sharing the computation of convolution between different proposals (feature map). The detection process is sped up but still depends on the region proposal method (Selective Search). Region proposals were generated by additional methods, i.e., Selective Search or Edge Box. To solve this problem, Ren et al. [17] introduced the Region Proposal Network (RPN).

Faster R-CNN consists of two main components, namely the RPN and the Fast R-CNN detector. RPN initializes squared reference boxes of aspect ratios and diverse scales at each convolutional feature map location. Each squared box is mapped to a feature vector. The feature vector is fed into two fully connected layers, an object category classification layer, and a box regression layer. Faster R-CNN enables highly efficient region proposal computation because RPN shares convolutional features with Fast R-CNN. With an image of arbitrary size as an input, RPN is trained end-to-end to generate high-quality region proposals as output. The Fast R-CNN detector also uses the ROI pooling layer to extract features from each candidate box and performs object classification and bounding-box regression. The entire system is a single, unified network for object detection.

3.2. RFCN: Region-Based Fully Convolutional Networks

A limitation of Faster RCN is that it does not share computations after ROI pooling. The amount of computation should be shared as much as possible. Faster R-CNN overcomes the limitations of Fast R-CNN but it still contains several non-shared fully connected layers that must be computed for each of hundreds of proposals. Region-based Fully Convolutional Network [19] was proposed as an improvement to Faster R-CNN. It consists of shared, fully convolutional architectures. In RFCN, fully connected layers after ROI pooling are removed, all other layers are moved prior to the ROI pooling to generate the score maps. RFCN infers 2.5 to 20 times faster than Faster R-CNN, yet it still maintains a competitive accuracy.

Backbone Architecture. The incarnation of RFCN in [19] is based on ImageNet pre-trained ResNet-101 model. To compute feature maps, the average pooling layer and the fc layer are removed. Instead, RFCN only uses the convolutional layers, and attaches a randomly initialized 1024-d 1×1 convolutional layer to reduce dimension at the last convolutional block in ResNet-101. In addition, RFCN uses the $k^2(C+1)$ channel convolutional layers to generate scope maps.

Position-sensitive score maps and Position-sensitive ROI pooling. RFCN regularly divides each ROI rectangle into $k \times k$ bins, then each bin has a size of $\approx \frac{w}{k} \times \frac{h}{k}$ for an ROI rectangle of a size $w \times h$. For each category, the last convolutional layer is built to create k^2 score maps. Inside the (i,j)-th bin $(0 \leq i, j \leq k-1)$, a position-sensitive ROI pooling operation pools only over the (i,j)-th score map:

$$r_c\left(i,j \mid \ominus\right) = \sum_{(x,y) \in bin(i,j)} z_{b,j,c}\left(x + x_0, y + y_0 \mid \ominus\right) / n \tag{1}$$

For the c-th category, $r_c(i,j)$ is the aggregated response in the (i,j)-th bin; $z_{b,j,c}$ is a score map among $k^2(C+1)$ maps; (x_0, y_0) represents the ROI's top left corner; and the number of pixels in the bin is marked as n. $\ominus$ is the set of learnable parameters; and the (i,j)-th bin is located at $\left[i\frac{w}{k}\right] \leq x \leq \left[(i+1)\frac{w}{k}\right]$ and $\left[j\frac{h}{k}\right] \leq y \leq \left[(j+1)\frac{h}{k}\right]$.

The k^2 position-sensitive scores are averaged to obtain a $(C+1)$-dimensional vector for each ROI $r_c\left(\ominus\right) = \sum_{i,j} r_c\left(i,j \mid \ominus\right)$. Then the SoftMax responses for categories are computed $s_c\left(\ominus\right) = e^{r_c(\ominus)} / \sum_{c'=0}^{C} e^{r_{c'}(\ominus)}$.

The RFCN resolves bounding-box regression similar to Fast R-CNN with $k^2(C+1)$-d convolutional layer, and appends a sibling $4k^2$-d convolutional layer additionally.

The position-sensitive ROI pooling makes a $4k^2$-d vector for each ROI. Then, a bounding box as $t = (tx, ty, tw, th)$ uses a 4-d vector aggregated by average voting.

Training. The loss function is defined on each ROI and calculated by the sum of cross-entropy losses and box regression loss:

$$L(s, t_{x,y,w,h}) = L_{cls}(s_c^*) + \lambda \left[c^* > 0\right] L_{reg}(t, t^*), \tag{2}$$

where c^* is the ground-truth label of ROI. For classification, $L_{cls}(s_c^*) = -\log(s_c^*)$ denotes the cross-entropy losses, L_{reg} denotes the bounding-box regression loss and t^* is the ground-truth box. If the argument is valid, $\left[c^* > 0\right]$ receives a value of 1, otherwise, 0.

Inference. The feature maps are the results of calculations on an image with a single scale of 600 shared between RPN and RFCN (as showed in Figure 2). Then, the RFCN part evaluates score maps and regresses bounding boxes based on ROIs which are proposed by the Region Proposal Network (RPN) part.

3.3. SNIPER: Scale Normalization for Image Pyramids with Efficient Resampling

SNIPER is an effective, multi-scale training method for identification, object detection and object separation [24]. Instead of processing pixels based on the pyramid (SN), SNIPER treats the context areas around the ground truths (called chips) at an appropriate scale. This greatly increases the speed during training when it operates on low-resolution chips. Relying on the memory efficient design, SNIPER benefits from mass standardization during the training process without having to synchronize standardized statistics on the GPU.

3.3.1. Chip Generation

SNIPER generates chips C_i at multiple scales $\{s_1, s_2, ..., s_i, ..., s_n\}$ in the image. For each scale, the image is first re-sized to width (W_i) and height (H_i). On this canvas, $K \times K$ pixel chips are placed at equal intervals of d pixels.

3.3.2. Chip Selection

All favorable chips are chosen greedily to cover the highest amount of valid ground-truth boxes. If it is completely enclosed inside a chip, a ground-truth box is said to be covered. Although positive chips cover all the positive instances, a significant portion of the background is not covered by them. In multi-scale training architecture, each pixel is processed at all scales in the picture. A naive strategy is to use object suggestions to define areas where objects are likely to present. If there are no region proposals in an image, it is considered to be background.

3.4. SSD: Single-Shot Detector

SSD [18] is a one-stage solution, which has tremendously reduced inference time and resulted in an accurate, high speed detector that can be used for real-time video processing.

3.4.1. Base Network VGG-16

SSD is built on top of VGG-16 base network [27] that focuses on simplicity and depth. In particular, the model uses 16 convolutional layers with only 3×3 filters to extract features. As the model goes deeper, the number of filter doubles after each max-pooling layer. Noticeably, the convolutional layers of the same type are combined as shown in Figure 2. At the end, three fully connected layers are followed by 4096 channels. The last one contains 1000 channels for each class and is concatenated with a SoftMax layer to return the detection results. The model works well on classification and localization tasks and has achieved 89% mAP on PASCAL-VOC 2007 dataset. VGG-16 has been one of the most interesting models to research even though it is not as fast as the newer ones. Its architecture has been reused in many models because of its valuable extracted features.

3.4.2. Model Architecture

SSD extends the pre-trained VGG-16 model (on ImageNet [10]) by adding new convolutional layers conv8_2, conv9_2, conv10_2, conv11_2 in addition to using the modified conv4_3 and fc_7 layers to extract useful features. Each layer is designed to detect objects at a certain scale using k anchor boxes, where $4k$ offsets and c class probabilities are computed by using 3×3 filters. Thus, given a feature map with a size of $m \times n$, the total number of filters to be used is $kmn(c+4)$. The anchor boxes are chosen manually. Here, we use the original formula, as follows, to calculate anchor box scales at different levels.

$$s_k = s_{min} + \frac{s_{max} - s_{min}}{m-1}(k-1) \tag{3}$$

$k \in [1, m]$ where $s_{min} = 0.2$ and $s_{max} = 0.9$.

3.5. RetinaNet

RetinaNet [22] is another one-stage detector. It aims to tackle the class imbalance problem between foreground and background remaining in one-stage detector. RetinaNet uses two main techniques: FPN backbone and focal loss as the loss function. FPN is built on top of a convolutional neural network and is responsible for extracting convolutional feature maps from the entire image. By using focal loss, RetinaNet changes weights in the loss function, focuses on hard, misclassified examples, which improves the prediction accuracy. With ResNet (FPN) as a backbone for feature extraction and two specific subnetworks for classification and bounding-box regression, RetinaNet has achieved state-of-the-art performance.

3.5.1. Class Imbalance

As a one-stage detector, RetinaNet has a much larger set of candidate object locations which is regularly sampled across an image ($\sim$100 k locations), covering spatial positions, scales and aspect ratios tightly. The easily classified background examples still dominate the training procedure. Bootstrapping or hard example mining is typically used as a solution for this problem. However, they are not efficient enough. To solve this, RetinaNet proposes a new loss function which can adaptively tune the contributed weights of object classes during training.

3.5.2. Focal Loss

Focal loss is computed by adding $(1 - p_i)^\gamma$ to cross-entropy loss as a modulating factor.

$$\sum_{i=1}^{k}(y_i log(p_i)(1 - p_i)^\gamma) + (1 - y_i)log(1 - p_i)p_i^\gamma) \tag{4}$$

3.5.3. RetinaNet Detector Architecture

RetinaNet adopts ResNet for deep feature extraction. ResNet builds a rich multi-scale feature pyramid from an input image of single resolution by using Feature Pyramid Network (FPN) [33]. It combines low-resolution, high-resolution, and semi-weak characteristics through a top-down pathway and lateral connections.

3.6. YOLO: You Only Look Once

YOLO is an object detection system targeted for real-time processing. Recently, the third version of YOLO has been published, YOLOv3 is extremely fast and accurate. In mAP measured at 0.5 IOU YOLOv3 is on par with focal loss but about $4 \times$ faster.

YOLOv3 takes an input image to predict 3D tensors respectively to three scales and each scale is divided into N $\times$ N grid cells. During training, each grid cell considers a class that it likely is and be responsible for detecting that class. Simultaneously, each grid cell is assigned with 3 initial prior boxes with various sizes. Finally, non-max suppression is applied to select the best boxes.

3.6.1. Feature Extraction

YOLOv3 uses a variant of Darknet, which originally has a 53-layer network trained on ImageNet. According to [23], Darknet-53 is better than ResNet-101 and 1.5 × faster. Darknet-53 has a performance similar to ResNet-152 and is 2 × faster.

3.6.2. Detection at Three Scales

YOLOv3 is different from its predecessors since it performs the detection process at three different scales. In YOLOv3, the detection is done by applying 1 × 1 detection kernels on feature maps of three different sizes at three different places in the network. YOLOv3 makes prediction at three scales, which are precisely given by down sampling the dimensions of the input image by 32, 16 and 8, respectively. Detections at different layer helps address the issue of detecting small objects, a common issue in YOLOv2.

3.6.3. Objective Score and Confidences

The object score illustrates the probability that an object is contained inside a bounding box and its value ranging from 1 to 0. A sigmoid is applied to compute the objectness scores. In terms of class confidences, they depict the probabilities of the detected object which belongs to a particular class. In YOLOv3, Non-maximum Suppression (NMS) is used to decide a class score and it is meant to alleviate the problem of multiple detections of the same object.

3.7. CenterNet

One-stage and two-stage detection have limitations: anchor box is designed with manual proportions that are easily affected by data and fixed during training. This requires a high computation cost, but the anchors are not always accurate. To address that, recently a series of anchor-free methods [34–36] are proposed. CenterNet [37] is a one-stage detector, anchor-free method. Reference [37] proposed a new center-based framework based on a single Hourglass network without FPN structure [38]. The object is represented by the central point of the bounding box. Other information is calculated by regression such as object size, dimension, and pose.

3.7.1. Object as Points

CenterNet considers the center point of an object as a prerequisite to localize the bounding box. As a result, Reference [37] use a keypoint estimator $\hat{y}$ to predict all center points and single a single size prediction for all object categories to alleviate the computational burden.

3.7.2. From Points to Bounding Boxes

CenterNet identifies the peak points in the heatmap before detecting all the values meet or greater than its 8-related neighbors and keep the top 100 points. Subsequently, each keypoint location is given by an integer coordinate (x_i, y_i) and the key point estimator value as a measure of its detection confidence is applied to produce the bounding box as below:

$$(\hat{x}_i + \delta\hat{x}_i - \hat{w}_i/2,\ \hat{y}_i + \delta\hat{y}_i - \hat{h}_i/2,\ \hat{x}_i + \delta\hat{x}_i + \hat{w}_i/2,\ \hat{y}_i + \delta\hat{y}_i + \hat{h}_i/2), \tag{5}$$

where $(\delta\hat{x}_i, \delta\hat{y}_i) = \hat{O}_{\hat{x}_i,\hat{y}_i}$ is the offset prediction and $(\hat{w}_i, \hat{h}_i) = \hat{S}_{\hat{x}_i,\hat{y}_i}$ is the size prediction [37].

4. Benchmark Experiments

In this section, we first introduce the benchmark dataset and the evaluation metrics. We then detail the model configuration and discuss the experimental results.

4.1. Dataset

VisDrone2019 dataset [39] consists of 288 videos with 261,908 frames and 10,209 static images that do not match the frames of videos. Data is collected from unmanned aerial vehicles such as DJI Mavic, Phantom series (3, 3A, 3SE, 3P, 4, 4A, 4P). The videos and images are collected at different times of day. The frames in the videos have the highest resolution of 3840 × 2160 and the still image is 2000 × 1500. Some images in the dataset are shown in the Figure 3. VisDrone2019 includes ten predefined categories of objects: *pedestrian, person car, van, bus, truck, motor, bicycle, awning-tricycle, and tricycle.* Only the training data is released by the contest organizers. In this paper, we use 56 clips of VID-train data, with 24,313 frames of the VisDrone2019 dataset as training data; and 7 clips of VID-val, with 2860 frames for model evaluation.

Figure 3. Some exemplary frames of videos in the VisDrone2019 dataset [39].

The number of unbalanced objects between classes is as follows. In VID-train, *pedestrian* (234,305), people (94,396), *bicycle* (40,255), car (505,301), *van* (46,940), *truck* (30,498), tricycle (28,338), *awning-tricycle* (13,011), *bus* (9653) and *motor* (102,819). In VID-val, *pedestrian* (32,404), *people* (17,908), bicycle (6842), *car* (31,821), van (6842), *truck* (1359), *tricycle* (3769), *awning-tricycle* (1718), *bus* (264) and *motor* (12,025). The class distribution of Visdrone VID is depicted in Figure 4. As a quick glimpse through the training set, there are the wide discrepancy in terms of the weather, the light source direction and the time interval as well as the drone motion in Figure 5.

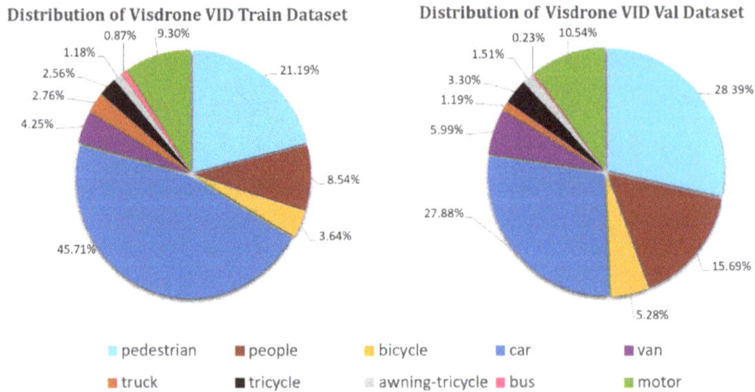

Figure 4. The class distribution in Visdrone VID-Train, Val Dataset.

Figure 5. Some example images of the challenge.

4.2. Evaluation Metrics

In this work, we use the Average Precision (AP) measurement [3,40], the commonly used metric to assess object detection accuracy. Given two bounding boxes, one for ground truth (the actual class label) and one for the detection result (the predicted class label), we use the Intersection over Union (IoU) to calculate the similarity between the two boxes and the score of the predicted box. It is computed as the intersected area (S_i) divided by the union (S_j) of the two areas. An IoU threshold η indicates whether the prediction is an object or not. If the actual class label is the predicted class label and IoU > η, it is considered a positive else it is considered a negative.

$$IoU = \frac{S_i \cap S_j}{S_i \cup S_j}$$

(6)

The AP computes the average precision value for recall value over 0 to 1. The mean Average Precision (mAP) is computed by taking the average over the AP of all classes. Precision is the proportion of the predicted bounding boxes matching actual ground truth. Recall is the proportion of ground-truth objects being correctly detected. For object detection, we report the performance results with AP (IoU = 0.50), AP (IoU = 0.75). The AP [3] summarizes the shape of the precision/recall curve, and is defined as the mean precision at a set of 11 equally spaced recall levels [0, 0.1, ... , 1]:

$$AP = \frac{1}{11} \times \sum_{r \in \{0,0.1,...,1\}} p_{interp}(r)$$

(7)

where:

$$p_{interp}(r) = max_{\tilde{r}:\tilde{r} \geq r} p(\tilde{r}) \tag{8}$$

4.3. Model Configuration

To make a fair comparison, namely each detection model at its best, we adjust the model's parameters following the recommendation of [41–43]. We provide the detailed configuration as below.

- For **YOLOv3**, we have trained a new model which adopts darknet53 (YOLOv3 code is available at: https://github.com/AlexeyAB/darknet) as pre-trained weights for the convolutional layers. In implementation, YOLO v3 uses a total of 9 anchors with three for each scale. It assigns the three biggest anchors for the first scale, the next three for the second scale, and the last three for the third. As we aim to detect effectively on 3 scales, we recalculate nine anchors correspondingly.
- For **RetinaNet**, we used ResNet152 as backbone model and changed anchor parameters (RetinaNet code is at: https://github.com/fizyr/keras-retinanet).
- **SNIPER** is the new architecture that allows us to improve AP by using negative chip mining. We adopt the available model (SNIPER code is at: https://github.com/mahyarnajibi/SNIPER). We extracted the required proposals for chip selection.
- For **SSD**, we used VGG16 as the pre-trained model and adjusted the aspect ratio used in the original SSD300 (SSD code is at: https://github.com/pierluigiferrari/ssd_keras).
- We trained **Faster R-CNN** (Faster R-CNN code is available at:https://github.com/rbgirshick/py-faster-rcnn), **RFCN** (RFCN code is at: https://github.com/YuwenXiong/py-RFCN), **CenterNet** (CenterNet code is at: https://github.com/xingyizhou/CenterNet) with default parameters.

Table 1 summarizes the detailed configuration. All models are trained or finetuned by using GeForce RTX 2080 Ti 11GB GPU run on Ubuntu 16.04.5 LTS OS.

Table 1. Configuration of SNIPER, RetinaNet, YOLOv3 and SSD.

Architecture	Configuration		
	Attribute	Default	Adjusted
RetinaNet	Base size	32 64 128 256 512	32 64 128 256 512
	Strides	8 16 32 64 128	8 16 32 64 128
	Ratios	0.5 1 2	0.5 1 2
	Scales	1 1.2 1.6	**0.5 1 1.2 1.6 2.0**
YOLOv3	Anchors	10, 13, 16, 30, 33, 23, 30, 61, 62, 45, 59, 119, 116, 90, 156, 198, 373, 326	**20, 310,** **35, 626,** **20, 3751,** **69, 1135,** **92, 2038,** **54, 6129,** **50, 8045,** **107, 12908,** **117, 15654**
SSD	Ratio	[2], [2, 3], [2, 3], [2, 3], [2], [2]	**[1.0, 2.0, 0.5]** **[1.0, 2.0, 0.5, 3.0, 1.0/3.0]** **[1.0, 2.0, 0.5, 3.0, 1.0/3.0]** **[1.0, 2.0, 0.5, 3.0, 1.0/3.0]** **[1.0, 2.0, 0.5]** **[1.0, 2.0, 0.5]**
SNIPER	Scales	[1400, 2000] [800, 1280] [480, 512]	**[1600, 2200]** [800, 1280] [480, 512]

4.4. Results

As aforementioned, we benchmark six state-of-the-art methods: Faster R-CNN, RFCN, SSD with default parameters and SNIPER, YOLOv3, RetinaNet with adjusted parameters which is presented in detail in Table 1. Table 2 shows the training time of the methods. RFCN and YOLOv3 take the least training time. Meanwhile, SSD requires a remarkable time for training. Table 3 shows the runtime performance of different methods. SSD and YOLOv3 achieve the fastest running time among the rest. In the meantime, RFCN only processes 1.75 frames per second. One-stage object detectors clearly run faster than the two-stage ones.

Table 2. Training time of Faster R-CNN, RFCN, SNIPER, RetinaNet, YOLOv3, SSD with VisDrone2018 on GeForce RTX 2080 Ti GPU.

Architecture	Training Time (hours)
Faster R-CNN	8.20
RFCN	4.30
SNIPER	62.50
RetinaNet	10.61
CenterNet	5.3
YOLOv3	4.70
SSD	251.18

Table 3. Runtime performance of Faster R-CNN, RFCN, SSD, YOLO, SNIPER, RetinaNet with VisDrone2019 on GeForce RTX 2080 Ti GPU.

Architecture	FPS (Frames Per Second)
Faster R-CNN	2.78
RFCN	1.75
SNIPER	7.6
SSD	76.9
YOLOv3	7.5
RetinaNet	5.9
CenterNet	7.1

Figure 6 visualizes the detection results of benchmarking methods. As a closer look, Tables 4 and 5 show the detailed results of six methods and the average performance with a threshold of IoU set as 0.5 and 0.75, respectively. In accordance with Table 4 (IoU = 0.5), CenterNet, RetinaNet and SNIPER are the only three algorithms achieving more than 25% mAP score. YOLOv3 ranks in the fourth with more than 20% mAP score. We observe that SSD is inferior, and RFCN performs better than Faster R-CNN. SSD performs the worst, only producing 10.80% mAP score. However, SSD ranks in the third with 9.10% on the *bus* class. As seen in Table 5 (IoU = 0.75), SNIPER, CenterNet and RetinaNet are the only three algorithms achieving more than 11% mAP score. RFCN ranks in the fourth with more than 8% mAP score. We observe that YOLOv3 performs the worst (3.20% mAP score). In the meantime, Faster R-CNN performs better than SSD.

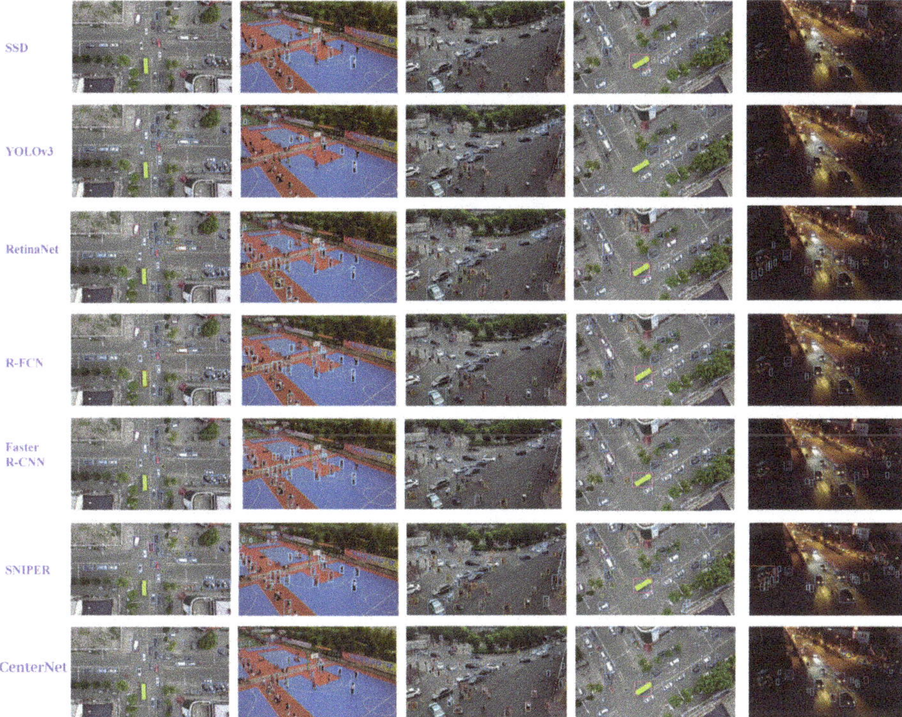

Figure 6. Visualization results of different object detection methods. Color legend: car, truck, bicycle, van, moto, pedestrian, people, bus, tricycle, awning-tricycle. Best view in high 400% resolution.

Regarding the performance, YOLOv3 has good performance with 7.5 FPS and 25.08 mAP (IoU = 0.5) then it drops rapidly to 3.2 mAP (IoU = 0.75) because YOLOv3 is does not perform well at localization. Instead, YOLOv3 is well-known for its runtime performance. We have performed the detection of YOLOv3 and realized the small confidence score for each object (<30%) due to the similarity of features. Therefore, we literally set the low confidence for the object detection demand with this YOLOv3. Meanwhile, CenterNet, RetinaNet and SNIPER achieve better detection results.

Table 4. The AP (IoU = 0.50) scores on the VisDrone2019 Validation set of each object category. The top three results are highlighted in red, blue and green fonts.

Detectors	Pedestrian	People	Bicycle	Car	Van	Truck	Tricycle	A-Tricycle	Bus	Motor	mAP
Faster R-CNN	21.36	11.02	16.97	46.85	18.72	16.96	12.22	10.81	9.09	11.13	17.51
RFCN	21.20	7.05	16.16	44.22	20.12	28.10	19.45	15.21	13.28	10.76	19.55
SNIPER	37.97	23.74	34.73	56.51	31.90	6.72	21.05	14.60	12.86	22.39	26.24
SSD	13.20	5.50	12.70	36.90	9.60	7.70	1.30	9.10	9.10	4.60	10.97
YOLOv3	32.64	16.92	36.94	51.67	33.65	12.83	8.97	26.01	4.87	26.36	25.08
RetinaNet	50.21	31.14	23.53	52.05	27.27	23.92	21.16	23.55	8.97	20.81	28.26
CenterNet	56.74	37.37	27.06	59.86	30	27.51	24.33	26.61	9.87	23.52	32.28

Table 5. The AP (IoU = 0.75) scores on the VisDrone2019 Validation set of each object category. The top three results are highlighted in red, blue and green fonts.

Detectors	Pedestrian	People	Bicycle	Car	Van	Truck	Tricycle	A-Tricycle	Bus	Motor	mAP
Faster R-CNN	3.30	0.71	9.09	19.12	9.09	9.09	1.64	3.03	9.09	0.95	6.51
RFCN	3.03	1.40	2.60	24.85	11.31	12.45	9.09	2.13	9.30	9.09	8.52
SNIPER	17.71	12.16	20.08	46.10	27.21	5.21	12.46	8.39	12.24	8.06	16.96
SSD	6.10	0.10	9.10	18.00	7.10	3.40	0.60	9.10	9.10	0.60	6.32
YOLOv3	4.03	1.22	1.27	9.44	1.94	4.67	2.36	4.78	1.43	0.95	3.20
RetinaNet	11.01	2.20	5.89	34.86	17.58	17.62	4.76	9.86	6.28	2.94	11.30
CenterNet	12.55	2.64	6.77	41.48	20.57	20.97	5.47	11.44	7.16	3.26	13.23

4.5. Analysis of Feature Maps Extraction

Figure 7 depicts feature maps of Faster R-CNN, RFCN, SSD, YOLO, SNIPER, RetinaNet, and CenterNet, on a real aerial image. We extracted feature maps at the final convolution layer. These feature maps offer deeper insights into how different methods capture object features from the aerial viewpoint.

As also seen from Figure 7, SNIPER, RetinaNet as well as CenterNet produce optimal feature maps. We observe that the object shapes are well captured with larger values, whereas the background is with smaller values. This is because focal loss is adept at learning imbalanced classes (foreground/background) while chip mining is extracted from a proposal network trained for a short training schedule, which identifies regions where objects are likely to be present. Simultaneously, keypoint estimation of CenterNet is considered to be the essential factor that facilitates finding center points and regresses to all other object properties, such as size, location, orientation.

Regarding YOLO and SSD: in the feature map obtained by SSD, the object shapes are not clear, and the edges of objects are not preserved. This explains why this method detects aerial objects inaccurately which is ascribed to shallow layers in a neural network. Simultaneously, the features maps extracted by YOLO is better than SSD's ones where object regions are more prominent from the background. However, the edges of the objects are still blurred.

As far as Faster R-CNN and RFCN feature maps are concerned, the object shapes are well preserved but are not clearly distinguished from the background. This is due to the variance of various angles of images, which is considered to be an obstacle of early feature extractors.

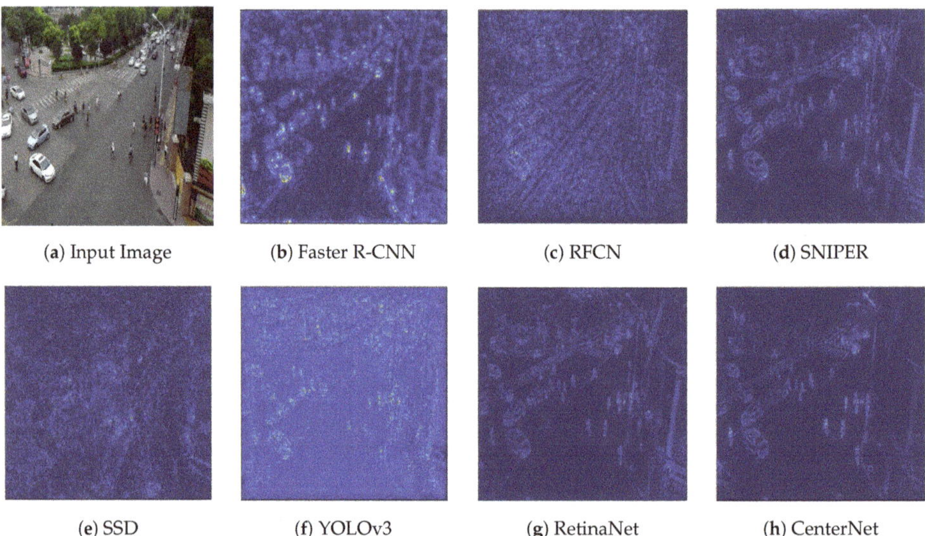

(**a**) Input Image   (**b**) Faster R-CNN   (**c**) RFCN   (**d**) SNIPER

(**e**) SSD   (**f**) YOLOv3   (**g**) RetinaNet   (**h**) CenterNet

Figure 7. Visualization of feature maps from different state-of-the-art object detection models.

4.6. Discussion

SSD is a unified object detector, which adopts a multi-scale approach. SSD uses a VGG16 network as a feature extractor and adds eight convolutional layers and ten layers separately, it also uses convolutional layers to reduce spatial dimension and resolution. To detect multi-scale objects, SSD makes independent object detections from multiple feature maps. Aspect ratios in SSD which are used as the anchor box scaling factors, so we widen the ratios range to ensure most objects could be captured. The higher resolution feature maps are responsible for detecting small objects, the first layer for object detection is conv4_3 which has a spatial dimension of 38×38, a pretty large reduction from the original input image. Furthermore, small objects can only be detected in left most feature maps. However, those maps contain low-level features, like edges or color patches that are less informative for classification. Shallow layers in a neural network may not generate enough high-level features to predict small objects [27]. Therefore, SSD usually performs worse for small objects compared to other detection methods. Although SSD is the penultimate detector, it achieves the second with 12.7% on the *bicycle* class, the third with 5.5% on the *people* class, 9.1% on the *awning-tricycle* class, 9.10% on the *bus* class and 4.6% on the *motor* class.

SSD is competitive with Faster R-CNN, RFCN on more substantial objects, which has poor performance on small objects. Faster R-CNN combines the Region Proposal Network (RPN) into Fast R-CNN. RPN produces box proposals based on the feature extractor. These box proposals are used to crop features from the same intermediate feature map. They fed to the remainder of the feature extractor to predict a class and refine box for each proposal. RFCN is similar to Faster R-CNN. RFCN crops features from the last feature layer before prediction to reduce the amount of computation. RFCN proposed a position-sensitive mechanism to keep translation variance for localization representations. Faster R-CNN has a mAP of 17.51%. Faster R-CNN ranks in the third with 16.96% on the *truck* class. RFCN has much better performance than Faster R-CNN and SSD, producing 19.55% AP. RFCN achieves the first with 28.10% on the *truck* class and 13.28% on the *bus* class, which ranks in the third with 19.45% on the *people* and 15.21% on the *awing-tricycle* class.

As far as the outstanding detectors are concerned, CenterNet, RetinaNet and SNIPER are the three algorithms that top the statistics in both IoU thresholds (0.5 and 0.75). CenterNet ranks first with 32.28%, followed by RetinaNet with 28.26% in case the threshold of IoU = 0.5. Paradoxically, in case the threshold of IoU = 0.75, which favors high accurate results, the figure for SNIPER overtakes that for CenterNet and achieves the best performance.

In particular, CenterNet achieves outstanding results in the benchmark dataset with both IoU thresholds. Please note that CenterNet object detector builds on keypoint estimation networks, finds object centers, and regresses to their size. The experimental results show that CenterNet works well with small IoU threshold, 0.5. Regarding SNIPER, the valid range, boxes, which the square root of their area lies in each range are marked as valid in that scale. Therefore, we increased the valid range to significantly detect objects in various sizes (small, medium and large objects) as shown in Table 1. Simultaneously, chip mining plays an important role in eliminating regions that are likely to contain the background and this measure could adapt to each viewpoint hence alleviating the drawback of diverse scales. As a result, these enhancements cooperate with the pyramid feature map to surpass other detectors in terms of average precision. In particular, at the 0.75 IOU threshold, SNIPER outperforms YOLOv3, with 16.96% and 3.2% respectively. This is mainly because YOLOv3 is inferior in terms of localization. Regarding Retina, by changing anchors, RetinaNet has an increment in terms of AP, 2.3% for *people* class. A scale adjustment has widened the variety of scaling factors to use per anchor location which could improve detection for the diverse size objects. Concurrently, the focal loss is designed to address a severe imbalance between foreground and background classes during training, as a result, this approach could tackle the problems of the unbalanced dataset, in which the number of training samples for car and bus classes outnumbers those from other classes.

5. Conclusion and Future Work

In this paper, we experimented the state-of-the-art object detection methods, namely Faster R-CNN, RFCN, SSD, YOLO, SNIPER, RetinaNet, and CenterNet, on aerial images. Among them, CenterNet, SNIPER and RetinaNet achieve the best performance in terms of average precision. Concurrently, YOLO is considered to be the optimal choice for real-time object detection applications which require the high FPS and moderate precision in detecting object. We notice the main challenges in the problem, for example, occlusion, scale, and class imbalance. From the aerial view, many objects are occluded, and their sizes are varied. We also notice the class imbalance issue during the training process. For example, most of the detectors perform much better on the car and pedestrian classes than on the awning-tricycle, tricycle, and bus classes due to more instances collected in the car and pedestrian classes.

In the future, we would like to investigate the fusion of different object detectors to even boost the state-of-the-art performance. In addition, we are interested in the task of aerial image segmentation. Obviously, the bounding boxes provided by the object detectors are very useful for the segmentation task. We also consider adopting the use of transfer learning [44] to consolidate and enhance the efficiency of training time.

Author Contributions: K.N. is responsible for discussion, paper writing and revising. N.T.H. and P.C.N. focus on training model, benchmark evaluation, and paper writing. K.-D.N. and N.D.V. are in charge of ideas, evaluation and paper writing. T.V.N. is responsible for ideas, discussion, paper structure, paper writing and revising. All authors have read and agreed to the published version of the manuscript.

Funding: This research is funded by Vietnam National University HoChiMinh City (VNU-HCM) under grant number C2018-26-03.

Acknowledgments: We are grateful to the NVIDIA corporation for supporting our research in this area. The authors would like to thank the editors and the reviewers for their professional suggestions.

Conflicts of Interest: The authors declare no conflict of interest.

Abbreviations

The following abbreviations are used in this manuscript:

PASCAL	Pattern Analysis, Statistical modelling and ComputAtional Learning
PASCAL-VOC	PASCAL Visual Object Classes Dataset
MS-COCO	Microsoft Common Objects in Context Dataset
ILSVRC	ImageNet Large Scale Visual Recognition Challenge
ROI	Region of Interests
VGGNet	Visual Geometry Group Network
ZFNet	Zeiler Fergus Networks
CNN	Convolutional Neural Networks
R-CNN	Regional CNN
SSD	Single-Shot Detector
YOLO	You Only Look Once Detector
RFCN	Region-based Fully Convolutional Networks
SNIPER	Scale Normalization for Image Pyramids with Efficient Resampling

References

1. Zou, Z.; Shi, Z.; Guo, Y.; Ye, J. Object detection in 20 years: A survey. *arXiv* **2019**, arXiv:1905.05055.
2. Russakovsky, O.; Deng, J.; Su, H.; Krause, J.; Satheesh, S.; Ma, S.; Huang, Z.; Karpathy, A.; Khosla, A.; Bernstein, M. Imagenet large scale visual recognition challenge. *Int. J. Comput. Vis.* **2015**, *115*, 211–252. [CrossRef]
3. Everingham, M.; Van Gool, L.; Williams, C.K.; Winn, J.; Zisserman, A. The PASCAL Visual Object Classes Challenge 2007 (VOC2007) Results 2007. Available online: http://www.pascal-network.org/challenges/VOC/voc2007/workshop/index.html/ (accessed on 5 October 2007)
4. Zhang, X.; Yang, Y.H.; Han, Z.; Wang, H.; Gao, C. Object class detection: A survey. *ACM Comput. Surv.* **2013**, *46*, 10. [CrossRef]

5. Everingham, M.; Van Gool, L.; Williams, C.K.; Winn, J.; Zisserman, A. The pascal visual object classes (voc) challenge. *Int. J. Comput. Vis.* **2010**, *88*, 303–338. [CrossRef]

6. Everingham, M.; Eslami, S.A.; Van Gool, L.; Williams, C.K.; Winn, J.; Zisserman, A. The pascal visual object classes challenge: A retrospective. *Int. J. Comput. Vis.* **2015**, *111*, 98–136. [CrossRef]

7. Lin, T.Y.; Maire, M.; Belongie, S.; Hays, J.; Perona, P.; Ramanan, D.; Dollár, P.; Zitnick, C.L. Microsoft coco: Common objects in context. *Comput. Vis. ECCV* **2014**, *8693*, 740–755.

8. Zhu, P.; Wen, L.; Du, D.; Bian, X.; Ling, H.; Hu, Q.; Wu, H.; Nie, Q.; Cheng, H.; Liu, C. VisDrone-VDT2018: The vision meets drone video detection and tracking challenge results. In Proceedings of the European Conference on Computer Vision (ECCV), Munich, Germany, 10 September 2018.

9. Zafeiriou, S.; Zhang, C.; Zhang, Z. A survey on face detection in the wild: Past, present and future. *Comput. Vis. Image Underst.* **2015**, *138*, 1–24. [CrossRef]

10. Dollar, P.; Wojek, C.; Schiele, B.; Perona, P. Pedestrian detection: An evaluation of the state of the art. *IEEE Trans. Pattern Anal. Mach. Intell.* **2011**, *34*, 743–761. [CrossRef]

11. Yang, Z.; Pun-Cheng, L.S. Vehicle detection in intelligent transportation systems and its applications under varying environments: A review. *Image Vis. Comput.* **2018**, *69*, 143–154. [CrossRef]

12. Krizhevsky, A.; Sutskever, I.; Hinton, G.E. Imagenet classification with deep convolutional neural networks. In *Advances in Neural Information Processing Systems 25 (NIPS 2012)*; Neural Information Processing Systems Foundation, Inc.: Vancouver, BC, Canada, 2012; pp. 1097–1105.

13. LeCun, Y.; Bengio, Y.; Hinton, G. Deep learning. *Nature* **2015**, *521*, 436–444. [CrossRef]

14. Sermanet, P.; Eigen, D.; Zhang, X.; Mathieu, M.; Fergus, R.; LeCun, Y. Overfeat: Integrated recognition, localization and detection using convolutional networks. *arXiv* **2013**, arXiv:1312.6229.

15. Girshick, R.; Donahue, J.; Darrell, T.; Malik, J. Rich feature hierarchies for accurate object detection and semantic segmentation. In Proceedings of the IEEE Conference on Computer Vision and Pattern Recognition, Columbus, OH, USA, 23–28 June 2014; pp. 580–587.

16. Girshick, R. Fast r-cnn. In Proceedings of the IEEE International Conference on Computer Vision (ICCV), Santiago, Chile, 7–13 December 2015; pp. 1440–1448.

17. Ren, S.; He, K.; Girshick, R.; Sun, J. Faster r-cnn: Towards real-time object detection with region proposal networks. *arXiv* **2015**, arXiv:1506.01497.

18. Liu, W.; Anguelov, D.; Erhan, D.; Szegedy, C.; Reed, S.; Fu, C.Y.; Berg, A.C. Ssd: Single shot multibox detector. *Comput. Vis. ECCV* **2016**, *9905*, 21–37.

19. Dai, J.; Li, Y.; He, K.; Sun, J. R-fcn: Object detection via region-based fully convolutional networks. In Proceedings of the 30th International Conference on Neural Information Processing Systems, Red Hook, NY, USA, 5–10 December 2016; pp. 379–387. Available online: https://arxiv.org/abs/1605.06409 (accessed on 21 June 2016)

20. Redmon, J.; Divvala, S.; Girshick, R.; Farhadi, A. You only look once: Unified, real-time object detection. In Proceedings of the IEEE Conference on Computer Vision and Pattern Recognition (CVPR), Las Vegas, NV, USA, 27–30 June 2016; pp. 779–788.

21. Redmon, J.; Farhadi, A. YOLO9000: Better, faster, stronger. In Proceedings of the IEEE Conference on Computer Vision and Pattern Recognition (CVPR), Honolulu, HI, USA, 21–26 July 2017; pp. 7263–7271.

22. Lin, T.Y.; Goyal, P.; Girshick, R.; He, K.; Dollár, P. Focal loss for dense object detection. In Proceedings of the IEEE International Conference on Computer Vision (ICCV), Venice, Italy, 22–29 October 2017; pp. 2980–2988.

23. Redmon, J.; Farhadi, A. Yolov3: An incremental improvement. *arXiv* **2018**, arXiv:1804.02767.

24. Singh, B.; Najibi, M.; Davis, L.S. SNIPER: Efficient multi-scale training. *arXiv* **2018**, arXiv:1805.09300.

25. Huynh, N.; Nguyen, P.; Vo, N.; Nguyen, K. Detecting Human from Space: An Evaluation of Deep Learning Modern Approaches. In Proceedings of the 15th International Conference on Multimedia Information Technology and Applications, University of Economics and Law, Vietnam National University, Ho Chi Minh City, Vietnam, 27 June 2019.

26. Zeiler, M.D.; Fergus, R. Visualizing and understanding convolutional networks. *Comput. Vis. ECCV* **2014**, *8689*, 818–833.

27. Simonyan, K.; Zisserman, A. Very deep convolutional networks for large-scale image recognition. *Int. Conf. Learn. Represent.* **2015**, arXiv:1409.1556.

28. Szegedy, C.; Liu, W.; Jia, Y.; Sermanet, P.; Reed, S.; Anguelov, D.; Erhan, D.; Vanhoucke, V.; Rabinovich, A. Going deeper with convolutions. In Proceedings of the IEEE Conference on Computer Vision and Pattern Recognition (CVPR) recognition, Boston, MA, USA, 7–12 June 2015; pp. 1–9.

29. He, K.; Zhang, X.; Ren, S.; Sun, J. Deep residual learning for image recognition. In Proceedings of the IEEE Conference on Computer Vision and Pattern Recognition (CVPR) recognition, Las Vegas, NV, USA, 27–30 June 2016; pp. 770–778.

30. Huang, G.; Liu, Z.; Van Der Maaten, L.; Weinberger, K.Q. Densely connected convolutional networks. In Proceedings of the IEEE Conference on Computer Vision and Pattern Recognition (CVPR) recognition, Honolulu, HI, USA, 21–26 July 2017; pp. 4700–4708.

31. He, K.; Zhang, X.; Ren, S.; Sun, J. Spatial pyramid pooling in deep convolutional networks for visual recognition. *IEEE Trans. Pattern Anal. Mach. Intell.* **2015**, *37*, 1904–1916. [CrossRef]

32. He, K.; Gkioxari, G.; Dollár, P.; Girshick, R. Mask r-cnn. In Proceedings of the IEEE International Conference on Computer Vision (ICCV), Venice, Italy, 22–29 October 2017; pp. 2961–2969.

33. Lin, T.Y.; Dollár, P.; Girshick, R.; He, K.; Hariharan, B.; Belongie, S. Feature pyramid networks for object detection. In Proceedings of the Conference on Computer Vision and Pattern Recognition (CVPR), Honolulu, HI, USA, 21–26 July 2017; pp. 2117–2125.

34. Law, H.; Teng, Y.; Russakovsky, O.; Deng, J. Cornernet-lite: Efficient keypoint based object detection. *arXiv* **2019**, arXiv:1904.08900.

35. Duan, K.; Bai, S.; Xie, L.; Qi, H.; Huang, Q.; Tian, Q. Centernet: Keypoint triplets for object detection. In Proceedings of the IEEE/CVF International Conference on Computer Vision (ICCV), Seoul, Korea, 27 October–2 November 2019; pp. 6569–6578.

36. Zhou, X.; Zhuo, J.; Krahenbuhl, P. Bottom-up object detection by grouping extreme and center points. In Proceedings of the IEEE/CVF Conference on Computer Vision and Pattern Recognition (CVPR), Long Beach, CA, USA, 15–20 June 2019; pp. 850–859.

37. Zhou, X.; Wang, D.; Krähenbühl, P. Objects as points. *arXiv* **2019**, arXiv:1904.07850.

38. Wu, X.; Sahoo, D.; Hoi, S.C. Recent advances in deep learning for object detection. *Neurocomputing* **2020**. [CrossRef]

39. Zhu, P.; Wen, L.; Bian, X.; Ling, H.; Hu, Q. Vision meets drones: A challenge. *arXiv* **2018**, arXiv:1804.07437.

40. Vo, N.D.; Nguyen, K.; Nguyen, T.V.; Nguyen, K. Ensemble of Deep Object Detectors for Page Object Detection. In Proceedings of the 12th International Conference on Ubiquitous Information Management and Communication (IMCOM '18), Langkawi, Malaysia, 5–7 January 2018; Association for Computing Machinery: New York, NY, USA, 2018. [CrossRef]

41. Redmon, J. Windows and Linux Version of Darknet Yolo v3 & v2 Neural Networks for Object Detection. 2019. Available online: https://github.com/AlexeyAB/darknet (accessed on 13 February 2020).

42. Gihub, F. Keras Implementation of RetinaNet Object Detection. 2019. Available online: https://github.com/fizyr/keras-retinanet. (accessed on 13 February 2020).

43. Ferrari, P. SSD300 Training Tutorial. 2019. Available online: https://github.com/pierluigiferrari/ssd_keras (accessed on 13 February 2020)

44. Chauhan, V.; Joshi, K.D.; Surgenor, B. Image Classification Using Deep Neural Networks: Transfer Learning and the Handling of Unknown Images. In Proceedings of the International Conference on Engineering Applications of Neural Networks, Crete, Greece, 24–26 May 2019; pp. 274–285. [CrossRef]

 *electronics*

Article

Image Text Deblurring Method Based on Generative Adversarial Network

Chunxue Wu [1], Haiyan Du [1], Qunhui Wu [2] and Sheng Zhang [1,*]

[1] School of Optical-Electrical and Computer Engineering, University of Shanghai for Science and Technology, Shanghai 200093, China; wcx@usst.edu.cn (C.W.); dhy5328@126.com (H.D.)
[2] Shanghai HEST Co. Ltd. Shanghai 201610, China; shhest@aliyun.com
* Correspondence: zhangsheng_usst@aliyun.com; Tel.: +86-13386002013

Received: 2 January 2020; Accepted: 23 January 2020; Published: 27 January 2020

Abstract: In the automatic sorting process of express delivery, a three-segment code is used to represent a specific area assigned by a specific delivery person. In the process of obtaining the courier order information, the camera is affected by factors such as light, noise, and subject shake, which will cause the information on the courier order to be blurred, and some information will be lost. Therefore, this paper proposes an image text deblurring method based on a generative adversarial network. The model of the algorithm consists of two generative adversarial networks, combined with Wasserstein distance, using a combination of adversarial loss and perceptual loss on unpaired datasets to train the network model to restore the captured blurred images into clear and natural image. Compared with the traditional method, the advantage of this method is that the loss function between the input and output images can be calculated indirectly through the positive and negative generative adversarial networks. The Wasserstein distance can achieve a more stable training process and a more realistic generation effect. The constraints of adversarial loss and perceptual loss make the model capable of training on unpaired datasets. The experimental results on the GOPRO test dataset and the self-built unpaired dataset showed that the two indicators, peak signal-to-noise ratio (PSNR) and structural similarity index (SSIM), increased by 13.3% and 3%, respectively. The human perception test results demonstrated that the algorithm proposed in this paper was better than the traditional blur algorithm as the deblurring effect was better.

Keywords: image deblurring; generative adversarial network; Wasserstein distance; adversarial loss; perceptual loss

1. Introduction

Image restoration [1] is an important research direction in image processing. It is a technique to study the cause of degradation and establish a mathematical model to restore high-quality images in response to the degradation in the image acquisition process. Image deblurring [2,3] is also a kind of image restoration. It is mainly aimed at the blurring effect caused by the relative displacement between the photographed object and the device due to camera shake or noise interference. The texture is also clearer and more natural. With this technology, low-quality blurred images can be restored to high-quality clear images. At present, there are many methods applied in the field of image deblurring. However, due to the high-quality requirements of image deblurring, image deblurring is still a very challenging research direction.

With the continuous development of human industrial technology, the application of characters on workpieces in the industrial field is also very important. The on-site environment in an industrial site is chaotic and complex. Factors such as camera shake, subject movement, light, and on-site noise will cause the image captured by the camera to appear blurred. The integrity of the information is urgently required in industrial sites. Solving this problem also has practical significance and application value. In the intelligent express sorting system, as long as the express code is obtained, you can know the area where the express was sent by the courier. If these courier slips are missing, guessing the information with the naked eye is not only cumbersome and heavy but also inefficient. If you can use a computer to automatically recover the blurred image information and automatically restore the blurred three-segment coding from the express order to a clear image, then you can reduce the manpower and material resources for manual data processing. Therefore, it is very important to deblur the fuzzy express image information in the automatic sorting system.

In recent years, with the rapid development of deep learning, a large number of scholars have researched image deblurring methods based on machine learning, and all have achieved good results [4]. In many studies, the image deblurring method based on generative adversarial network (GAN) has been widely recognized. This not only takes into account the rationality of image texture details but also considers the uniformity of the overall image structure. Therefore, this paper proposes an image deblurring method based on the generative adversarial network. Firstly, in order to eliminate the difference between the real blurred image and the generated blurred image, we established an unpaired dataset and solved the image deblurring problem based on this. Then, a GAN model was established, which consisted of two generative adversarial networks. For the conversion of blurred images to clear images and the conversion of clear images to blurred images, the loss function was optimized by combining adversarial loss and perceptual loss. Finally, a stable network model was obtained by iteratively training the GAN model on unpaired datasets. During each training process, the model was updated to achieve a better deblurring effect.

The specific sections of the article are as follows. Section 2 briefly introduces the related work on image deblurring. Section 3 presents the network model proposed in this paper and explains each part of the model in detail. Section 4 gives the experimental demonstration, and Section 5 gives the summary of this article and future expectations.

2. Related Works

With the popularization of handheld devices and multimedia communication technology, image deblurring technology avoids further increasing the high cost of the device as well as noise interference between redundant components. It has broad application scenarios and strong technical advantages. If we can first use the deblurring method to restore the corresponding clear image and then use the restored image as the input of the subsequent neural network, the accuracy of the output of the algorithm will be greatly improved. Therefore, image deblurring technology, as an important part of computer vision data preprocessing, has also become a research hotspot in the field of computer vision and computer graphics.

At present, there are two main research methods for image deblurring. One is the nonblind deblurring method, which uses a known blur kernel function that directly deconvolves the degraded model of the blurred image to obtain a restored high-definition image. The other is the blind deblurring method, which is used when the fuzzy process is unknown. A brief introduction of these two methods is as follows.

The nonblind deblurring method is a more traditional image deblurring method. It first obtains the blur kernel information through a certain technique and then deconvolves the blur image according to the obtained blur kernel to restore a high-definition image. The classic deconvolution algorithms include the Lucy–Richardson algorithm, the Wiener filter, and the Tikhonov filter.

In reality, in most cases, the fuzzy function is unknown. Therefore, it is necessary to make assumptions on the fuzzy source and parameterize the fuzzy function. The most common assumption is that the blur is uniformly distributed on the image. For example, the method proposed by Fergus et al. [5] achieved groundbreaking results, and the literature [6–8] has been optimized based on it. In addition, there are some methods for dealing with cases where the blur is unevenly distributed on the image, but this type of algorithm also simplifies the problem from different angles. For example, Whyte et al. [9] used a parametric geometric model to model camera motion, and Gupta et al. [10]

assumed that blur was caused solely by 3D camera motion. These traditional methods have achieved certain effects. However, because the model makes too many assumptions, they have a lot of limitations in the application scene and cannot solve the problem of image blur caused by various complicated factors in actual life.

With the development of deep learning in the field of computer vision, scholars everywhere have begun to use deep learning to deal with image deblurring. Earlier works were still based on the idea of nonblind deblurring, allowing neural networks to estimate fuzzy kernel information. For example, Sun et al. [11] used a convolutional neural network (CNN) to estimate the fuzzy kernel and then restored the image based on the estimated fuzzy kernel. Chakrabarti et al. [12] used de-CNN to predict the Fourier coefficients of the fuzzy kernel and deblurred the image in the frequency domain. Gong et al. [13] used a full convolutional network (FCN) to estimate the motion flow of the entire image and restored a blurred image based on it. Due to the use of a nonblind deblurring algorithm, the above methods need to obtain a clear image after obtaining the estimated fuzzy kernel through CNN and then use a traditional deconvolution algorithm to deconvolve the blurred image. This leads to slow running speed of the algorithm, and the restoration effect depends entirely on the estimation results of the blur kernel.

In recent years, with the deep prospect of deep learning in the areas of image semantic repair and image compression [14], more and more scholars have discovered that the work that neural networks can cover is far more than just estimating fuzzy kernels. In 2017, Nah et al. [15] proposed the use of multiscale convolutional neural networks to directly deblur images. They used an end-to-end training method to allow the network to directly reproduce clear images without first estimating the blur function. This type of method is called the blind deblurring method. Compared with the previous method, this method greatly improves the model effect and running speed. Other similar methods are those proposed by Noroozi et al. [16], Ramakrishnan et al. [17], and Yao et al. [18,19]. Later, Kupyn et al. [20] proposed the use of conditional generative adversarial networks (CGAN) to deblur images. They followed the basic structure of pix2pix, a general framework for image translation tasks proposed by Isola et al. [21], and modified it to obtain the DeblurGAN image deblurring algorithm model. This model obtained better image deblurring effect than the multiscale convolutional neural network used by Nah et al. At the same time, the network structure was simpler and faster. To some extent, this reflects the fact that the generative adversarial network really performs well on image deblurring tasks.

In this paper, an image deblurring method based on GAN is proposed for unpaired datasets. Because there is no blur–clear image pair in unpaired datasets, a single GAN cannot directly calculate the loss function. Therefore, the proposed model uses two generations. Adversarial networks can realize the mutual conversion between blur and clear images and indirectly calculate the loss function between the input and output images. Therefore, the model has the ability to train and learn on unpaired datasets. At the same time, a loss function that combines adversarial loss and perceptual loss is used for training, making the image generated by the model clearer and more real.

3. Image Deblurring Model

In order to eliminate the difference between the real blurred image and the blurred image synthesized by the algorithm, as well as to achieve a better image deblurring effect in the real industrial scene, we used the CycleGAN structure [22] on the premise of having an unpaired dataset. An image deblurring model based on a generative adversarial network was established. The overall structure of the model is shown in Figure 1. This model consists of two generative adversarial networks A and B, which are used to achieve the conversion from blurred images to clear images and from clear images to blurred images. A and B networks are composed of their respective generators and discriminators. The model also adds a loss function to the network that combines adversarial loss and perceptual loss. Such a model just constitutes a cyclic structure consisting of clear-> blur-> clear and blur-> clear->

blur, which better constrains the content of the generated sample. This article will introduce each part of the model separately.

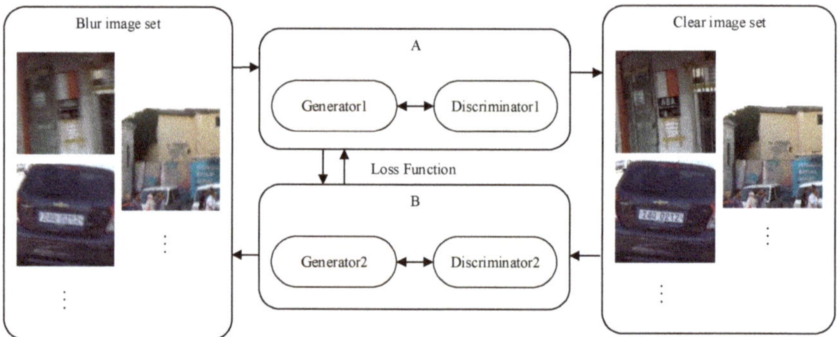

Figure 1. The overall structure of the model.

3.1. Structure of GAN

3.1.1. Generator Model

In generative adversarial networks, generative networks are the key. In previous studies, some classic network structures [23,24] have also achieved outstanding results. Among them, the deep residual network structure has achieved better performance in generating high-definition image tasks because it can greatly increase the number of layers in the network, as shown in [25,26]. The deep residual network was proposed by He et al. [24] in 2016. It is composed of several residual block (ResBlock) and other layers. After using the residual module, the number of layers of the network can be greatly deepened without the phenomenon of difficult model convergence, model degradation, and disappearance of gradients. Therefore, we used the deep residual network structure as the network structure of the generator. In addition, we added global skip connection [27] to the generative network. Global skip connection can directly connect the input of the entire network to the output. Therefore, the intermediate network only learns the residual between the output and the input, thereby reducing the amount of network learning and making the network converge faster and fit to improve the generalization ability of the model.

On the specific network structure, combined with CycleGAN, deep residual network, and global skip connection, the network contains 22 convolutional layers and two deconvolutional layers. A batch normalization (BN) layer [28,29] is added after each convolutional layer. The activation function uses a linear rectification (ReLU) function. The network structure of the generator is shown in Figure 2.

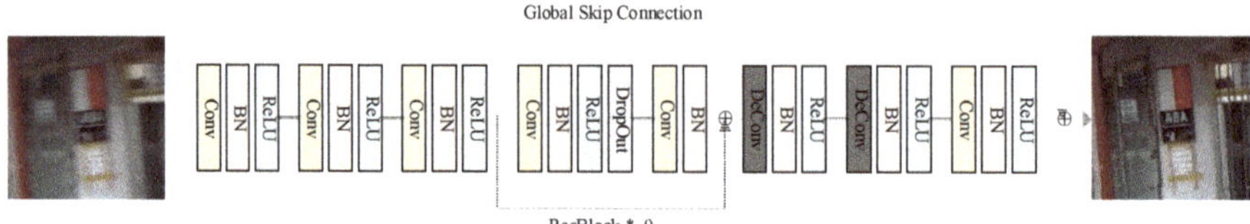

Figure 2. Generator network.

3.1.2. Discriminator Model

Compared with the generator, the task of the discriminator is to learn the difference between the generated sample and the real sample, including whether the structure in the picture is natural and the content is clear. In order to more accurately measure whether the samples are clear and natural, we used PatchGAN, proposed by Isola et al. [21], as the discriminative network structure. Because PatchGAN pays more attention to the local information of the image, the generated image details are more abundant, and the visual effect is more realistic. Unlike the original PatchGAN, we removed the last Sigmoid function activation layer of the original PatchGAN and used Wasserstein distance instead of the original loss function. PatchGAN is unique in that it pays more attention to the local information of the image, which makes the generated image richer with detailed information and the visual effect more realistic.

Because the input of PatchGAN is an image block, we used a sliding window with a size of 70×70 to traverse the entire generated sample. Each image block can output a value through PatchGAN, and the average value of all image blocks can then be obtained with the authenticity of the entire image. The structure of the entire discrimination network is shown in Figure 3.

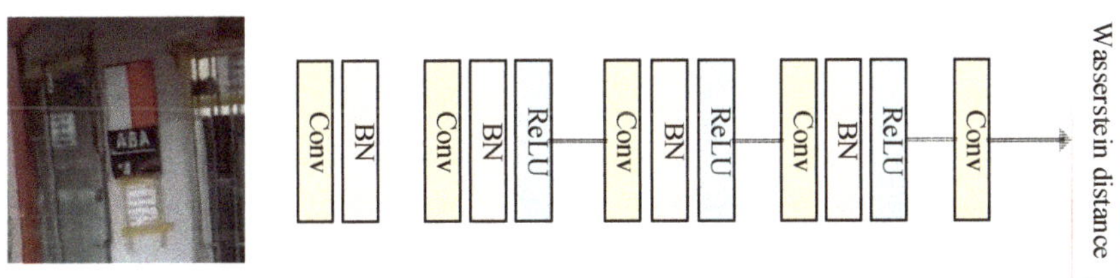

Figure 3. Discriminator network.

3.2. Loss Function of Network

The loss function is the most basic and critical factor in deep learning. By rationally designing the loss function and continuously optimizing it, the network can learn what they should learn without a clear image, thereby achieving a deblurring effect. In this work, the loss function of the entire network is a combination of adversarial loss [30] and perceptual loss. With these two loss functions, the generator can produce clear and realistic images. For the convenience of description, in the following content, Z is used to represent the samples in the clear image set, T is the samples in the blur image set, and N is the number of samples.

3.2.1. Adversarial Loss

Adversarial loss refers to the loss function between two generative adversarial networks A and B. For A, its role is to make the generated image as realistic and clear as possible, and for B, its role is to make the generated sample have as realistic motion blur as possible. In the development of generative adversarial networks, various adversarial loss functions have appeared, including cross-entropy loss functions [31], squared loss functions [32], and Wasserstein distance loss functions [33]. Because WGAN-GP [33,34] uses the Wasserstein distance loss function as the adversarial loss of the network and increases the gradient penalty term for discriminating the network, it has achieved the most stable training effect at present. Therefore, we used the Wasserstein distance loss function for the confrontation loss. The calculation process of the adversarial loss is shown in Figure 4.

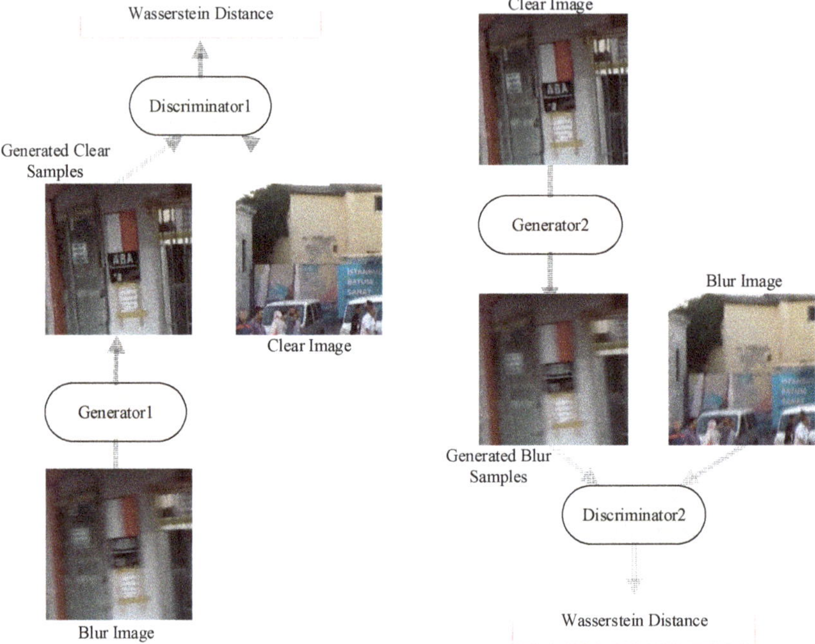

Figure 4. The calculation process of adversarial loss.

The formulas of the two generative adversarial networks are shown in Equations (1) and (2).

$$L_{GAN}(A) = \frac{1}{N}\sum_{n=1}^{N}[D_A(T) - D_A(G_A(Z))] \tag{1}$$

$$L_{GAN}(B) = \frac{1}{N}\sum_{n=1}^{N}[D_B(Z) - D_B(G_B(T))] \tag{2}$$

In the above formula, $G_A(Z)$ represents the sample generated by the generator in network A on the clear image set Z, and $G_B(T)$ represents the sample generated by the generator in network B on the blurred image set T. $D_A(T)$ represents the probability that the discriminator in network A judges whether the blurred image set T is a real image. $D_B(Z)$ represents the probability that the discriminator in network B judges whether the clear image set Z is a real image.

3.2.2. Perceptual Loss

Perceptual loss has the ability of visual perception close to the human eye. Compared with other pixel-level loss functions, it can make the generated image look more realistic and natural. Perceptual loss was originally proposed by Johnson et al. [35], and it has achieved good results in multiple application areas, such as image style transfer [35], image segmentation [36,37], image super-resolution reconstruction [26,38], and image deblurring [20]. The calculation of the perceptual loss depends on the visual geometric group (VGG) network [39]. The specific calculation steps are as follows. First, input two images to be tested, namely, a real image and a generated image, into a pretrained VGG network. Then, extract the feature map output by one or several convolutional layers from the VGG network. Finally, calculate the mean square error (MSE) on the feature maps corresponding to the two images to be tested. In this work, the feature map output from the eighth convolution layer in the

VGG-16 network was selected to calculate the perceptual loss. The calculation process is shown in Figure 5. The formula is shown in Equation (3).

$$L_{per}(G_A, G_B) = \frac{1}{N} \sum_{n=1}^{N} \left[\frac{1}{shw} \varphi(G_B(G_A(Z))) - \varphi(Z)_2^2 + \frac{1}{shw} \varphi(G_A(G_B(T))) - \varphi(T)_2^2 \right] \tag{3}$$

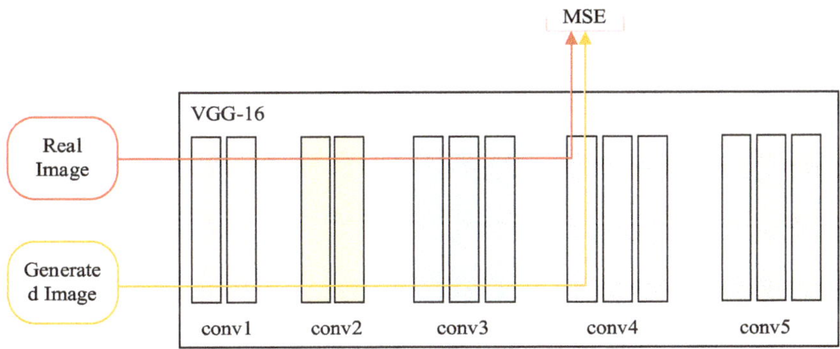

Figure 5. The calculation process of perceptual loss.

Here, φ represents the feature map output from the eighth convolutional layer of VGG-16, and s, h, and w represent the number of channels, height, and width of the feature map, respectively.

In summary, the overall loss function of the network is the result of the weighted summation of the above two loss functions, and the formula is shown in Equation (4).

$$L(A, B, Z, T) = L_{GAN}(G_A, D_A, Z, T) + L_{GAN}(G_B, D_B, T, Z) + \mu L_{per}(G_A, G_B) \tag{4}$$

In the above formula, μ represents the weight of the perceptual loss function. As with the original generative adversarial network, the generator needs to minimize the loss function, and the discriminator needs to maximize the loss function. Therefore, the result of the optimal generator is as follows:

$$G_A', G_B' = arg \min_{G_A, G_B} \max_{D_A, D_B} L(A, B, Z, T) \tag{5}$$

3.3. Algorithm Implementation

In the algorithm implementation process of the entire network model, the role of Discriminator1 and Discriminator2 is to provide gradients for Generator1 and Generator2 to guide its optimization process. The role of generative network B is to realize the conversion from clear images to blurred images and assist generative network A to complete the learning process so that A can generate and input consistent samples. After Discriminator1, Generator2, and Discriminator2 complete their respective auxiliary work, Generator1 is responsible for restoring the input blurred image into a clear image as the output result of the entire algorithm model. In the model training process, we used the buffer pool strategy proposed by Shrivastava et al. [40] to reduce model oscillation. When updating the parameters of the discriminant network, the historical samples generated by the generator and the new samples generated in this iteration were used as the input of the discriminative network, thereby increasing the stability of the model. The size of the generated sample buffer pool was 50, the batch size of all model training was 1, and the number of iterations was 300 epochs. The optimization algorithm used the Adam algorithm [41], and the initial learning rate was 0.0002. The specific implementation process was as Algorithm 1.

Algorithm 1: The algorithm flow of this model.

1: Initialize the input shape, h = 256, w = 256, s = 3, and the output shape of PatchGAN, pa = h/2**4, and the
loss weight, μ = 10, and the optimizer, Adam(0.0002, 0.5)

2: Input (img_1,h,w,s), Input (img_2,h,w,s)

3:    Combined model trains generator to discriminator

4:    **for** epoch in range(300):

5:    **for** batch_i, img_1, img_2 **in** enumerate (dataloader.loadbtach(1)):

6:      fake_2 = generator(img_1), fake_1 = generator(img_2)

7:      recon_1 = generator(fake_2), recon_2 = generator(fake_1)

8:      vali_1 = discriminator(fake_1), vali_2 = discriminator(fake_2)

9:    **if** loss **is** $arg \min\limits_{G_A,G_B} \max\limits_{D_A,D_B} L(A, B, Z, T)$**:**

10:    clear_image = concatenate (img_1, fake_2, recon_1, img_2, fake_1, recon_2)

11: **return** clear_image

---

4. Experiments and Results

The GAN model proposed in this paper is implemented based on the Python language and the
PyTorch deep learning framework, and it can run on an Intel Xeon computer with a 2.40 GHz CPU and
32 GB RAM. The model idea was established by looking at the references for one month. The model
training took 15 days, and the model finally reached a stable state.

4.1. Datasets

Because there are no large-scale unpaired image deblurring datasets yet to be disclosed, it is
meant for comparison with algorithms on paired datasets. Therefore, we still used paired datasets for
training but used unpaired training methods. We used two datasets: the GOPRO dataset and a small
unpaired dataset built in this work. All models were trained on the GOPRO dataset. The self-built
unpaired dataset was only used as a test dataset due to its small number. These two datasets are
described separately.

4.1.1. GOPRO Dataset

The GOPRO dataset is currently the largest and the highest-resolution open-paired dataset in the
field of image deblurring. It consists of 3214 pairs of blurred and clear images. In order to obtain a
more realistic blurred image, all the blurred images in this dataset are fused from multiple clear images
taken in the real scene rather than synthesized by means of clear image convolution blur kernel. For
this work, a simplified version of this dataset was downloaded from the internet. The dataset was
used as the training dataset, and the images in all datasets were cropped from the original GOPRO
dataset. The size was 256 × 256. An image of a partial dataset is shown in Figure 6.

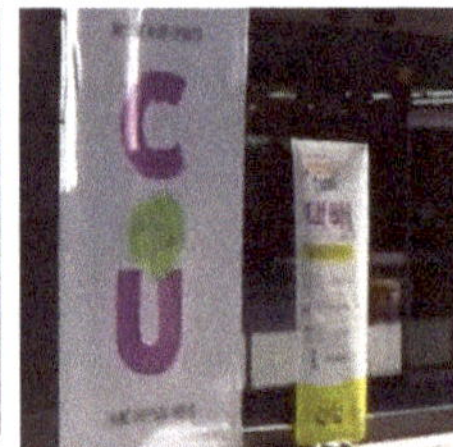

Figure 6. Partial GOPRO dataset.

4.1.2. Unpaired Dataset

In order to eliminate the difference between the real blurred image and the blurred image synthesized by the algorithm, as well as to achieve a better image deblurring effect in the real industrial scene, we trained the network model by unpaired training. In order to test the effectiveness of the model in real scenarios, a small unpaired test dataset was also established. The data contained only 70 blurred images, which were collected or photographed by the author from different sources, and they all originated from real scenes. An image of a partial dataset is shown in Figure 7.

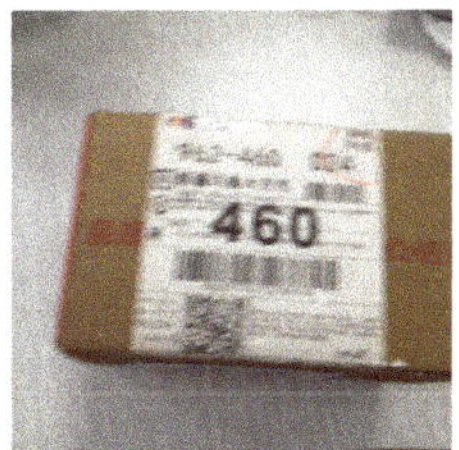

Figure 7. Partial unpaired dataset.

4.2. Experimental Results and Comparison

Based on the CycleGAN, we established an image deblurring algorithm model based on a generative adversarial network. First, the model was trained on the GOPRO dataset, and the number of iterations was 300 epochs. Then, in order to verify the validity of this model, the CycleGAN and DeblurGAN methods were selected to make comparisons with the algorithm used in this work, and they were tested on GOPRO test dataset and self-built unpaired dataset. In order to objectively and comprehensively analyze the deblurring effect of the algorithm, we used two quantitative evaluation indicators of peak signal-to-noise ratio (PSNR) and structural similarity index (SSIM) as well as human perception test methods for evaluation. PSNR evaluates the quality of an image by measuring the error between corresponding pixels in two images, while SSIM measures the degree of image distortion from brightness, contrast, and structural information. The human perception test refers to evaluation of the test results of the model with the human eye. The test subject needs to choose the one that is considered better from the two given images or indicate whether the choice is uncertain. The two images given were reconstructed from two randomly selected models among the three contrasting models. After the subjects completed the test, we combined the pass rate and the TrueSkill evaluation system to measure the deblurring ability of the three models.

First, this paper will show the test results of each model on the GOPRO test dataset. Table 1 shows the quantitative evaluation results, Figure 8 shows the perceptual test results of each model on the GOPRO test set, and Figure 9 shows the deblurring results of each model on the GOPRO test set.

Table 1. Quantitative evaluation results on GOPRO dataset.

Models / Parameters	PSNR	SSIM
CycleGAN	22.6	0.902
Ours	25.6	0.929
DeblurGAN	26.8	0.943

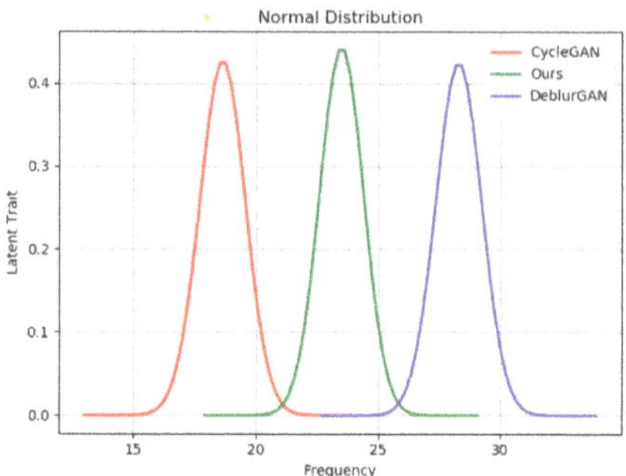

Figure 8. Perceptual test results on the GOPRO test set.

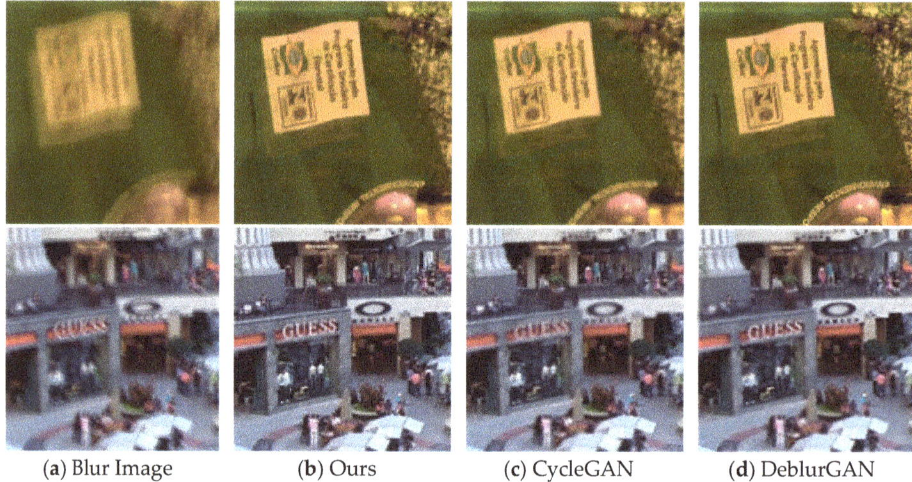

| (**a**) Blur Image | (**b**) Ours | (**c**) CycleGAN | (**d**) DeblurGAN |

Figure 9. Experimental results of different algorithms on GOPRO dataset.

As can be seen from Table 1, the results of the method used in this paper were significantly improved compared to CycleGAN, and the PSNR and SSIM indicators increased by 13.3% and 3%, respectively. Compared with DeblurGAN, our method achieved similar results.

In Figure 8, the results of each model are represented by a Gaussian curve. The mean μ of the curve represents the average deblurring ability of the model, and the variance σ represents the stability of the model. The higher the average value of the curve and the smaller the variance, the better the model. Therefore, it can be seen from the above figure that the model in this paper had better deblurring effect compared to CycleGAN, but it was slightly worse than DeblurGAN.

From the results in Figure 9, it can be seen that, compared with CycleGAN, the effect of the proposed model was significantly improved. Not only was the deblurring ability enhanced, but the disadvantages of chromatic aberration, edge distortion, and unnatural sharpness were also eliminated, making the repair. The resulting image looked more real and natural. Compared with DeblurGAN, the method used in this paper obtained similar results on images with low blurring degree, and the images after deblurring were all natural and clear. However, on the images with a high degree of blurring, the deblurring effect of the method used in this paper was not thorough enough, and it was not as good as DeblurGAN.

Next, this paper will show the test results of each model on a self-built unpaired dataset. Figure 10 shows the perceptual test results of each model on the unpaired dataset, and Figure 11 shows the deblurring results of each model on the unpaired dataset.

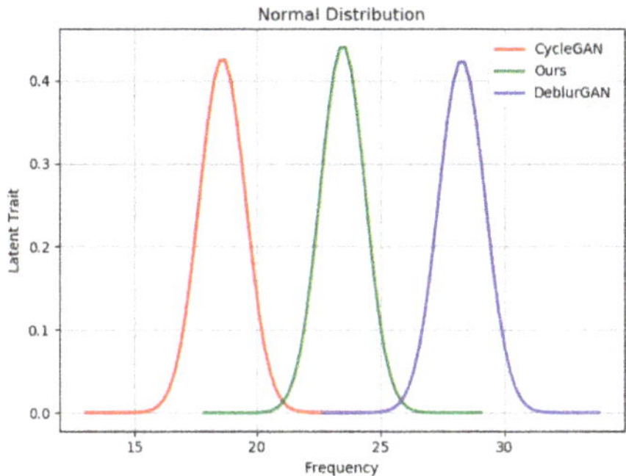

Figure 10. Perceptual test results on unpaired dataset.

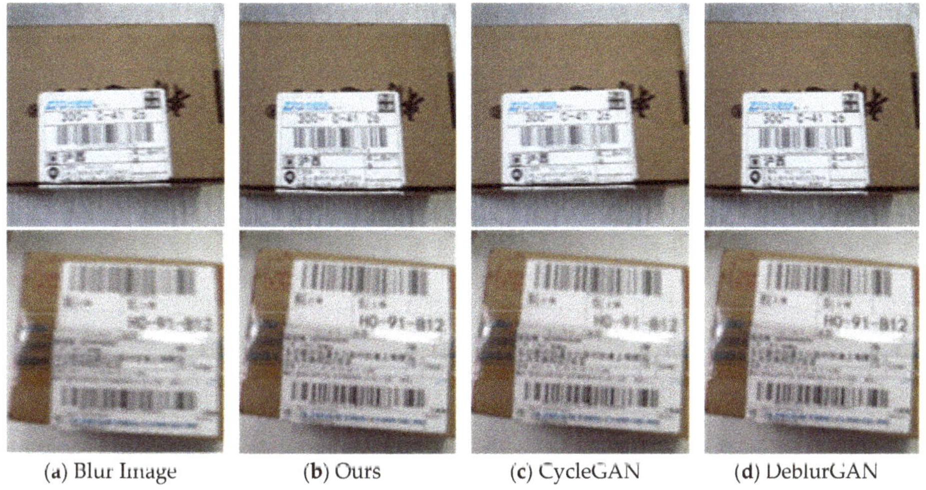

(a) Blur Image (b) Ours (c) CycleGAN (d) DeblurGAN

Figure 11. Experimental results of different algorithms on unpaired dataset.

From the results in Figures 10 and 11, we can see that, on the unpaired dataset, the proposed model had better deblurring effect than CycleGAN, and the repaired image looked more realistic and natural. However, its performance was slightly worse than DeblurGAN.

Combining the experimental results on the above two datasets, it can be seen that the loss function, which combines the adversarial loss and perceptual loss, can play a certain role in constraining the content of the generated image. However, because the generated image was not directly constrained, the image was generated by the constraint. The reconstructed image was used to indirectly constrain the generated image, so the effect was limited, and the effect achieved was not as good as that on the paired dataset. However, in general, the method achieved certain results on highly difficult unpaired datasets. Compared to traditional CycleGAN, the deblurring effect was significantly improved. If a large-scale image dataset in a real scene can be obtained, the effect will be better.

In order to show the importance of the combination of various parts of the model to the deblurring effect, ablation research was also performed in this work. The ablation of the model was achieved by reducing the number of convolution layers of the generator network to 16 layers and removing one of the perceptual loss functions. Finally, through model training, the evaluation index results on the GOPRO paired dataset are shown in Table 2.

Table 2. Quantitative evaluation results after ablation.

Parameters Models	PSNR	SSIM
Ours	25.6	0.929
Ablation	18.5	0.767

As can be seen from the above table, after reducing the generator convolutional layer and removing the perceptual loss, the PSNR and SSIM evaluation indexes were reduced too much, by 27% and 17%, respectively, resulting in poor model performance. Therefore, the various components of the model proposed in this work are particularly important in the field of image deblurring.

5. Conclusions and Future Works

With the widespread use of handheld devices and digital images on multimedia communication networks, image deblurring technology has more and more application value. In particular, in an intelligent express sorting system, using a computer to restore the fuzzy three-segment coded information on the courier slip to a clear image can improve the recognition effect of the subsequent three-segment code. Therefore, this paper proposes an image deblurring method based on generative adversarial networks. First, in view of the shortcomings of the existing algorithms, we dealt with image deblurring on unpaired datasets to solve motion deblurring in actual scenes. Then, based on CycleGAN, an image deblurring model based on a generative adversarial network was established to realize the conversion of blurred images to clear images and the conversion of clear images to blurred images. The process of network learning was also constrained by combining adversarial loss and perceived loss so that the network could better learn the motion-blurred data distribution characteristics in the actual scene. Finally, an evaluation was performed on the GOPRO dataset and the self-built unpaired dataset. The experimental results showed that the proposed method could obtain good deblurring effect on both datasets and that it was better than CycleGAN. However, some improvements are still required. For example, in future work, we may try to introduce a multiscale network structure into the model and deepen the network layers at the same time to improve the capacity of the model. There are also loss functions of other structures designed to strengthen their constraints on the generated sample content, which can be used to achieve a complete deblurring effect.

Author Contributions: Conceptualization, C.W.; methodology, H.D. and Q.W.; software, H.D.; validation, H.D.; formal analysis, C.W. and H.D.; investigation, Q.W.; resources, S.Z.; data curation, H.D.; writing—original draft preparation, H.D.; writing—review and editing, H.D.; visualization, H.D.; supervision, S.Z. All authors have read and agree to the published version of the manuscript.

Funding: This research was funded by SHANGHAI SCIENCE AND TECHNOLOGY INNOVATION ACTION PLAN PROJECT, grant number 19511105103.

Acknowledgments: The authors would like to thank all anonymous reviewers for their insightful comments and constructive suggestions to polish this paper to high quality. This research was supported by the Shanghai Science and Technology Innovation Action Plan Project (19511105103) and the Shanghai Key Lab of Modern Optical System.

Conflicts of Interest: The authors declare no conflict of interest. The funders had no role in the design of the study; in the collection, analyses, or interpretation of data; in the writing of the manuscript, or in the decision to publish the results.

References

1. Vyas, A.; Yu, S.; Paik, J. Image Restoration. In *Multiscale Transforms with Application to Image Processing*; Springer: Singapore, 2018; pp. 133–198.

2. Mahalakshmi, A.; Shanthini, B. A survey on image deblurring. In Proceedings of the 2016 International Conference on Computer Communication and Informatics (ICCCI), Coimbatore, India, 7–9 January 2016.

3. Dawood, A.; Muhanad, F. Image deblurring tecniques2019. *J. Sci. Eng. Res.* **2019**, *6*, 94–98.

4. Leke, C.A.; Marwala, T. Deep Learning Framework Analysis. In *Deep Learning and Missing Data in Engineering Systems. Studies in Big Data*; Springer: Cham, Switzerland, 2019; Volume 48, pp. 147–171.

5. Fergus, R.; Singn, B.; Hertzmann, A.; Roweis, S.T.; Freeman, W.T. Removing Camera Shake from a Single Photograph. *ACM Trans. Graphics* **2006**, *25*, 787–794. [CrossRef]

6. Xu, L.; Jia, J. Two-Phase Kernel Estimation for Robust Motion Deblurring. In Proceedings of the 11th European Conference on Computer Vision, Heraklion, Greece, 5–11 September 2010.

7. Babacan, S.D.; Molina, R.; Do, M.N.; Katsaggelos, A.K. Bayesian Blind Deconvolution with General Sparse Image Priors. In Proceedings of the 12th European conference on Computer Vision, Florence, Italy, 7–13 October 2012.

8. Li, X.; Zheng, S.; Jia, J. Unnatural L0 Sparse Representation for Natural Image Deblurring. In Proceedings of the 2013 IEEE Conference on Computer Vision and Pattern Recognition, Portland, OR, USA, 23–28 June 2013.

9. Whyte, O.; Sivic, J.; Zisserman, A.; Ponce, J. Non-uniform Deblurring for Shaken Images. *Int. J. Comput. Vision* **2012**, *98*, 168–186. [CrossRef]

10. Gupta, A.; Joshi, N.; Zitnick, C.; Micheael, C.; Curless, B. Single Image Deblurring Using Motion Density Functions. 6311. In Proceedings of the 11th European Conference on Computer Vision, Heraklion, Greece, 5–11 September 2010; pp. 171–184.

11. Sun, J.; Cao, W.; Xu, Z.; Ponce, J. Learning a Convolutional Neural Network for Non-uniform Motion Blur Removal. In Proceedings of the 2015 IEEE Conference on Computer Vision and Pattern Recognition (CVPR), Boston, MA, USA, 7–12 June 2015.

12. Chakrabarti, A. A Neural Approach to Blind Motion Deblurring. In Proceedings of the 14th European Conference on Computer Vision, Amsterdam, The Netherlands, 11–14 October 2016; pp. 221–235.

13. Gong, D.; Yang, J.; Liu, L.; Zhang, Y.; Reid, I.; Shen, C.; van Den Hengel, A.; Shi, Q. From Motion Blur to Motion Flow: A Deep Learning Solution for Removing Heterogeneous Motion Blur. In Proceedings of the 2017 IEEE Conference on Computer Vision and Pattern Recognition (CVPR), Honolulu, HI, USA, 21–26 July 2017; pp. 3806–3815.

14. Hu, Q.; Wu, C.; Wu, Y.; Xiong, N. UAV Image High Fidelity Compression Algorithm Based on Generative Adversarial Networks Under Complex Disaster Conditions. *IEEE Access* **2019**, *7*, 91980–91991. [CrossRef]

15. Nah, S.; Kim, T.H.; Lee, K.M. Deep Multi-scale Convolutional Neural Network for Dynamic Scene Deblurring. In Proceedings of the 2017 IEEE Conference on Computer Vision and Pattern Recognition (CVPR), Honolulu, HI, USA, 21–26 July 2017.

16. Noroozi, M.; Paramanand, C.; Favaro, P. Motion Deblurring in the Wild. In Proceedings of the 39th German Conference on Pattern Recognition, GCPR 2017, Basel, Switzerland, 12–15 September 2017; pp. 65–77.

17. Ramakrishnan, S.; Pachori, S.; Gangopadhyay, A.; Raman, S. Deep Generative Filter for Motion Deblurring. In Proceedings of the 2017 IEEE International Conference on Computer Vision (ICCV), Venice, Italy, 22–29 October 2017; pp. 2993–3000.

18. Jiang, X.; Yao, H.; Zhao, S. Text image deblurring via two-tone prior. *Neurocomputing* **2017**, *242*, 1–14. [CrossRef]

19. Huang, Y.; Yao, H.; Zhao, S.; Zhang, Y. Towards more efficient and flexible face image deblurring using robust salient face landmark detection. *Multimedia Tools Appl.* **2017**, *76*, 123–142. [CrossRef]

20. Kupyn, O.; Budzan, V.; Mykhailych, M.; Mishkin, D.; Matas, J. DeblurGAN: Blind Motion Deblurring Using Conditional Adversarial Networks. In Proceedings of the 2018 IEEE/CVF Conference on Computer Vision and Pattern Recognition, Salt Lake City, UT, USA, 18–23 June 2018.

21. Isola, P.; Zhu, J.-Y.; Zhou, T.; Efros, A.A. Image-to-Image Translation with Conditional Adversarial Networks. In Proceedings of the 2017 IEEE Conference on Computer Vision and Pattern Recognition (CVPR), Honolulu, HI, USA, 21–26 July 2017; pp. 5967–5976.

22. Zhu, J.-Y.; Park, T.; Isola, P.; Efros, A.A. Unpaired Image-to-Image Translation using Cycle-Consistent Adversarial Networks. In Proceedings of the 2017 IEEE International Conference on Computer Vision (ICCV), Venice, Italy, 22–29 October 2017; pp. 2223–2232.

23. Szegedy, C.; Liu, W.; Jia, Y.; Sermanet, P.; Reed, S.; Anguelov, D.; Erhan, D.; Vanhoucke, V.; Rabinovich, A. Going Deeper with Convolutions. In Proceedings of the 2015 IEEE Conference on Computer Vision and Pattern Recognition (CVPR), Boston, MA, USA, 7–12 June 2015.

24. He, K.; Zhang, X.; Ren, S.; Sun, J. Deep Residual Learning for Image Recognition. In Proceedings of the 2016 IEEE Conference on Computer Vision and Pattern Recognition (CVPR), Las Vegas, NV, USA, 27–30 June 2016; pp. 770–778.

25. Wang, T.-C.; Liu, M.-Y.; Zhu, J.-Y.; Tao, A.; Kautz, J.; Catanzaro, B. High-Resolution Image Synthesis and Semantic Manipulation with Conditional GANs. In Proceedings of the 2018 IEEE/CVF Conference on Computer Vision and Pattern Recognition, Salt Lake City, UT, USA, 18–23 June 2018; pp. 8798–8807.

26. Ledig, C.; Theis, L.; Huszar, F.; Caballero, J.; Cunningham, A.; Acosta, A.; Aitken, A.; Tejani, A.; Totz, J.; Wang, Z.; et al. Photo-Realistic Single Image Super-Resolution Using a Generative Adversarial Network. In Proceedings of the 2017 IEEE Conference on Computer Vision and Pattern Recognition (CVPR), Honolulu, HI, USA, 21–26 July 2017.

27. Shi, J.; Li, Z.; Ying, S.; Wang, C.; Liu, Q.; Zhang, Q.; Yan, P. MR Image Super-Resolution via Wide Residual Networks With Fixed Skip Connection. *IEEE J. Biomed. Health Inf.* **2018**, *23*, 1129–1140. [CrossRef] [PubMed]

28. Cai, J.; Chang, O.; Tang, X.-L.; Xue, C.; Wei, C. Facial Expression Recognition Method Based on Sparse Batch Normalization CNN. In Proceedings of the 2018 37th Chinese Control Conference (CCC), Wuhan, China, 25–27 July 2018; pp. 9608–9613.

29. Rajeev, R.; Samath, J.A.; Karthikeyan, N.K. An Intelligent Recurrent Neural Network with Long Short-Term Memory (LSTM) BASED Batch Normalization for Medical Image Denoising. *J. Med. Syst.* **2019**, *43*, 234. [CrossRef] [PubMed]

30. Kim, D.-W.; Chung, J.-R.; Kim, J.; Lee, D.Y.; Jeong, S.Y.; Jung, S.-W. Constrained adversarial loss for generative adversarial network-based faithful image restoration. *ETRI J.* **2019**, *41*, 415–425. [CrossRef]

31. Goodfellow, I.; Pouget-Abadie, J.; Mirza, M.; Xu, B.; Warde-Farley, D.; Ozair, S.; Courville, A.; Bengio, Y. Generative Adversarial Nets. In Proceedings of the Neural Information Processing Systems 2014, Montreal, QC, Canada, 8–13 December 2014.

32. Mao, X.; Li, Q.; Xie, H.; Lau, R.Y.K.; Wang, Z.; Smolley, S.P. Least Squares Generative Adversarial Networks. In Proceedings of the 2017 IEEE International Conference on Computer Vision (ICCV), Venice, Italy, 22–29 October 2017; pp. 2813–2821.

33. Arjovsky, M.; Chintala, S.; Bottou, L. Wasserstein GAN. *arXiv* **2017**, arXiv:1701.07875.

34. Gulrajani, I.; Ahmed, F.; Arjovsky, M.; Dumoulin, V.; Courville, A. Improved Training of Wasserstein GANs. *arXiv* **2017**, arXiv:1704.00028.

35. Johnson, J.; Alahi, A.; Li, F.-F. Perceptual Losses for Real-Time Style Transfer and Super-Resolution. In Proceedings of the 14th European Conference on Computer Vision, Amsterdam, The Netherlands, 11–14 October 2016.

36. Xiong, N.N.; Shen, Y.; Yang, K.; Lee, C.; Xu, C. Color sensors and their applications based on real-time color image segmentation for cyber physical systems. *J. Image Video Proc.* **2018**, *23*. [CrossRef]

37. Liu, C.; Zhou, A.; Xu, C.; Zhang, G. Image segmentation framework based on multiple feature spaces. *IET Image Proc.* **2015**, *9*, 271–279. [CrossRef]

38. Xi, S.; Wu, C.; Jiang, L. Super resolution reconstruction algorithm of video image based on deep self encoding learning. *Multimedia Tools Appl.* **2019**, *78*, 4545–4562. [CrossRef]

39. Simonyan, K.; Zisserman, A. Very Deep Convolutional Networks for Large-Scale Image Recognition. *arXiv* **2015**, arXiv:1409.1556.

40. Shrivastava, A.; Pfister, T.; Tuzel, O.; Susskind, J.; Wang, W.; Webb, R. Learning from Simulated and Unsupervised Images through Adversarial Training. In Proceedings of the 2017 IEEE Conference on Computer Vision and Pattern Recognition (CVPR), Honolulu, HI, USA, 21–26 July 2017.

41. Kingma, D.; Ba, J. Adam: A Method for Stochastic Optimization. *arXiv* **2015**, arXiv:1412.6980.

Article

Exploring Impact of Age and Gender on Sentiment Analysis Using Machine Learning

Sudhanshu Kumar [1,*], Monika Gahalawat [1], Partha Pratim Roy [1] and Debi Prosad Dogra [2] and Byung-Gyu Kim [3,*]

[1] Department of Computer Science & Engineering, Indian Institute of Technology Roorkee, Uttarakhand 247667, India; monikagahalawat62@gmail.com (M.G.); proy.fcs@iitr.ac.in (P.P.R.)

[2] School of Electrical Sciences, Indian Institute of Technology Bhubaneswar, Odisha 752050, India; dpdogra@iitbbs.ac.in

[3] Department of IT Engineering, Sookmyung Women's University, Seoul 04310, Korea

* Correspondence: skumar2@cs.iitr.ac.in (S.K.); bg.kim@ivpl.sookmyung.ac.kr (B.-G.K.)

Received: 25 January 2020; Accepted: 17 February 2020; Published: 22 February 2020

Abstract: Sentiment analysis is a rapidly growing field of research due to the explosive growth in digital information. In the modern world of artificial intelligence, sentiment analysis is one of the essential tools to extract emotion information from massive data. Sentiment analysis is applied to a variety of user data from customer reviews to social network posts. To the best of our knowledge, there is less work on sentiment analysis based on the categorization of users by demographics. Demographics play an important role in deciding the marketing strategies for different products. In this study, we explore the impact of age and gender in sentiment analysis, as this can help e-commerce retailers to market their products based on specific demographics. The dataset is created by collecting reviews on books from Facebook users by asking them to answer a questionnaire containing questions about their preferences in books, along with their age groups and gender information. Next, the paper analyzes the segmented data for sentiments based on each age group and gender. Finally, sentiment analysis is done using different Machine Learning (ML) approaches including maximum entropy, support vector machine, convolutional neural network, and long short term memory to study the impact of age and gender on user reviews. Experiments have been conducted to identify new insights into the effect of age and gender for sentiment analysis.

Keywords: sentiment analysis; social media; machine learning; lexicon

---

1. Introduction

The growth of the internet has led to a huge influx of data that holds vast and valuable insights about the public opinion. Every internet user who expresses an opinion on the web becomes a part of this information circuit where other users benefit from these public reviews and hence can make an informed decisions. With the data collected (reviews, posts, comments) from different social media platform such as Facebook, Twitter, Amazon, Goodreads, IMDb or blogs, the task of using these reviews to find the polarity of public (positive, negative or neutral) opinion is called Sentiment analysis. Sentiment analysis is generally performed on movie reviews [1,2], restaurant or food reviews [3,4], along with data from microblogs [5,6], providing some useful insights to different organizations to improve business strategies by attracting new customers. The categorization of customers based on age and gender present an important information that can make products more effectively fullfill the demands of different age and gender group persons. This fine-grain information about customers are value-added to enhance the revenue of the company and its reputation in the global market. E-commerce companies want to know the mindset of the customers. For example, females do more shopping in comparision to male in the E-commerce site and certain portals such as firstcry website are more popular for various products of different age groups including newborn, infant, toddler, etc. Sentiment analysis [7–10] has evolved over the years with different dictionary-based and machine learning techniques implemented to obtain better accuracy. With the advent of deep learning techniques [11–13] in sentiment analysis, prior information has also played a big role in adequately expressing the polarity of opinions. Kahaki et al. [14] proposed an age estimation system based on orthopantomographs images. The orthopantomogram is a dental X-ray of the upper and lower jaw. The geometric mean projection transform was used on Malaysian children dental development dataset with 456 patient's X-ray images to extract and classify the 3rd molar teeth in the orthopantomography images. The proposed system results showed a reliable age estimation method. The importance of age estimation may also be useful for civil, criminal, law enforcement, airport security and for forensic purposes. In a similar study on the same Malaysian children dataset, the automatic age assessment [15] was proposed based on pre-trained deep convolution neural network. The results of this approach concluded that the method can efficiently classify the images with high accuracy and precision.

Li et al. [16] proposed a framework providing an abstract of the opinions using sentiment analysis. The authors have taken into consideration the opinion subjectivity and user credibility in their proposed approach. Lockenhoff et al. [17] analyzed how different age groups express their emotions. The authors found that older adults describe the positive emotions better than their negative emotions as compared to younger adults. Another research by Zimmermann et al. [18] for emotion regulation based on age and gender [19] of the person was able to handle emotions in a better way. The results showed that people in middle adolescence showed the least emotional regulation as compared to other age groups. Even gender differences were also encountered as either under or over estimating a particular emotion. Thus, based on these established psychological differences between people of different age and gender, we aim to find out whether these differences can also be observed in the opinions that the individuals express on on-line platforms.

This study explores the differences in sentiment expressing abilities of different groups and their subsequent impact on the sentiment reviews. This can be extremely helpful for commercial applications as they can focus directly on a particular audience that is more receptive towards their product rather than making a generalized marketing strategy. It will further help in differentiating their brand from other leading companies to provide better customer support. Oh et al. [20] studied the market segmentation on basis of gender and age of users to find the travel potential of different groups based on their incomes and available leisure time. Keshari et al. [21] analyzed the effectiveness of advertising appeals on different gender and age groups based on how the consumers respond to these advertisements. Figure 1 demonstrates the basic flow diagram of the framework used for sentiment analysis, where the data collected from social media is used to extract two datasets on basis of age and gender. The different sentiment analysis approaches have been implemented on this data. The main contributions of the paper are as follows:

1. Exploring the impact of user expression based on age and gender using different feature extraction methods.
2. We create a dataset that contains user reviews along with the user's age and gender information.
3. A detailed analysis on the impact of user expression is presented through extensive experiments.
4. Finally, a comparison with different machine learning and a dictionary-based classifier is also discussed.

The rest of the paper is organized as follows. In Section 2, we discuss the existing research work in sentiment analysis. In Section 3, the methodologies implemented on the dataset have been discussed, along with a comparison of different approaches. Section 4 describes the experimental results with dataset description. Finally, in Section 5 the work has been concluded along with discussion of some future possibilties.

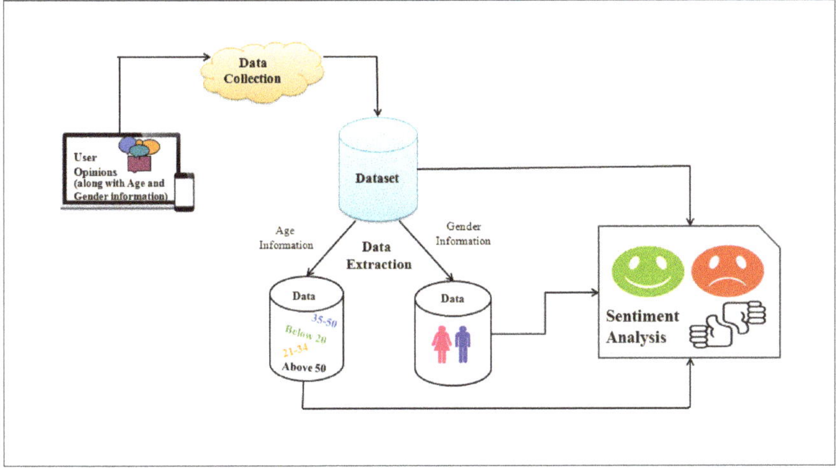

Figure 1. Data points are collected from the social media along with the user's age and gender information. Sentiment analysis is then performed on the newly created data sets.

2. Related Work

In this section, we discuss the recent works of sentiment analysis as researchers try to find a better approach to predict the sentiment polarity. Twitter and Facebook have been the most popular social media platforms as people express their opinion about every topic on these social networking sites, which helps in understanding public sentiment. Appel et al. [22] used twitter sentiment and movie review datasets to implement a hybrid approach based on ambiguity management, semantic rules, and sentiment lexicon. The authors compared this proposed hybrid system results with the standard supervised algorithms such as Naive Bayes (NB) and Maximum Entropy (ME). The proposed system achieves higher precision score and accuracy than the supervised algorithms. Similarly, Zainuddin et al. [23] used a twitter dataset of aspect-based sentiment analysis to perform a fine-grained analysis. They proposed a hybrid approach using a feature selection method that performs better than the standard methods.

Blogs have been a relevant source of data in sentiment analysis with posts containing reviews and comments. Fan et al. [24] analyzed blog text to improve the quality of advertisements in the blogs that were more relevant to the user. To find the blogger's overall emotions towards any particular topic, Kuo et al. [25] create a social opinion graph as generally every blogger is somewhat influenced by its social circle. So their social interactions can be used to find the overall sentiment orientation of the blogger. Li et al. [26] used opinions expressed on the web such as blogs, reviews and comments to design a new technique to further enhance the accuracy of clustering based approaches. This approach is proven to more suitable in finding neutral opinions. The authors [27] proposed a new extraction and opinion mining system based on a type-2 fuzzy ontology called T2FOBOMIE. The proposed system received input from a user, extracts the relevant features from an input query and then converts into to a search query with hotel reviews. The feature opinions, user requirements and hotel information were integrated in a T2FOBOMIE system to achieve high performance.

Apart from using products, movie, restaurants or book reviews for sentiment analysis, researchers have also focused on analyzing sentiment in other languages than English. Pak et al. [28] have proposed a technique that works quite well for other languages as well, though they have not tested their algorithm on multilingual data . The author [29] has implemented a methodology to find sentiment polarity within a multilingual framework and the testing was performed using movie reviews in German language collected from amazon. Similarly, Zhou et al. [30] translated Chinese reviews to English language and then used English language corpus to perform sentiment analysis on these translated reviews. The authors presented that translated reviews outperform original reviews. Another study on Chinese public figures has been performed in [31] to analyze the opinion polling of public figures.

The analysis of opinions expressed by people from different genders or different age groups should align with their psychological differences, as is illustrated by different research groups. There have been multiple research studies on how different individuals handle different emotions and the way these individuals express their emotions even before the advent of internet. The authors [32] examined gender differences in conducting a study on 400 college students in five age groups from preschoolers to adults. The study aligned with the stereotypes of gender and age emotional expressiveness. Stoner et al. [33] considered people of both genders and in different age groups to study their anger expressing ability. The research showed that young adult group expressed anger more as compared to old adult age group. In this study, the author did not find out much differences on basis of gender in this aspect.

A research by Davis [34] on gender differences in negative emotions showed that boys expressed a greater negative affect as compared to girls when they were disappointed. Brody et al. [35] researched more on gender and emotional expression and showed that gender differences in emotional expressiveness were culturally specific in asian international students. Another study by Kring et al. [36] in which they showed emotional videos to a group of students and reaffirmed that women are generally more expressive than men even in case of experienced emotions. A study by Birditt [37] examined age and gender differences in description of emotional reactions. It contained 185 individuals as 85 males and 100 female aged from 13 to 99 which showed that adolescents and young adults were reported more likely to describe anger and giving more intensive aversive responses as opposed to the male adult group.

3. Methodology

To process the reviews, the steps in the Figure 2 are followed. Firstly, the dataset segregated into two sets on basis of age and gender and then separated into categories based on the specific age and gender. Secondly, each particular data group is divided in training data and testing reviews.

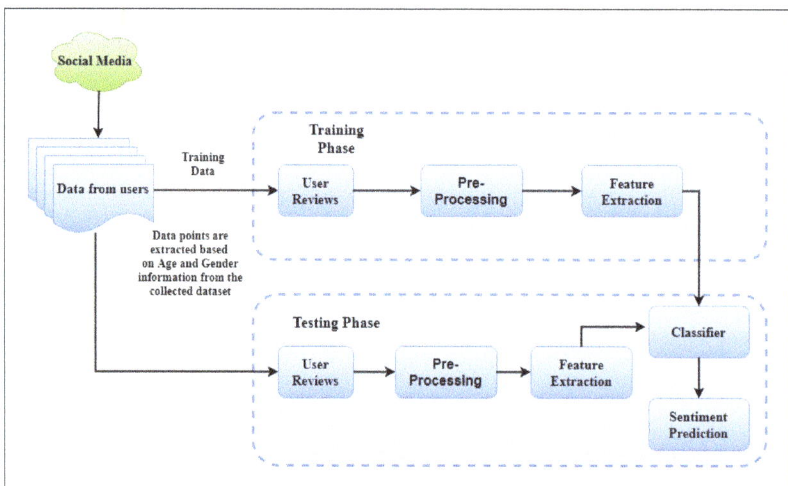

Figure 2. Flow Diagram representing the steps taken for sentiment analysis where the classifier algorithm is implemented at the end of training after the data pre-processing and feature extraction and it used in testing step to produce the final results. The user reviews need to go through pre-processing and feature extraction in the testing phase as well before being passed on to the classifier algorithm.

The reviews [38] are pre-processed to remove the unnecessary information from the reviews that has no effect on the polarity of the sentence. So, we perform data cleansing through the steps as shown in Table 1. Then, the feature extraction steps are performed as explained in Section 3.1.2. Finally, the classifier algorithm predicts the label which when compared to the ground truth gives the accuracy of the classifier. We have collected data regarding people's preference for the books (hard cover, kindle ebooks or audio books) along with their age and gender information. We implement different algorithms for sentiment analysis on each set of data separately and the results are then compared to identify the respective differences between the groups. Also, a dictionary-based approach has been implemented on the collected dataset.

Table 1. Pre-processing steps that have been performed on the user reviews for doing data cleansing and removing uninformative parts that has no effect on the sentiment score of the sentence.

S.No.	Description of Noisy and Uninformative Parts in Reviews
1.	Removing punctuations, numbers and symbols since they do not add any substantial meaning to the sentence that may affect it's sentiment score.
2.	Removing stop words as they make no impact on the sentiment score of the expressed opinion.
3.	Replacing the acronyms of a word with the actual word.
4.	Transforming the text to lowercase.
5.	Replacing emoticons with the sentiment that the emoticon expresses.
6.	Tokenize the review.

3.1. Feature Extraction

Bag of words feature [39] extraction is used in NB, ME and SVM methods, while word2vec creates a feature vector using either Continuous bag of words or Skip gram model which is further used in LSTM and CNN. The methods are explained below.

3.1.1. Bag-of-Words

Bag of words model is a very flexible and simple model used for feature extraction. This model keeps a track of number of occurrences, also called term frequency of every word that appears in the sentence. Also, a specific subjectivity score is assigned to each word of the sentence. The score for each word is added up to find the total score. Depending upon this total score, the polarity of each sentence is decided.

3.1.2. Word2Vec

Word2Vec model is used for forming word embeddings. It is a two-layer neural network created by Tomas Mikolov at google to process text. It takes the text dataset as an input and then outputs a set of vectors [40]. Word2Vec is a combination of two techniques, i.e., Skip-gram model and Continuous bag of words (CBOW) model. This model is very useful as it detects similarities of words in its vector form rather than textual format. These similarities are detected on the basis of word's meaning guessed through its past appearances and association with other words.

3.2. Dictionary-Based Classifier

Valence Aware Dictionary and Sentiment Reasoner [41] (VADER) is a dictionary-based approach that maps words to sentiment by building a or a 'dictionary of sentiment'. In this approach, each word present in the sentence is assigned a score as per the meaning of that word in the dictionary. A final compound score of the sentence is calculated which varies from -1 to 1. This score represents whether the sentence is positive or negative. The compound score for each sentence in the dataset is combined and an average score for the whole document is analyzed. To compare it with the other machine learning approaches, we convert the average score to accuracy by dividing the score of the whole document by the total number of reviews in that particular data set. VADER focuses on the words used in the sentence and then assigns score to each word based on the dictionary.

3.3. Machine Learning Based Classifiers

We discuss in detail five machine learning based algorithms to determine the sentiment accuracy of the dataset.

3.3.1. Naive Bayes

This is a probabilistic model based on the Bag-of-words module to store only the frequencies of each word and ignore their positioning with respect to each other. By using Bayes Theorem, it estimates the probability that a feature set will belong to a particular predefined label. Naive Bayes classification model [42], based on the distribution of words present in the document or sentence, computes the posterior probability that this document or sentence will belong to a particular class. The probability is based on the distribution and frequency of the words rather than their positioning with respect to each other.

$$P(label|features) = \frac{P(label) * P(features|label)}{P(features)},\tag{1}$$

where P(label|features) determines the probability that a feature set belongs to a particular label. P(label) is the prior estimate of the label. P(features|label) is the probability that the given feature set belongs to this particular label and P(features) is the prior estimate that this given feature set occurred. However, this classification system makes one fundamental assumption, i.e., words in a reviews, category pair occur independent of other words.

3.3.2. Maximum Entropy

Maximum Entropy (ME) [43] belongs to the class of exponential models. Its polarity is more based on the positioning of words rather than their frequencies. It does not assume that all the features are independent of each other like Naive Bayes. Based on the principle of ME, from all the models, we pick the one that has the largest entropy. The ME classifier uses encoding to convert the feature sets into vectors. Then for computation of most likely labels for each feature set, we combine the calculated weight for each feature [44].

The Maximum Entropy modeling technique provides a probability distribution that is as close to the uniform distribution, so its result is better than Naive Bayes.

3.3.3. Support Vector Machine (SVM)

Support Vector Networks works for multiple machine learning problems such as regression and classification. The main principle that works behind SVM is finding a particular linear classifier that separates all the classes in the search space in the best possible manner. After the pre-processing of the reviews, the improved feature sets were used for sentiment classification, i.e., positive and negative reviews. With the help of hyper plane in support vector machine the data is divided into two classes such as positive and negative. This hyperplane used to map the new examples or the data in the test cases in the same search plane and predict the class to which the data example has more probability of belonging [45].

3.3.4. Long Short Term Memory (LSTM)

Recurrent Neural Networks (RNN) focus on the issue of considering the past information so as to understand the meaning of current and next words. LSTM network [46] is a type of RNN that is capable of handling long term dependencies as otherwise it was difficult for RNN to connect multiple long term dependencies [47]. After being first introduced by Hochreiter and Schmidhuber in 1997, LSTM has gone through multiple changes over the years. LSTM solves the problem of vanishing and exploding gradient [48], which is a severe limitation for RNN.

The steps of LSTM are defined as: The first step is to decide the information that is going to be deleted from the memory cell. A sigmoid layer executes this decision after looking at prior information i_{t-1} and current input c_t. This sigmoid layer outputs a number between 0 and 1 that determines the amount of information that needs to be retained based on weight W_o. o_t represents the output of the current cell, and b_o is the bias for this particular cell.

$$o_t = \sigma(W_o * [i_{t-1}, c_t] + b_o). \tag{2}$$

Next, it decides the new information that is to be updated into the memory cell. It is done through two steps, a sigmoid layer to decide the values to update and a *tanh* layer to create a vector of new values. n_t denotes the information that is to be updates based on weight W_n and bias b_n and $\tilde{V}_t$ is the data to be included in the current state information. An LSTM cell is shown in Figure 3.

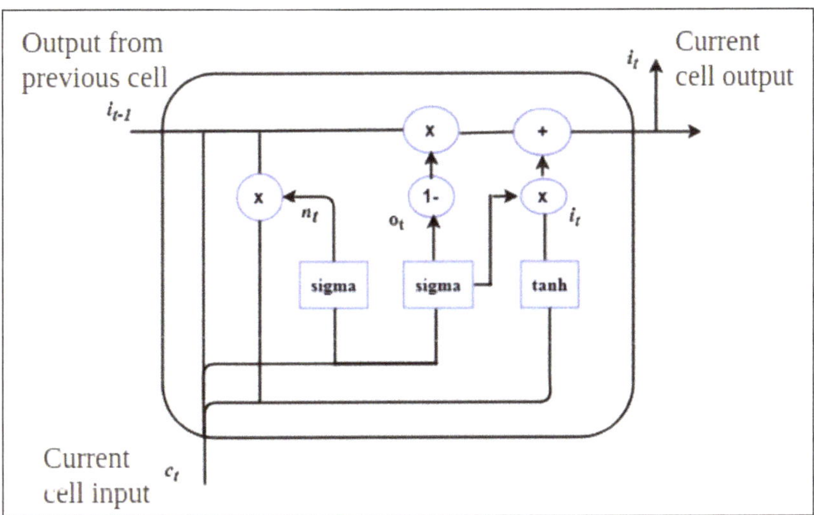

Figure 3. Long Short Term Memory cell, the data flow is from left to right where the current cell input parameter is c_t, i_{t-1} is the output from the previous LSTM cell containing prior information, which is forwarded to the current cell. Both these values are concatenated based on the parameters n_t which denotes the information that is to be updated, o_t which represents the output within the current cell giving the final output value for this layer as i_t that serves as prior information to the next LSTM cell.

$$n_t = \sigma(W_n * [i_{t-1}, c_t] + b_n), \tag{3}$$

$$\tilde{V}_t = tanh(W_V * [i_{t-1}, c_t] + b_V). \tag{4}$$

Now this information is updated into the next cell V_t by multiplying the old state with o_t.

$$V_t = o_t * V_{t-1} + n_t * \tilde{V}_t \tag{5}$$

In the last step, we again implement a sigmoid layer to find f_t that denotes the information which will be given as output based on weight W_f and bias b_f. The *tanh* layer updates the required parts and gives i_t as the output of the cell.

$$f_t = \sigma(W_f * [i_{t-1}, c_t] + b_f), \tag{6}$$

$$i_t = f_t * tanh(V_t). \tag{7}$$

The final output i_t from this cell will serve as prior information for the next cell to find out its subsequent cell state. Nowdays, LSTM are increasingly used to classify test data over other classification algorithms. It is trained on book review dataset with 32 neurons per layer followed by a sigmoid activation function. The netwok has been trained on different epochs and achieved good accuracy compare to other algorithms.

3.3.5. Convolution Neural Network (CNN)

CNN was originally developed for computer vision and its applications, it makes use of local features of the image on which multiple layers with convolving features can be implemented. To implement CNN on the textual reviews, we train a CNN model [49] on book reviews dataset with a single layer on top of the features extracted from the sentences using the word2vec model. First layer is the convolution layer where we slide multiple filters of different sizes over the 128 word embeddings dimensions to produce a feature map based on the particular filter. Max-pooling layer follows this by convolving the results of previous layer into one long feature vector. Max pooling layer finds the most prominent feature vector from the feature map belonging to every filter, which is then passed on to fully connected softmax layer. Dropout regularization is performed before we use softmax layer to classify the result. Regularization randomly drops out some hidden units from the layer to prevent the co-adaptation on training data which may lead to over-fitting. This network is shown in Figure 4.

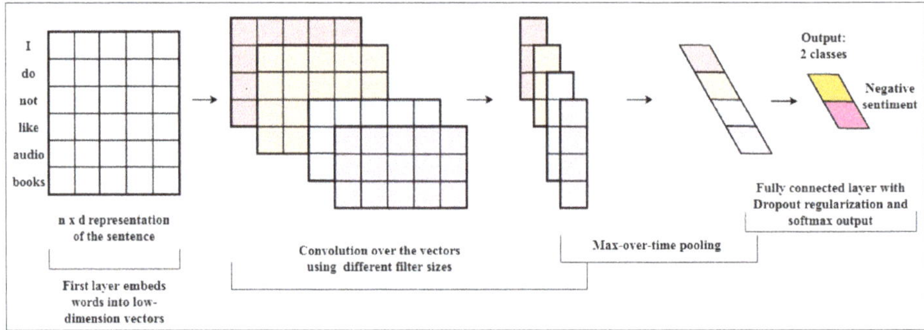

Figure 4. First layers of the model form low-dimensional vectors from the sentence words. The convolution is done by the next layer, using multiple filter sizes such as sliding over 3 or 4 words at a time. Next, the result is max-pooled into a long feature vector and the final results is given using a softmax layer after adding dropout regularization.

4. Experiments and Discussion

In this section, we first describe the dataset, explaining the process of data collection and its further processing that we have done in our experiment. We present the results (see Sections 4.2.1 and 4.2.2) obtained from the feature extraction methods and different classifiers.

4.1. Dataset Description

One of the most crucial parts of this study is data collection. Generally, datasets for sentiment analysis are easily available on the internet which can not be used here as along with the expressed opinion. The micro-blogging and other sites like twitter, Facebook, Amazon, Goodreads, and IMDb do not divulge their user's personal information due to privacy concerns so we create a new dataset that contains all the required information.

The dataset for this experiment is created by collecting opinions of nearly 900 users from the social networking site Facebook. The users have answered a questionnaire containing multiple questions that ask their reviews on preferences of book medium as a Google Form. The questionnaire consisted of questions based on the user's opinions regarding kindles, paperbacks, hardcover, picture, and audiobooks. Further, the questionnaire discusses if the user's thought that digital mediums such as kindle or ebooks could replace hardcover or paperbacks for them. The questions elaborated on whether the user liked audiobooks better than other formats and a short description of their opinions. The form registers the user's opinion, along with the gender and age groups to which they belonged. Along with the user opinions, they have also stated their preference as a positive/negative opinion that serves as the ground truth for the classifiers.

We have selected this domain because we intended to avoid topics with unbalanced spectrum of audience like sports, fashion or television that leaned more towards a particular gender or age group. The responses given by the users to the questionnaire is shown in Figure 5, from the overall reviews we have received, 60% are positive, while the other 40% are negative. From this dataset, we have also segregated the reviews into separate groups, first based on gender, where we have data in a 70% to 30% division to more opinions expressed by the female users. Based on age demographics, the dataset has four age groups into which the users have identified themselves. From the total reviewers, 40% of them belong to the age group of Below 20. The age group 21–34 has nearly 30%, while 20% are in the 35–50 age group. The rest of the users belong to the oldest age group of Above 50.

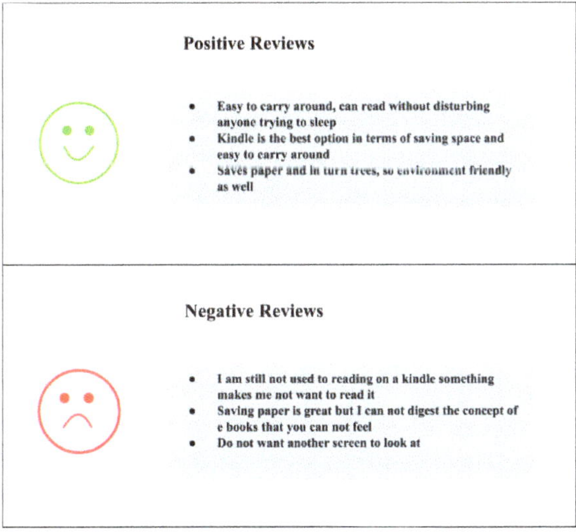

Figure 5. The collected reviews segregated into positive and negative reviews.

4.2. Result Analysis

We have shown the result of machine learning and dictionary-based approaches on the basis of age and gender information. The results of these classifiers are expressed in terms of accuracy [50].

$$Accuracy = \frac{Correctly\ Predicted\ Observations}{Total\ number\ of\ observations}. \tag{8}$$

4.2.1. Effect of Age

The extracted dataset based on age is divided into four groups: one group with age below 20, second with age from 21 to 34, third from 35 to 50 and the last one with age above 50. Thus, a total four groups are created containing positive and negative responses from people of that particular age group. Another group (without age information) containing reviews from all the age groups is formed to compare its results to the other groups as shown in Figure 6.

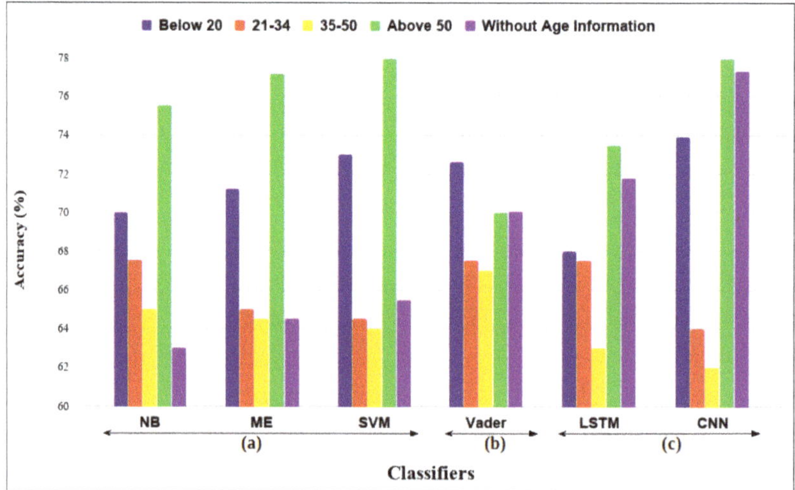

Figure 6. Comparison on basis of Age between different Feature Extraction and Classifier techniques: (**a**) Machine Learning classifiers using Bag-of-words feature extraction method; (**b**) Dictionary-based approach; (**c**) Machine Learning approaches using word2vec feature extraction method.

Pre-processing of all the reviews is performed individually by removing the punctuations, symbols and the stop words from the user reviews as explained in Section 3.1.1. Bag-of-words model on pre-processed data is used to create feature vector which is then used in different classifiers such as NB, ME and SVM. The low dimensional feature vectors are formed from sentences using word2vec model which are then used in LSTM and CNN methods. VADER is also implemented on the pre-processed data. After these approaches are implemented on the separated groups of data individually, the results are recorded.

The 'Above 50' age group performs better as compared to all other age groups in all the classifiers with the highest accuracy of 78% in CNN and SVM classifier. 'Below 20' age group has better accuracy compared to the other two middle age groups where the age group '21–34' performs better than the other age group in all instances, even though the difference between these two age groups are not considerable. Better performance of the eldest age groups shows that the sentiment analysis approaches are able to predict the sentiment in this age group more easily as compared to others groups. The group of data without any age information performs better in LSTM and CNN as compared to other machine learning approaches, where it performs worse than the groups with age information.

4.2.2. Effect of Gender

We label the full dataset into two groups (Male and Female) based on gender containing their positive and negative reviews. Pre-processing, feature extraction and different classifiers are implemented on these data groups similarly as in Section 3. The results are represented in Figure 7.

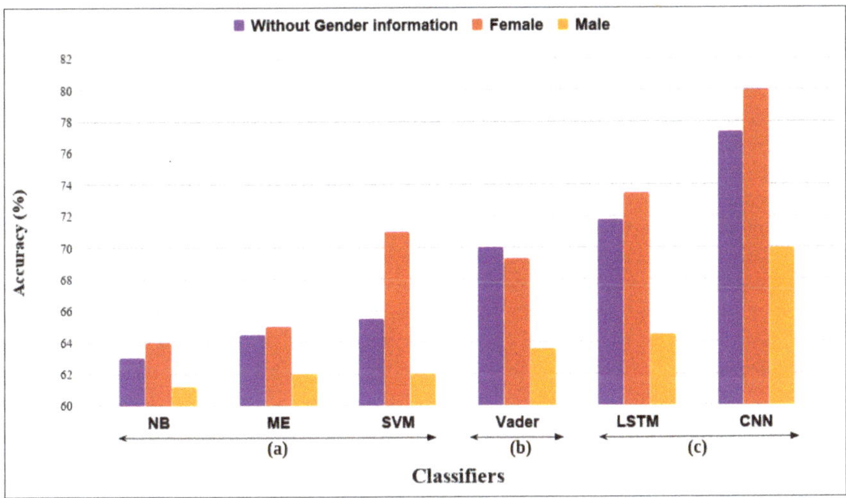

Figure 7. Comparison on basis of Gender between different Feature Extraction and Classifier techniques: (**a**) Machine Learning classifiers using Bag-of-words feature extraction method; (**b**) Dictionary-based approach; (**c**) Machine Learning approaches using word2vec feature extraction method.

It can be clearly seen that female data generates better accuracy as compared to the data without gender information and the male data. Female data has the best accuracy in CNN classifier of 80%, which is better than the other classifiers. This result aligns with the psychological studies that females express their opinion better as compared to their male counterparts. The sentiment in female data is easier to predict, hence giving a better accuracy. This pattern of female data having better accuracy can be observed in all the machine learning approaches.

5. Conclusions and Future Work

In this paper, we have compared multiple sentiment analysis techniques on the dataset collected from nearly 900 users from Facebook along with the users' age and gender information. We extracted this dataset into four groups to analyze the impact of age and gender on the way the user expresses his/her opinion. Machine learning and Dictionary-based techinques have been performed to know the sentiment analysis of the reviews. With respect to gender, female data recorded the best accuracy while for age, the Above 'Age 50' group has the better accuracy as compared to all other age groups. The results can be further improved by collecting more data for both male and female and different age groups.

In future work, we can also include exploration of reviews in audio and visual format to detect emotions from the way of speech and facial expressions of the user to provide more comprehensive investigations from different aspects.

Author Contributions: All authors have contributed to this paper. M.G. and S.K. proposed the main idea, worked on the introduction and data collection. M.G., S.K. and P.P.R. were involved in the methodology and M.G. performed the analyses. M.G. and S.K. drafted the manuscript. B.-G.K., P.P.R. and D.P.D. contributed to the final version of the paper. All authors have read and agreed to the published version of the manuscript.

Funding: This research was supported by Basic Science Research Program through the National Research Foundation of Korea(NRF) funded by the Ministry of Education (NRF-2016R1D1A1B04934750) and the APC was funded by (NRF-2016R1D1A1B04934750).

Acknowledgments: This research was supported by Basic Science Research Program through the National Research Foundation of Korea(NRF) funded by the Ministry of Education (NRF-2016R1D1A1B04934750).

Conflicts of Interest: The authors declared that they have no conflicts of interest to this work.

References

1. Manek, A.S.; Shenoy, P.D.; Mohan, M.C.; Venugopal, K. Aspect term extraction for sentiment analysis in large movie reviews using gini index feature selection method and svm classifier. *Worldw. Web* **2017**, *20*, 135–154. [CrossRef]
2. Dos Santos, C.; Gatti, M. Deep convolutional neural networks for sentiment analysis of short texts. In Proceedings of the COLING, the 25th International Conference on Computational Linguistics: Technical Papers, Dublin, Ireland, 23–29 August 2014; pp. 69–78.
3. Kiritchenko, S.; Zhu, X.; Cherry, C.; Mohammad, S. Nrc-canada-2014: Detecting aspects and sentiment in customer reviews. In Proceedings of the 8th International Workshop on Semantic Evaluation, Dublin, Ireland, 23–24 August 2014; pp. 437–442.
4. Pontiki, M.; Galanis, D.; Papageorgiou, H.; Manandhar, S.; Androutsopoulos, I. Semeval-2015 task 12: Aspect based sentiment analysis. In Proceedings of the 9th International Workshop on Semantic Evaluation, Denver, CO, USA, 4–5 June 2015; pp. 486–495.
5. Cao, D.; Ji, R.; Lin, D.; Li, S. A cross-media public sentiment analysis system for microblog. *Multim. Syst.* **2016**, *22*, 479–486. [CrossRef]
6. Ghosh, R.; Zhang, L.; Dekhil, M.E.; Liu, B. Performing sentiment analysis on microblogging data, including identifying a new opinion term therein. US Patent 9,275,041, 1 March 2016.
7. Ullah, M.A.; Islam, M.M.; Azman, N.B.; Zaki, Z.M. An overview of multimodal sentiment analysis research: Opportunities and difficulties. In Proceedings of the 2017 IEEE International Conference on Imaging, Vision & Pattern Recognition, Himeji, Japan, 1–3 September 2017; pp. 1–6.
8. Cambria, E. Affective computing and sentiment analysis. *IEEE Intell. Syst.* **2016**, *31*, 102–107. [CrossRef]
9. Liu, B. Sentiment analysis and subjectivity. *Handb. Nat. Lang. Proc.* **2010**, *2*, 627–666.
10. Kumar, S.; Yadava, M.; Roy, P.P. Fusion of eeg response and sentiment analysis of products review to predict customer satisfaction. *Inf. Fus.* **2019**, *52*, 41–52. [CrossRef]
11. Kim, J.H.; Kim, B.G.; Roy, P.P.; Jeong, D.M. Efficient facial expression recognition algorithm based on hierarchical deep neural network structure. *IEEE Access* **2019**, *7*, 41273–41285. [CrossRef]
12. Yoo, S.M.; Cho, C.; Lee, K.H.; Park, J.; Jin, S.; Lee, Y.; Kim, B.G. Structure of deep learning inference engines for embedded systems. In Proceedings of the IEEE 2019 International Conference on Information and Communication Technology Convergence, Kuala Lumpur, Malaysia, 24–26 July 2019; pp. 920–922.
13. Kim, J.H.; Hong, G.S.; Kim, B.G.; Dogra, D.P. Deepgesture: Deep learning-based gesture recognition scheme using motion sensors. *Displays* **2018**, *55*, 38–45. [CrossRef]
14. Kahaki, S.M.M.; Ismail, W.; Nordin, M.J.; Ahmad, N.S.; Ahmad, M. Automated age estimation based on geometric mean projection transform using orthopantomographs. *J. Adv. Technol. Eng. Stud.* **2017**, *3*, 6–10.
15. Kahaki, S.M.; Nordin, M.J.; Ahmad, N.S.; Arzoky, M.; Ismail, W. Deep convolutional neural network designed for age assessment based on orthopantomography data. *Neural Comput. Appl.* **2019**, *3*, 1–12. [CrossRef]
16. Li, Y.M.; Li, T.Y. Deriving market intelligence from microblogs. *Decis. Support Syst.* **2013**, *55*, 206–217. [CrossRef]
17. Lockenhoff, C.E.; Costa, P.T.; Lane, R.D. Age differences in descriptions of emotional experiences in oneself and others. *J. Gerontol. Ser. B Psychol. Sci. Soc. Sci.* **2008**, *63*, 92–99. [CrossRef] [PubMed]
18. Zimmermann, P.; Iwanski, A. Emotion regulation from early adolescence to emerging adulthood and middle adulthood. *Int. J. Behav. Dev.* **2014**, *38*, 182–194. [CrossRef]
19. Kaur, B.; Singh, D.; Roy, P.P. Age and gender classification using brain–computer interface. *Neural Comput. Appl.* **2019**, *31*, 5887–5900. [CrossRef]

20. Oh, H.; Parks, S.C.; Demicco, F.J. Age-and gender-based market segmentation: A structural understanding. *Int. J. Hosp. Tour. Adm.* **2002**, *3*, 1–20. [CrossRef]

21. Keshari, P.; Jain, S. Effect of age and gender on consumer response to advertising appeals. *Paradigm* **2016**, *20*, 69–82. [CrossRef]

22. Appel, O.; Chiclana, F.; Carter, J.; Fujita, H. Successes and challenges in developing a hybrid approach to sentiment analysis. *Appl. Intell.* **2018**, *48*, 1176–1188. [CrossRef]

23. Zainuddin, N.; Selamat, A.; Ibrahim, R. Hybrid sentiment classification on twitter aspect-based sentiment analysis. *Appl. Intell.* **2018**, *48*, 1218–1232. [CrossRef]

24. Fan, T.K.; Chang, C.H. Sentiment-oriented contextual advertising. *Knowl. Inf. Syst.* **2010**, *23*, 321–344. [CrossRef]

25. Kuo, Y.H.; Fu, M.H.; Tsai, W.H.; Lee, K.R.; Chen, L.Y. Integrated microblog sentiment analysis from users' social interaction patterns and textual opinions. *Appl. Intell.* **2016**, *44*, 399–413. [CrossRef]

26. Li, G.; Liu, F. Sentiment analysis based on clustering: a framework in improving accuracy and recognizing neutral opinions. *Appl. Intell.* **2014**, *40*, 441–452. [CrossRef]

27. Ali, F.; Kim, E.K.; Kim, Y.G. Type-2 fuzzy ontology-based opinion mining and information extraction: A proposal to automate the hotel reservation system. *Appl. Intell.* **2015**, *42*, 481–500. [CrossRef]

28. Pak, A.; Paroubek, P. Twitter as a corpus for sentiment analysis and opinion mining. In *LREC*; University of Paris: Paris, France, 2010; pp. 1320–1326.

29. Denecke, K. Using sentiwordnet for multilingual sentiment analysis. In Proceedings of the IEEE 24th International Conference on Data Engineering Workshop, Cancun, Mexico, 7–12 April 2008; pp. 507–512.

30. Zhou, G.; Zhu, Z.; He, T.; Hu, X.T. Cross-lingual sentiment classification with stacked autoencoders. *Knowl. Inf. Syst.* **2016**, *47*, 27–44. [CrossRef]

31. Cheng, J.; Zhang, X.; Li, P.; Zhang, S.; Ding, Z.; Wang, H. Exploring sentiment parsing of microblogging texts for opinion polling on chinese public figures. *Appl. Intell.* **2016**, *45*, 429–442. [CrossRef]

32. Fabes, R.A.; Martin, C.L. Gender and age stereotypes of emotionality. *Personal. Soc. Psychol. Bull.* **1991**, *17*, 532–540. [CrossRef]

33. Stoner, S.B.; Spencer, W.B. Age and gender differences with the anger expression scale. *Educ. Psychol. Meas.* **1987**, *47*, 487–492. [CrossRef]

34. Davis, T.L. Gender differences in masking negative emotions: Ability or motivation? *Dev. Psychol.* **1995**, *31*, 660–667. [CrossRef]

35. Brody, L.R. Gender and emotion: Beyond stereotypes. *J. Soc. Issues* **2010**, *53*, 369–393. [CrossRef]

36. Kring, A.M.; Gordon, A.H. Sex differences in emotion: Expression, experience, and physiology. *J. Personal. Soc. Psychol.* **1998**, *74*, 686–703 . [CrossRef]

37. Birditt, K.S.; Fingerman, K.L. Age and gender differences in adults' descriptions of emotional reactions to interpersonal problems. *J. Gerontol. Ser. B Psychol. Sci. Soc. Sci.* **2003**, *58*, 237–245. [CrossRef]

38. Kharde, V.A.; Sonawane, S.S. Sentiment analysis of Twitter data: A survey of techniques. *Int. J. Comput. Appl.* **2016**, *139*, 5–15.

39. Saini, R.; Kumar, P.; Roy, P.P.; Pal, U. Trajectory classification using feature selection by genetic algorithm. In *Proceedings of the 3rd International Conference on Computer Vision and Image Processing*; Springer: Berlin/Heidelberg, Germany, 2020; pp. 377–388.

40. Xue, B.; Fu, C.; Shaobin, Z. A study on sentiment computing and classification of Sina Weibo with Word2vec. In Proceedings of the IEEE International Congress on Big Data, Anchorage, AK, USA, 27–July 2 June 2014; pp. 358–363.

41. Hutto, C.J.; Gilbert, E. Vader: A parsimonious rule-based model for sentiment analysis of social media text. In Proceedings of the 8th International Conference on Weblogs and Social Media, Ann Arbor, MI, USA, 1–4 June 2014; pp. 216–255.

42. Zhang, H. The optimality of naive bayes. *AA* **2004**, *1*, 1–6.

43. Nigam, K.; Lafferty, J.; McCallum, A. Using maximum entropy for text classification. In *IJCAI-99 Workshop on Machine Learning for Information Filtering*; IJCAI: Stockholom, Sweden, 1999; Volume 1, pp. 61–67.

44. Kaufmann, J.M. JMaxAlign: A maximum entropy parallel sentence alignment tool. In Proceedings of the COLING 2012: Demonstration Papers, Mumbai, India, 8–15 December 2012; pp. 277–288.

45. Mullen, T.; Collier, N. Sentiment analysis using support vector machines with diverse information sources. In Proceedings of the 2004 Conference on Empirical Methods in Natural Language Processing, Barcelona, Spain, 25–26 July 2004; pp. 412–418.

46. Saini, R.; Kumar, P.; Kaur, B.; Roy, P.P.; Dogra, D.P.; Santosh, K. Kinect sensor-based interaction monitoring system using the blstm neural network in healthcare. *Int. J. Mach. Learn. Cybern.* **2019**, *10*, 2529–2540. [CrossRef]

47. Hochreiter, S.; Schmidhuber, J. Long short-term memory. *Neural Comput.* **1997**, *9*, 1735–1780. [CrossRef] [PubMed]

48. LeCun, Y.; Bottou, L.; Bengio, Y.; Haffner, P. Gradient-based learning applied to document recognition. *Proc. IEEE* **1998**, *86*, 2278–2324. [CrossRef]

49. Kim, Y. Convolutional neural networks for sentence classification. In Proceedings of the 2014 Conference on Empirical Methods in Natural Language Processing (EMNLP), Doha, Qatar, 25–29 October 2014; pp. 1746–1751.

50. Saito, T.; Rehmsmeier, M. The precision-recall plot is more informative than the roc plot when evaluating binary classifiers on imbalanced datasets. *PLOS ONE* **2015**, *10*, e0118432. [CrossRef] [PubMed]

Article

Real-Time Detection and Recognition of Multiple Moving Objects for Aerial Surveillance

Wahyu Rahmaniar [1,2], Wen-June Wang [1,*] and Hsiang-Chieh Chen [3]

[1] Department of Electrical Engineering, National Central University, Zhongli 32001, Taiwan; wahyurahmaniar@yahoo.com
[2] Department of Computer Science and Electronics, Universitas Gadjah Mada, Yogyakarta 55281, Indonesia
[3] Department of Electrical Engineering, National United University, Miaoli 36063, Taiwan; chc@nuu.edu.tw
* Correspondence: wjwang@ee.ncu.edu.tw

Received: 14 October 2019; Accepted: 17 November 2019; Published: 20 November 2019

Abstract: Detection of moving objects by unmanned aerial vehicles (UAVs) is an important application in the aerial transportation system. However, there are many problems to be handled such as high-frequency jitter from UAVs, small size objects, low-quality images, computation time reduction, and detection correctness. This paper considers the problem of the detection and recognition of moving objects in a sequence of images captured from a UAV. A new and efficient technique is proposed to achieve the above objective in real time and in real environment. First, the feature points between two successive frames are found for estimating the camera movement to stabilize sequence of images. Then, region of interest (ROI) of the objects are detected as the moving object candidate (foreground). Furthermore, static and dynamic objects are classified based on the most motion vectors that occur in the foreground and background. Based on the experiment results, the proposed method achieves a precision rate of 94% and the computation time of 47.08 frames per second (fps). In comparison to other methods, the performance of the proposed method surpasses those of existing methods.

Keywords: moving object; image stabilization; object detection; optical flow; surveillance; UAVs

---

1. Introduction

There has been increased worldwide interest in unmanned aerial vehicles (UAVs) used for surveillance in recent years due to their high mobility and flexibility. In general, the UAV with a camera attached for surveillance flying over the mission area can be controlled manually by an operator or automatically by using computer vision. One of the most important tasks of aerial surveillance is the detection of moving objects that can be used to convey essential information in images, such as pedestrian detection and tracking [1–3], vehicle detection and tracking [4,5], object counting [6], estimation and recognition of object activity [7–9], human and vehicle interactions [10], intelligent transportation systems [11,12], traffic management [13,14], and autonomous robot navigation [15,16].

Several studies have proposed some methods to detect moving objects using stationary cameras, such as Gaussian Mixture Model (GMM) [17], Bayesian background model [18], Markov Random Field (MRF) [19,20], and frame differences [21,22]. These methods extract and identify moving objects by seeking the changes in pixels in each frame. However, these techniques rely on static pixels in the images and are not suitable for processing images from moving cameras that have dynamic pixels. Therefore, stationary cameras limit the application of image processing on videos from moving cameras, e.g., aerial vehicles, mobile robots, and handheld cameras. Thus, the problem for detecting moving objects using a moving camera attracted the attention of researchers in recent years [23].

Detecting moving objects using UAVs has many difficulties to implement in real time and in real environments. These difficulties include camera movements, dynamic background, abrupt motion of the objects or camera, rapid illumination changes, camouflage of stationary objects as moving objects, moving object appearance changes, noise from low-quality images, and so on. Several approaches have been proposed to detect moving objects by moving cameras using object segmentation techniques. Saif et al. [24] presented a dynamic motion model using moment invariant and segmentation which extracts one frame in one second but it is not fast enough for the real-time detection. Their result has some false detection, such as a parked car recognized as a moving object. Maier et al. [25] used the deviations between all pixels of the anticipated geometry of two or more consecutive frames to distinguish moving and static objects, but the result depended on the accuracy of the optical flow calculation and the amount of radial distortion. Kalantar et al. [26] proposed a moving object detection framework without explicitly overlaying frame pairs, where each frame is segmented into regions and subsequently represented as a regional adjacency graph (RAG).

In our propose method, we do not only want to achieve the accuracy of the moving objects detection using a moving camera but also to do it in real-time processing. Some previous studies used an optical flow schemes approach to define the movement path of pixels that are tracked on two consecutive frames. Wu et al. [27] used a coarse-to-fine threshold scheme on particle trajectories in the sequence of images to detect moving objects. The background movement is subtracted using the adaptive threshold method to get fine foreground segmentation. Then, mean-shift segmentation is used to refine the detected foreground. Cai et al. [28] combined procedures of the brightness constancy relaxation and intensity normalization within the optical flow to extract the moving objects in the background based on the growing region of the velocity field. In this case, the images are obtained from the robot competition arena which has a homogeneous background. Minaeian et al. [29] used the foreground estimation to segment moving targets through the integration of spatiotemporal differences and local motion history. However, these previous methods did not adequately prove reliability in real-time processing.

This paper proposes a method for detecting multiple moving objects from the sequence of images taken by a UAV which can be applied for real-time applications. The detection and recognition are performed through this method for different objects, such as people and cars. In addition, the image sequences to be tested by this method may contain a complex background. This paper proposes a reliable method for object detection in images where the processing time to get the foreground is shorter than that of segmentation method employed in previous studies [24–26]. Aerial image stabilization is proposed to reduce the mixing of camera and object movements, where the background moves due to the camera movement and the foreground moves due to camera and object movement. Furthermore, the unwanted camera movements make the motion vectors field estimation between two consecutive frames incompatible with the actual situation. This situation differentiates the direction of the motion vectors of static objects from the background, even though the objects are a part of the background. Thus, the static objects tend to be recognized as moving objects. To solve such problems, the proposed method provides the motion vectors classification to distinguish static and dynamic (moving) objects.

The remainder of the paper is organized as follows. Section 2 introduces materials and the main algorithm. Section 3 illustrates performance results using multiple videos taken from a UAV. Finally, conclusions are drawn in Section 4.

2. Materials and Method

2.1. Materials

The experiment was executed using Visual Studio C++ in the 3.40 GHz CPU with 8 GB RAM. The performance of the proposed method is evaluated using three types of aerial image sequences (action1, action2, and action3) obtained from the UCF (http://crcv.ucf.edu/data/UCF_Aerial_Action.php) aerial action dataset with the resolution of 960 × 540. These image sequences were recorded at different flying altitudes ranging from 400–450 feet. Action1.mpg and action2.mpg were taken by the UAV at similar altitudes, where people and cars are the main objects in the image. Action3.mpg was taken at a higher altitude than other videos, so the objects look smaller when compared to other videos.

2.2. The Proposed Method

The challenge to the moving object detection by a moving camera is obvious. The application of the proposed framework can be used to distinguish the foreground from a dynamic background into a simpler formulation. The systematic approach starts with image stabilization to reduce unwanted movement in the sequence of images. The unwanted movements are the motion of the camera as well as any vibration of the UAV. Inaccuracies in motion compensation can cause failure on the estimation of the background and foreground pixels [30]. However, despite using image stabilization, motion vectors in the static objects (background) and moving objects (foreground) are still difficult to distinguish.

Additionally, in order to detect several moving objects with different sizes and speeds we require the correct calculation of motion vector fields. Furthermore, static and dynamic objects are distinguished based on their movement direction (MD). There are two kinds of MD to be estimated: The direction of the object's movement (foreground) and the direction of the background's movement. It should be noted that the background motion is affected by the camera's movement. Figure 1 shows an illustration of the movement of a UAV affecting camera movement. The background movement corresponding to the motion of a moving camera is affected by UAV movements on the yaw, pitch, and roll axis. So, efficient affine transformation is needed.

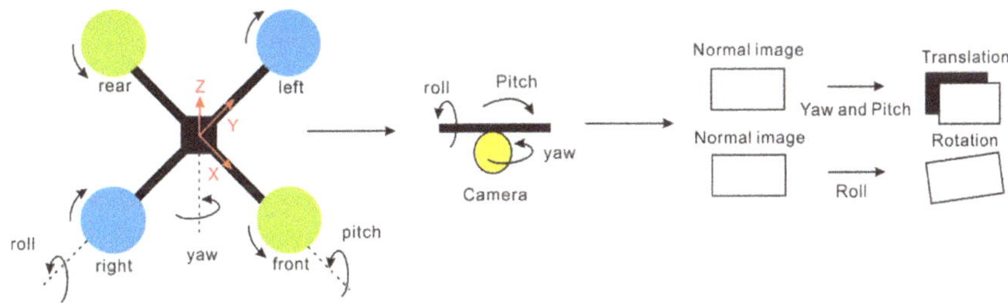

Figure 1. Unmanned aerial vehicles (UAV) movement modeling.

Figure 2 shows an overview of the structure of the system. The algorithm consists of three steps to accomplish the main task: Step 1 is the aerial image stabilization, step 2 is the object detection and recognition, and step 3 is the classification of the motion vectors. The proposed algorithm handles each frame for the moving objects detection and recognition so that it can be used in real-time applications with online image processing.

Step 1: Image stabilization is performed to handle unstable UAV platforms. This step aligns each frame with the adjacent frame in a sequence of aerial images to eliminate the effect of camera movement. This stabilization method consists of motion estimation and compensation. We used the methods of speeded-up robust features (SURF) [31–33] and affine transformation [34] to estimate the camera movement based on the position of features which are similar between the previous $(t-1)$ and current (t) frames. Then use the Kalman filter [35,36] to overcome the changes in frame position due to UAV movement such that the camera movement is compensated for each frame. This image transformation is applied to the frame t, so it affect the results of MD in the background and foreground.

Step 2: People and cars are detected in the images as the moving objects candidates or foreground. In this step, Haar-like features [37] and cascade classifiers [38,39] are used to detect and recognize the objects in the images and determine the region of interest (ROI) for the objects. This is followed by labeling the background and foreground.

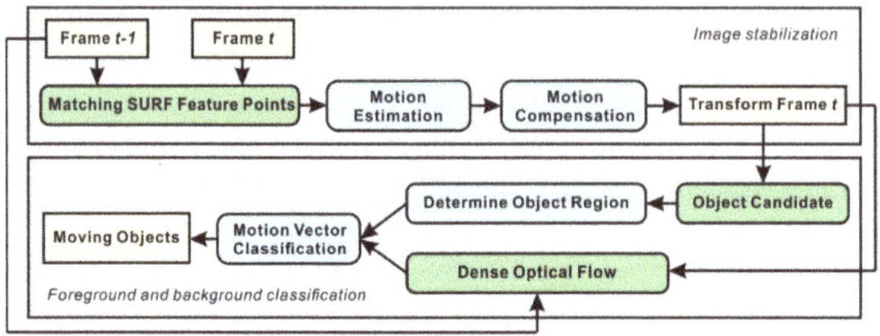

Figure 2. System overview of the real-time moving object detection and recognition using UAV.

Step 3: Calculate the motion vectors from two consecutive images based on the dense optical flow [40]. Background modeling is sometimes incompatible with actual camera movements due to UAV movements and camera transitions. It is noted that Step 1 makes the MD between static and dynamic objects clearer to be distinguished. MD is specified as the value of a highly repetitive motion vector in frame t, which is calculated in the background and each foreground. If the foreground has the same MD as the background, then the object is omitted from the foreground. Thus, the final result is the ROI in the image showing the moving objects.

The details of each step are explained as follows.

2.3. Step 1: Aerial Image Stabilization

This step uses an affine motion model to handle rotation, scaling, and translation. The affine model can be used to estimate movement between frames under certain conditions in the scene [41,42]. For every two successive frames, the previous frame is defined as $f(t-1)$ and the current frame is defined as $f(t)$. In order to reduce the computation time, let the image size be reduced to 75% of the original size and the color is changed into a gray-scale, where $\hat{f}(t)$ denotes the new image with the above size and color of $f(t)$. The local features on each frame are found using SURF [31] as the feature detector and descriptor. SURF uses an integral image [43] to compute different box filters to detect feature points in the image. If $f(t)$ is an input image and $f_{(x,y)}(t)$ is the pixel value of the location (x,y) at $f(t)$, the value $P(i,j,t)$ is defined as

$$P(i,j,t) = \sum_{x=0}^{x \leq i} \sum_{y=0}^{y \leq j} f_{(x,y)}(t).$$

(1)

Haar wavelet [30] corresponding to d_x and d_y are calculated in the x-direction and y-direction, respectively, around each feature point to form a descriptor vector presented as

$$v = \left(\sum d_x, \sum d_y, \sum |d_x|, \sum |d_y| \right).$$

(2)

Then, a 4×4 array with each vector having four orientations is constructed and centered on the feature point. Therefore, there will be a total of 64-length vectors for each feature point.

Fast Library for Approximate Nearest Neighbor (FLANN) [44] is used to select a set of feature point pairs between $\hat{f}(t-1)$ and $\hat{f}(t)$. Then, the minimum distance for all pairs of feature points is calculated using the Euclidean distance. The matching pair is determined as a feature point pair with a distance less than 0.6. If the total number of matching pairs are more than three, then the selected feature points are used for the next step. Otherwise, the previous trajectory is used as an estimate of the current movement.

In homogenous coordinates, the relationship between a pair of feature points in the $\hat{f}(t-1)$ and $\hat{f}(t)$ is given by

$$\begin{bmatrix} x(t) \\ y(t) \end{bmatrix} = H \begin{bmatrix} x(t-1) \\ y(t-1) \\ 1 \end{bmatrix}, \tag{3}$$

where H is the homogeneous affine matrix given by

$$H = \begin{bmatrix} 1 + a_{11} & a_{12} & T_x \\ a_{21} & 1 + a_{22} & T_y \end{bmatrix}, \tag{4}$$

where a_{ij} is the parameter from the rotation angular θ, T_x, and T_y are parameters of the translation T on the x-axis and y-axis, respectively. An affine matrix can be represented as a least squares problem by

$$\begin{aligned} L &= mh, \\ L &= \begin{bmatrix} x(1)' & y(1)' & \dots & x(\bar{q})' & y(\bar{q})' \end{bmatrix}^T, \\ m &= \begin{bmatrix} M_0(1) & M_1(1) & \dots & M_0(\bar{q}) & M_1(\bar{q}) \end{bmatrix}^T, \\ h &= \begin{bmatrix} 1 + a_{11} & a_{12} & T_x & 1 + a_{21} & a_{22} & T_y \end{bmatrix}^T, \end{aligned} \tag{5}$$

where $q = 1, \dots, \bar{q}$ is the number order of features, $M_0(q) = (\ x(q)\ \ y(q)\ \ 1\ \ 0\ \ 0\ \ 0\ )$, and $M_1(q) = (\ 0\ \ 0\ \ 0\ \ x(q)\ \ y(q)\ \ 1\ )$.

The optimal estimation h in Equation (5) can be found by using Gaussian elimination to minimize Root Mean Squared Errors (RMSE) calculated by

$$RMSE = \frac{1}{Q}\|L - mh'\| = \sqrt{\frac{\sum\limits_{q=1}^{Q} \left(L_q - (mh')_q\right)^2}{Q^2}}. \tag{6}$$

Because the affine transform cannot represent the three-dimensional motion which occurs in the image, the outliers are generated in motion estimation. To solve this problem, Random Sample Consensus (RANSAC) [45] is used to filter outliers during the estimation.

Next, the translation and rotation trajectories are compensated to generate a new set of transformations for each frame using the Kalman filter. The Kalman filter consists of two essential parts, prediction and measurement correction. The prediction step estimates the state of the trajectory $\hat{z}(t) = \left[T_x(t), T_y(t), \hat{\theta}(t) \right]$ at $\hat{f}(t)$ as

$$\hat{z}(t) = z(t-1), \tag{7}$$

where the initial state is defined by $z(0) = [0, 0, 0]$ and the error covariance can be estimated by

$$\hat{e}(t) = e(t-1) + \Omega_p, \tag{8}$$

where the initial error covariance is defined by $e(0) = [1, 1, 1]$ and Ω_p is the noise covariance of the process. Optimum Kalman gain can be computed as follows

$$K(t) = \frac{\hat{e}(t)}{\hat{e}(t) + \Omega_m}, \tag{9}$$

where Ω_m is the noise covariance of the measurement. The error covariance can be compensated by

$$e(t) = (1 - K(t))\hat{e}(t). \tag{10}$$

Then, the measurement correction step compensates the trajectory state at $\hat{f}(t)$, which can be computed as

$$z(t) = z(t) + K(t)(\Gamma(t) - z(t)), \tag{11}$$

where the new state contains the compensated trajectory defined by $z(t) = \left[T'_x(t), T'_y(t), \theta'(t)\right]$ and $\Gamma(t)$ is the accumulation of the trajectory measurement that can be calculated as follows

$$\Gamma(t) = \sum_{\tau=1}^{t-1}\left[\left(\overline{T}_x(\tau) + T_x(t)\right), \left(\overline{T}_x(\tau) + T_x(t)\right), \left(\overline{\theta}(\tau) + \theta(t)\right)\right] = \left[\Gamma_x(t), \Gamma_y(t), \Gamma_\theta(t)\right]. \tag{12}$$

Therefore, a new trajectory can be obtained by

$$\left[\overline{T}_x(t), \overline{T}_y(t), \overline{\theta}(t)\right] = \left[T_x(t), T_y(t), \theta(t)\right] + \left[\sigma_x(t), \sigma_y(t), \sigma_\theta(t)\right], \tag{13}$$

where $\sigma_x(t) = T'_x(t) - \Gamma_x(t)$, $\sigma_y(t) = T'_y(t) - \Gamma(t)$, and $\sigma_\theta(t) = \theta'(t) - \Gamma_\theta(t)$.

Then, warp $f(t)$ is in the new image plane and let us apply the new trajectory in Equation (13) to get the transformation $\overline{f}(t)$ in the current frame

$$\overline{f}(t) = f(t)\left[\begin{array}{cc} \Phi(t)\cos\overline{\theta}(t) & -\Phi(t)\sin\overline{\theta}(t) \\ \Phi(t)\sin\overline{\theta}(t) & \Phi(t)\cos\overline{\theta}(t) \end{array}\right] + \left[\begin{array}{c} \overline{T}_x(t) \\ \overline{T}_y(t) \end{array}\right], \tag{14}$$

where $\Phi(t)$ is a scale factor computed by

$$\Phi(t) = \frac{\cos\overline{\theta}(t)}{\cos\left(\tan^{-1}\left(\frac{\sin\overline{\theta}(t)}{\cos\overline{\theta}(t)}\right)\right)(t)}. \tag{15}$$

2.4. Step 2: Object Detection and Recognition

In this step, the background and foreground are determined in each frame that has been transformed in Step 1. The foreground is made up of the moving object candidates, which are people and cars, in the image. The foreground is detected and recognized using Haar-like features and a boosted cascade of classifiers with training and detection stages. The basic idea behind Haar-like features is to detect objects of various sizes in the images. Figure 3 shows the template of the Haar-like features where each feature consists of two or three adjacent rectangular groups and can be scaled up or down. The pixel intensity values in the white and black groups are accumulated separately. So, the distinction between adjacent groups gives light and dark regions. Therefore, Haar-like features are suitable for defining information in images to find objects on different scales in which some simple patterns are used to identify the existence of objects.

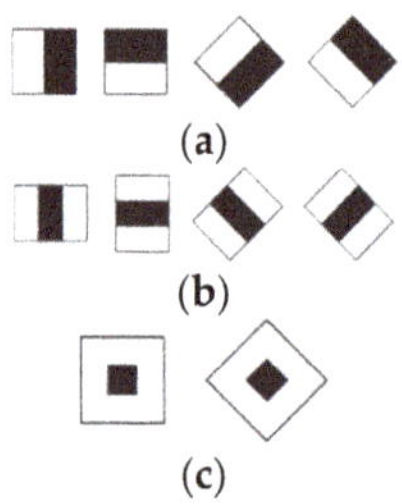

Figure 3. Haar-like features: (**a**) Edge, (**b**) line, (**c**) center-surround.

The Haar-like feature value is calculated as the weighted sum of the pixel gray level values which are summed over the black rectangle and the entire feature area. Then, an integral image [41] is used to minimize the number of array references in the sum of the pixels in a rectangular area of an image. Figure 4a,b show the example of the main objects to be selected. Figure 4c shows the example of the additional objects to be selected which are non-moving objects, i.e., road signs, fences, boxes, road patterns, grass patterns, power lines, roadblocks, and so on. This additional object has the purpose to reduce false detection where the type of object often tends to be recognized as foreground. Negative images are the images of landscapes and roads taken by a UAV without containing cars or people. In this study, the minimum and maximum sizes of positive images to be trained are 16×35 and 136×106, respectively.

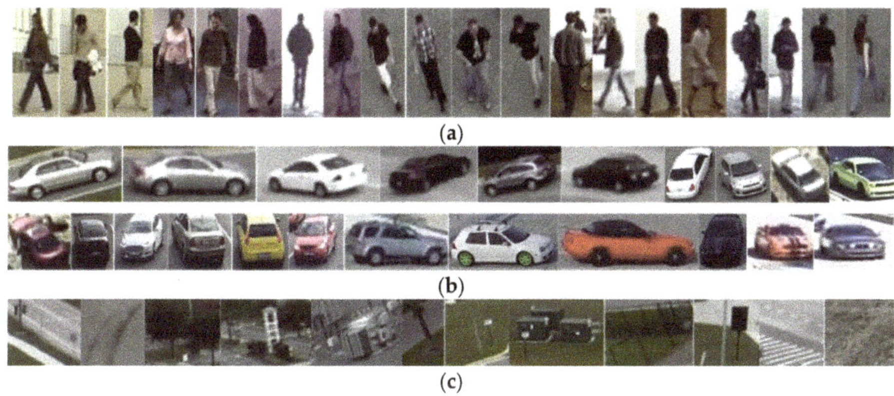

Figure 4. Examples of positive images: (**a**) Person, (**b**) car, (**c**) non-moving object.

The AdaBoost algorithm [46] is used to combine features of the selected classifier. A classifier is chosen as the threshold to determine the best classification function for each feature. A training sample is set as (α_s, β_s), $s = 1, 2, \ldots, N$, where $\beta_s = 0$, or 1 for negative or positive labels, respectively, and it is the class label for the sample α_s. Each sample is converted to a gray-scale then scaled down to the base resolution of the detector. The AdaBoost algorithm creates a weight vector which is distributed over all training samples in the iteration. The initial weight vector for all samples $(\alpha_1, \beta_1), \ldots, (\alpha_N, \beta_N)$ is set as $\omega_1(s) = 1/N$. The error associated with the selected classifier is evaluated as

$$\varepsilon_i = \sum_{s=1}^{N} \omega_i(s), \text{ if } \left| \lambda_i(\alpha_s) \neq \beta_s \right|. \tag{16}$$

The $\lambda_i(\alpha_s) = 0$, or 1 is a selected classifier for negative or positive labels, respectively, and $i = 1, 2, \ldots, I$ is the iteration number. The selected classifier is used to update the weight vector as

$$\omega_{i+1}(s) = \omega_i(s)\delta_i^{1-r_s}$$
$$\text{where } r_s = \begin{cases} 0, \text{ if } \alpha_s \text{ classified correctly,} \\ 1, \text{ otherwise} \end{cases} \tag{17}$$

and δ_i is the weighting parameter set by

$$\delta_i = \frac{\varepsilon_i}{1 - \varepsilon_i}. \tag{18}$$

The final classifier stage $W(\alpha)$ is the labeled result of each region represented as

$$W(\alpha) = \begin{cases} 1, \text{ if } \sum_{i=1}^{I} \left[\log\left(\frac{1}{\delta_i}\right) \times \lambda_i(\alpha) \right] \geq \frac{1}{2} \sum_{i=1}^{I} \log\left(\frac{1}{\delta_i}\right). \\ 0, \text{ otherwise} \end{cases} \tag{19}$$

Figure 5 shows a sub-window that slides over the image to identify the region containing the object. The region is labeled at each classifier stage either as positive (1) or negative (0). The classifier passes to the next stage if the region is labeled as positive, which means that the region is recognized as an object. Otherwise, the region is labeled as negative and is rejected. The final stage shows the region of the moving object candidates. The region of the non-moving object is not to be displayed in the image and is used to evaluate the detected object. If the region of a moving object candidate is the same as the non-moving object, then the region is eliminated as a foreground. Let the n-th foreground region be represented as

$$Obj[n] = [(x_{\min}(n), y_{\min}(n)), (x_{\max}(n), y_{\max}(n))], \tag{20}$$

where $(x_{\min}(n), y_{\min}(n))$ and $(x_{\max}(n), y_{\max}(n))$ are the minimum and maximum positions of the rectangular foreground pixel locations, respectively.

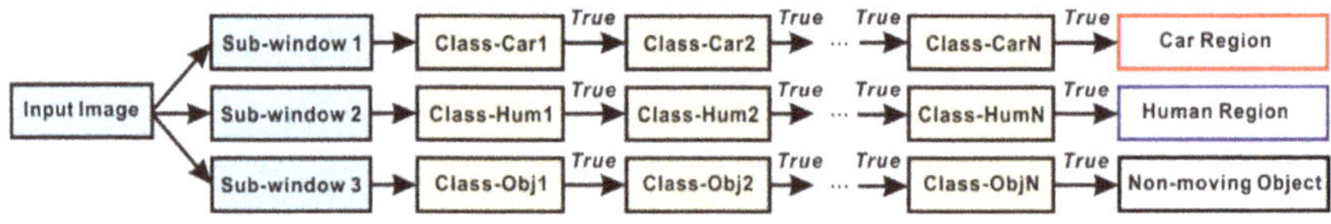

Figure 5. Cascade classifier for object detection and recognition.

False detection of the moving object candidates is eliminated immediately using a comparison of the region with non-moving objects. This will speed up the computation time in the next step.

2.5. Step 3: Motion Vector Classification

The Farneback optical flow [40] is adopted to obtain motion vectors of two consecutive images. The Farneback optical flow uses a polynomial expansion to provide high speed and accuracy for field estimation. Suppose there is a 10×10 window $G(j)$ and the pixel j is chosen inside the window. By using polynomial expansion, each pixel in $G(j)$ can be approximated by a polynomial so called "local coordinate system" at $f(t-1)$ which can be computed as follows

$$f^p_{lcs}(t-1) = p^T A(t-1)p + b^T(t-1)p + c(t-1), \tag{21}$$

where p is a vector, $A(t-1)$ is a symmetric matrix, $b(t-1)$ is a vector, and $c(t-1)$ is a scalar. The local coordinate system at $f(t)$ can be defined by

$$f^p_{lcs}(t) = p^T A(t)p + b^T(t)p + c(t). \tag{22}$$

Then, a new signal is constructed at $f(t)$ by a global displacement $\Delta(t)$ as $f^p_{lcs}(t) = f^{p-\Delta(t)}_{lcs}(t-1)$. The relation between the local coordinate systems of two input images will be

$$\begin{aligned} f^p_{lcs}(t) &= (p - \Delta(t))^T A(t-1)(p - \Delta(t)) + b^T(t-1)(p - \Delta(t)) + c(t-1) \\ &= p^T A(t-1)p^T + (b(t-1) - 2A(t-1)\Delta(t))^T p + \Delta^T(t)A(t-1)\Delta(t) - b^T(t-1)\Delta(t) + c(t-1) \end{aligned} \tag{23}$$

The coefficients can be equated in Equations (22) and (23) as

$$A(t) = A(t-1), \tag{24}$$

$$b(t) = b(t-1) - 2A(t-1)\Delta(t), \tag{25}$$

and

$$c(t) = \Delta(t)\left(\Delta^T(t)A(t-1) - b^T(t-1)\right) + c(t-1). \tag{26}$$

Therefore, the total displacement with the extraction in the ROI can be solved by

$$\Delta(t) = -\frac{1}{2}A^{-1}(t-1)(b(t) - b(t-1)). \tag{27}$$

The displacement in Equation (27) is a translation from each corresponding ROI consisting of the x-axis $(\Delta_x(t))$ and y-axis $(\Delta_y(t))$, so the angular value of the motion vector can be calculated by

$$\Delta_\theta(t) = \tan^{-1}\left(\frac{\Delta_{(x+1,y)}(t) - \Delta_{(x-1,y)}(t)}{\Delta_{(x,y+1)}(t) - \Delta_{(x,y-1)}(t)}\right) \times \frac{180}{\pi}. \tag{28}$$

Since the motion vector is calculated for each 10×10 pixels neighborhood, the total displacement is the matrix of size $(image\_width/10) \times (image\_height/10)$. Thus, the new n-th foreground region is determined by

$$Obj2[n] = \left[\frac{(x_{\min}(n), y_{\min}(n))}{10}, \frac{(x_{\max}(n), y_{\max}(n))}{10}\right]. \tag{29}$$

Figure 6a shows regions marked with red and blue ROI, representing the moving objects candidate (foreground), identified as a person and a car, respectively. Figure 6b shows an example of the estimate motion vector distribution. In images taken by a static camera, the motion vectors in the background are zero, signifying MD value is zero. This means that there is no movement (represented by the direction of the arrows) between two consecutive frames. In our case (images were taken by a moving camera), motion vectors in the background have several different directions as shown in Figure 6b. The red ROI is a parked car classified as a non-moving object, where the motion vectors are similar to most motion vectors in the background. The blue ROI shows a person walking, classified as a moving object, where the motion vectors are different from most motion vectors in the background. Thus, MD on each moving object candidate is obtained as the most occurrence of motion vectors in each ROI. In the background, MD can be obtained as the most occurrence motion vectors in images other than the foreground.

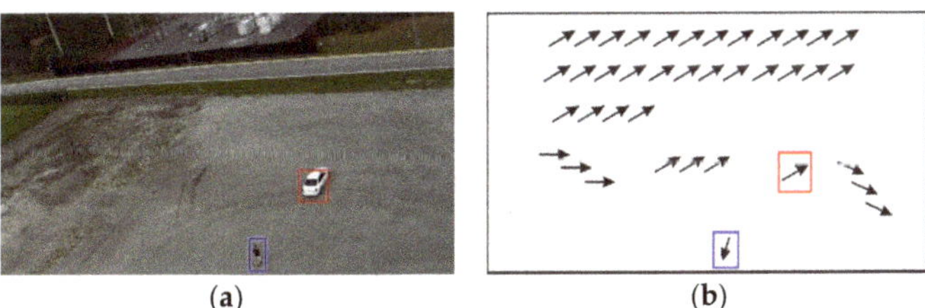

Figure 6. Optical flow estimation: (**a**) Original image, (**b**) motion vectors.

Figure 7 shows a flowchart of the classification of motion vectors and the selection of moving objects, which are implemented in Algorithm 1. In each ROI, motion vectors with angular values that are equal to or greater than zero are grouped into the same class. If the motion vector is in the background, it is classified as $\Delta_B[B]$ where B is the number order of classes in the background and the total number of members of each class is denoted by $N_B[B]$. If the motion vector is in the foreground, it is classified as $\Delta_F[n, F[n]]$ where $F[n]$ is the number order of classes in the foreground and the total number of members of each class is $N_F[n, F[n]]$. Then, MD of the background $\overline{\Delta}_B$ and n-th foreground $\overline{\Delta}_F[n]$ are determined as the biggest Δ_B and $\Delta_F[n]$, respectively. If $\overline{\Delta}_F[n]$ has a value on the threshold of $\overline{\Delta}_B$, then the object is identified as a non-moving object and is not considered as a moving object candidate. Otherwise, the object is identified as a moving object. Finally, the image will only show the ROI of the selected object. The minimum and maximum MD threshold values in the background are -5 and $+5$, respectively. We choose these values because the MD between background and static objects may have little difference which is not out of the threshold range $[-5, +5]$.

Algorithm 1. The proposed classification for selecting moving objects.

---

1. **Input:**

    -   Motion vector : Δ_θ
    -   Number of foregrounds : n
    -   Foreground region : $obj2[n]$

2. **Initial:**
3. $B = 0$, $F[n] = 0$//*number of classes*
4. $N_B[B] = 0$, $N_F[n, F[n]] = 0$//*number of members in each class*
5. $\Delta_B[B] = 0$, $\Delta_F[n, F[n]] = 0$//*motion vector in each class*
6. **FOR** i = 1 *to* end of the column
7. **FOR** j = 1 *to* end of the row
8. **IF** region of $\Delta_\theta = obj2[n]$
9. The motion vectors are classified in each foreground based on $\Delta_F[n, F[n]]$ to get $F[n]$ and $N_F[n, F[n]]$.
10. **ELSE**
11. The motion vectors are classified in the background based on $\Delta_B[B]$ to get B and $N_B[B]$.
12. **END**
13. Select the class with the most members in the background to determine $\overline{\Delta}_B$.
14. Select the class with the most members in each foreground to determine $\overline{\Delta}_F[n]$.
15. Eliminate the object-n which has similar movement direction with the background.
16. **Output:** Moving object region

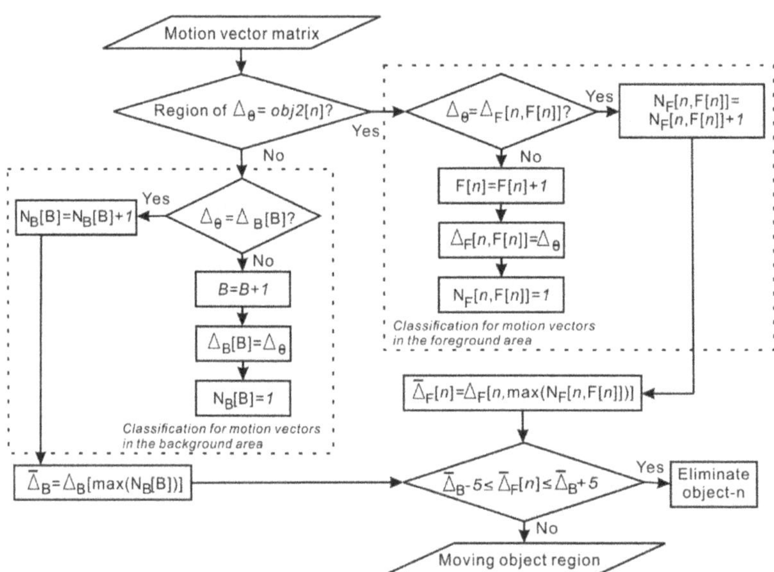

Figure 7. Flowcharts to classify motion vectors and select moving objects.

3. Results and Discussion

3.1. Result of Motion Vectors

The tested images were unstable due to the movement of the UAV. This caused their motion vectors with regards to static (non-moving) and dynamic (moving) objects to be unsuitable to distinguish.

Figures 8 and 9 show the results of the motion vectors without and with image stabilization, respectively. Figures 8a and 9a show the motion vectors in the background. Figures 8b and 9b show the motion vectors in the ROI as a static object car. Figures 8c and 9c show the motion vectors in the ROI as dynamic objects people. Figure 8 shows that the motion vectors of the dynamic and static objects are almost the same with slight difference from the motion vectors in the background. Thus, the result of the motion vectors without image stabilization was incorrect.

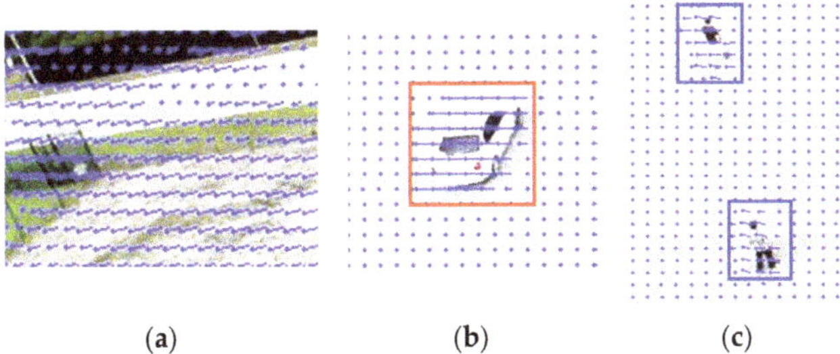

(a)   (b)   (c)

Figure 8. Result of motion vectors without image stabilization: (**a**) Background, (**b**) car, (**c**) people.

Figure 9b shows that the motion vectors in the car (static object) are almost the same as the background. Figure 9c shows that the motion vectors in the people (dynamic objects) are very different from the background. Thus, the results of the motion vectors with image stabilization were very suitable to distinguish between static and dynamic objects.

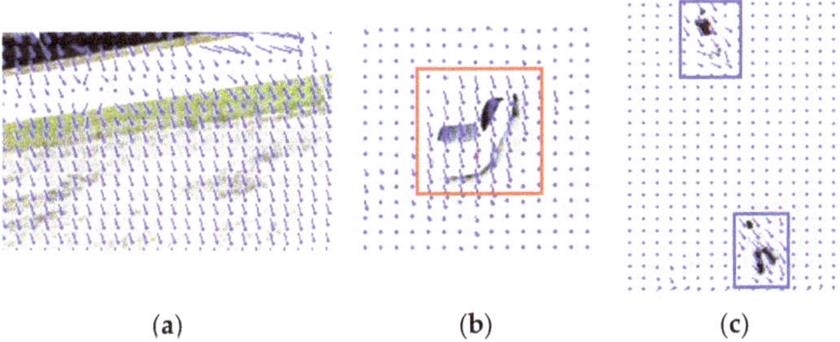

(a)   (b)   (c)

Figure 9. Result of motion vectors with image stabilization: (**a**) Background, (**b**) car, (**c**) people.

3.2. Result of Moving Objects Detection

Figures 10–12 show the results of detection and recognition of moving objects. In some cases, there were false detections on moving objects candidates because motion vectors classified these objects as undesirable and so omitted them. Figures 10 and 11 show the sequence of images obtained from Action1 and Action2, respectively. Sometimes the algorithm did not detect a small object in the image. For example, a small car in Figure 11a was not detected as the foreground. Although the classification result of the motion vector showed the car as a moving object, the final result eliminated the car because the object region was not recognized as the foreground.

Figure 12 shows result of the sequence of images obtained from Action3 which contains five people playing together and making little movements every once in a while. The detection result showed that if there were only slight displacements on an object, it was difficult to distinguish the motion vector. So, the object tended to be detected as a non-moving object.

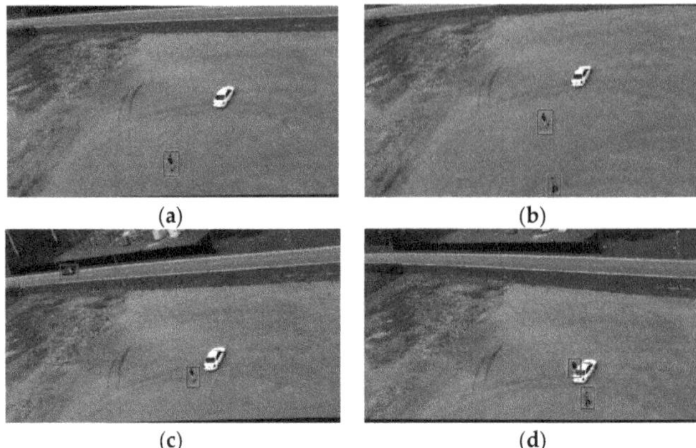

Figure 10. The result of moving object detection in Action1: (**a**) Frame 25, (**b**) frame 100, (**c**) frame 210, (**d**) frame 405.

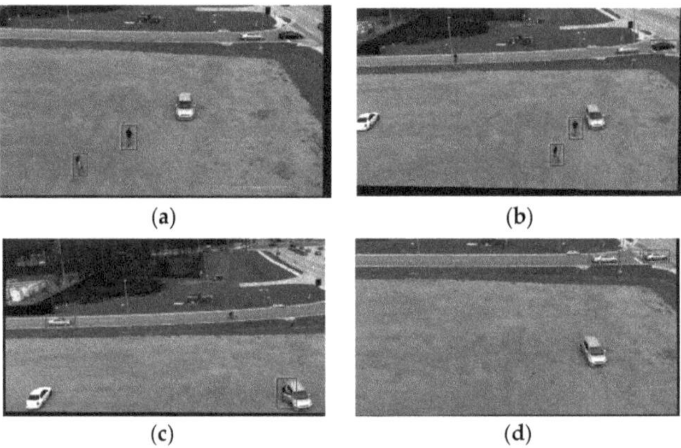

Figure 11. The result of moving object detection in Action2: (**a**) Frame 25, (**b**) frame 100, (**c**) frame 170, (**d**) frame 440.

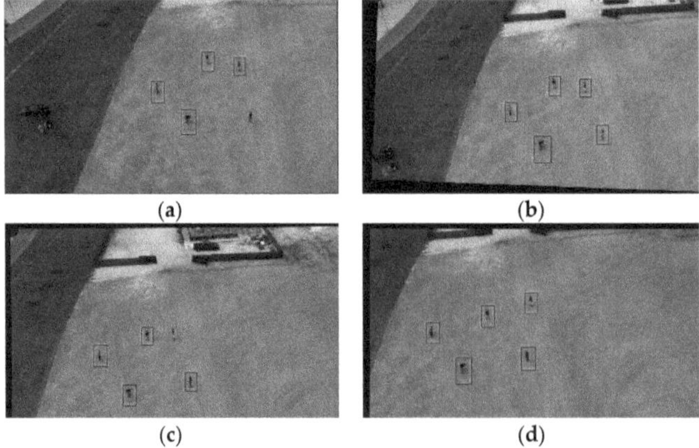

Figure 12. The result of moving object detection in Action3: (**a**) Frame 5, (**b**) frame 60, (**c**) frame 120, (**d**) frame 300.

The results of computation performance are summarized in Table 1 in respect of frames per second (fps). The average time cost is about 47.08 fps which is faster than previous methods in [23–28]. Table 2 shows the performance accuracy in terms of True Positive (TP), False Positive (FP), False Negative (FN), Precision Rate (PR), recall, and f-measure. TP is the detected region that corresponds to the moving object. FP is the detected region that is not related to the moving object. FN is the region associated with the moving object that is not detected. The performance accuracy can be computed as

$$PR = \frac{TP}{TP + FP},\tag{30}$$

$$Recall = \frac{TP}{TP + FN},\tag{31}$$

$$F\text{-measure} = 2 \times \frac{PR \times Recall}{PR + Recall}.\tag{32}$$

Although many articles have tried to solve the same problem (moving object detection using a moving camera), the proposed method has performed well for real-time computation time in a real environment with a complex background. The detection results also showed that the proposed method detected moving objects with high accuracy, although the UAV had some unwanted motion and vibration. The comparison of computation time and accuracy between the results of the proposed method with those methods in [23–26,28] are reported in Table 3. The proposed method achieved an average precision rate of 0.94 and a recall of 0.91. Action1 had the highest PR and recall compared to other videos because there were only a few objects and their sizes were quite large. Action2 had the lowest PR because there were a lot of objects that were similar to the person and car such as trees, fences, road signs, houses, and bushes. Action3 had the lowest recall due to some small objects in the video with displacement.

Table 1. Computation time performance in frames per second (fps).

Video Name	Average Fps
Action1	49.9
Action2	42.16
Action3	49.2
Average	47.08

Table 2. Detection results performance.

Video Name	TP	FP	FN	PR	Recall	F-Measure
Action1	124	7	6	0.95	0.95	0.95
Action2	245	19	23	0.92	0.91	0.91
Action3	184	12	25	0.94	0.88	0.90
Average				0.94	0.91	0.92

Table 3. Comparison of performance results.

Method	Computation Time (fps)	Accuracy		
		PR	Recall	F-Measure
Proposed	47.08	0.94	0.91	0.92
[23]	1	0.7	0.76	0.72
[24]	-	0.66	0.86	0.74
[25]	-	0.94	0.89	0.91
[26]	1.6	-	-	0.73
[28]	5	-	-	0.76

The method in [27] did not discuss the accuracy of the detected moving object as well as the computation time performance. It focused on the optical flow to describe the direction of pixel movement. However, this method, i.e., [27], is suitable for application on an image with a homogeneous background. In our case, a moving camera produced several objects in the background that had no correlation with moving objects but had pixel movements. This condition occurs in image sequences that have complex backgrounds such as our datasets. Thus, the method in [27] is not suitable to be applied to our datasets. In addition, we used a simple dense optical flow which is sufficient to calculate the motion vector fields between two consecutive frames and has a fast computation time. Then, we used the classification, which is feasible to distinguish the motion vectors between static and dynamic objects, to determine MD in the background and foreground.

The proposed method can be used for various moving objects, not only for people and cars. In this work, we used people and car objects to test the performance of the method, because these objects are often investigated as moving objects using moving cameras [23–29]. High frequency jitter, small size objects, and low-quality images make detection of moving objects using UAVs a difficult task. But, using the framework that we propose, we can resolve the problem. Furthermore, a machine learning approach is used to detect and recognize the foreground because it can be applied to almost all processors without GPU. This method is proposed for use on a PC or on-board system. In other words, if the image capture by UAVs can be transmitted to a ground station such as a PC using a wireless camera or transmitted to an additional board such as Raspberry Pi on a UAV, then the image can be processed online and in real time.

Based on information from datasets and previous studies [23–29], we can conclude that the proposed algorithm will be applicable under the conditions: UAV altitude is less than 500 feet and speed is less than 15 m/s. In addition, based on our experiment results, our algorithm had the best results at a video frame rate of less than 50 fps.

4. Conclusions

A novel method for multiple moving objects detection using UAVs is presented in this paper. The main contribution of the proposed method is to detect and recognize moving objects by using a UAV with moving camera with excellent accuracy and can be used in real-time applications. An image stabilization method was used to handle unwanted motion in aerial images so that a significant difference in motion vectors can be obtained to distinguish between static and dynamic objects. The object detection that was used to determine the region of the moving object candidate had a fast computation time and good accuracy on complex backgrounds. Some false detections can be handled using a motion vector classification, in which the object that has a movement direction similar to the background will be removed as a moving object candidate. Comparing the results on various sequences of aerial images, the proposed method can be a potential real-time application in the real environment.

Author Contributions: W.R. contributed to the conception of the study and wrote the manuscript, performed the experiment and data analyses, and contributed significantly to algorithm design and manuscript preparation. W.-J.W. and H.-C.C. helped perform the analysis with constructive discussions, writing, review, and editing.

Funding: We would like to thank the Ministry of Science and Technology of Taiwan for supporting this work by the grant 108-2634-F-008-001.

Conflicts of Interest: The authors declare no conflict of interest.

References

1. Zhou, X.; Yang, C.; Yu, W. Moving object detection by detecting contiguous outliers in the low-rank representation. *IEEE Trans. Pattern Anal. Mach. Intell.* **2013**, *35*, 597–610. [CrossRef] [PubMed]
2. Kang, B.; Zhu, W.-P. Robust moving object detection using compressed sensing. *IET Image Process.* **2015**, *9*, 811–819. [CrossRef]
3. Chen, B.-H.; Shi, L.-F.; Ke, X. A robust moving object detection in multi-scenario big data for video surveillance. *IEEE Trans. Circuits Syst. Video Technol.* **2018**, *29*, 982–995. [CrossRef]

4. Liu, K.; Mattyus, G. Fast Multiclass vehicle detection on aerial images. *IEEE Geosci. Remote Sens. Lett.* **2015**, *12*, 1938–1942.

5. Wu, Q.; Kang, W.; Zhuang, X. Real-time vehicle detection with foreground-based cascade classifier. *IET Image Process.* **2016**, *10*, 289–296.

6. Chen, Y.W.; Chen, K.; Yuan, S.Y.; Kuo, S.Y. Moving object counting using a tripwire in H.265/HEVC bitstreams for video surveillance. *IEEE Access* **2016**, *4*, 2529–2541. [CrossRef]

7. Wang, H.; Oneata, D.; Verbeek, J.; Schmid, C. A Robust and efficient video representation for action recognition. *Int. J. Comput. Vis.* **2016**, *119*, 219–238. [CrossRef]

8. Lin, Y.; Tong, Y.; Cao, Y.; Zhou, Y.; Wang, S. Visual-attention-based background modeling for detecting infrequently moving objects. *IEEE Trans. Circuits Syst. Video Technol.* **2017**, *27*, 1208–1221. [CrossRef]

9. Hammoud, R.I.; Sahin, C.S.; Blasch, E.P.; Rhodes, B.J. Multi-source multi-modal activity recognition in aerial video surveillance. In Proceedings of the IEEE Conference on Computer Vision and Pattern Recognition Workshops, Columbus, OH, USA, 23–28 June 2014; pp. 237–244.

10. Ibrahim, A.W.N.; Ching, P.W.; Gerald Seet, G.L.; Michael Lau, W.S.; Czajewski, W.; Leahy, K.; Zhou, D.; Vasile, C.I.; Oikonomopoulos, K.; Schwager, M.; et al. Recognizing human-vehicle interactions from aerial video without training. *IEEE Robot. Autom. Mag.* **2012**, *19*, 390–405.

11. Liang, C.W.; Juang, C.F. Moving object classification using a combination of static appearance features and spatial and temporal entropy values of optical flows. *IEEE Trans. Intell. Transp. Syst.* **2015**, *16*, 3453–3464. [CrossRef]

12. Nguyen, H.T.; Jung, S.W.; Won, C.S. Order-preserving condensation of moving objects in surveillance videos. *IEEE Trans. Intell. Transp. Syst.* **2016**, *17*, 2408–2418. [CrossRef]

13. Lee, G.; Mallipeddi, R. A genetic algorithm-based moving object detection for real-time traffic surveillance. *Signal. Process. Lett.* **2015**, *22*, 1619–1622. [CrossRef]

14. Chen, B.H.; Huang, S.C. An advanced moving object detection algorithm for automatic traffic monitoring in real-world limited bandwidth networks. *IEEE Trans. Multimed.* **2014**, *16*, 837–847. [CrossRef]

15. Minaeian, S.; Liu, J.; Son, Y.J. Vision-based target detection and localization via a team of cooperative UAV and UGVs. *IEEE Trans. Syst. Man Cybern. Syst.* **2016**, *46*, 1005–1016. [CrossRef]

16. Gupta, M.; Kumar, S.; Behera, L.; Subramanian, V.K. A novel vision-based tracking algorithm for a human-following mobile robot. *IEEE Trans. Syst. Man Cybern. Syst.* **2017**, *47*, 1415–1427. [CrossRef]

17. Mukherjee, D.; Wu, Q.M.J.; Nguyen, T.M. Gaussian mixture model with advanced distance measure based on support weights and histogram of gradients for background suppression. *IEEE Trans. Ind. Inf.* **2014**, *10*, 1086–1096. [CrossRef]

18. Zhang, X.; Zhu, C.; Wang, S.; Liu, Y.; Ye, M. A bayesian approach to camouflaged moving object detection. *IEEE Trans. Circuits Syst. Video Technol.* **2017**, *27*, 2001–2013. [CrossRef]

19. Xu, Z.; Zhang, Q.; Cao, Z.; Xiao, C. Video background completion using motion-guided pixels assignment optimization. *IEEE Trans. Circuits Syst. Video Technol.* **2016**, *26*, 1393–1406. [CrossRef]

20. Benedek, C.; Szirányi, T.; Kato, Z.; Zerubia, J. Detection of object motion regions in aerial image pairs with a multilayer markovian model. *IEEE Trans. Image Process.* **2009**, *18*, 2303–2315. [CrossRef]

21. Wang, Z.; Liao, K.; Xiong, J.; Zhang, Q. Moving object detection based on temporal information. *IEEE Signal. Process. Lett.* **2014**, *21*, 1403–1407. [CrossRef]

22. Bae, S.-H.; Kim, M. A DCT-based total JND profile for spatiotemporal and foveated masking effects. *IEEE Trans. Circuits Syst. Video Technol.* **2017**, *27*, 1196–1207. [CrossRef]

23. Yazdi, M.; Bouwmans, T. New trends on moving object detection in video images captured by a moving camera: A survey. *Comput. Sci. Rev.* **2018**, *28*, 157–177. [CrossRef]

24. Saif, A.F.M.S.; Prabuwono, A.S.; Mahayuddin, Z.R. Moving object detection using dynamic motion modelling from UAV aerial Images. *Sci. World J.* **2014**, *2014*, 1–12. [CrossRef] [PubMed]

25. Maier, J.; Humenberger, M. Movement detection based on dense optical flow for unmanned aerial vehicles. *Int. J. Adv. Robot. Syst.* **2013**, *10*, 146–157. [CrossRef]

26. Kalantar, B.; Mansor, S.B.; Halin, A.A.; Shafri, H.Z.M.; Zand, M. Multiple moving object detection from UAV videos using trajectories of matched regional adjacency graphs. *IEEE Trans. Geosci. Remote Sens.* **2017**, *55*, 5198–5213. [CrossRef]

27. Wu, Y.; He, X.; Nguyen, T.Q. Moving object detection with a freely moving camera via background motion subtraction. *IEEE Trans. Circuits Syst. Video Technol.* **2017**, *27*, 236–248. [CrossRef]

Electronics **2019**, *8*, 1373

28. Cai, S.; Huang, Y.; Ye, B.; Xu, C. Dynamic illumination optical flow computing for sensing multiple mobile robots from a drone. *IEEE Trans. Syst. Man Cybern. Syst.* **2017**, *48*, 1370–1382. [CrossRef]
29. Minaeian, S.; Liu, J.; Son, Y.J. Effective and efficient detection of moving targets from a UAV's camera. *IEEE Trans. Intell. Transp. Syst.* **2018**, *19*, 497–506. [CrossRef]
30. Leal-Taixé, L.; Milan, A.; Schindler, K.; Cremers, D.; Reid, I.; Roth, S. Tracking the trackers: An analysis of the state of the art in multiple object tracking. *arXiv* **2017**, arXiv:1704.02781. Available online: https://arxiv.org/abs/1704.02781 (accessed on 10 April 2017).
31. Bay, H.; Ess, A.; Tuytelaars, T.; Van Gool, L. Speeded-Up Robust Features (SURF). *Comput. Vis. Image Underst.* **2008**, *110*, 346–359. [CrossRef]
32. Shene, T.N.; Sridharan, K.; Sudha, N. Real-time SURF-based video stabilization system for an FPGA-driven mobile robot. *IEEE Trans. Ind. Electron.* **2016**, *63*, 5012–5021. [CrossRef]
33. Rahmaniar, W.; Wang, W.-J. A novel object detection method based on Fuzzy sets theory and SURF. In Proceedings of the International Conference on System Science and Engineering, Morioka, Japan, 6–8 July 2015; pp. 570–584.
34. Kumar, S.; Azartash, H.; Biswas, M.; Nguyen, T. Real-time affine global motion estimation using phase correlation and its application for digital image stabilization. *IEEE Trans. Image Process.* **2011**, *20*, 3406–3418. [CrossRef] [PubMed]
35. Wang, C.; Kim, J.; Byun, K.; Ni, J.; Ko, S. Robust digital image stabilization using the Kalman filter. *IEEE Trans. Consum. Electron.* **2009**, *55*, 6–14. [CrossRef]
36. Ryu, Y.G.; Chung, M.J. Robust online digital image stabilization based on point-feature trajectory without accumulative global motion estimation. *IEEE Signal. Process. Lett.* **2012**, *19*, 223–226. [CrossRef]
37. Viola, P.; Jones, M.J. Robust real-time face detection. *Int. J. Comput. Vis.* **2004**, *57*, 137–154. [CrossRef]
38. Ludwig, O.; Nunes, U.; Ribeiro, B.; Premebida, C. Improving the generalization capacity of cascade classifiers. *IEEE Trans. Cybern.* **2013**, *43*, 2135–2146. [CrossRef]
39. Rahmaniar, W.; Wang, W. Real-Time automated segmentation and classification of calcaneal fractures in CT images. *Appl. Sci.* **2019**, *9*, 3011. [CrossRef]
40. Farneback, G. Two-frame motion estimation based on polynomial expansion. In Proceedings of the Scandinavian Conference on Image Analysis, Halmstad, Sweden, 29 June–2 July 2003; pp. 363–370.
41. Cayon, R.J.O. Online Video Stabilization for UAV. Master's Thesis, Politecnico di Milano, Milan, Italy, 2013.
42. Li, J.; Xu, T.; Zhang, K. Real-Time Feature-Based Video Stabilization on FPGA. *IEEE Trans. Circuits Syst. Video Technol.* **2017**, *27*, 907–919. [CrossRef]
43. Hong, S.; Dorado, A.; Saavedra, G.; Barreiro, J.C.; Martinez-Corral, M. Three-dimensional integral-imaging display from calibrated and depth-hole filtered kinect information. *J. Disp. Technol.* **2016**, *12*, 1301–1308. [CrossRef]
44. Muja, M.; Lowe, D.G. Fast approximate nearest neighbors with automatic algorithm configuration. In Proceedings of the International Conference on Computer Vision Theory and Applications, Lisboa, Portugal, 5–8 February 2009; pp. 331–340.
45. Fischler, M.A.; Bolles, R.C. Random sample consensus: A paradigm for model fitting with applications to image analysis and automated cartography. *Commun. ACM* **1981**, *24*, 381–395. [CrossRef]
46. Yu, H.; Moulin, P. Regularized Adaboost learning for identification of time-varying content. *IEEE Trans. Inf. Forensics Secur.* **2014**, *9*, 1606–1616. [CrossRef]

 *electronics*

Article

CNN-Based Ternary Classification for Image Steganalysis

Sanghoon Kang [1], Hanhoon Park [1,*] and Jong-Il Park [2]

[1] Department of Electronic Engineering, Pukyong National University, 45 Yongso-ro, Nam-gu, Busan 48513, Korea; totohoon01@naver.com

[2] Department of Computer Science, Hanyang University, 222, Wangsimni-ro, Seongdong-gu, Seoul 04763, Korea; jipark@hanyang.ac.kr

* Correspondence: hanhoon.park@pknu.ac.kr; Tel.: +82-51-629-6225

Received: 11 September 2019; Accepted: 23 October 2019; Published: 26 October 2019

Abstract: This study proposes a convolutional neural network (CNN)-based steganalytic method that allows ternary classification to simultaneously identify WOW and UNIWARD, which are representative adaptive image steganographic algorithms. WOW and UNIWARD have very similar message embedding methods in terms of measuring and minimizing the degree of distortion of images caused by message embedding. This similarity between WOW and UNIWARD makes it difficult to distinguish between both algorithms even in a CNN-based classifier. Our experiments particularly show that WOW and UNIWARD cannot be distinguished by simply combining binary CNN-based classifiers learned to separately identify both algorithms. Therefore, to identify and classify WOW and UNIWARD, WOW and UNIWARD must be learned at the same time using a single CNN-based classifier designed for ternary classification. This study proposes a method for ternary classification that learns and classifies cover, WOW stego, and UNIWARD stego images using a single CNN-based classifier. A CNN structure and a preprocessing filter are also proposed to effectively classify/identify WOW and UNIWARD. Experiments using BOSSBase 1.01 database images confirmed that the proposed method could make a ternary classification with an accuracy of approximately 72%.

Keywords: image steganalysis; WOW; UNIWARD; ternary classification; convolutional neural network (CNN)

---

1. Introduction

Interest in information security technologies, such as image steganography/steganalysis, has significantly grown because of the universalization of digital multimedia and communication. Image steganography is a technique in which a secret message is embedded into an image, called *cover image,* and the message-embedded image, called *stego image,* is transmitted through a public channel without gaining the attention of a third party, thereby implementing covert communication. The image steganalysis is the reverse process of image steganography, which aims to determine whether or not the image to be tested contains a secret message and then finds out the hidden message.

The performance of image steganographic methods depends on two conflicting parameters: embedding capacity, which represents how many messages we can hide, and the image quality after embedding, which is closely related to message concealment. Therefore, most image steganographic methods have achieved a high embedding capacity at the expense of low image quality after embedding, and vice versa.

Early image steganographic methods include the least significant bit (LSB) substitution method [1], which replaces the least significant bits of image pixels by secret messages, and the pixel value differencing (PVD) methods [2–4] that determine the amount of secret messages to be embedded in proportion to the difference between adjacent pixels. These early image steganographic methods sequentially embed secret messages into all pixels of an image, although they have been recently extended to embed messages in randomly selected pixels using pseudo-random generators for secure message hiding [5,6].

Sequentially embedding secret messages into all pixels of an image is well known to change the statistical characteristics of the image. In Figure 1, the solid line refers to a probability density function (PDF) of the differences between two adjacent pixels on a cover image. The dot lines refer to the different PDFs on the stego images created by different image steganographic methods. The PDF of the LSB stego image is significantly different from that of the cover image in the section where the differences are small. This statistical difference is easily detected by statistical attacks, such as the RS analysis in [7]. Thus, image steganographic methods have come to consider more how not to be detected by steganalytic attacks than how many messages to embed. To avoid statistical attacks, image steganographic methods began to consider where the message would be embedded. Methods such as HUGO [8], WOW [9], and UNIWARD [10] tried to embed a message into only pixels with a small distortion, mainly on image edges, by analyzing the distortion caused by embedding a message into each pixel. For example, HUGO measured the embedding distortion by reverse-engineering the processes of the subtractive pixel adjacency matrix (SPAM) [11], a steganalytic method that calculated a co-occurrence matrix for the differences of the adjacent pixels in eight directions of vertical, horizontal, and diagonal to analyze the statistical changes in the pixel values caused by the message embedding. HUGO could reduce the probability of being detected by the SPAM by 1/7.

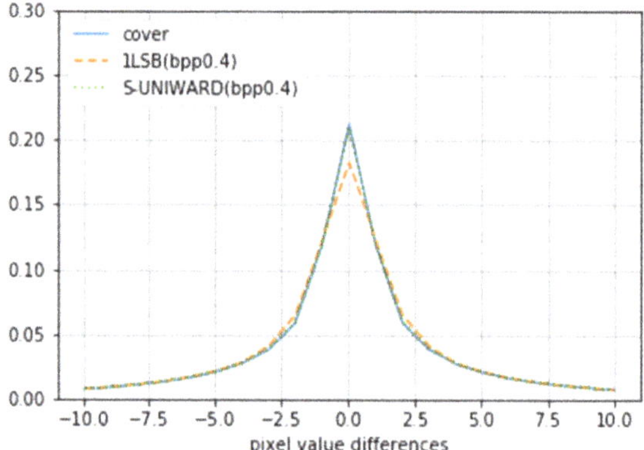

Figure 1. Probability density functions of the differences between the adjacent pixels on a cover image and its stego images.

The performance of image steganalysis in detecting image steganography has greatly improved with the development of image steganography to more covertly and skillfully hide a message. Image staganalytic methods generally try to extract traces of image steganography in the image by using high-pass filters (HPF) and identify images to which image steganography has been applied through classification. Early steganalytic methods extracted image features using manually designed HPFs (those features are called handcrafted features hereafter) and detected image steganography using classifiers based on machine learning algorithms, such as support vector machines (SVM) [12] and random forest [13]. A representative method using handcrafted features is the spatial rich model (SRM) [14].

With the great success of convolutional neural networks (CNN) in object detection and recognition [15,16], using CNNs for steganalysis has been actively investigated [17–27]. Unlike handcrafted feature-based methods, a CNN can automatically extract and learn the features that are optimal or well suited for identifying steganographic methods. Therefore, CNN-based steganalytic methods have demonstrated a better performance compared to handcrafted feature-based methods.

However, most existing image steganalytic methods, regardless of whether or not CNNs are used, have focused on identifying whether or not a secret message is hidden in an image (i.e., the binary classification between a normal (or cover) image in which any message has not been embedded and a stego image in which a message has been embedded). Discriminating stego images created by different steganographic methods has been less considered; thus, the binary classifiers are not suitable for discriminating these stego images. Discriminating the stego images created by WOW and UNIWARD that embed a message in a similar and skillful manner is very difficult.

The classification of stego images created by different steganographic methods plays an important role in restoring embedded messages beyond judging whether or not a message is embedded. In this study, as the first step to restore messages embedded by steganographic methods, a CNN-based steganalytic method is proposed to classify the stego images created by different steganographic methods. The structure of a ternary classifier is specially designed to distinguish between the stego images created by WOW and UNIWARD and the normal images without messages. Through comparative experiments with the existing binary classifiers, the reason why multiple steganographic methods should be classified in a single ternary classifier, and various methods for improving the performance of the proposed ternary classifier are presented.

Compared to existing image steganalytic methods, the primary contributions of this study are as follows:

- a single framework is provided for identifying multiple steganographic methods;
- a CNN-based ternary classifier is proposed for image steganalysis; and
- effective methods for extending a CNN to discriminate similar WOW and UNIWARD stego images are proposed and evaluated.

This study is an extension of [28] and differs from the previous study in the following respect:

- a CNN-based ternary classifier with a new preprocessing filter is proposed;
- more details for designing it are provided; and
- the performance of the proposed classifier is intensively evaluated.

The remainder of this paper is organized as follows: Section 2 briefly reviews the conventional image steganographic and steganalytic methods; Section 3 explains the proposed steganalytic method; Section 4 experimentally evaluates its performance using images from a database available online; and Section 5 presents the conclusions and suggestions for future work.

2. Related Work

2.1. WOW and UNIWARD

WOW and UNIWARD calculate the degree of distortion when a message is embedded in an image, and then embed a small amount of message in regions where the distortion is small. We refer herein to such methods as adaptive steganographic methods. This makes it more difficult to detect hidden messages by embedding messages only in high-frequency regions with relatively little distortion and makes it possible to avoid steganalytic attacks using statistical analysis because the change in the statistical characteristics of the images caused by message embedding is very small (Figure 1).

Adaptive steganographic methods have suggested different approaches for quantifying the image distortion caused by message embedding. The image distortion function for WOW is defined as follows:

$$D(X, Y) = \rho_{ij}(X, Y_{ij})|X_{ij} - Y_{ij}|. \tag{1}$$

Here, X and Y are a cover and its stego images, respectively, and ρ is a function that examines the detectability in all neighboring directions of each pixel using the HPFs in Figure 2. Thus, a message is not embedded if the detectability is high even in one direction. The message is embedded into the pixels for which the detectability is low in all directions.

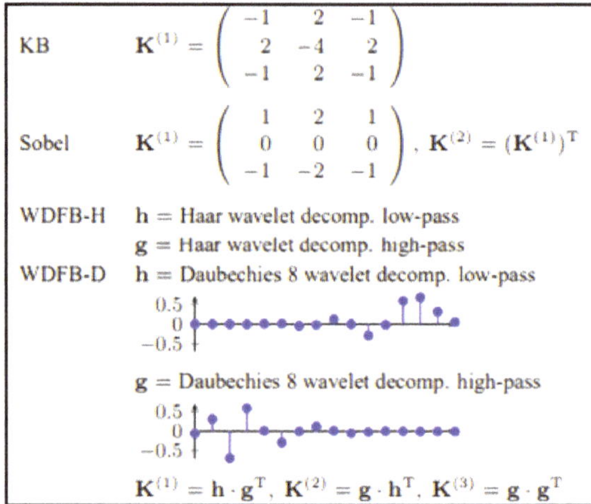

Figure 2. HPFs and wavelet filters used in WOW [9].

For UNIWARD, the residual images were calculated using the wavelet filters in Figure 2. The image distortion function is defined as follows by the sum of the absolute difference between the cover and the stego residual images:

$$D(X, Y) = \sum_{k=1}^{3} \sum_{u=1}^{n_1} \sum_{v=1}^{n_2} \frac{|W_{uv}^k(X) - W_{uv}^k(Y)|}{\sigma + |W_{uv}^{(k)}(X)|}. \tag{2}$$

Here, $W^{(k)}$ represents the residual image calculated using the kth filter; n_1 and n_2 are the image width and height, respectively, and σ is a constant stabilizing the numerical calculations.

Consequently, WOW and UNIWARD have different image distortion functions, but their approaches to embedding messages are very similar.

2.2. SRM

The SRM [14] is a handcrafted feature-based steganalytic method that uses various types of linear and nonlinear HPFs (Figure 3) to extract a number of meaningful features from the images. The features are then classified using an ensemble classifier (i.e., a random forest) that uses Fisher linear discriminants as the base classifiers.

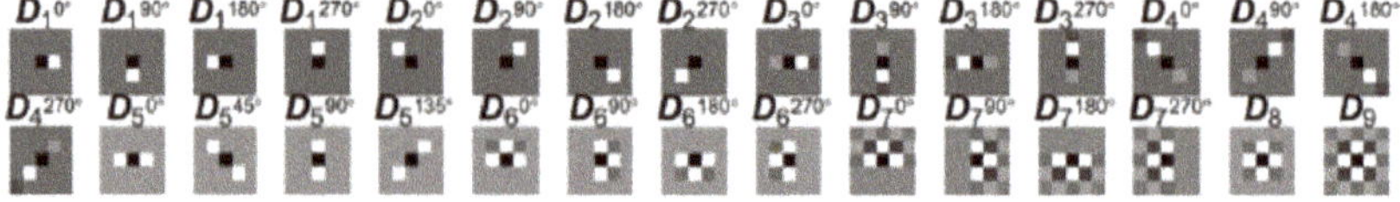

Figure 3. Thirty linear and nonlinear 5 × 5 SRM filters [19]. The filters are padded with zeros to obtain a unified size of 5 × 5.

The SRM was the most effective method used to detect image steganography before the CNN-based image steganalytic methods emerged. The SRM is highly accurate compared to

CNN-based methods. The method of extracting many features using various types of HPFs has also been widely used in CNN-based ones [19,20,25–27].

2.3. CNN-Based Image Steganalysis

CNNs can automatically extract the optimal features required to detect and recognize objects in images, and can classify features with high accuracy [15,16]. Therefore, studies using CNNs are greatly increasing in the image steganalysis field. However, unlike other deep learning problems, the CNN-based image steganalysis has a preprocessing process of applying HPFs to input images. This process enhances the pixel variation caused by embedding messages such that the CNN can detect it well while also removing the low-frequency area, where the messages are less likely to be embedded.

Xu and Wu proposed a simple yet effective initial CNN for image steganalysis [17]. They used a network comprising five convolutional layers and a single fully connected layer (Figure 4). They also used a 5 × 5 HPF in a preprocessing stage, generated eight feature maps in the first convolutional layer, and doubled the number of feature maps and halved the size of the feature maps in the subsequent convolutional layers. Each convolutional layer comprised the processes of convolution, batch normalization, activation, and pooling. They improved the steganalytic performance of the network by adding the absolute layer to the first convolutional layer and by using the tanh activation function in the first two convolutional layers. Yuan et al. used the same network structure as the initial CNN, but utilized three HPFs in a preprocessing stage [18].

Figure 4. Initial CNN for image steganalysis [17]. The CNN extracts 128 1 × 1 feature maps from a 512 × 512 input image.

ReST-Net [19] uses three different filter sets, namely 16 simplified linear SRM, 14 nonlinear SRM, and 16 Gabor filters (Figures 3 and 5) in the preprocessing stage to extract much more features from the input images. In addition, ReST-Net constructs three subnetworks (Figure 6). After separately training the subnetworks using each preprocessing filter, it trains a new fully connected layer using transfer learning while fixing the parameters of three subnetworks.

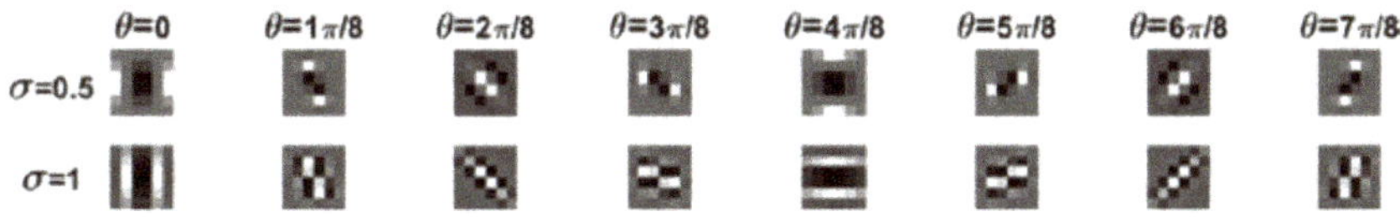

Figure 5. Sixteen 6 × 6 Gabor filters with different orientations and scales [19].

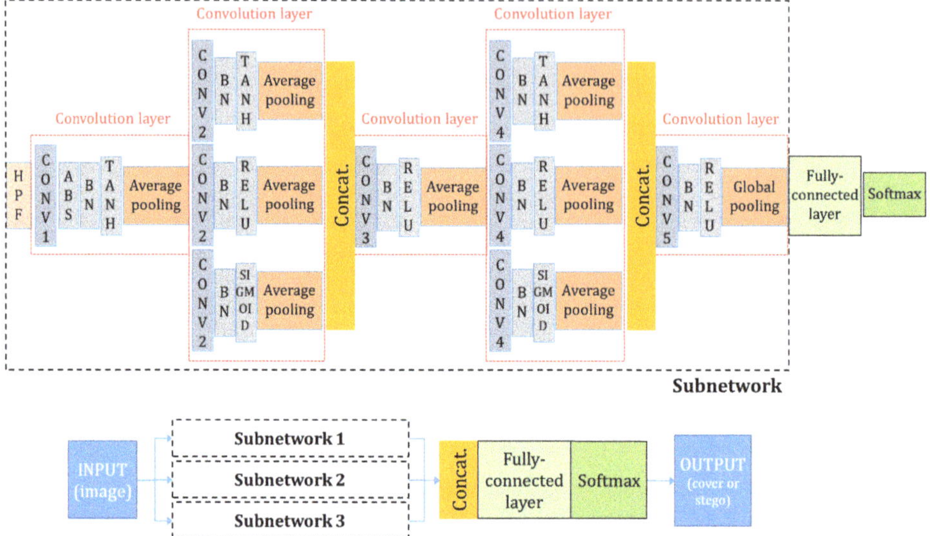

Figure 6. Structure of Rest-Net [19], which comprises three three subnetworks that are a modification of the initial CNN [17] and uses transfer learning.

Yedroudj-Net [20] has a similar structure with the initial CNN [17], but uses linear SRM filters in the preprocessing stage, and has two additional fully connected layers (Figure 7). It removes the average pooling process in the first convolutional layer to prevent loss of information caused by pooling. In the first two convolutional layers, it uses the TLU function instead of the tanh function to remove the strong, but statistically insignificant information. It has an additional scaling process after batch normalization. Yedroudj-Net has achieved approximately 4–5% improvement in accuracy in binary classification in comparison with the initial CNN [17].

Figure 7. Structure of Yedroudj-Net [20].

Deep residual networks for image steganalysis have also been proposed [22,23]. These networks could be made much deeper by employing residual shortcuts (Figure 8). In [22], without fixing the preprocessing filters or initializing the filter coefficients with the SRM filters, the preprocessing process is significantly expanded using several convolutional and residual layers to realize a completely data-driven steganalysis.

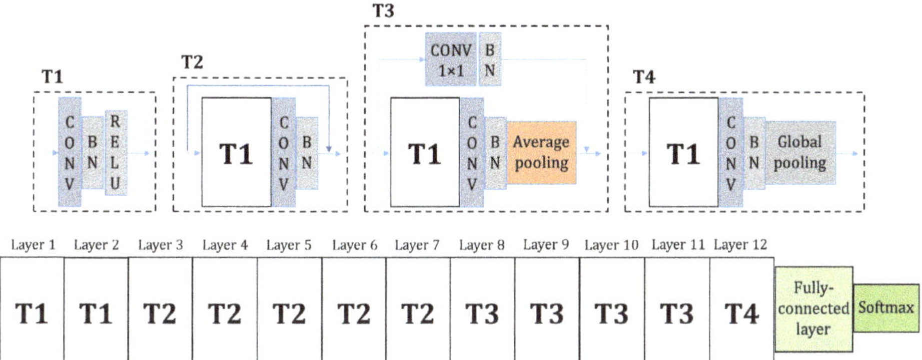

Figure 8. Structure of a deep residual network used in [22].

Ke et al. proposed a multi-column CNN that extracts various features using filters of different sizes in convolutional layers and allows the input image to be of an arbitrary size or resolution [24]. As a multi-task learning approach, Yu et al. extended a CNN by adding fully convolutional networks that take the output of each convolutional layer as the input for a pixel binary classification that estimates whether or not each pixel in an image has been modified because of steganography [26].

Wu et al. proposed a new normalization, called shared normalization, that uses the same mean and standard deviation, instead of the minibatch mean and standard deviation, for all training and test batches to normalize each input batch and address the limitation of batch normalization for image steganalysis [21]. Meanwhile, Ni et al. proposed a selective ensemble method that can choose to join or delete a base classifier by reinforcement learning to reduce the number of base classifiers while ensuring the classification performance [25].

As such, the existing CNN-based staganalytic methods could successfully increase the classification accuracy by deepening or widening the CNNs and using various types of preprocessing filters. However, these methods aimed for the binary classification of cover and stego images, and, thus, may not be available for the N-ary ($N > 2$) classification. Two adaptive steganographic methods, namely WOW and UNIWARD, embed a small amount of messages in a similar manner (Section 2.1); hence, the binary classifiers are very likely to misclassify the WOW and UNIWARD stego images.

3. Proposed Method

3.1. Similarity between WOW and UNIWARD

The adaptive steganographic methods, namely WOW and UNIWARD, use directional filters to analyze how different the differences from the neighboring pixels (i.e., the degree of image distortion) are when a message is embedded into each pixel of an image, and then selectively embed the message into a pixel with a small degree of image distortion. WOW and UNIWARD use different functions to measure the image distortion, but their processes of embedding the message are very similar; thus, the existing CNN-based binary classifiers become confused when discriminating WOW and UNIWARD, and are very likely to make an incorrect classification.

We conducted an experiment in which UNIWARD stego images were input to a binary classifier that had been trained for WOW and vice versa to demonstrate the difficulty of discriminating WOW and UNIWARD using binary classifiers. The CNN used in the literature [17] (Figure 4) was used for the experiment. The other experimental conditions were the same as those given in Section 4.

Table 1 shows that, even when two different steganographic methods (i.e., WOW and UNIWARD) were used in the training and testing phases, respectively, the classification rates for the stego images were still high. For example, the classification rates were 67.13% when the UNIWARD stego images were input into the classifier trained using the WOW stego images. In other words, it is very likely that they are confused with each other because they are too similar to discriminate. Therefore, using existing binary classifiers to classify WOW and UNIWARD is ineffective.

Table 1. Cross identification between WOW and UNIWARD ($bpp = 0.4$).

Training	Testing	Classification Rates (%)			
		For Cover	For UNIWARD Stego	For WOW Stego	Total
UNIWARD	UNIWARD	84.02	73.56	-	78.80
WOW	WOW	77.63	-	78.25	77.94
UNIWARD	WOW	84.02	-	60.48	72.25
WOW	UNIWARD	77.63	67.13	-	72.38

3.2. Combining Pre-Trained Binary Classifiers to Discriminate WOW and UNIWARD

We attempted to train two CNN-based binary classifiers for WOW and UNIWARD and simply combine the two classifiers in parallel to determine the result of the classifier with a higher probability as a final result (Figure 9). This was based on the assumption that the results of the classifier with a greater probability would be right if different classification results are obtained by the two classifiers. Table 2 presents the classification results for the cover, WOW stego, and UNIWARD stego images (the details for the experimental conditions are given in Section 4). The classification rates for the WOW and UNIWARD stego images significantly decreased because of the similarity between WOW and UNIWARD. In other words, the simple combination of two binary classifiers is not useful for discriminating WOW and UNIWARD.

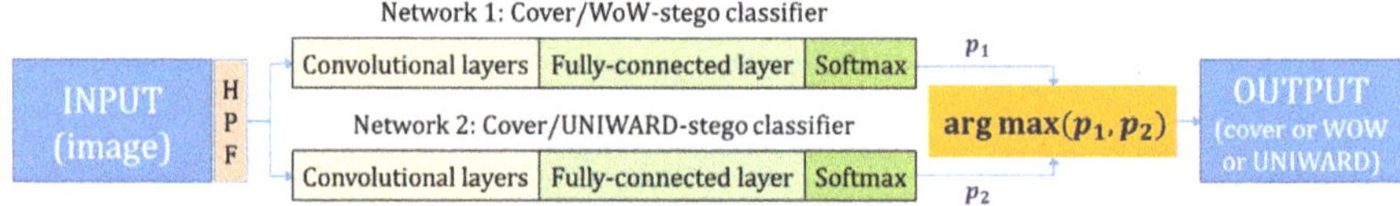

Figure 9. Combining two binary classifiers in parallel for the ternary classification.

Table 2. Ternary classification rates obtained by simply combining two binary classifiers separately trained for WOW and UNIWARD ($bpp = 0.4$).

For Cover (%)	For UNIWARD Stego (%)	For WOW Stego (%)	Total (%)
85.23	50.13	46.24	60.53

We also conducted an experiment for ternary classification through transfer learning. The network parameters of classifiers were fixed after training each binary classifier for WOW and UNIWARD. The fully connected layer was then removed from each binary classifier, and a common fully connected layer was added and trained for the ternary classification (Figure 10). Table 3 shows the classification results for the cover, WOW stego, and UNIWARD stego images (the details for the experimental conditions are given in Section 4). The classification rates for the WOW and UNIWARD stego images were very low, and were lower than those obtained by simply combining two binary classifiers in parallel. This result indicates that the network parameters in the fully connected layer were not correctly trained because of the similarity between WOW and UNIWARD.

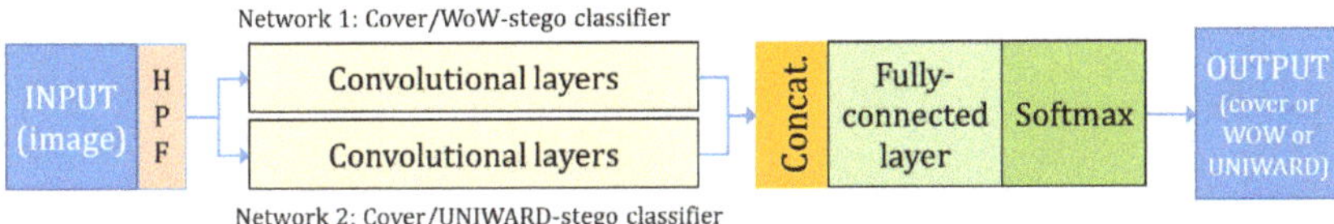

Figure 10. Ternary classification through transfer learning.

Table 3. Ternary classification rates obtained by transfer learning ($bpp = 0.4$).

For Cover (%)	For UNIWARD Stego (%)	For WOW Stego (%)	Total (%)
72.23	45.56	42.90	53.56

Referring to the abovementioned experiments, a new single network should be designed to simultaneously learn the cover, WOW stego, and UNIWARD stego images from the beginning and correctly classify the similar steganographic methods, WOW and UNIWARD.

3.3. Designing a CNN for Ternary Classification

The CNN used in [17] is the most basic CNN for image steganalysis, and most conventional CNNs are its modifications. Therefore, it was used as the base CNN herein. First, the base CNN was tested for ternary classification without modification. The cover, WOW stego, and UNIWARD stego images were simultaneously learned in a single network (Figure 4). Table 4 presents the classification rates (the details for the experimental conditions are given in Section 4), which are better than those obtained by combining pre-trained binary classifiers. However, the cover images were relatively well classified at approximately 84%, but the WOW and UNIWARD stego images were rarely classified as expected. In conclusion, the network structure should be extended, and the preprocessing filter for extracting the steganalytic features should be more carefully designed to make the classifier originally developed for the binary classification between the cover and stego images available for ternary classification.

Table 4. Ternary classification rates when simultaneously learning the cover, WOW stego, and UNIWARD stego images using the conventional classifier [17] ($bpp = 0.4$).

For Cover (%)	For UNIWARD Stego (%)	For WOW Stego (%)	Total (%)
84.05	56.45	50.39	63.63

We tried to extend the based CNN by attempting to add more convolutional layers (each comprising convolution, normalization, activation, and pooling operations) because more classification power would be required for the ternary classification compared to the binary classification. Figure 11 shows the structure of the networks extended with additional convolutional layers. Table 5 displays the classification rates of the extended networks (the details for the experimental conditions are given in Section 4). As a result, adding convolutional layers improved the classification rates by 2–4%; however, the classification rates rather became lower with two or more additional convolutional layers, indicating that the network needs to be deeper for ternary classification, but the depth should be properly adjusted.

Table 5. Ternary classification rates of deeper networks in Figure 11 ($bpp = 0.4$).

	For Cover (%)	For UNIWARD Stego (%)	For WOW Stego (%)	Total (%)
Figure 11a	75.81	67.48	59.83	67.70
Figure 11b	72.40	58.18	69.80	66.79
Figure 11c	61.24	60.18	75.27	65.56

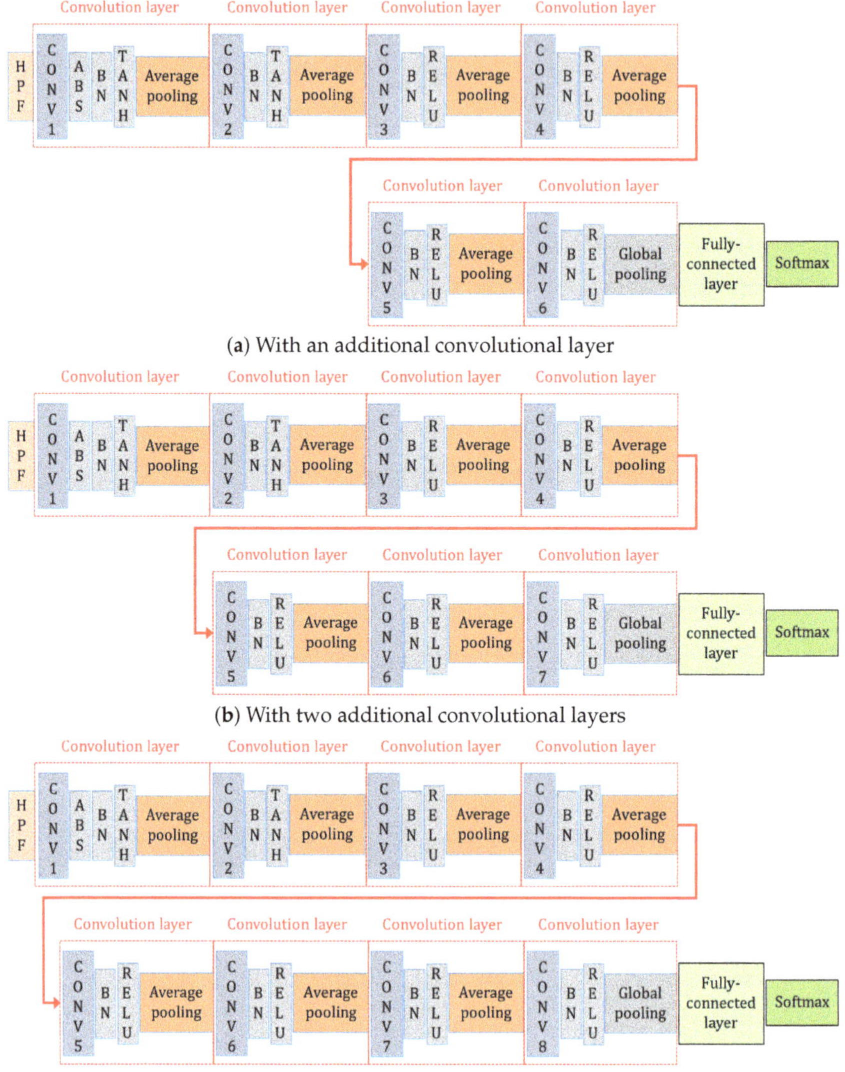

(**a**) With an additional convolutional layer

(**b**) With two additional convolutional layers

(**c**) With three additional convolutional layers

Figure 11. Extending the conventional network [17] with additional convolutional layers.

We also attempted to use a deep residual network (Figure 12a) or a convolution-stacked network (Figure 12b), where the convolutional blocks were stacked as done in [29] because those residual or convolution-stacked networks demonstrated a significantly improved performance in image recognition. However, as shown in Table 6, the classification rates were not good, indicating that these networks were not suitable for image steganalysis or for ternary classification.

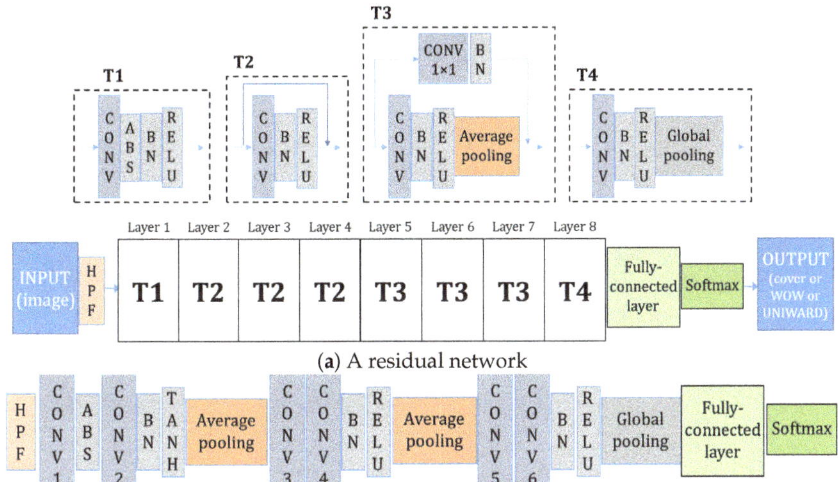

Figure 12. Deep residual network and convolution-stacked network for ternary classification.

Table 6. Ternary classification rates of the residual and convolution-stacked networks of Figure 12 ($bpp = 0.4$).

	For Cover (%)	For UNIWARD Stego (%)	For WOW Stego (%)	Total (%)
Figure 12a	80.03	23.52	4.30	35.95
Figure 12b	83.87	62.43	32.87	59.72

As explained in Section 2.3, the CNN-based classifiers for image steganalysis have preprocessing filters to facilitate the extraction of steganalytic features from images. Many conventional methods tried to use various preprocessing filters for performance improvement. For the ternary classification, we decided to use the SRM filters mostly used in conventional methods and conducted an experiment to determine their performance. The base CNN was used with three different preprocessing filter sets: 30 SRM filters (Figure 3), three groups of 10 SRM filters, and 10 selected SRM filters (Figure 13). The second filter set was obtained by dividing 30 SRM filters into three groups of 10 (using different numbers of groups was worse [28]). The filters of each group were applied to the input image. Ten filtered results were generated by performing the element-wise sum between the filtered results of each group [28]. The third filter set is a new one proposed herein. More effective filters were selected from 30 SRM filters. Each of the 30 SRM filters was applied to the arbitrary cover and stego images. The differences between the filtered cover and stego images were then computed (Figure 14). Subsequently, 10 filters with higher differences were selected, assuming that those filters would extract steganalytic features from the images well. For all of the filter sets, eight feature maps were generated in the first convolutional layer and doubled in the subsequent convolutional layers. Tables 4 and 7 (the details for the experimental conditions are given in Section 4) show that the classification rates of the base CNN did not increase as the number of filters increased, unlike expected. The results of the three groups of 10 SRM filters were better than those of the others, indicating that simply increasing the number of filters does not guarantee performance improvements, and finding the appropriate filters for a given CNN is necessary.

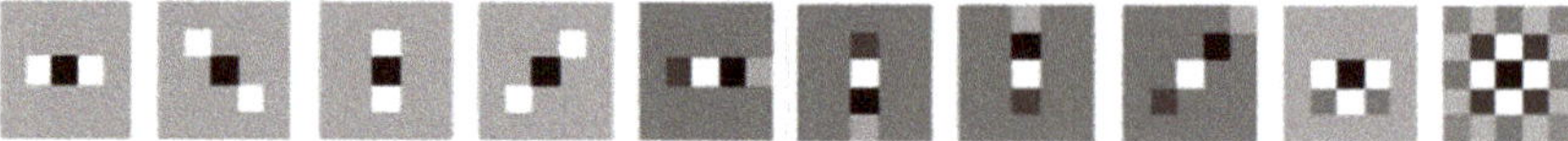

Figure 13. Ten selected SRM filters. They can better detect tiny variation on images, among 30 SRM filters of Figure 3.

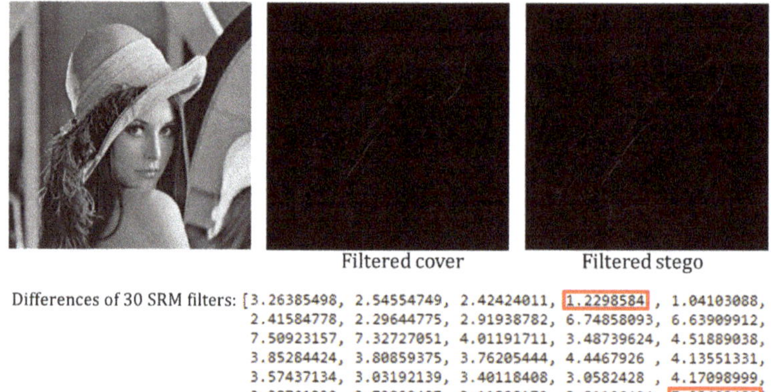

Filtered cover   Filtered stego

Differences of 30 SRM filters: [3.26385498, 2.54554749, 2.42424011, 1.2298584 , 1.04103088,
2.41584778, 2.29644775, 2.91938782, 6.74858093, 6.63909912,
7.50923157, 7.32727051, 4.01191711, 3.48739624, 4.51889038,
3.85284424, 3.80859375, 3.76205444, 4.4467926 , 4.13551331,
3.57437134, 3.03192139, 3.40118408, 3.0582428 , 4.17098999,
2.38761902, 3.79829407, 3.11508179, 3.81126404, 7.23495483]

Figure 14. Selection of more effective SRM filters: a large difference (e.g., 1.229 and 7.234) depending on the filter between the filtered cover and stego images is found after each SRM filter is applied.

Table 7. Ternary classification rates of the base CNN with different preprocessing filters ($bpp = 0.4$).

Preprocessing Filters	For Cover (%)	For UNIWARD Stego (%)	For WOW Stego (%)	Total (%)
30 SRM	50.43	19.98	90.90	53.70
Three groups of 10 SRM	65.44	51.01	75.26	63.90
10 selected SRM	68.55	73.15	42.41	61.37

Together with increasing the number of filters, we also attempted to increase the feature maps in the first convolutional layers from 8 to 60. Table 8 shows that the classification rates of the base CNN became significantly lower, except for the 10 selected SRM filters, when the number of feature maps increased. Unlike most conventional CNNs that achieve performance improvement by using more filters or feature maps, the base CNN had a better performance with a small number of filters maybe because the base CNN failed to learn a large amount of information extracted by many filters or feature maps. From these results, we conclude that the base CNN should be deeper such that more filters or feature maps can be used.

Table 8. Ternary classification rates of the base CNN with different preprocessing filters when increasing the feature maps in the first convolutional layers to 60 ($bpp = 0.4$).

Preprocessing Filters	For Cover (%)	For UNIWARD Stego (%)	For WOW Stego (%)	Total (%)
5×5 HPF	90.44	40.15	44.89	58.49
30 SRM	4.54	87.39	26.04	39.32
Three groups of 10 SRM	79.17	32.52	67.39	59.69
10 selected SRM	46.08	45.27	88.99	60.11

3.4. Proposed Classifier for Ternary Classification

We proposed a CNN-based classifier for the ternary classification. The base CNN [17] was extended with an additional convolutional layer. The feature maps were increased to 60 in the first convolutional layer and doubled in the subsequent convolutional layers: thus, 1920 feature maps were fed into the fully connected layer. Ten selected SRM filters were used as the preprocessing filters.

4. Experimental Results and Discussion

All the experiments presented in the previous sections and in this section were conducted with the following conditions: 10,000 gray scale images of 512×512 in BOSSBase 1.01 [30] were quartered, and the resulting 40,000 images were divided into the training and testing sets, each comprising 30,000 and 10,000 images, respectively. The stego images for both sets were generated with a random payload of $bpp = 0.4$ (In most steganalytic studies, 0.1, 0.2, and 0.4 bpp have been used for testing steganalytic methods. However, when using adaptive steganographic methods, 0.1 and 0.2 bpp are too small to identify the stego images, even in binary classification [31]. The average PSNRs of the WOW and UNIWARD stego images of 0.4 bpp are 58.76 and 59.36 dB, respectively; thus, the image quality of the stego images of 0.4 bpp is still very high.) using WOW and UNIWARD. As a result, 90,000 (30,000 for cover, WOW stego, and UNIWARD stego images each) training images of 256×256 and 30,000 (10,000 for cover, WOW stego, and UNIWARD stego images each) testing images were used. For training, a momentum optimizer [32] with a momentum value of 0.9 was used. The learning rate started at 0.001 and decreased to 90% in every 5000 iterations. The minibatch size was 64 (32 pairs of cover and stego images). The other hyperparameters were set the same as in the conventional method [17]. All CNNs were implemented using the TensorFlow library [33].

The proposed classifier was evaluated with different preprocessing filters. As a new preprocessing filter set, 16 Gabor filters were used together with the 10 selected SRM filters, as has been done in [19]. The results in Table 9 are the classification rates for the cover, WOW stego, and UNIWARD stego images obtained using different preprocessing filters.

Table 9. Ternary classification rates of the network of Figure 11a with different preprocessing filters ($bpp = 0.4$).

Preprocessing Filters	For Cover (%)	For UNIWARD Stego (%)	For WOW Stego (%)	Total (%)
5×5 HPF	68.52	46.07	61.01	58.53
30 SRM	75.49	51.85	71.46	66.26
Three groups of 10 SRM	76.22	59.66	77.45	70.78
10 selected SRM	75.65	69.71	71.32	72.22
10 selected SRM + 16 Gabor	76.23	56.26	62.10	64.86

Unlike the base CNN, using more filters and feature maps increased the classification rates; however, utilizing too many and different types of filters was not good. The results of the 10 selected SRM filters (i.e., the proposed one) were the best. The experimental results demonstrated that the cover, WOW stego, and UNIWARD stego could be classified with an accuracy of approximately 72% through the single CNN-based ternary classifier proposed herein.

We also attempted to change the tanh functions of the first two convolutional layers to TLU functions, as has been done in [20], and the ReLU functions of the subsequent convolutional layers to leaky ReLU functions, but the classification rates were not good (Table 10).

Table 10. Ternary classification rates when changing the activation functions of the proposed CNN ($bpp = 0.4$).

For Cover (%)	For UNIWARD Stego (%)	For WOW Stego (%)	Total (%)
39.60	72.29	91.87	67.93

5. Conclusions and Future Works

This study proposed a CNN-based ternary classifier to identify cover, WOW stego, and UNIWARD stego images. The existing binary classifiers were designed to learn and detect a specific steganographic method; hence, they were not suitable for discriminating different steganographic methods. Adaptive steganographic methods, such as WOW and UNIWARD, embed a small amount of the secret message in a similar manner; therefore, discriminating their stego images using the existing binary classifiers or combining them was very difficult. However, the proposed ternary classifier could effectively learn the difference between both steganographic methods and discriminate them with high accuracy. The classification between different steganographic methods using the proposed ternary classifier was the first step in restoring the embedded message instead of simply determining whether or not a message has been embedded.

It was experimentally confirmed that, in designing a CNN-based ternary classifier for image steganalysis, simply expanding the width or depth of the CNN does not guarantee performance improvements. In other words, the CNN width and depth need experimental optimization. This study demonstrated the results of such an experimental optimization.

The proposed method had an accuracy of approximately 72%, which is not very high. Therefore, ways to improve the accuracy by further highlighting the differences between WOW and UNIWARD must be explored in the future. Ways to design a CNN-based classifier suitable for classifying a larger number (≥ 3) of steganographic methods, including those with other embedding domains (e.g., DCT and wavelet domains), must also be explored.

Author Contributions: Conceptualization, S.K. and H.P.; Funding acquisition, H.P. and J.-I.P.; Methodology, S.K. and H.P.; Software, S.K.; Supervision, H.P. and J.-I.P.; Validation, S.K. and H.P.; Writing—original draft, S.K.; Writing—review and editing, H.P. and J.-I.P.

Funding: This work was supported by the research fund of the Signal Intelligence Research Center supervised by the Defense Acquisition Program Administration and Agency for the Defense Development of Korea.

Conflicts of Interest: The authors declare no conflict of interest.

References

1. Chan, C.K.; Cheng, L.M. Hiding data in images by simple LSB substitution. *Pattern Recognit.* **2004**, *37*, 469–474. [CrossRef]
2. Wu, D.C.; Tsai, W.H. A steganographic method for images by pixel-value differencing. *Pattern Recognit.* **2003**, *24*, 1613–1626. [CrossRef]
3. Chang, K.C.; Chang, C.P.; Huang, P.S.; Tu, T.M. A novel image steganographic method using tri-way pixel-value differencing. *J. Multimed.* **2008**, *3*, 37–44. [CrossRef]
4. Darabkh, K.A.; Al-Dhamari, A.K.; Jafar, I.F. A new steganographic algorithm based on multi directional PVD and modified LSB. *J. Inf. Technol. Control* **2017**, *46*, 16–36. [CrossRef]
5. Kordov, K.; Stoyanov, B. Least significant bit steganography using Hitzl-Zele chaotic map. *Int. J. Electron. Telecommun.* **2017**, *63*, 417–422. [CrossRef]
6. Stoyanov, B.P.; Zhelezov, S.K.; Kordov, K.M. Least significant bit image steganography algorithm based on chaotic rotation equations. *C. R. Acad. Bulg. Sci.* **2016**, *69*, 845–850.
7. Fridrich, J.; Goljan, M.; Du, R. Detecting LSB steganography in color and gray-scale images. *IEEE Multimed. Mag.* **2001**, *8*, 22–28. [CrossRef]
8. Pevny, T.; Filler, T.; Bas, P. Using high-dimensional image models to perform highly undetectable steganography. In Proceedings of the 12th International Conference on Information Hiding, Calgary, AB, Canada, 28–30 June 2010; pp. 161–177.
9. Holub, V.; Fridrich, J. Designing steganographic distortion using directional filters. In Proceedings of the IEEE Workshop on Information Forensic and Security, Tenerife, Spain, 2 December 2012.
10. Holub, V.; Fridrich, J.; Denemark, T. Universal distortion function for steganography in an arbitrary domain. *EURASIP J. Inf. Secur.* **2014**, *2014*, 1. [CrossRef]
11. Penvy, T.; Bas, P.; Fridrich, J. Steganalysis by subtractive pixel adjacency matrix. *IEEE Trans. Inf. Forensics Secur.* **2010**, *5*, 215–224.
12. Cortes, C.; Vapnik, V. Support-vector networks. *Mach. Learn.* **1995**, *20*, 273–297. [CrossRef]
13. Ho, T.K. Random decision forests. In Proceedings of the 3rd International Conference on Document Analysis and Recognition, Montreal, QC, Canada, 14–16 August 1995; pp. 275–282.

14. Fridrich, J.; Kodovský, J. Rich models for steganalysis of digital images. *IEEE Trans. Inf. Forensics Secur.* **2012**, *7*, 868–882. [CrossRef]

15. Krizhevsky, A.; Sutskever, I.; Hinton, G.E. ImageNet classification with deep convolutional neural networks. In Proceedings of the 25th International Conference on Neural Information Processing Systems, Lake Tahoe, NV, USA, 3–6 December 2012; pp. 1097–1105.

16. He, K.; Zhang, X.; Ren, S.; Sun, J. Deep residual learning for image recognition. In Proceedings of the IEEE Conference on Computer Vision and Pattern Recognition, Las Vegas, NV, USA, 27–30 June 2016; pp. 770–778.

17. Xu, G.; Wu, H. Structure design of convolution neural networks for steganalysis. *IEEE Signal Process. Lett.* **2016**, *23*, 708–712. [CrossRef]

18. Yuan, Y.; Lu, W.; Feng, B.; Weng, J. Steganalysis with CNN using multi-channels filtered residuals. *LNCS* **2017**, *10602*, 110–120.

19. Li, B.; Wei, W.; Ferreira, A.; Tan, S. ReST-Net: Diverse activation modules and parallel subnets-based CNN for spatial image steganalysis. *IEEE Signal Process. Lett.* **2018**, *25*, 650–654. [CrossRef]

20. Yedroudj, M.; Comby, F.; Chaumont, M. Yedroudj-Net: An efficient CNN for spatial steganalysis. In Proceedings of the IEEE International Conference on Acoustics, Speech and Signal Processing, Calgary, AB, Canada, 15–20 April 2018; pp. 15–20.

21. Wu, S.; Zhong, S.; Liu, Y. A novel convolutional neural network for image steganalysis with shared normalization. *IEEE Trans. Multimed.* **2019**. [CrossRef]

22. Boroumand, M.; Chen, M.; Fridrich, J. Deep residual network for steganalysis of digital images. *IEEE Trans. Inf. Forensics Secur.* **2019**, *14*, 1181–1193. [CrossRef]

23. Wu, S.; Zhong, S.; Liu, Y. Deep residual learning for image steganalysis. *Multimed. Tools Appl.* **2018**, *77*, 10437–10453. [CrossRef]

24. Ke, Q.; Ming, L.D.; Daxing, Z. Image steganalysis via multi-column convolutional neural network. In Proceedings of the 14th IEEE International Conference on Signal Processing, Beijing, China, 12–16 August 2018; pp. 550–553.

25. Ni, D.; Feng, G.; Shen, L.; Zhang, X. Selective ensemble classification of image steganalysis via deep Q network. *IEEE Signal Process. Lett.* **2019**, *26*, 1065–1069. [CrossRef]

26. Yu, X.; Tan, H.; Liang, H.; Li, C.T.; Liao, G. A multi-task learning CNN for image steganalysis. In Proceedings of the 10th IEEE International Workshop on Information Forensics and Security, Hong Kong, China, 11–13 December 2018.

27. Zhang, T.; Zhang, H.; Wang, R.; Wu, Y. A new JPEG image steganalysis technique combining rich model features and convolutional neural networks. *Math. Biosci. Eng.* **2019**, *16*, 4069–4081. [CrossRef]

28. Kang, S.; Park, H.; Park, J.-I. Toward ternary classification in CNN-based image steganalysis. In Proceedings of the 15th International Conference on Multimedia Information Technology and Applications, Ho Chi Minh City, Vietnam, 27 June–1 July 2019.

29. Simonyan, K.; Zisserman, A. Very Deep Convolutional Networks for Large-Scale Image Recognition. Available online: https://arxiv.org/pdf/1409.1556.pdf (accessed on 4 September 2019).

30. Bas, P.; Filler, T.; Pevny, T. Break our steganographic system—The ins and outs of organizing BOSS. In Proceedings of the International Workshop on Information Hiding, Prague, Czech Republic, 18–20 May 2011; pp. 59–70.

31. Kim, J.; Park, H.; Park, J.-I. CNN-based image steganalysis using additional data embedding. *Multimed. Tools Appl.* **2019**, in press.

32. Qian, N. On the momentum term in gradient descent learning algorithms. *Neural Netw.* **1999**, *12*, 145–151. [CrossRef]

33. TensorFlow. Available online: https://www.tensorflow.org/ (accessed on 4 September 2019).

 *electronics*

Article

False Positive Decremented Research for Fire and Smoke Detection in Surveillance Camera using Spatial and Temporal Features Based on Deep Learning

Yeunghak Lee [1] and Jaechang Shim [2],*

[1]   Department of Multimedia Engineering, Andong National University, Andong 36729, Korea; yhyi@anu.ac.kr
[2]   Department of Computer Engineering, Andong National University, Andong 36729, Korea
*   Correspondence: jcshim@anu.ac.kr; Tel.: +82-10-9770-5645

Received: 17 September 2019; Accepted: 7 October 2019; Published: 15 October 2019

Abstract: Fire must be extinguished early, as it leads to economic losses and losses of precious lives. Vision-based methods have many difficulties in algorithm research due to the atypical nature fire flame and smoke. In this study, we introduce a novel smoke detection algorithm that reduces false positive detection using spatial and temporal features based on deep learning from factory installed surveillance cameras. First, we calculated the global frame similarity and mean square error (MSE) to detect the moving of fire flame and smoke from input surveillance cameras. Second, we extracted the fire flame and smoke candidate area using the deep learning algorithm (Faster Region-based Convolutional Network (R-CNN)). Third, the final fire flame and smoke area was decided by local spatial and temporal information: frame difference, color, similarity, wavelet transform, coefficient of variation, and MSE. This research proposed a new algorithm using global and local frame features, which is well presented object information to reduce false positive based on the deep learning method. Experimental results show that the false positive detection of the proposed algorithm was reduced to about 99.9% in maintaining the smoke and fire detection performance. It was confirmed that the proposed method has excellent false detection performance.

Keywords: deep learning; fire and smoke detection; spatial and temporal; wavelet transform; coefficient of variation

1. Introduction

Many civilian fire injuries and civilian fire deaths occur each year due to intentionally-set fires and naturally occurring fires, which causes much property damage. Fires are classified into structure fires (home structures which include one-and two-family, manufactured homes, and apartments), non-residential structure fires (public assembly, school and college, store and office, industrial facilities, and other structures), and outdoor fires (bush, grass, forest, rubbish, and vehicle fires) [1]. Research on automatic fire detection or monitoring has long been the focus of the interior structure fires and non-residential structure fires to protect casualties and property damage from fires.

Smoke is very important because it indicates the start of a fire. However, sometimes the flames start first; thus, both smoke and flames require early detection to extinguish the fire early. Many methods of detecting smoke and flames to extinguish a fire early have been studied. In order to reduce the damage caused by fire, many early fire detection systems using heat sensors, smoke sensors, and flame detection sensors that detect flames by infrared rays (spectrum) and ultraviolet rays (spectrum) are frequently used [2,3]. Sensors used in buildings, factories, and interior spaces detect the presence of particles produced by fire flames and smoke in close proximity using a chemical reaction by ionization that requires proximity. Traditional fire alarm systems using sensors show good detection results in close proximity for activation or very narrow spaces [4,5]. However, sensor-based sensing systems are expensive because many devices need to be installed for fast detection. The disadvantage of the thermal sensor is that the detection is slow because it uses the temperature difference from the surroundings. The smoke sensor may be delayed depending on the speed of the smoke or may not be detected depending on the air flow. In addition, sensor-based detection systems cannot provide users with information about the location or size of a fire. The main disadvantage of the sensor based system is that it is difficult to install outdoors. As mentioned earlier, fires can occur anywhere and anytime, and must be detected at various locations.

In order to overcome the shortcomings of the sensor-based detection systems, many methods of detecting smoke and fire using camera sensors (image-based) have been studied [6,7]. Compared to sensor based fire detectors, video fire detectors have many advantages, such as fast response, long range detection and large protected areas. However, most of the recent video fire detections have a high rate of false alarms [8].

Vision based fire detection includes short range fire detection and long range fire detection. Long-distance forest or wildfire smoke and fire detection system using fixed CCD (Charge-Coupled Device) cameras is the monitoring of smoke and fire from distant mountains or fields [9–11]. In addition, Zhao et al. [12] described wildfire identification based on deep learning using unmanned aerial equipment. To extract local extremal regions of smoke, they used the rapidly growing Maximally Stable Extremal Region detection method in the field of initial smoke region detection.

More research has been conducted on short distance fire and smoke detection than on long distance forest of wildfire. Early fire detection using cameras detected fires in tunnels and mountains using black and white images [13,14]. Early feature extraction detected flames by measuring histogram changes using the temporal change characteristics of flames from black and white images. Recently, image-based flame detection methods using motion, color, shape, texture, and frequency analysis have been studied for the last 20 years [15–21].

Conventional flame detection methods include a method using RGB (Red, Green, Blue) HSV (Hue, Saturation, Value), YCbCr color models, etc., wavelet transform after detecting moving areas and flame color pixels, flame intensity changes over time, the shape of contour of fire flame in HSV color models and time-space domain, and a method using infrared image.

The color image fire detection algorithms determine cases where the flame's color level exceeded a certain threshold in the brightness information of the color space such as RGB, YCbCr, HIS (Hue, Saturation, Intensity), and CIEL * a * b * (CIELAB, Commission Interationale De L'éclairage) [22–26]. Algorithms using spatial domain analysis are algorithms for distinguishing between the flame color and the non-flame color. There are algorithms for determining the fire or analyzing the frequency components of the flame region by analyzing the texture of the flame candidate area [15,26,27]. The algorithm using the frequency analysis of the time domain determines the fire by analyzing the frequency of a specific level value of the flame candidate region that changes over time [22,23,25,26].

Chen et al. [27] studied the fire detection system using RGB and HSI color model and rule-based by using the characteristic that the flame movement is spread in irregular shape when fire occurs. Toreyin et al. [28] proposed a system that detects fire and non-fire using temporal and spatial wavelet analysis of input images as a feature of high frequency components, based on the fact that smoke appears translucent in the early stages of fire. Yuan [7] proposed an algorithm which is fast estimated the motion orientation of smoke and an accumulative motion model which is used the integral image. This is a method of generating a direction histogram for a motion vector by using a feature of upward moving of smoke, and determining that smoke is a case when there are a lot of motion vectors in a relatively upward direction. Yuan [29] proposed a smoke detection algorithm based neural network classification to train using feature vectors, which are generated by LPB (Local Binary pattern) and LBPV (Local Binary pattern Variance) histograms for rotation and lighting in multi-scale pyramid images. Celik and Demirel [30] presented the experimental results using YCbCr color space and proposed a pixel classification algorithm for flames. To this end, they suggested a very innovative algorithm that separates the chrominance from luminance components. However, this method used heuristic membership and did not produce good results for the new data. Fujiwara [31] proposed a smoke detection algorithm for smoke shapes using a fractal encoding method using the self-organism of smoke in grayscale images. Liu and Ahuja [16] detected the fire region based on the area expansion method using the fire initial region that has high brightness. They asserted that the fire zones and non-fire zones are classified by Fourier coefficients change over time. Philips [32] classified the fire region using the changes in status over time for candidate region, after the fire flame candidate region is dedicated by the color histogram adapted Gaussian filter. Tian et al. [33] detected smoke regions by image separation. After the background model was created, the smoke was detected by gray color and partial transparency. The limitation of the vision-based method is that it fails to detect transparent smokes. Moreover, it often mistakenly detects many natural objects, for example, the sun, various artificial lights or light reflects on various surfaces, dust particles, as well as flame and smoke. Additionally, scene complexity and low-video quality can affect the robustness of vision-based flame detection algorithms, thus increasing the false alarm rate. Barmpoutis et al. [34,35] also asserted that high false alarm rates are caused by natural objects, which have similar characteristics with flame, and by the variation of flame appearance. Other causes have claimed environmental changes that complicate fire detection including clouds, movement of rigid body objects in the scene, and sun and light reflections. Hence, the difficulty of fire flame detection from digital images is due to the chaotic and complex nature of fire phenomena. Lee et al. [36] proposed smoke detection algorithm based on the Histogram of Oriented Gradients and LBP. Adaboost, which is constructing a strong classifier as linear combination, was used to classify trained object.

In contrast, the deep learning based fire flame and smoke detection systems have automatic feature extraction; thus, making the process much more reliable and efficient than the conventional feature extraction methods. However, such a deep learning approach requires tremendous computational power, not only during training periods, but also when deploying trained models to hardware to perform specific tasks. As a fire detection method using a security surveillance camera, fire detection techniques using real-time image analysis and deep learning have been proposed.

Recently, several kinds of deep learning algorithms for fire flame and smoke detection have been proposed. Frizzi et al. [37] researched the Convolution Neural Network (CNN) based smoke and flame detection, Sang [38] studied the classification of smoke image and flame image feature using composite product neural network, Wu et al. [39] Studied the detection of fire and smoke regions by extracting dynamic and static features using ViBe algorithm, Shen et al. [40] detected the fire flame using the YOLO (You Look Only Once) model, and Khan et al. [41] also researched a disaster management system to respond to early fire detection and automatic reaction within the inside and outside environment using CNN. Zhang et al. [42] researched forest fire detection utilizing fire patches detection using two joined deep CNNs to detect fire in forest images. However, these models have many parameters to render, which require a large computing space. Thus, these models are unsuitable for onfield fire detection applications using low-cost low-performance hardware. Muhammad et al. [43] used Foggia's dataset [44]. They fine-tuned various variants of CNNs: AlexNet [41], SqueezeNet [43], GoogleNet [44], and MobileNetV2 [45]. They used Foggia's dataset [46] as the major portion of their train dataset. Although Foggia's dataset includes 14 fire and 17 non-fire videos with multiple frames, the dataset contains a lot of similar images, which restricts the performance of the model trained on this dataset to a very specific range of images. Recently, much research has been conducted on Faster R-CNN, which shows higher performance than other network models, like as R-CNN (Region-based Convolutional Network) and Fast R-CNN. Barmpoutis et al. [39] studied higher-order linear dynamical systems based multidimensional texture analysis as the deep learning networks. They classified the fire using the Faster R-CNN model based on the spatial analysis on Grassmann manifold. Wildland forest fire and smoke detection algorithm with Faster R-CNN was suggested by the Zhang et al. [47] to avoid the complex process.

As mentioned above, malfunctions of smoke and flame detection using image processing have been drastically reduced due to the development of deep learning, but the malfunction still exist due to problems of deep learning. The goal of most of the existing approaches is detecting either smoke or fire from images, but as explained, they suffer from a variety of limitations. To solve the problem of these limitations, in this paper, Faster R-CNN model is proposed with object attribution for increasing the smoke and fire flame detection and decreasing the false positive rate. This method is capable of detecting both smoke and fire flame images at the same time, and offers many advantages and exhibits better performance than other existing visual recognition CNN models for the recognition of fire flame and smoke in images. Additionally, we researched a novel algorithm on rigid change of natural environment to reduce the false positive smoke detection based on advanced deep learning, as shown Figure 1.

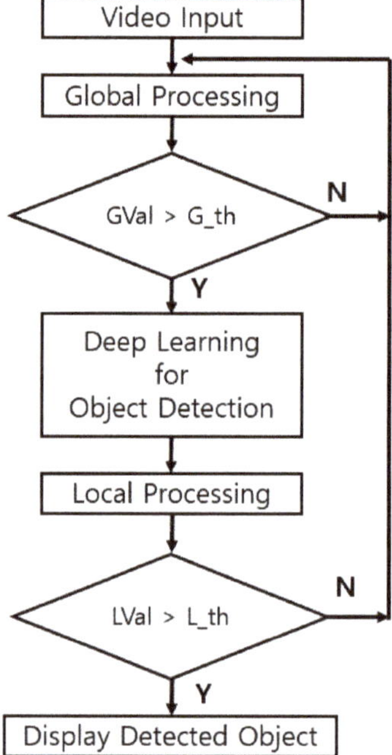

Figure 1. Flowchart of the proposed algorithm.

This paper is organized as follows. We propose deep learning model architecture for flame and smoke detection in surveillance camera in Section 2. This paper explains several theories to reduce the rate of false alarms and improve the detection rate in Section 3. Our experimental results and discussion are implemented in Section 4. Finally, the manuscript presents a brief conclusion and future research directions in Section 5.

2. Deep Learning (Faster R-CNN)

It is often more difficult to distinguish objects within an image than to classify images. Deep learning using the R-CNN method takes several steps. Once the R-CNN creates a region proposal or a bounding box for an area where an object exists, it unifies the size of the extracted bounding box to use as input to CNN. Next, the model uses SVM (Support Vector Machine) to classify the selected region. Finally, it uses a linear regression model so that the bounding box of the categorized object sets the exact coordinates. CNN for training data is divided into three parts. Figure 2 depicts the full flow of the proposed system. In Figure 2, RPN (Region Proposal network) was used to find a predefined number of regions (bounding boxes) that can contain objects using features computed by CNN. The next step is to get a list of possible related objects and their locations in the original image. We apply region of interest pooling (ROIP), using boundary boxes for features and related objects extracted from CNN, and extract the features corresponding to related objects as new tensors. Finally, this information is used to classify the contents of the bounding box and the bounding box coordinates are adjusted in the R-CNN module. As a result of the Faster R-CNN, a bounding box of related objects is displayed on the screen. The proposed algorithm part is added at end of Faster R-CNN. We finally select the case where FD (Final Decision) is greater than threshold (TH) using several features in the bounding box.

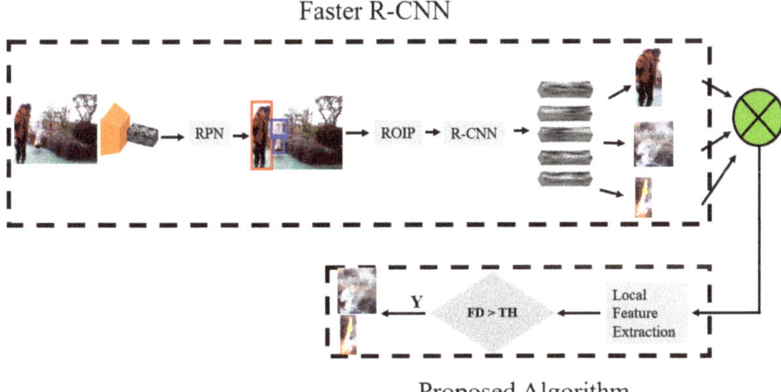

Figure 2. Faster R-CNN system flow.

2.1. Labeling Dataset

Labeling of the fire flame and smoke in the images was done using the LabelImg program. This paper used a variety size of labeling including fire flame and smoke to train the images, as shown in Figure 3. The labeling results are stored in the .xml file with the image file name along with the four-point coordinates of each rectangle. For labeling dataset, there are two things to be considered. First, a list of class is necessary for the dataset. Second, bounding boxes (Xmin, Ymin, Xmax, Ymax) will be generated by the labeling program according to the classes for images.

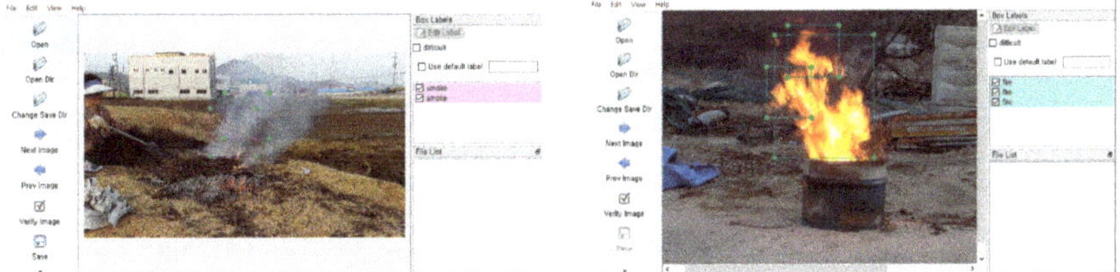

Figure 3. Example of labeling area for fire and smoke dataset image.

2.2. Training Data with Faster R-CNN

Faster R-CNN [48] is a method of applying a new method called Region Proposal Network (RPN) that merely integrates the part that generates the region proposal within the model. This is a new application of the RPN network for object detection. The function of RPN is to output the rectangle and object score of the part that proposes the object in the input image. It is a fully connected network and is designed to share a convolutional layer with Faster R-CNN. Trained RPN improves the quality of the proposed area and improves the accuracy of object detection. In general, Faster R-CNN searches external slow selections by CPU calculations but speeds them up by using internal fast RPNs by GPU calculations. The RPN comes after the last convolutional layer, followed by ROIP, classification, and bounding boxes are located, as shown in Figure 4. RPN extracts 256 or 512 features from the input image by convolution calculation using 3 × 3 window. This is then used as a box classifier layer and a box regress layer. The predefined reference box name used as the bounding box candidate at each position of the sliding window is used as the box regression. It extracts features by applying predefined anchor boxes of various ratios/sizes using the center position, moving the sliding window of the same size. In our model, we used nine anchor boxes (three sizes and three proportions), and each box is considered as a candidate for the bounding box at each position of the sliding window in the image.

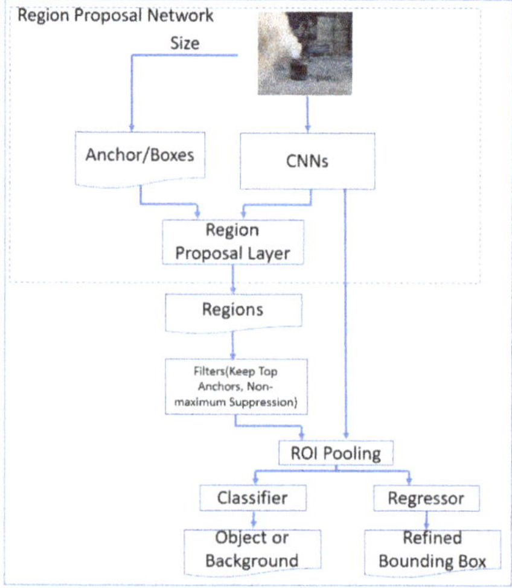

Figure 4. The architecture of faster R-CNN.

2.3. Creating Inference Graph

An inference graph is also known as a freezing model that is saved for further process. While training the dataset with the model, each pair at different time steps, one is holding the weights ".data", and another is holding the graph ".meta". The labeled image information is progressed using the Faster R-CNN model described above, and the ".meta" file is generated as a training result. The next step is making the graph file (".pb file") which is using the ".meta" file generated in the previous step. Finally, when we use the ".pb" file to detect the objects in the images, the result image including the bounding box and object score will be displayed on the monitor.

3. Feature Extraction Methods

3.1. Structural Similarity

SSIM (Structural Similarity) [49] is a measure of the similarity of the original image and distortion due to compression and transformation. This is more widely used in signal processing because it has higher accuracy than the Mean Square Error (MSE) method, which uses a measure of the difference between pixel values of two images. We used the evaluation of the test image (X) against the original image (Y) to measure the quantification of visual similarity. The more similar the test image to the original image, the closer the value is to 1.0, and the more different the test image is to the original image, the closer the value is to 0.0. The SSIM formulas are defined as follows:

$$L(x,y) = \frac{2\mu_x\mu_y + K1}{\mu_x^2 + \mu_y^2 + K1} \tag{1}$$

$$M(x,y) = \frac{2\sigma_x\sigma_y + K2}{\sigma_x^2\sigma_y^2 + K2} \tag{2}$$

$$N(x,y) = \frac{\sigma_{xy} + K3}{\sigma_x\sigma_y + K3} \tag{3}$$

where μ_x and μ_y are the mean of the pixels, σ_x and σ_y are the standard deviations, and σ_{xy} is covariance. *K1*, *K2*, and *K3* are constants for preventing the denominator and numerator from becoming zero. $L(x, y)$ is the relationship of the brightness difference, $M(x, y)$ is the contrast difference, and $N(x, y)$ is the similarity of the structural change between x and y. The structural similarity is shown in Equation (4):

$$\text{SSIM} = [L(x,y)]^\alpha [M(x,y)]^\beta [N(x,y)]^\gamma \tag{4}$$

where α, β, and γ represent the importance of each term; 1.0 was used in this paper.

3.2. RGB Color Histogram

Generally, smoke is grayish (dark gray, gray, light gray, and white). Black smoke occurs by unburned materials or a combustion at high temperatures; this means that a certain time has passed since the fire occurred. This paper focuses on the smoke of the initial generation, and sets the conditions as shown in Equation (5) to use smoke colors ranging from gray to white:

$$C = (R + G + B)/3, \quad \tau1 < C_L < \tau2, \quad \tau3 < C_H < \tau4 \tag{5}$$

where C is the output image, R is the red image, G is the green image, and B is the blue image. This research set the C_L to a minimum value between 80 ($\tau1$) and 150 and the C_H ($\tau2$) an upper range value between 180 ($\tau3$) and 250 ($\tau4$). The average image C is histogrammed into 256 bins (0 to 255) for each pixel. The values stored in each bin of the histogram are normalized using the input image size, and the sum is obtained, as in Equation (6):

$$H_S = \sum_{i=0}^{255} \frac{b_i}{(h \times w)} \tag{6}$$

where H_S is the RGB color histogram result value, b_i means the histogram bins from 0 to 255, which is only included Equation (5) range, and h and w is height and width for an input image. The grayish color is distributed intensively between 80 and 250.

Fire flames are usually bright orange or red (red -> orange -> yellow -> white -> mellow). This paper used HSV color instead of RGB color. The range of HSV color used in the paper is as follows:

- H: 0 to 40
- S: 100 to 255
- V: 80 to 255

As shown in smoke color extraction, HSV color image is also calculated for the average value for the filtered range image. The HSV histogram is obtained by Equation (6).

3.3. Coefficient of Variation (CV)

The coefficient of variation is a type of statistic that represents the precision or scatter of a sample, such as variance and standard deviation, in that it shows how scattered the distribution is relative to the mean. CV is a measure of how large the standard deviation is relative to the mean. These coefficients of variation are useful for comparing the spread in two types of data and for comparing variability when the differences between data are large. It is also used to determine the volatility of economic models and securities of economists and investors, as well as areas such as engineering or physics, when conducting quality assurance research and ANOVA gauge R & R [50].

The coefficient of variation is the standard value divided by the mean, as shown Equation (7):

$$CV = \sigma/m \tag{7}$$

where σ is standard deviation and m is mean. It showed that the image with smoke and fire flame region has lower *CV* value. In the contrast, the region with false alarm showed higher *CV* value, as shown Figure 5. This paper adapted as the coefficient value (weighting value) of wavelet transform to remove the false alarm cases. In case of Fire flame, we used the R color in RGB color space, and adapted Y color in YCbCr color space for the smoke region.

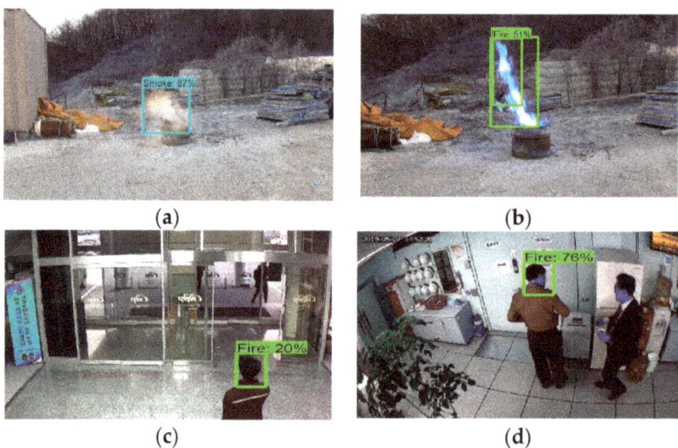

Figure 5. The result of coefficient variation values for detected area, (**a**) smoke area Coefficient of Variation (CV) value: 1.5 (87%), (**b**) fire area CV value: 1.9 (51%), (**c**) false alarm area CV value: 6.2 (20%), and (**d**) false alarm area CV value: 13.6 (76%).

3.4. Wavelet Transform

In general, smoke is blurry and uneven, thus, it is difficult to detect the contour using the contour detection method. DWT (Discrete Wavelet Transform) [51,52] supports multiple resolutions, and can express contour information of vertical, horizontal, and diagonal components, respectively. Using this feature to represent smoke in DWT energy, it is more apparent than in conventional edge detection methods.

When smoke with translucent characteristics occurs, the smoke part of the image frame is less sharp and the high frequency component is reduced in the area. Wavelet algorithms are generally suitable for expressing image textures and edge characteristics of smoke and fire flames. Background images generally have lower wavelet energy and few moving objects. In contrast, the edge of smoke images becomes less visible, and may disappear from the scene after a certain time. It means that the high frequency energy of the background scene is decreasing. In order to identify smoke in a scene, any decrease in high frequency from the detected blob images in the frame was monitored by a spatial wavelet transform algorithm.

As shown in Figure 6, if the smoke spreads to the edges of the image, it may be difficult to see initially and the smoke may darken over time, causing part of the background to disappear. [53,54]. This means that there is a high probability that smoke will be present and smoke detection will be easier, as shown in Figure 6. Therefore, this paper used the spatial energy to evaluate the sub-image energy by dividing the image into first stage wavelet transform and summing the squared from each coefficient images in Equation (8):

$$E(x,y) = \sqrt{\left[LH(x,y)^2 + HL(x,y)^2 + HH(x,y)^2\right]} \tag{8}$$

where x and y represent positions within the image, and LH, HL, and HH each contain contour information of the high frequency component of the DWT (Discrete Wavelet Transform). LH is horizontal low-band vertical high-band, HL is horizontal high-band vertical low-band, and HH is horizontal high-band vertical high-band. $E(x, y)$ is wavelet energy at each pixel in the candidate region which is detected by deep learning algorithm within each frame.

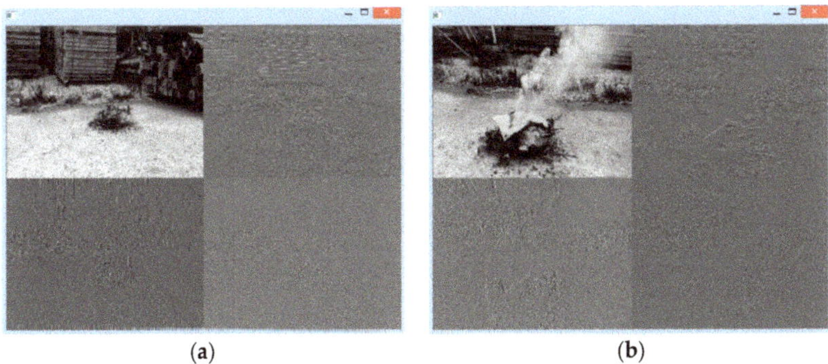

(a)   (b)

Figure 6. Single level of wavelet transform results, (**a**) non-smoke sub-images and (**b**) smoke sub-images.

4. Experimental Results

We proposed a new algorithm using similarity and color histogram of global and local area in the frame to reduce smoke false positive rate generated by fire detection systems using Onvif camera based on deep learning. In this paper, we used a computer with an Intel Core i7-7700 (3.5 GHz) CPU, 16 GB of memory, and Geforce TITAN-X to perform the experiment. The flame and smoke databases used in this study was obtained from the internet, and general direct ground and factory recorded video. The video recording device was a mobile phone camera, a Canon G5 camera, and a Raspberry pi camera. Python 3.5, Tensorflow, and Opencv were used in this paper.

In order to implement the proposed algorithm, the following process was carried out. The first step is labeling dataset from training database. The first task is labeling data using the LabelImg program, as shown Figure 4. The labeling categories used in this paper are flame, smoke, Grinder, Welding, and human. The result of labeling data is stored in an .xml file that contains the object type name and the four-point coordinates of the object area.

The second step is training process with labeled images. In the training process, the input image is a JPEG or PNG file. The .xml file should be converted to the learning data format of the Tensorflow. Since the meta data and labels of these images are stored in a separate file and must be read separately from the meta data and label file, the code becomes complicated when reading the training data. Additionally, performance degradation can occur if the image is read in JPEG or PNG format and decoded each time. However, the TFRecord file format avoids the above performance degradation and makes it easier to develop. The TFRecord file format stores the height and width of the image, the file name, the encoding format, the image binary, and the rectangle coordinate of the object in the image. Through this process, the entire training data is classified and stored as 70% training data and 30% validation data. We used the FASTER-CNN ResNet (Deep Residual Network) as the primary model for training, and it is characterized by the smallest number of objects and the highest detection rate. The fire images used in the training consisted of 21,230 pieces.

Finally, we extracted the training model. The learning process stores a check pointer that represents the learning result for the predetermined pointer. Each check pointer has meta information about the Tensorflow model file format and can be learned again. However, because there is a lot of unnecessary information in the ".meta" file, the .meta file needs to be improved to use the actual model. Finally, a ".pb" file is generated that combines the weights except for the unnecessary data in the ".meta" file.

In this paper, we used factory recorded video images, mobile camera, Raspberry pi camera, and general camera as the experimental data. Figure 7 shows an example of continuous frames of video used in the experiment. Fire detection experiment was performed using ".pb" file based on Fater R-CNN model. Figure 8 shows fire and smoke detection results included true positive and false positive using general deep learning.

Figure 7. Example of the frame sequence of test video.

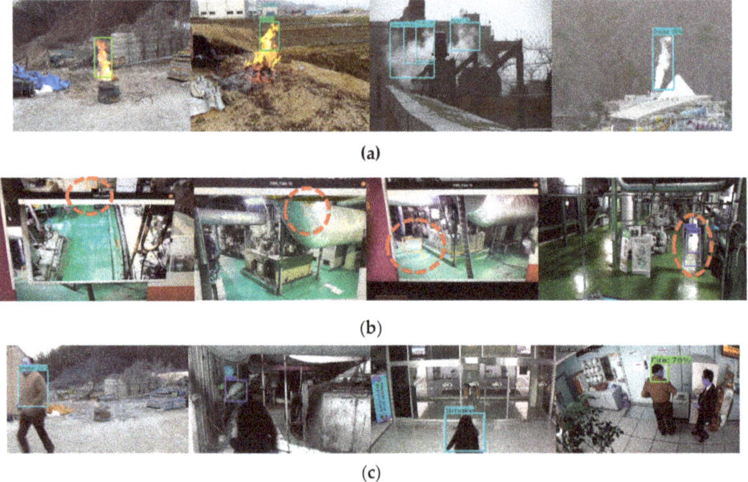

(a)

(b)

(c)

Figure 8. The experimental results using the Faster R-CNN: (**a**) the results of true positive, (**b**) the results of false positive (similar shape and color and reflection of sun and light), (**c**) the results of false positive (moving objects and similar color).

Figure 8a shows the result of the experiment to detect fire and smoke using various videos. The detection threshold of Faster R-CNN was 30% or higher. Figure 8b,c shows the result of false positive detection by applying deep learning training results. Although false positives have appeared in many places, there are two types of false positives. First, smoke or flame is detected by reflection of sunlight. Second, facilities inside and outside the factory show similar shapes and colors like smoke and fire. Third, when objects are moving around, deep learning system recognize them as fire flame or smoke for the similar shape of trained fire flame and smoke, as shown in Figure 8c. Table 1 shows the fire and smoke detection results for several videos.

Table 1. The results of video test using general Faster R-CNN (frame).

Videos	Ground Truth	True Positive	True Negative	False Positive
Video 1 (F/S)	85	85	0	0
Video 2 (F/S)	102	102	0	0
Video 3 (F/S)	890	890	0	890
Video 4 (F/S)	1159	1159	0	0
Video 5 (F/S)	1477	1477	0	0
Video 6 (F/S)	1112	1112	0	0
Video 7 (F/S)	544	544	0	7
Video 8 (F/S)	1940	1940	0	0
Video 9 (NON)	12112	0	11984	128
Video 10 (NON)	15015	0	15009	6
Video 11 (NON)	6745	0	6639	106
Video 12 (NON)	14949	0	14943	6
Video 13 (NON)	14891	0	14875	16
Video 14 (NON)	14975	0	14965	10
Video 15 (NON)	4402	0	4387	15
Video 16 (NON)	13454	0	13448	6

Videos 1 to 8 contain smoke and fire flame and Videos 9 to 16 contain non-fire (factory and office) scenes. Video 3, Video 7, and non-fire Video included a number of false positive frames. Especially, Video 3 showed the same number of true positive frames and false positives. It means that each frame has False Positive object in the images. In Table 1, Ground Truth represents the total number of frames in the video, True Positive (TP) indicates when a fire flame and smoke is detected as fire flame and smoke. True Negative (TN) indicates that non-fire objects are not detected as fire flame and smoke. False Positive (FP) is a case where non-fire objects are detected as a fire. NON signifies a non-fire video and F/S signifies a fire flame and smoke video. In Table 1, F/S means including fire and smoke frames and NON means without fire and smoke frames.

In the case of Videos, they is not generated in a continuous frame. Since the video is 30 fps, it can be sufficiently compensated. However, in the case of Video 3, Video 7, and non-fire video, the alarm continues to ring and the stress of the worker becomes higher. In order to reduce false positives generated in False Positive Videos, we use the following characteristics. The first is a global check. We checked the motion characteristics before performing deep learning using mean square error (8) and three frame differences (9) [55]. Since there is motion when a fire occurs, if a block of moving pixels is generated, it is registered as a fire candidate state. If the fire candidate frame status is True, a deep learning process is performed, as shown Figure 1.

$$S_k = SSIM\left(f_i,\ f_j\right),\ M_k = MSE\left(f_i,\ f_j\right),\ A_k = diff\left(f_i,\ f_j\right) \tag{9}$$

$$FS_G = \begin{cases} 1 & if\ S_k < th1,\ M_k < th2,\ A_k < th3 \\ 0 & else \end{cases} \tag{10}$$

where FSG is global decision parameter.

The second is a local check for the detected area (bounding box) by deep learning. If there is a trained class in the input frame image, a bounding box is created and stored as a local area of interest. The next step is to verify the local area of interest again. In this paper, we determine the final fire region using the color histogram H, $SSIM$ index, and mean square error (MSE), coefficient variant, and wavelet transform with other frames as the following equation:

$$F_L = \begin{cases} 1 & if\ M_k < fth1,\ A_k < fth2,\ H_{sum\_F} < fth3,\ WE_k < fth4 \\ 0 & else \end{cases}$$
$$WE_k = \sqrt{FWV^2 + C\_R\_HH^2 \times (R\_H_{sum} + Y\_H_{sum})}$$
$$FWV = \sqrt{C\_R\_HH^2 \times CV + C\_Y\_HH^2 \times CV} \tag{11}$$

where k means frames, from *fth1* to *fth4* are threshold value by experiment. C_R_HH and C_Y_HH is the wavelet transform coefficient HH for RGB and $YCbCr$ color. Moreover, R_H_{sum} and C_H_{sum} are the result of R color and Y color histogram for the local region. We compared the local region (bounding box area) of interest using the three frame difference algorithm (first, middle, and last frames) from the stored 10 frame images.

The final smoke region, in common with fire detection, we also adapted same sequence as the following equation:

$$S_L = \begin{cases} 1 & if\ M_k > sth1,\ A_k > sth2,\ H_{sum\_F} > sth3,\ WE_k < sth4 \\ 0 & else \end{cases} \tag{12}$$

This paper added the following conditions to remove false positives:

$$SD1 = \sqrt{C\_Y\_HH^2 \times FWV}$$
$$SD2 = SD1 \times FWV$$
$$SD3 = \{(CV_S + CV_F)/2\} \times SD2$$
$$SD4 = CV_S \times C\_Y\_HL\_LH \tag{13}$$
$$C\_Y\_HL\_LH = \sqrt{C\_Y\_HL^2 \times FWV + C\_Y\_LH^2 \times FWV}$$
$$S_{SD} = \begin{cases} 1 & if\ SD1 > sth5,\ SD2 > sth6,\ SD3 > sth7,\ SD4 < sth8 \\ 0 & else \end{cases}$$

where CVS and CVF are the coefficient variance of local smoke and fire region, respectively. In this paper, it is regarded as a fire if FD is satisfied as shown in the following equation:

$$FS_G = \begin{cases} 1 & if\ F_L > 0,\ S_L > 0,\ F_{SD} > 0 \\ 0 & else \end{cases} \tag{14}$$

We described the result of the experiment applying the proposed algorithms in Table 2.

Table 2 shows the experimental results using the proposed algorithm. In the Videos, the false positive rate dropped to 0% and the fire detection of Video 1 to Video 6 persisted. Even though the Video 3 and Video 4 missed a few fire images, it has no problem because it is not continuously generated and the alarm system has no problem sending a warning signal to operator if it misses one or two frames. As shown in Table 2, the proposed algorithm using color histogram, wavelet transform, and coefficient variant was able to eliminate false positives (similar shape and color objects, sun and light reflection, moving objects, etc.) shown in Figure 8b,c. The results of the proposed algorithm using color histogram performance, high frequency components of wavelet transform, which is background discrimination of smoke and fire flame, and coefficient variant coefficients showed higher ratio of false alarm removal than the traditional deep learning method. However, in the case of Video 7 and Video 8, we must seriously consider the case of the missing frames. Additionally, we tested other factory and office videos. It also marked zero false positive rate for the proposed method. The false positive rate for the additional 16 videos was 99%, and the image examples used in the video experiment are shown in Figure 9. Figure 9a is office and factory videos and Figure 9b is fire and smoke videos. Since this involves a lot of movement, it is likely that it has affected the frames missing in Video 7 and Video 8.

Table 2. The results of video test using proposed algorithm.

Videos	Ground Truth	True Positive	True Negative	False Positive
Video 1 (F/M)	85	85	0	0
Video 2 (F/M)	102	102	0	0
Video 3 (F/M)	890	888	0	0
Video 4 (F/M)	1159	1158	0	0
Video 5 (F/M)	1477	1477	0	0
Video 6 (F/M)	1112	1112	0	0
Video 7 (F/M)	544	502	0	0
Video 8 (F/M)	1040	949	0	0
Video 9 (NON)	12112	0	12112	0
Video 10 (NON)	15015	0	15015	0
Video 11 (NON)	6745	0	6745	0
Video 12 (NON)	14949	0	14949	0
Video 13 (NON)	14891	0	14891	0
Video 14 (NON)	14975	0	14975	0
Video 15 (NON)	4402	0	4402	0
Video 16 (NON)	13454	0	13454	0

Figure 9. Experimental videos for proposed algorithm test: (**a**) factory and office videos and (**b**) fire and smoke videos.

5. Conclusions

Fires resulting from small sparks can cause terrible natural disasters that can lead to both economic losses and the loss of human lives. In this paper, we describe a new fire flame and smoke detection method to remove false positive detection using spatial and temporal features based on deep learning from surveillance cameras. In general, a deep learning method using the shape of an object frequently generate false positives, where general object is detected as the fire or smoke. To solve this problem, first, we used motion detection using the three frame difference algorithm as the global information.

We then applied the frame similarity using SSIM and MSE. Second, we adapted the Faster R-CNN algorithm to find smoke and fire candidate region for the detected frame. Third, we determined the final fire flame and smoke area using the spatial and temporal features; wavelet transform, coefficient of variation, color histogram, frame similarity, and MSE for the candidate region. Experiments have shown that the probability of false positives in the proposed algorithm is significantly lower than that of conventional deep learning method.

For future work, it is necessary to study the analysis for the moving videos and the experiment using the correlation of the frame and the deep learning model to further reduce false positives and missing fire and smoke frames.

Author Contributions: We provide our contribution as follow, conceptualization, Y.L. and J.S.; methodology, Y.L. and J.S.; software, Y.L.; validation, Y.L.; formal analysis, Y.L.; investigation, Y.L.; resources, Y.L.; data curation, Y.L.; writing—original draft preparation, Y.L.; writing—review and editing, Y.L. and J.S.; visualization, Y.L.; supervision, J.S.; project administration, J.S.; funding acquisition, J.S.

Funding: This research was supported by the MSIP (Ministry of Science, ICT & Future Planning), Korea, under the National Program for Excellence in SW) (IITP-2019-0-01113) supervised by the IITP (Institute for Information & communications Technology Planning & Evaluation).

Conflicts of Interest: The funders had no role in the design of the study; in the collection, analyses, or interpretation of data; in the writing of the manuscript, or in the decision to publish the results.

References

1. Evarts, B. Fire Loss in the United States during 2017. Available online: https://www.nfpa.org/~{}/media/FD0144A044C84FC5BAF90C05C04890B7.ashx (accessed on 14 October 2019).
2. Lee, B.; Kwon, O.; Jung, C.; Park, S. The development of UV-IR combination flame detector. *J. KIIS* **2001**, *16*, 1–8.
3. Kang, D.; Kim, E.; Moon, P.; Sin, W.; Kang, M. Design and analysis of flame signal detection with the combination of UV/IR sensors. *J. Korean Soc. Int. Inf.* **2013**, *14*, 45–51.
4. Lee, D.H.; Lee, S.W.; Byun, T.; Cho, N.I. Fire detection using color and motion models. *IEIE Trans. Smart Process. Comput.* **2017**, *6*, 237–245. [CrossRef]
5. Yuan, F.; Shi, J.; Xia, X.; Fang, Y.; Mei, Z.F.T. High-order local ternary patterns with locality preserving projection for smoke detection and image classification. *Inf. Sci.* **2016**, *372*, 225–239. [CrossRef]
6. Chen, J.; Wang, Y.; Tian, Y.; Huang, T. Wavelet based smoke detection method with RGB contrast-image and shape constrain. In Proceedings of the 2013 Visual Communications and Image Processing (VCIP), Kuching, Malaysia, 23–27 November 2013.
7. Yuan, F. A fast accumulative motion orientation model based on integral image for video smoke detection. *Pattern Recognit.* **2008**, *29*, 925–932. [CrossRef]
8. Yu, C.; Mei, Z.; Zhang, X. A real-time video fire flame and smoke detection algorithm. *Proc. Eng.* **2013**, *62*, 891–898. [CrossRef]
9. Zhou, Z.; Shi, Y.; Gao, Z.; Li, S. Wildfire smoke detection based on local extremal region segmentation and surveillance. *Fire Saf. J.* **2016**, *85*, 55–58. [CrossRef]
10. Zhao, Y.; Li, Q.; Gu, Z. Early smoke detection of forest fire video using CS Adaboost algorithm. *Optik* **2015**, *126*, 2121–2124. [CrossRef]
11. Nowzad, A.; Jok, A.; Reulke, R.; Jackel, K. False Alarm Reduction in Image-Based Smoke Detection Algorithm Using Color Information FWS 2015. Available online: https://www.researchgate.net/publication/296483761_False_Alarm_Reduction_in_Image-Based_Smoke_Detection_Algorithm_Using_Color_Information (accessed on 14 October 2019).
12. Zhao, Y.; Ma, J.; Li, X.; Zhang, J. Saliency detection and deep learning-based wildfire identification in UAV imagery. *Sensors* **2018**, *18*, 712. [CrossRef]
13. Noda, S.; Ueda, K. Fire detection in tunnels using an image processing method. In Proceedings of the Vehicle Navigation & Information Systems Conference, Yokohama, Japan, 31 August–2 September 1994; pp. 57–62.
14. Breejen, E.D.; Breuers, M.; Cremer, F.; Kemp, R.; Roos, M.; Schutte, K.; de Vries, J.S. Autonomous forest fire detections. In Proceedings of the International Conference on Forest Fire Research 14th Conference on Fire and Forest Meteorology, Coimbra, Portugal, 16–20 November 1998; Volume 2, pp. 2003–2012.

15. Yamagish, H.; Yamaguchi, J. Fire flame detection algorithm using a color camera. In Proceedings of the International Symposium on Micromechatronics and Human Science, Nagoya, Japan, 23–26 November 1999.
16. Beall, K.; Grosshadler, W.; Luck, H. Smoldering fire detection by image-processing. In Proceedings of the 12th International Conference on Automatic Fire Detection, Gaithersburg, MD, USA, 26–28 March 2001.
17. Qiu, T.; Yan, Y.; Lu, G. An autoadaptive edge-detection algorithm for flame and fire image processing. *IEEE Trans. Instrum. Meas.* **2012**, *61*, 1486–1493. [CrossRef]
18. Liu, C.B.; Ahuja, N. Vision based fire detection. In Proceedings of the 17th International Conference on Pattern Recognition (ICPR 2004), Cambridge, UK, 26 August 2004; pp. 134–137.
19. Celik, T.; Demirel, H.; Ozkaramanli, H.; Uyguroglu, M. Fire detection using statistical color model in video sequences. *J. Vis. Commun. Image Represent.* **2007**, *18*, 176–185. [CrossRef]
20. Ko, B.C.; Ham, S.J.; Nam, J.Y. Modeling and formalization of fuzzy finite automata for detection of irregular fire flames. *IEEE Trans. Circ. Syst. Video Technol.* **2011**, *21*, 1903–1912. [CrossRef]
21. Ye, W.; Zhao, J.; Wang, S.; Wang, Y.; Zhang, D.; Yuan, Z. Dynamic texture based smoke detection using Surfacelet transform and HMT model. *Fire Saf. J.* **2015**, *73*, 91–101. [CrossRef]
22. Li, J.; Qi, Q.; Zou, X.; Peng, H.; Jiang, L.; Liang, Y. Technique for automatic forest fire surveillance using visible light image. *IGSNNRR CAS* **2005**, 3135–3138.
23. Celik, T.; Ma, K.K. Computer vision based fire detection in color images. In Proceedings of the 2008 IEEE Conference on Soft Computing in Industrial Applications, Muroran, Japan, 25–27 June 2008; pp. 258–263.
24. Toreyin, B.U.; Dedeoglu, Y.; Cetin, A.E. Flame detection In Video Using Hidden Markov Models. In Proceedings of the IEEE International Conference on Image Processing, Piscataway, NJ, USA, 14 September 2005; pp. 1230–1233.
25. Zhang, Z.; Zhao, J.; Zhang, D.; Qu, C.; Ke, Y.; Cai, B. Contour based forest fire detection using FFT and wavelet. In Proceedings of the International Conference on Computer Science and Software Engineering, Wuhan, China, 12–14 December 2008; pp. 760–763.
26. Celik, T. Fast and efficient method for fire detection using image processing. *ETRI J.* **2010**, *32*, 881–890. [CrossRef]
27. Chen, T.H.; Wu, P.H.; Chiou, Y.C. An early fire-detection method based on image processing. In Proceedings of the 2004 International Conference on Image Processing, Singapore, 24–27 October 2004; pp. 1707–1710.
28. Töreyin, B.U.; Dedeoglu, Y.; Gudukbay, U.; Cetin, A.E. Computer vision based method for real-time fire and flame detection. *Pattern Recognit. Lett.* **2006**, *27*, 49–58. [CrossRef]
29. Yuan, F. Video-based smoke detection with histogram sequence of LBP and LBPV pyramids. *Fire Saf. J.* **2011**, *46*, 132–139. [CrossRef]
30. Celik, T.; Demirel, H. Fire detection in video sequences using a generic color model. *Fire Saf. J.* **2009**, *44*, 147–158. [CrossRef]
31. Fujiwara, N.; Kenji Terada, K. Extraction of a smoke region using fractal coding. In Proceedings of the IEEE International Symposium on Communications and Information Technology, Sapporo, Japan, 26–29 October 2001.
32. Phillips, W.; Shah, M.; Lobo, N.D.V. Frame recognition in video. *Pattern Recognit. Lett.* **2002**, *23*, 319–327. [CrossRef]
33. Tian, H.; Li, W.; Wang, L.; Ogunbona, P. A novel video-based smoke detection method using image separation. In Proceedings of the 2012 IEEE International Conference on Multimedia and Expo, Melbourne, Australia, 9–13 July 2012; pp. 532–537.
34. Barmpoutis, P.; Dimitropoulos, K.; Kaza, K.; Grammalidis, N. Fire detection from image using faster R-CNN and multidimensional texture analysis. In Proceedings of the 2019 IEEE International Conference on Acoustics, Speech and Signal Processing, Brighton, UK, 12–17 May 2019.
35. Jadon, A.; Omama, M.; Varshney, A.; Ansari, M.S.; Sharma, R. FireNet: A Specialized Lightweight Fire & Smoke Detection Model for Real-Time IoT Applications. In Proceedings of the IEEE Region 10 Conference (TENCON 2019), Kochi, Kerala, India, 17–20 October 2019.
36. Lee, Y.; Kim, T.; Shim, J. Smoke detection system research using fully connected method based on adaboost. *J. Multimed. Inf. Syst.* **2017**, *4*, 479–482.
37. Frizzi, S.; Kaabi, R.; Bouchouicha, M.; Ginoux, J.; Moreau, E.; Fnaiech, F. Convolutional neural network for video fire and smoke detection. In Proceedings of the Annual Conference of the IEEE Industrial Electronics Society, Florence, Italy, 23–26 October 2016; pp. 877–882.

38. Sang, W. Implementation of image based fire detection system using convolution neural Network. *J. Korea Inst. Electron. Commun. Sci.* **2017**, *12*, 331–336.

39. Wu, X.; Lu, X.; Leung, H. An adaptive threshold deep learning method for fire and smoke detection. In Proceedings of the International Conference on Systems, Man, and Cybernetics, Banff, AB, Canada, 5–8 October 2017; pp. 1954–1959.

40. Shen, D.; Chen, X.; Nguyen, M.; Yan, W.Q. Flame detection using deep learning. In Proceedings of the International Conference on Control, Automation and Robotics, Singapore, 18–21 November 2018.

41. Muhammad, K.; Ahmad, J.; Baik, S.W. Early fire detection using convolutional neural networks during surveillance for effective disaster management. *J. Neurocomput.* **2018**, *288*, 30–42. [CrossRef]

42. Zhang, Q.; Xu, J.; Xu, L.; Guo, H. Deep convolutional neural networks for forest fire detection. In Proceedings of the 2016 International Forum on Management, Education and Information Technology Application, Guangzhou, China, 30–31 January 2016.

43. Muhammad, K.; Ahmad, J.; Lv, Z.; Bellavista, P.; Yang, P.; Baik, S.W. Efficient deep CNN-based fire detection and localization in video surveillance applications. *IEEE Trans. Syst. Man Cybern. Syst.* **2018**, *49*, 1419–1434. [CrossRef]

44. Muhammad, K.; Ahmad, J.; Mehmood, I.; Rho, S.; Baik, S.W. Convolutional neural networks based fire detection in surveillance videos. *IEEE Access* **2018**, *6*, 18174–18183. [CrossRef]

45. Muhammad, K.; Khan, S.; Elhoseny, M.; Ahmed, S.H.; Baik, S.W. Efficient fire detection for uncertain surveillance environment. *IEEE Trans. Ind. Inf.* **2019**, *15*, 3113–3122. [CrossRef]

46. Foggia, P.; Saggese, A.; Vento, M. Real-time fire detection for video-surveillance applications using a combination of experts based on color, shape, and motion. *IEEE Trans. Circ. Syst. Video Technol.* **2015**, *25*, 1545–1556. [CrossRef]

47. Zhang, Q.X.; Lin, G.H.; Zhang, Y.M.; Xu, G.; Wang, J.J. Wildland forest fire smoke detection based on faster RCNN using synthetic smoke images. *Procedia Eng.* **2018**, 441–446. [CrossRef]

48. Ren, S.; He, K.; Girshick, R.; Sun, J. Faster R-CNN: Towards real-time object detection with region proposal networks. In Proceedings of the Advances in Neural Information Processing Systems 28 (NIPS 2015), Montreal, QC, Canada, 7–12 December 2015.

49. Lee, Y.; Ansari, I.; Shim, J. Rear-approaching vehicle detection using frame similarity base on faster R-CNN. *Int. J. Eng. Technol.* **2018**, *7*, 177–180. [CrossRef]

50. Coefficient Variance. Available online: https://en.wikipedia.org/wiki/Coefficient_of_variation (accessed on 12 August 2019).

51. Chui, C.K. *An Introduction to Wavelets*; Academic Press: Cambridge, MA, USA, 1992.

52. Wei, Y.; Chunyu, Y.; Yongming, Z. Based on wavelet transformation fire smoke detection method. In Proceedings of the 2009 9th International Conference on Electronic Measurement & Instruments, Beijing, China, 16–19 August 2009.

53. Toreyin, B.U.; Dedeoglu, Y.; Cetin, A.E. Wavelet based real-time smoke detection in video. In Proceedings of the 13th European Signal Processing Conference, Antalya, Turkey, 4–8 September 2005.

54. Toreyin, B.U.; Dedeoglu, Y.; Cetin, A.E. Contour based smoke detection in video using wavelets. In Proceedings of the 14th European Signal Processing Conference, Florence, Italy, 4–8 September 2006.

55. Zhang, Y.; Wang, X.; Qu, B. Three-Frame Difference Algorithm Research Based on Mathematical Morphology. *Procedia Eng.* **2012**, *29*, 2705–2709. [CrossRef]

 *electronics*

Brief Report

Image Classification with Convolutional Neural Networks Using Gulf of Maine Humpback Whale Catalog

Nuria Gómez Blas [1,†,‡], Luis Fernando de Mingo López [1,*,†,‡], Alberto Arteta Albert [2,†] and Javier Martínez Llamas [1,†,‡]

1   Escuela Técnica Superior de Ingeniería de Sistemas Informáticos, Universidad Politécnica de Madrid, 28031 Madrid, Spain nuria.gomez.blas@upm.es (N.G.B.); javier.martinez.llamas@alumnos.upm.es (J.M.L.)
2   Department of Computer Science, College of Arts and Sciences, Troy University, Troy, AL 36082, USA; aarteta@troy.edu
*   Correspondence: fernando.demingo@upm.es; Tel.: +34-91-067-3566
†   These authors contributed equally to this work.
‡   Current address: Escuela Técnica Superior de Ingeniería de Sistemas Informáticos, Calle Alan Turing s/n, 28031 Madrid, Spain.

Received: 19 March 2020; Accepted: 27 April 2020; Published: 29 April 2020

Abstract: While whale cataloging provides the opportunity to demonstrate the potential of bio preservation as sustainable development, it is essential to have automatic identification models. This paper presents a study and implementation of a convolutional neural network to identify and recognize humpback whale specimens by processing their tails patterns. This work collects datasets of composed images of whale tails, then trains a neural network by analyzing and pre-processing images with TensorFlow and Keras frameworks. This paper focuses on an identification problem, that is, since it is an identification challenge, each whale is a separate class and whales were photographed multiple times and one attempts to identify a whale class in the testing set. Other possible alternatives with lower cost are also introduced and are the subject of discussion in this paper. This paper reports about a network that is not necessarily the best one in terms of accuracy, but this work tries to minimize resources using an image downsampling and a small architecture, interesting for embedded system.

Keywords: convolutional neural networks; pattern recognition; machine learning

---

1. Introduction

Humpback whales have patterns of black and white pigmentation and scars on the underside of their tails that are unique to each whale, just as fingerprints are to humans. Researchers document the marks or flukes on the right and left lobes of the tail and rate the percentage of dark vs. light skin pigmentation in a range (0–100, 0–100) (100 percent white to 100 percent black).

While whale cataloging provides the opportunity to demonstrate the potential of bio preservation as sustainable development, and at the same time honoring the principles of conservation, it is essential to have automatic identification models.

For scientific purposes, each humpback whale sighted in the North Atlantic is assigned to a catalog number. The unique scarring and shading patterns also provide the inspiration for common names. For Gulf of Maine humpbacks, researchers and naturalists work together each year to name new adult whales and young animals sighted in a second year. New calves are not named because their coloring and scarring often dramatically change during that first year.

Information collected for humpbacks in the sanctuary constitutes the longest and most detailed data set for baleen whales in the world. Photographs in the Gulf of Maine Humpback Whale Catalog, maintained by the Provincetown Center for Coastal Studies, and the North Atlantic Humpback Whale Catalog, maintained by the College of the Atlantic in Maine, allow scientists and naturalists to identify and monitor individual animals and gather valuable information about population sizes, migration, health, sexual maturity and behavior patterns. Photographing individual whales and their calves each year helps to identify family relationships. Four generations of humpback whales have been documented in certain maternal lines or matrilines.

The computing performance of Artificial Intelligence has increased remarkably in recent years. While it has been available to more and more people at the same time, its technological and social impact will grow exponentially. This fact has included Artificial Intelligence in almost any field of Information Technology, with massive companies such as Google, Facebook, Amazon or Microsoft that offer services, solutions and tools based on Artificial Intelligence such as: Video Games, Virtual Assistants and Financial Services.

One of the main areas that has the most development is the field of image/pattern recognition. This is mainly due to the utility and social precision (compared to the human eye) offered by this field. This type of technology is used in a wide range of systems, from surveillance tools and facial recognition to medical applications such as the early identification of tumors.

Their potential and social involvement is so high that their development must be deeply linked to ethical responsibility. It is notable that Artificial Intelligence and field recognition are not subject to controversy, as long as they receive criticism due to abusive practices or lack of ethics/privacy.

From the first years of Information Technology, the possibility that the machines were able to think has been really attractive and has reached the minds of several writers and artists along with history. They imagined androids that were absolutely indistinguishable from humans and artificial intelligence with capabilities that the human mind is incapable of understanding, among many others. However, to understand the feasibility of these examples, it is necessary to understand what Artificial Intelligence is and how it works.

The term Artificial Intelligence has been raised historically from different points of view: the ability to think or the ability to intelligently act [1]. On the one hand, the first focuses on the approach of a human idea of intelligence, in which machines think and are rational. On the other hand, the second approach is based not so much on the process as on the result, considering the Artificial Intelligence to the ability to act and emulate what would be the result of a strictly rational action. A fundamental part of intelligence lies in learning, a process through which, through information, study and experience, a certain amount of training is achieved. That is why the need arises within the framework of Artificial Intelligence to adequately equip the knowledge systems.

With this objective, automatic learning or automatic learning was born, which, thanks to data processing, seeks to identify common patterns that allow the elaboration of increasingly precise and improved predictions. However, these algorithms have historically required complex statistical knowledge. Following the evolution of machine learning, recent years have seen the birth of a new concept, known as deep learning. Unlike machine learning, deep learning understands the world as a hierarchy of concepts [2], diluting the information in different layers through the use of modules, which transform their representation into a higher and more abstract level. This allows the amplification of the relevant information and eliminates the superfluous one [3].

2. Convolutional Neural Networks

Artificial neural networks or neural networks are mathematical models that try to emulate the natural behavior of biological neural networks. In these models, a network of logical units or neurons interconnected with each other is established. With this connection, they can process the received information and issue a result to the next layer determined by an activation function that takes into account the weight of each input, see Figure 1b. This behavior adds more importance to specific incoming connections.

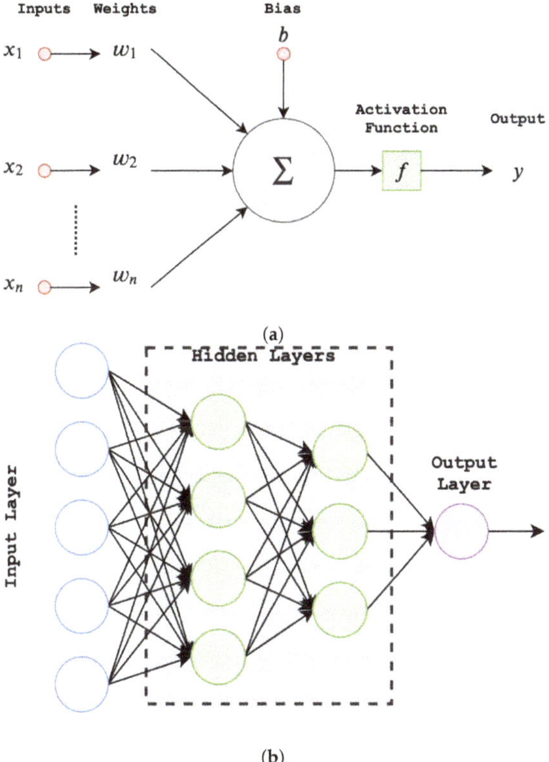

Figure 1. Illustration of an artificial neuron and a simple neural network with 2 hidden layers. In practice, there could be many layers and many neurons per layer. Note that the output of a neuron depends on a non-linear combination of the inputs, provided $f(x)$ is a non-linear function. (**a**) Artificial neuron. (**b**) Multilayer perceptron.

In this model, output obtained in neuron y (Figure 1a) is given by Equation (1), where $\bar{x} = \{x_1, \cdots, x_n\}$ represents the input data, $\bar{w} = \{w_1, \cdots, w_n\}$ is the weight matrix and b is the so-called the bias term.

$$y = f(\sum_{i=1}^{n} w_i x_i + b) \tag{1}$$

During the training phase of a neural network the weight and bias parameters are readjusted in order to adapt the model to a specific task and improve the predictions. The activation function f will be selected according to the problem to solve (a sigmoid or hyperbolic function).

Multilayer neural networks organize and group artificial neurons into levels or layers. These have an input layer and an output layer and might have a variable number of hidden layers between them.

The input layer is formed by neurons that introduce the information into the network, but they do not produce processing, so they only act as a receiver and propagator. The hidden layers are formed by those neurons where both the entrance and the exit connect with other layers of neurons. The output layer is the last level of the network and produces a set of results out of it. The connections of the multilayer neural networks usually move forward, connecting the neurons with their next layer. They are called feed forward networks.

Convolutional neural networks (ConvNet or CNNs) [4,5] are a class of multilayer feed forward neural networks specially designed for the recognition and classification of images. Classification is simply a more general term than pattern recognition. In both cases, you have a set of classes K and a collection of observations, where each observation is represented by a set of features. The problem is to find a mapping of features to a member of K such that you minimize some measure/estimate of out-of-sample classification error. In pattern recognition, you are simply using a very complex/large feature space. For example, facial recognition will have an input space equal to the number of pixels in the image (this is no different than classifying a loan application as high or low risk based on several measures of creditworthiness).

Computers perceive images in different ways to humans, as long as for these an image consists of a two-dimensional vector with the values relative to the pixels (Figure 2). They have got a channel for greyscale images or three of them for color (RGB).

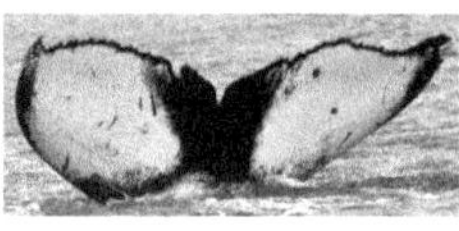

182	179	79	$\cdots$	102	101	104
202	202	195	$\cdots$	95	99	102
172	175	179	$\cdots$	114	118	122
			$\cdots$			
64	64	179	$\cdots$	95	100	126
64	64	64	$\cdots$	30	81	114
68	72	68	$\cdots$	135	106	151

Figure 2. Whale tail image vector. Left image is the picture taken and right one is the coded image using a matrix of real values $\in [0, 255]$, where 0 means a white pixel and 255 a black one.

Convolutional networks follow a certain structure, with three main types of layers: Convolutional layer, pooling layer and fully-connected layer, see Figure 3. A series of alternate conversions and subsamples or reductions are made, until finally through a series of completely connected layers (multilayer perception) the desired output is obtained, equivalent to the number of classes.

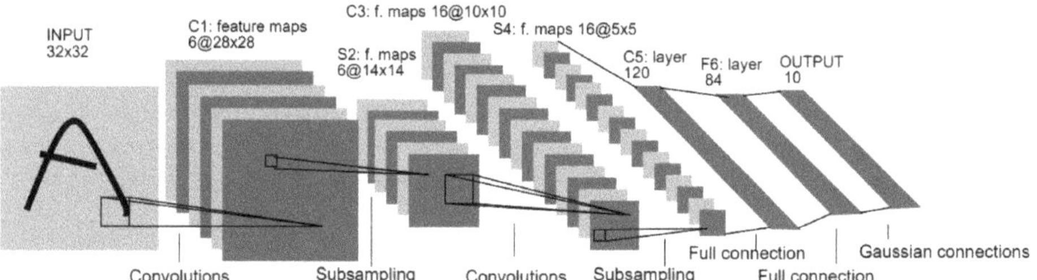

Figure 3. A classic convolutional neural network (CNN) model: LeNet-5 architecture (original image published by LeCun Y. et al. [6]). It consists of two sets of convolutional and average pooling layers, followed by a flattening convolutional layer, then two fully-connected layers and finally a softmax classifier.

The convolution makes a series of products and sums between the starting matrix and a kernel matrix or filter of size n. On the other hand, the sub-sampling reduces the dimension of the input matrix by dividing it into sub-regions and allowing the generalization of the characteristics (Figure 4).

There are different architectures that are currently considered as the state of the art, such as AlexNet, Inception or VGGNet, highlighting residual neural networks or ResNet [7] among them.

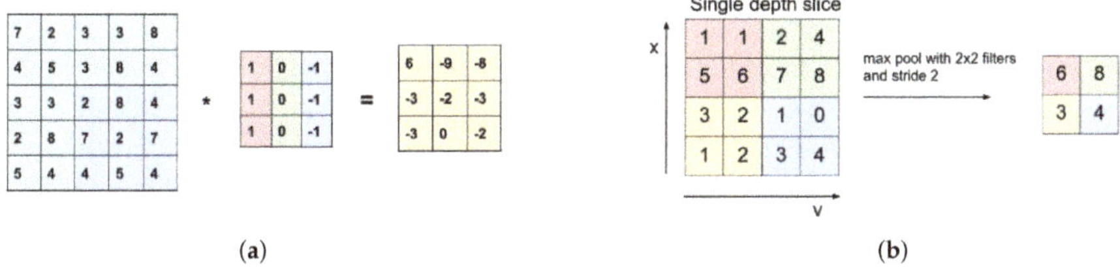

(a)   (b)

Figure 4. Matrix operators in convolutional neural networks. Convolution involves a sum of element by element multiplication, which in turn is the same as a dot product on multidimensional matrices which machine learning researchers call tensors. (**a**) Convolution. (**b**) Max Pooling.

2.1. Evolution of Convolutional Networks: ResNet, AlexNet, VGGNet, Inception

The task of training the whole network from the scratch can be carried out using a large dataset like ImageNet using convolutional neural networks (CNN). The reason behind this is, sharing of parameters between the neurons and sparse connections in convolutional layers. It can be seen in Figure 3. In the convolution operation, the neurons in one layer are only locally connected to the input neurons and the set of parameters are shared across the 2-D feature map.

Most CNNs have huge memory and computation requirements, especially while training. Hence, this becomes an important concern. Similarly, the size of the final trained model becomes important to consider if you are looking to deploy a model to run locally on mobile. As you can guess, it takes a more computationally intensive network to produce more accuracy. Therefore, there is always a trade-off between accuracy and computation.

Apart from these, there are many other factors like ease of training, the ability of a network to generalize well, etc. There are other architectures that are the most popular ones and are presented in the order that they were published and they also had increasingly better accuracy from the earlier ones, see Table 1.

- AlexNet [8]: This architecture was one of the first deep networks to push ImageNet Classification accuracy by a significant stride in comparison to traditional methodologies. It is composed of 5 convolutional layers followed by 3 fully connected layers, as depicted in Figure 1. AlexNet, proposed by Alex Krizhevsky, uses Rectified Linear Unit (ReLu) for the non-linear part, instead of a Tanh or Sigmoid function which was the earlier standard for traditional neural networks. The advantage of the ReLu over sigmoid is that it trains much faster than the latter because the derivative of sigmoid becomes very small in the saturating region and therefore the updates to the weights almost vanish. This is called vanishing gradient problem. In the network, ReLu layer is put after each and every convolutional and fully-connected layer (FC). Another problem that this architecture solved was reducing the over-fitting by using a Dropout layer after every FC layer. Dropout layer has a probability associated with it and is applied at every neuron of the response map separately. It randomly switches off the activation with the probability. The idea behind the dropout is similar to the model ensembles. Due to the dropout layer, different sets of neurons which are switched off, represent a different architecture and all these different architectures are trained in parallel with weight given to each subset and the summation of weights being one. For n neurons attached to DropOut, the number of subset architectures formed is 2^n. It amounts to prediction being averaged over these ensembles of models. This provides a structured model regularization which helps in avoiding the over-fitting. Another view of DropOut being helpful is that since neurons are randomly chosen, they tend to avoid developing co-adaptations among themselves thereby enabling them to develop meaningful features, independent of others.

- VGG16 [9]: This architecture is from VGG group, Oxford. It was an improvement over AlexNet by replacing large kernel-sized filters (11 and 5 in the first and second convolutional layer, respectively) with multiple 3×3 kernel-sized filters one after another. With a given receptive field (the effective area size of input image on which output depends), multiple stacked smaller size kernel is better than the one with a larger size kernel because multiple non-linear layers increases the depth of the network which enables it to learn more complex features, and that too at a lower cost. For example, three 3×3 filters on top of each other with stride 1 ha, a receptive size of 7, but the number of parameters involved is $3(9C^2)$ in comparison to $49C^2$ parameters of kernels with a size of 7. Here, it is assumed that the number of input and output channel of layers is C. In addition, 3×3 kernels help in retaining finer level properties of the image.

- GoogLeNet/Inception [9]: While VGG achieves a phenomenal accuracy on ImageNet dataset, its deployment on even the most modest sized GPUs is a problem because of huge computational requirements, both in terms of memory and time. It becomes inefficient due to large width of convolutional layers. In a convolutional operation at one location, every output channel is connected to every input channel, and so we call it a dense connection architecture. The GoogLeNet builds on the idea that most of the activations in a deep network are either unnecessary (value of zero) or redundant because of correlations between them. Therefore the most efficient architecture of a deep network will have a sparse connection between the activations. There are techniques to prune out such connections which would result in a sparse weight/connection. Kernels for sparse matrix multiplication are not optimized in BLAS or CuBlas (CUDA for GPU) packages which render them to be even slower than their dense counterparts. Therefore, GoogLeNet devised a module called inception module that approximates a sparse CNN with a normal dense construction. Since only a small number of neurons are effective, the width/number of the convolutional filters of a particular kernel size is kept small. In addition, it uses convolutions of different sizes to capture details at varied scales. Another salient point about the module is that it has a so-called bottleneck layer. It helps in the massive reduction of the computation requirement.

- ResNet [9]: According to the evolution of CNN, increasing the depth should increase the accuracy of the network, as long as over-fitting is taken care of. The problem with increased depth is the signal required to change the weights, which arises from the end of the network by comparing ground-truth and prediction becomes very small at the earlier layers, because of increased depth. It essentially means that earlier layers are almost negligibly learned. This is called vanishing gradient. The second problem with training the deeper networks is performing the optimization on huge parameter space and therefore naively adding the layers leading to higher training error. Residual networks allow training of such deep networks by constructing the network through modules called residual models as shown in the figure. This is called degradation problem. The architecture is similar to the VGGNet consisting mostly of 3×3 filters. From the VGGNet, a shortcut connection as described above is inserted to form a residual network. This can be seen in the figure which shows a small snippet of earlier layer synthesis from VGG-19.

Table 1. Convolutional neural networks architectures evolution (The top-1 accuracy rate is the ratio of images whose ground truth category is exactly the prediction category with maximum probability, while the top-5 accuracy rate indicates the ratio of images whose ground-truth category is within the top-5 prediction categories sorted by the probabilities, accuracy is obtained using ImageNet dataset).

Architecture	Top-1 Accuracy	Top-5 Accuracy	Year
Alexnet	57.1	80.2	2012
Inception-V1	69.8	89.3	2013
VGG	70.5	91.2	2013
Resnet-50	75.2	93	2015
Inception-V3	78.8	94.4	2016

2.2. Programming Language

Among the most popular programming languages in the field of machine learning, two of them stand out: Python and R, being the latter largely oriented towards statistical analysis. In terms of deep learning, the clear dominator is Python, thanks in part to the large number of libraries and frameworks developed for this language, such as PyTorch, Caffe, Theano, TensorFlow or Keras. That is why Python has been chosen in its version 3.6. In addition, different libraries implemented for this language will be used to facilitate the development process. Some of the most relevant packages are the following:

- NumPy, focused on the scientific computation, provides vectors or arrays as well as powerful mathematical tools on them. Pandas for the analysis of data, mainly the reading process of the different CSV files needed. Pandas allows quick access to the data, as well as a powerful treatment of it.
- Scikit-learn, a machine learning library, with powerful statistical tools that will be used mainly during the pre-processing of the data, prior to the implementation of the neural network.
- Python Imaging Library (PIL) used for reading and converting images. In the development phase, although eliminated in the final version, the Matplotlib library has been used in order to create the needed graphs. More specifically, it has been used only for visualization and to obtain the images, both original and processed, during the elaboration of this document. Therefore, it lacks relevance and usefulness in the final versions.
- Developed by Google in order to meet their needs along the machine learning environment, TensorFlow is a library for numerical calculations that mainly uses data flow diagrams. Published under a license of open code in 2015, it has since become one of the most popular benchmarks in the development of deep learning systems and neural networks.
- Born with flexibility and usability in mind, Keras is a library for high level neural networks that was developed for Python. One of the biggest advantages when it comes to Keras usage is that it seeks to greatly simplify the development-related tasks. It is possible to run it with libraries such as TensorFlow, Theano or CNTK as a backend, understanding each other as an interface rather than as a framework. In 2017, Keras was integrated into the source code of TensorFlow allowing its development with a higher level of abstraction. Deep learning and its development have been greatly facilitated by the use of Graphics Processing Unit (GPUs). The high number of calculations that are carried out and their high complexity, especially during the training of a neural network, make high-performance hardware an actual need. GPUs come into play in this current scenario, as long as its usage in the training of neural networks allows to reduce considerably the time taken, compared to the exclusive use of CPUs. This is due to its high number of cores that allow parallel processing.
- çDIA parallel calculus architecture, next to cuDNN, its library for deep neural networks, allows the use of GPUs in TensorFlow.

The performance of the application can vary considerably depending on the available specifications, being critical in the final results and, especially, in the total execution time.

The system used has an Intel Core i7-8700 6-core processor with a base frequency of 3.2 GHz up to 4.6 GHz and 16 GB of DDR4 RAM. It also has a NVIDIA GeForce GTX 1060 graphics card with 3 GB of VDR DDR5 memory with a total of 1152 CUDA cores. This hardware is able to perform the training of the neural network and the processing of images in a comfortable way, however different approaches to the problem would require more memory allocated in the GPU. Please note that all code, datasets and samples are available at https://github.com/javiermzll/CCN-Whale-Recognition.

3. State-of-the-Art in Kaggle Competition: Humpback Whale Identification Methods

After centuries of intense whaling, recovering whale populations still have a hard time adapting to warming oceans and struggle to compete every day with the industrial fishing industry for food.

To aid whale conservation efforts, scientists use photo surveillance systems to monitor ocean activity. They use the shape of whales' tails and unique markings found in footage to identify what species of whale they are analyzing and meticulously log whale pod dynamics and movements. For the past 40 years, most of this work has been done manually by individual scientists, leaving a huge trove of data untapped and underutilized.

"In this competition, you're challenged to build an algorithm to identify individual whales in images. You will analyze Happywhale's database and Gulf of Maine Humpback Whale Catalog of over 25,000 images, gathered from research institutions and public contributors. By contributing, you will help to open rich fields of understanding for marine mammal population dynamics around the globe. Note, this competition is similar in nature to this competition with an expanded and updated dataset. We'd like to thank Happywhale and Gulf of Maine Humpback Whale Catalog for providing this data and problem. Happywhale is a platform that uses image process algorithms to let anyone to submit their whale photo and have it automatically identified."

Next, find a brief state-of-the-art of more interesting and catchy methods used by other researchers in terms of accuracy, see Table 2 for a quick overview.

- SIFT-Based [10]. This is one of the most beautiful and, at the same time, unusual. David, now a Kaggle Grandmaster (Rank 12), was 4th on the Private LeaderBoard and shared his solution as a post on Kaggle Discussions forum. He worked with full-resolution images and used traditional key point matching techniques, utilizing SIFT and ROOTSIFT. In order to deal with false positives, David trained a U-Net to segment the whale from the background. Interestingly, he used smart post-processing to give classes with only one training example more chance to be in the TOP-1 prediction. The takeaway is that we should never be blinded by the power of deep learning and underestimate the abilities of traditional methods.
- Pure Classification [11,12]. The team Pure Magic thanks radek (7th place), consisting of Dmytro Mishkin, Anastasiia Mishchuk and Igor Krashenyi, pursued approach that was a combination of metric learning (triplet loss) and classification, as Dmytro described in his post. They tried Center Loss to reduce overfitting when training classification models for a long time, along with temperature scaling before applying softmax. Among the numerous backbone architectures that were used, the best one was SE-ResNeXt-50, which was able to reach 0.955 LeaderBoard. Their solution is way more diverse than that, and I highly suggest you to refer to the original post.
- CosFace, ArcFace [13]. As it was mentioned in the post by Ivan Sosin (his team BratanNet was 9th in this competition), they used CosFace and ArcFace approaches. From the original post: Among others Cosface and Arcface stand out as newly discovered state-of-the-art (SOTA) for face recognition task. The main idea is to bring examples of the same class close to each other in cosine similarity space and to pull apart distinct classes. Training with cosface or arcface generally is classification, so the final loss was CrossEntropy. When using larger backbones like InceptionV3 or SE-ResNeXt-50, they noticed overfitting, so they switched to lighter networks like ResNet-34, BN-Inception and DenseNet-121.The team also used carefully selected augmentations and numerous network modification techniques like CoordConv and GapNet. What was particularly interesting in their approach is the way they dealt with new whales. From the original post: Starting from the beginning we realized that it is essential to do something with new whales in order to incorporate them into the training process. Simple solution was to assign each new whale a probability of each class equal to 1/5004. With the help of weighted sampling technique it gave us some boost. Then we realized that we could use softmax predictions for new whales derived from the trained ensemble. Therefore, we came up with distillation. We choose distillation instead of pseudo labels, because new whale is considered to have different labels from the train labels. Though it might not really be true. To further boost the model capability we added

test images with pseudo labels into the train dataset. Eventually, our single model could hit 0.958 with snapshot ensembling. Unfortunately, ensembling trained this way did not give any score improvement.

- Siamese Networks [14]. One of the first architecture was a siamese network with numerous branch architectures and custom loss, which consisted of a number of convolutional and dense layers. The branch architectures that were used included ResNet-18, ResNet-34, Resnet-50 and SE-ResNeXt-50. A progressive learning was used, with the resolution strategy $229 \times 229 \rightarrow 384 \times 384 \rightarrow 512 \times 512$. That is, first the network is trained on 229×229 images with little regularization and larger learning rate. After convergence, the net resets the learning rate and increased regularization, consequently training the network again on images of higher resolution. The models were optimized using Adam optimizer with an initial learning rate of 1^{-4}, reducing 5 times on plateau. The best-performing single model with ResNet-50 scored 0.929.

- Metric Learning [15]. Another approach that was used was metric learning with Margin Loss. Numerous ImageNet-pretrained backbone architectures were used, which included: ResNet-50, ResNet-101, ResNet-152, DenseNet-121 and DenseNet-169. The networks were trained progressively mostly using $448 \times 448 \rightarrow 672 \times 672$ strategy. Adam optimizer is used, decreasing the learning rate 10 times after 100 epochs. We also used a batch size of 96 for the whole training. The most interesting part is a 2% boost right away. It is a metric learning method that was developed by Sanakoyeu, Tschernezki, et al. [16]. What it does is every n epochs it splits the training data as well as the embedding layer into k clusters. After setting up the bijection between the training chunks and the learners, the model trains them separately while accumulating the gradients for the branch network. As a result of the huge class imbalance, heavy augmentations were used, which included random flips, rotate, zoom, blur, lighting, contrast and saturation change. During inference, dot products between the query feature vector and the train gallery feature vectors were calculated and a class with the highest dot product value was selected as the TOP-1 prediction. It is noteworthy to mention that the best-performing single model with a DenseNet-169 backbone scored 0.931.

- Classification on Features [17]. It trains the classification model using the features extracted from all previous models and concatenated together (after applying PCA). The head for the classification consisted of two dense layers with dropout in between. The model trained very quickly because precomputed features were used. This approach allowed to get a 0.924 score and brought even more diversity in the overall ensemble.

- New Whale Classification. One of the most complicated parts of this competition was to correctly classify the new whales (as about 30% of all images belonged to the new_whale class). The popular strategy to deal with this was to use a simple threshold. That is, if the maximum probability that the given image X belongs to some known whale class is smaller than the threshold, it was classified as the new_whale. For each best-performing model and ensemble, its TOP-4 predictions are taken, sorted in descending order. Then, for every other model, probabilities for the selected 4 classes are used. The goal was to predict whether the whale is new or not based on these features. A combination of LogRegression, Support Vector Machines (SVM), several k-NN models and LightGBM is used. The combination of all scores is 0.9655.

This paper reports about a network that is not necessarily the best competitor in Kaggle challenge, but this work tries to minimize resources using an image downsampling and a small architecture, interesting for embedded systems, see accuracy results in Table 2.

Table 2. State-of-the-art of Kaggle competition models vs. proposed model. Note that Kaggle submissions are evaluated according to the Mean Average Precision @ 5 (MAP@5). Proposed method uses a straight forward convolutional neural network, main advantages are the low resources needed and fast training process to be embedded in a small computing device. The method is not the best one but it has a good accuracy.

Method	Score/Accuracy
SIFT-Based	0.967
New Whale Classification	0.965
Pure Classification	0.955
CosFace, ArcFace	0.958
Metric Learning	0.931
Siamese Networks	0.929
Classification on Features	0.924
Proposed method	0.785

4. Design

Over the last five years, convolutional neural nets have offered more accurate solutions to many problems in computer vision, and these solutions have surpassed a threshold of acceptability for many applications. Convolutional neural networks have supplanted other approaches to solving problems in these areas, and enabled many new applications. While the design of convolutional neural nets is still something of an art form, in our work we have try to minimize the storage amount for CNN model parameters and processor load using an image downsampling and a small architecture, interesting for embedded systems, such as: cameras mounted on the ship, single board computers, etc. in order to classify images online.

4.1. Dataset

The key aspect to face the construction of a neural network is the study and analysis of the dataset on which it is intended to work. The dataset is constituted by a set of images of humpback whales, more specifically their tails, see Figure 5.

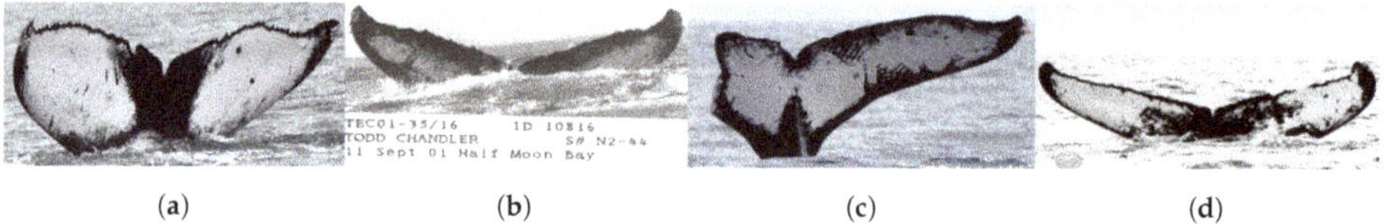

(a) (b) (c) (d)

Figure 5. Humpback whale tail images in the dataset. (**a**) 1a36b244. (**b**) 1c5d333f. (**c**) 1efa630a. (**d**) 2c77045a.

As the images show, the tails present different characteristics. The most notable difference a priori is the variation of color between them, which can be presented both in greyscale or in color. The contrast of resolutions, or the presence of elements external to the whale itself, such as annotations, is also a relevant aspect.

There are approximately 25,000 images divided into two sets, one training 9851 images while the other one evaluates 15,600 images. This distribution hinders the subsequent training of the network since the size of the training set is notably smaller to its homologous. Alongside the images, the training set provides a CSV file that collects the labels of each image, check Table 3. Our neural network will deal with the following problem: Identify humpback whale specimens and if there is no record of it, catalogue it as a new whale with the label `new_whale`, counting in total with 4251 different classes or individuals. This paper focuses on an identification problem, that is, since it is an identification challenge, each whale is a separate class and whales were photographed multiple times and one attempts to identify a whale class in the testing set.

Table 3. Header of the training file.

	Image	Id
0	00022e1a.jpg	w_d15442c
1	000466c4.jpg	w_1287fbc
2	00087b01.jpg	w_da2efe0
3	001296d5.jpg	w_19e5482
4	0014cfdf.jpg	w_f22f3e3

Data training samples is the focus of every CNN algorithm whether the training process can achieve effective convergence or whether it will produce overfitting. In this study, the discussion on setting the number of training samples is quite meaningful.

The origin of the 1000-image magic number comes from the original ImageNet classification challenge, where the dataset had 1000 categories, each with a bit less than 1000 images for each class. This was good enough to train the early generations of image classifiers like AlexNet, and so proves that around 1000 images is enough. It seems to get trickier to train a model from scratch in some cases until you get into the low hundreds. The biggest exception is when a transfer learning on an already-trained model is used. It is using a network that has already seen a lot of images and learned to distinguish between the classes, therefore it could usually teach it new classes in the same domain with as few as ten or twenty examples.

What does in the same domain mean? It is a lot easier to teach a network that has been trained on photos of real world objects to recognize other objects, but taking that same network and asking it to categorize completely different types of images like x-rays, faces or satellite photos is likely to be less successful, and at least require a lot more training images.

Another key point is the representative modifier. That is there because the quality of the images is important, not just the quantity. What is crucial is that the training images are as close as possible to the inputs that the model will see when it is deployed. ImageNet consists of photos taken from the web, so they are usually well-framed and without much distortion. A smaller amount of training images that were taken in the same environment that it will produce better end results than a larger number of less representative images.

Augmentations are important too. The training data can be augmented by randomly cropping, rotating, brightening, or warping the original images. TensorFlow controls this with command line flags like flip-left-to-right and random-scale. This has the effect of effectively increasing the size of your training images, and is standard for most ImageNet-style training pipelines. It can be very useful for helping out transfer learning on smaller sets of images as well though. Distorted copies are not worth quite as much as new original images when it comes to overall accuracy, but if dataset only has a few images it is a great way to boost the results and will reduce the overall number of images needed.

The real situation is to try, so if a dataset has fewer images than the rule suggests, do not let it stop you, but this rule of thumb will be a good starting point for planning an approach at least.

4.2. Image Processing

It is necessary to process each image in advance in order to facilitate the extraction of the present features present in it. If all the data is homogenized, this effect is achieved.

Firstly, the image is converted to a grey scale (Figure 6), ranging from three different channel colors to a single one. There is a double reason behind this conversion—the existence of original images in the black and white dataset and the absence of color characteristics and information. This fact provokes that only the pattern of the tail stands out as useful information.

(a) (b) (c) (d) (e)

Figure 6. RGB image to grey image conversion. (**a**) Original image. (**b**) Red component. (**c**) Green component. (**d**) Blue component. (**e**) Grey image.

The neural network requires the dimension of the input vectors to be fixed, and that is why each image must be rescaled beforehand, resulting, in this case, in 100×100 size matrices with a single channel. The proposal of such downsampling method for inferring the resolutions of images and downsampling images of higher resolution as a preprocessing step in whale recognition. Some authors have demonstrated that image downsampling increases the identification performance of recognition and re-identification [18].

Each pixel has a value ranging between 0 and 255. This amplitude of range, due to the operation of the convolutional networks, allows the incorrect identification of the characteristics for each vector. To correct this disadvantage, it is convenient to normalize the image previously, in a process known as zero mean and unit variance normalization (Figure 7).

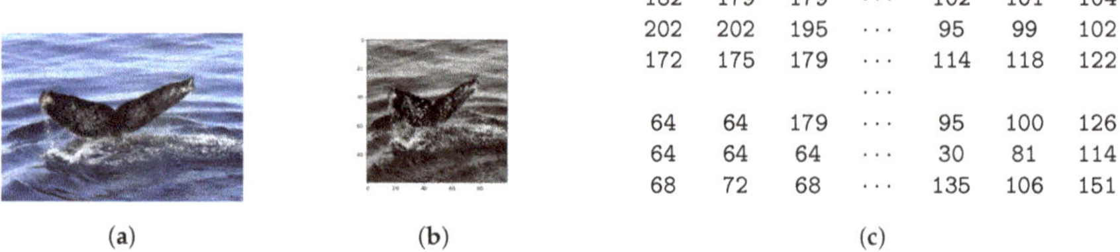

(a) (b) (c)

Figure 7. Scale image process. From an HD image to a resized and gray image represented as a matrix (such representation is needed to train the model). (**a**) HD image. (**b**) Gray and scaled image. (**c**) Matrix representation.

The first step is to center the image on the value 0 by subtracting its mean from each value of the matrix.

$$M = \begin{Bmatrix} 67.8595 & 64.8595 & \cdots & -13.140503 & -10.140503 \\ 87.8595 & 87.8595 & \cdots & -15.140503 & -12.140503 \\ & & \cdots & & \\ -50.140503 & -50.140503 & \cdots & -33.140503 & -0.14050293 \\ -46.140503 & -42.140503 & \cdots & -8.140503 & 36.859497 \end{Bmatrix} \tag{2}$$

Subsequently, the range is compressed dividing each value between matrix's standard deviation.

$$M = \begin{Bmatrix} 1.4328924 & 1.3695457 & \cdots & -0.2774693 & -0.21412258 \\ 1.855204 & 1.855204 & \cdots & -0.31970045 & -0.25635374 \\ & & \cdots & & \\ -1.0587456 & -1.0587456 & \cdots & -0.6997808 & -0.0029668 \\ -0.97428334 & -0.88982105 & \cdots & -0.17189142 & 0.7783096 \end{Bmatrix} \tag{3}$$

4.3. Data Augmentation and Imbalance Class

Neural networks are primarily used for classification tasks where the network learns by looking at data points belonging to different classes. Imagine you have a classification problem where you have to identify whether a picture shown to the network has a dog in it or a cat. Now assume that your training set has a total of 10,000 images, 9998 images are of dogs and there is only one image which has a cat and the remaining image has none of them. This is largely what class imbalance looks like [19], when you have unequal distribution of labeled data in different classes, see Figure 8. A common practice within the classification of images through neural networks is the increase of data present in the dataset (data augmentation). This is especially useful where the proportion of classes is not balanced and there are numerous categories with only one sample.

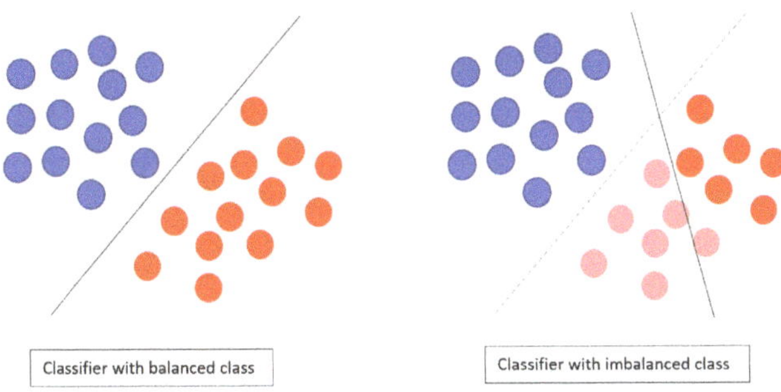

Figure 8. General representation of the problem of classifier trained with imbalance class.

In order to increase the number of data available for the training, a series of processes are performed on an original image, thus generating a series of derived images. Among the possible modifications are the rotation (Figure 9), translation, noise reduction, etc.

Figure 9. Image rotation.

The effectiveness of this technique lies in the way neural networks understand the images and their characteristics. If an image is slightly modified, it is perceived by the network as a completely different image belonging to the same class.

This decreases the chances of the network to focus on irrelevant orientations or positions while maintaining the relevance of the desired characteristics such as the pattern of the tail in this paper.

An increase of the training set has been made following the following criteria: If a class has less than 10 images, the difference with its original sample number is generated. Therefore, if a specimen had 4 samples, an additional one would be generated from the images associated with a whale, see Figure 10.

Figure 10. Augment over specimen w_964c1b3.

4.4. Neural Network Architecture

The implemented network presents the structure shown in Table 4. A convolutional layer is established, followed by batch normalization [20] and a max pool reduction. Following this fact, another convolutional layer is defined, followed by a reduction by mean or average pooling. This firstly allows extracting the most important characteristics of the image. Secondly, at the next level of depth, smoothing the extraction ensures the lack of loss of relevant information. In both layers, a Rectified Linear Unit (ReLU) activation function is used, defined by:

$$f(x) = x^{+} = \max(0, x) \tag{4}$$

where x is the input of the neuron, returning the value x if it is positive or 0 if it is negative. This allows to considerably accelerate the training to be easily computable in comparison to other activation functions such as softmax.

Finally, a conventional multilayer neural network is connected. This network is formed by 450 neurons with dropout (Figure 11), followed by a dense output layer which is completely connected to softmax as an activation function, as well as 4251 neurons that are equivalent to the total number of classes to classify. The first layer receives as input a vector of one dimension, which is achieved compressing the output of two dimensions obtained from the convolutional layers.

The dropout allows ignoring a percentage of input units or neurons during training, in this case 80%. This ignorance disembogues in a neural network smaller than the original and with therefore fewer parameters, which decreases the dependencies between neurons and thus avoids overtraining.

The softmax activation function is used in classification problems with multiple options, as in this case, and it returns the probabilities of each class. The aforementioned function is given by:

$$f(x_i) = \frac{e^{x_i}}{\sum_{j=1}^{4251} e^{x_j}} \tag{5}$$

It compresses the values between 0 and 1 and makes the sum of all the resulting values equal to 1, being x an input vector of size equal to the number of units or neurons of the layer (4251).

Table 4. Architecture of the implemented network.

Layer (Type)	Output Shape	Param
conv0 (Conv2D)	(none, 94, 94, 32)	1600
bn (BatchNormalization)	(None, 94, 94, 32)	128
activation (Activation)	(None, 94, 94, 32)	0
max_pool (maxPooling2D)	(None, 47, 47, 32)	0
conv1 (Conv2D)	(None, 45, 45, 64)	18496
activation_1 (Activation)	(None, 45, 45, 64)	0
avg_pool (AveragePooling2D)	(None, 15, 15, 64)	0
flatten (Flatten)	(None, 14400)	0
ReLU (Dense)	(None, 450)	6480450
dropout (Dropout)	(None, 450)	0
softmax (Dense)	(None, 4251)	1917201

Total params: 8,417,875
Trainable params: 8,417,811
Non-trainable params: 64

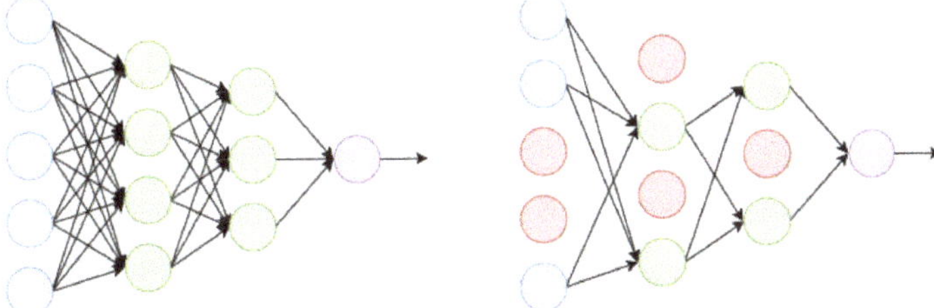

Figure 11. Dropout schema used in convolutional neural networks (Input layer in blue, hidden layers in green, output layer in purple and dropout neurons in red).

4.5. Training

TensorBoard tool (belonging TensorFlow) is used in this section to monitor and visualize our neural network during the training phase, performing a simulation while obtaining the precision and loss graphs.

A key aspect of neural networks is the loss function that reveals about how far away the predictions of a model are ($\hat{y}$) from the real labels y. It is a positive value that improves the performance of the model while decreasing.

The loss function used on the implemented model is Categorical Cross-Entropy (CCE) [21,22], which is especially useful in problems where several excluding classes exist. CCE is preferred for training deep networks with softmax outputs, since: It puts more emphasis on correct classification and can distinguish better between "almost correctly classified" and "totally wrongly classified" than 0-1-loss, It corresponds to maximum likelihood for networks with softmax output and It has an information-theoretical interpretation based on probability and therefore makes sense to use with probabilities (which is the nature of our predictions). The function is given by:

$$H(y, \hat{y}) = \sum_x y_x \log \frac{1}{\hat{y}_x} = -\sum_x y_x \log \hat{y}_x \tag{6}$$

where x is a discrete variable and $\hat{y}$ is the prediction for the real distribution y. The accuracy of the model during the training increases while the value of the loss function decreases, see Figure 12.

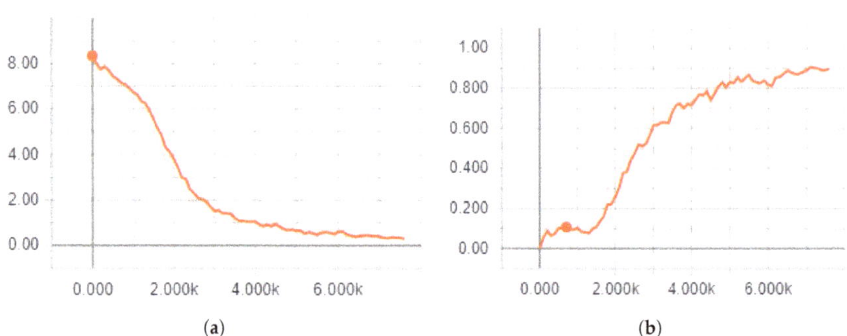

Figure 12. Loss/Accuracy during the >7000 iterations in the training process. (**a**) Loss during training.
(**b**) Accuracy during the training.

The simulation of the training takes as reference the exit of the first dense layer before proceeding
to the classification. This does not allow an exact representation of what would be the final output
with the total classes, but easily exemplifies the behavior of the neural network and the data during
the training phase.

Apart from the performance aspect, it is also important to know what the model has learned.
This is necessary so as to ensure that the model has not learned something discriminatory or biased.
One way of approaching this problem is from a data visualization perspective. By visualizing how the
model groups the data, we can get an idea of what the model thinks are similar and dissimilar data
points. This is beneficial in order to understand why a model makes certain predictions and what kind
of data is fed into the algorithm.

An embedding is essentially a low-dimensional space into which a high dimensional vector can
be translated. During translation, an embedding preserves the semantic relationship of the inputs by
placing similar inputs close together in the embedding space. Multidimensional space helps to group
semantically related items together while keeping the dissimilar items apart. This can prove to be
highly useful in a machine learning task. Consider the following visualizations of real embeddings,
see Figures 13 and 14.

The Embedding Projector is open-sourced and integrated into the TensorFlow platform or can be
used as a standalone tool at projector.tensorflow.org. It offers four well-known methods for reducing
the dimensionality of the data. Each method can be used to create either a two- or three-dimensional
view for exploration.

- Principal Component Analysis (PCA) is a technique which extracts a new set of variables
  called Principal Components from the existing data. These Principal Components are a linear
  combination of original features and try to capture as much information from the original dataset.
  The Embedding Projector computes the top 10 principal components for the data, from which we
  can choose two or three to view.
- t-SNE or T-distributed stochastic neighbor embedding visualizes high-dimensional data by giving
  each data point a location in a two or three-dimensional map. This technique finds clusters in
  data thereby making sure that an embedding preserves the meaning in the data.
- A custom projection is a linear projection onto the horizontal and vertical axes which have been
  specified using the data labels. The custom projections mainly help to decipher the meaningful
  directions in data sets.
- UMAP stands for Uniform Manifold Approximation and Projection for Dimension Reduction.
  t-SNE is an excellent technique to visualize high dimensional datasets but has certain drawbacks
  like high computation time and loss of large scale information. UMAP, on the other hand, tries to
  overcome these limitations as it can handle pretty large datasets easily while also preserving the
  local and global structure of data.

Four situations are differentiated during the execution in the case of study. The beginning, where the network is not trained (Figure 13). The first iterations (Figure 14) where most samples are considered belonging to the new whale class due to their common characteristics among all the images and to their greater proportion in comparison to the other classes. The phase where features from other classes are starting to be noticed (Figure 14a). The final phase, where the network completes its training and differentiates with greater precision (Figure 14c). A group with similar characteristics has been selected to follow its grouping using a t-SNE projection.

The t-SNE algorithm comprises two main stages. First, t-SNE constructs a probability distribution over pairs of high-dimensional objects in such a way that similar objects have a high probability of being picked while dissimilar points have an extremely small probability of being picked. Second, t-SNE defines a similar probability distribution over the points in the low-dimensional map, and it minimizes the Kullback–Leibler divergence (KL divergence) between the two distributions with respect to the locations of the points in the map. Note that while the original algorithm uses the Euclidean distance between objects as the base of its similarity metric, this should be changed as appropriate (see [23]).

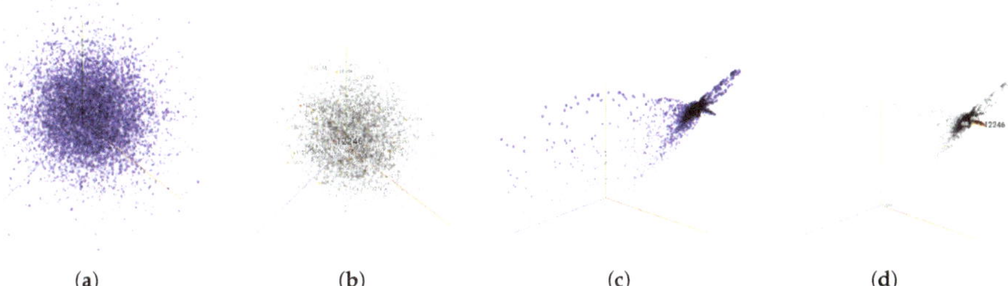

(a)   (b)   (c)   (d)

Figure 13. Different states in the simulation (Figures 13 and 14 are obtained in the training phase, and Figure 13b,d are obtained in testing phase) using a T-distributed stochastic neighbor embedding (t-SNE) projector. (**a**) Init state. (**b**) Init state. (**c**) First iterations. (**d**) First iterations.

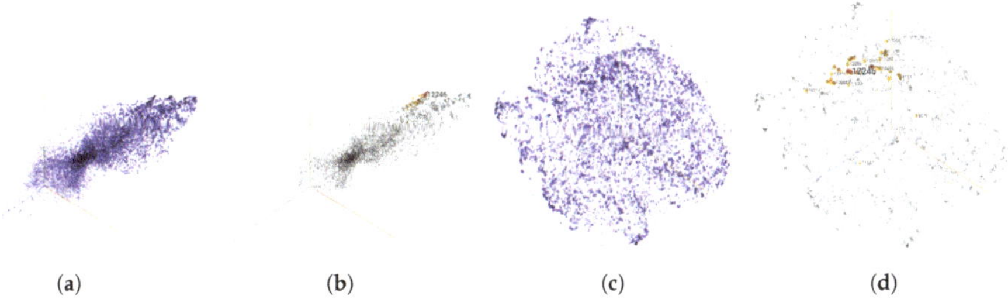

(a)   (b)   (c)   (d)

Figure 14. Different states in simulation (Figure 14a,c in training, and Figure 14b,d in testing) using a t-SNE projector. (**a**) Last iterations. (**b**) Last iterations. (**c**) Final stage. (**d**) Final state.

4.6. Optimization Algorithm

The learning process is summarized in one idea: the minimization of the loss function by updating the parameters, w, of the network, turning it into an optimization problem. Adam or Adaptive Moment Estimation [24] is an optimization algorithm that allows modifying the learning rate as the training is performed. Specifically, it is a variant of the stochastic gradient descent.

The neural network has a series of hyper parameters among them the learning rate α. This parameter allows the gradient descent algorithms to determine the following point, so:

$$w_j = w_j - \alpha \frac{\partial f(w_j)}{\partial w_j} \tag{7}$$

where w_j is one of the parameters, f the loss function and α the learning rate.

If its value is too small, the convergence, and therefore the learning, will take too long, in the same way as if the rate is sufficiently large the next point will exceed the minimum, making it impossible to reach it. It is therefore true that an adequate value of the learning rate is the key for the correct performance during the training.

The descent of the gradient, see Figures 15 and 16, is an iterative process where the weights are updated in each iteration, which is the reason why, in order to obtain the best result, it is necessary to process the dataset more than once. As long as processing a complete dataset in an iteration is difficult by memory restrictions, it is divided into parts called batches of a certain size, so if the dataset contains 1000 images and a batch size of 100, there will be 10 divisions in total.

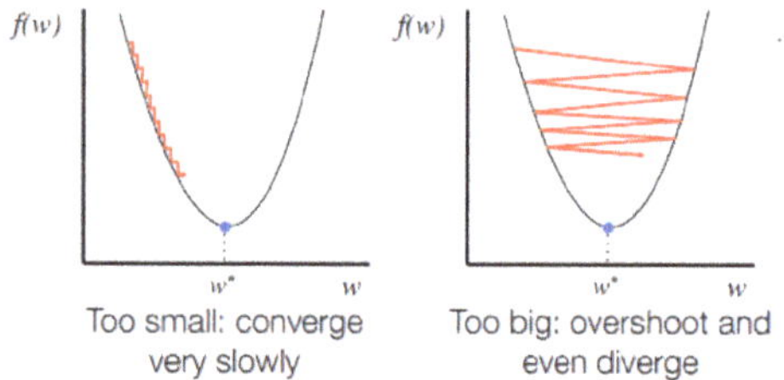

Figure 15. Learning rate α values. A small value produces slow convergence while a high value could overshoot and even diverge.

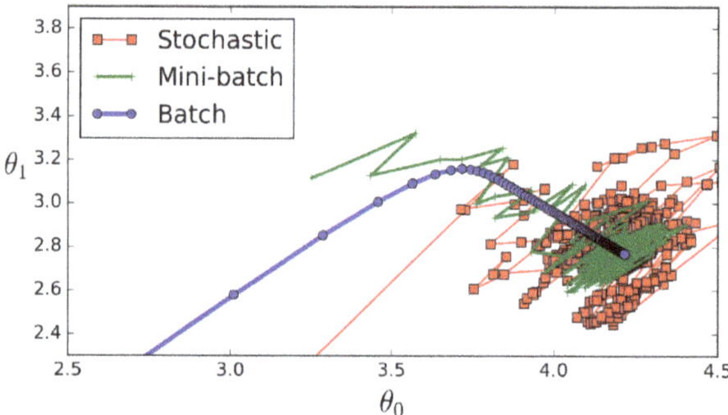

Figure 16. Learning trend using different gradient descent methods (batch, mini-batch and stochastic).

Thus, the learning process is carried out through iterations over the different batches. Depending on the size, the frequency with which the weights of the network are updated varies, so the smaller the size, the higher the frequency of update. Smaller sizes are generally used for the size of the dataset, mini-batch. If the size is equal it is called batch, and if it is equal to 1 it is known as stochastic.

A batch size of 128 over a total of 9850 images of the training set has been defined, which produces a total of 77 iterations for the achievement of the totality of the data. In addition, 100 epochs or tours on the complete dataset have been established.

The dataset subjected to an increase during the intermediate stages of development has a total of 124,065 images, a massive number compared to the original 9850 images. This fact produces a total of 970 iterations per epoch, having been limited to 50.

4.7. Label Coding

Due to the mathematical operations they perform, most of machine learning algorithms do not allow working on categorical data (labels), requiring numerical values. The encoding technique known as One-Hot has been used for the implementation, where a vector is generated for each category where only one of the values can be 1, being the rest 0.

The coding process involves two operations:

$$V = \begin{bmatrix} Blue & Red & Green & White & Black \end{bmatrix} \tag{8}$$

First off, the categories are converted into a whole value $V = [0, 3, 4, 1, 2]$. Being subsequently codified as One-Hot vectors.

$$M = \begin{Bmatrix} 1 & 0 & 0 & 0 & 0 \\ 0 & 0 & 0 & 1 & 0 \\ 0 & 0 & 0 & 0 & 1 \\ 0 & 1 & 0 & 0 & 0 \\ 0 & 0 & 1 & 0 & 0 \end{Bmatrix} \tag{9}$$

Kaggle submissions are evaluated according to the Mean Average Precision @ 5 (MAP@5).

$$MAP@5 = \frac{1}{U} \sum u = 1^U \sum k = 1^{min(n,5)} P(k) \times rel(k) \tag{10}$$

where U is the number of images, $P(k)$ is the precision at cutoff k, n is the number predictions per image and $rel(k)$ is an indicator function equaling 1 if the item at rank k is a relevant (correct) label, zero otherwise.

Once a correct label has been scored for an observation, that label is no longer considered relevant for that observation, and additional predictions of that label are skipped in the calculation. For example, if the correct label is A for an observation, the following predictions all score an average precision of 1.0.

5. Results and Evaluation

One of the main drawbacks during training is the overfitting or under fitting of the data. If the training data does not contain enough information, the system will not be able to correctly recognize images. On the contrary, if a model is over trained it will increase the precision on the training set but it will be incapable of generalizing when new data is received.

That is why the training data is usually split into two; a training set and an evaluation set. The test set must be representative of the total dataset and must be large enough to be effective, usually between 70% and 80% of the original size. There may also be a third set of tests to check the final version of a model, providing new data and ensuring an unbiased evaluation.

This allows to obtain closer results to reality since the data of the validation set remains hidden from the network during training and thus avoids overfitting.

In the implemented application, the validation set has been established at 10% of the total, while the training set was 90%. It has only been divided during the phase of development, by increasing the data set as described, to adjust the network's hyper parameters and study their performance, using the entire original dataset in advanced versions of the model and not the increased dataset. This is due to two factors: the considerable increase in both the time of image pre-processing and in the training time of the neural network as well as the improvement in the percentage of final success of only 1%. However, if the goal is to achieve the maximum precision, the data increase in the final version is to be incorporated, in addition to a more aggressive processing on the data.

The decisions that justify the need to establish a process of image enhancement are several. First, the asymmetrical distribution of the data set, existing categories with a single example, which would result in the presence of values exclusively in one of the sets when dividing them during the evaluation results in counterproductive. The second and most important reason is the existence of a set of test data, with unknown values, provided by Kaggle [25], allowing us to generate a CSV file with the results, upload them to the competition and compare the performance of the model. Therefore, this set of tests will be used as a validation set in the final versions of the application.

The whole process of designing, implementing and training a neural network is summarized in the prediction stage, where it is expected that the network responds with a certain precision to a series of entries. The model implemented reaches a precision of 0.4407 that can be considered promising, due to the characteristics of the selected dataset (4251 asymmetric classes) and to the hardware limitations that could exist in terms of memory. If we compare the public classification of the Kaggle competition, the developed model would be placed in the 54th position out of a total of 528 participants, with the highest probability being 0.78563 and the second 0.64924, see Table 2.

To extract the results and the resulting classes it is necessary to reverse the labels coding process. In order to achieve this goal, the vector of predictions of length 4251 for each input is obtained, where a probability is established for each class. Then, the five highest probabilities are selected and their positions in the array are reverted to the original label.

This paper tried to minimize resources using an image downsampling and a small architecture. To perform calculations efficiently and save resources, the algorithm process must be simplified, or the size of the data file must be strictly controlled. Sensitivity analysis of convolutional neural networks is related to the concept of adversarial and rubbish examples. Adversarial examples are slightly modified images which show a significant change in the model output compared to the original image. The idea of rubbish examples studies random noise images that lead to arbitrary classification decisions although their appearance can not be related to the particular object category or any natural image at all. See Figure 17 corresponding to the sensitivity analysis of proposed model.

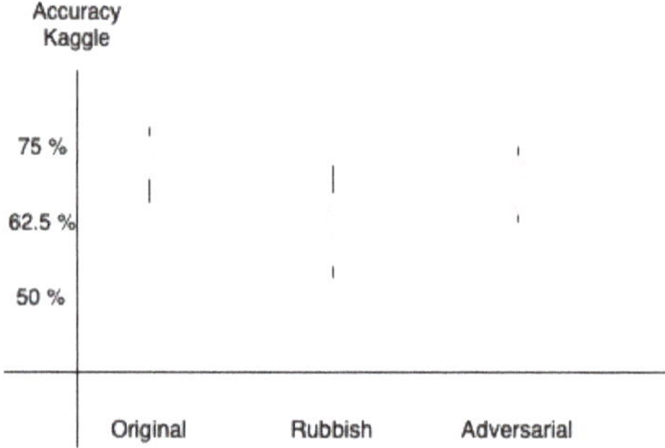

Figure 17. Sensitivity analysis showing candle chart of original method versus rubbish and adversarial modifications.

As for the input data preprocessing, image conversion, feature extraction or band combination are all effective ways to increase the value of the input data and reduce noise. Collect labeled data is expensive. For images, proposed method applies data augmentation with simple techniques like rotation, random cropping, shifting, shear and flipping to create more samples from existing data (other color distortion includes hue, saturation and exposure shifts could be also applied). It is possible to add training data with non-labeled classes. For samples with high confidence prediction, they have been added to the training dataset with the corresponding label predictions. Another interesting approach consists of building a hierarchy of classifiers for image classification with a large number of classes Subcategory classification is another rapidly growing subfield of image classification. Using object part information is beneficial for fine-grained classification. Generally, the accuracy can be improved by localizing important parts of objects and representing their appearances discriminatively. Part annotation information is used to learn a compact pose normalization space. All the above-mentioned methods make use of part annotation information for supervised training. However, these annotations are not easy to collect and these systems have difficulty in scaling up and to handle many types of fine-grained classes. To avoid this problem, it is possible to find localized parts or regions in an unsupervised manner.

6. Conclusions and Further Work

The original approach to this paper has been the study and deepening in the fields of artificial intelligence, deep learning and convolutional neural networks. That is why, despite the fact that the accuracy achieved does not imply a value close to the state of the art, the different methods and techniques used during the development of systems that allow recognizing images have to be explored. Likewise, the influence and importance of previous work on the neural networks themselves has been analyzed, such as the processing of images and the comprehension of data sets.

During the development of this work, several of the original design decisions were altered. The usage of a single library for neural networks, TensorFlow, was raised, but in later versions Keras was also due to two factors: its previous inclusion in the TensorFlow library and the possibility of achieving a greater extent to offer a higher abstraction approach at a programming level compared to TensorFlow.

It has been possible to verify how the importance of the performance of a neural network does not reside exclusively in the architecture that it may have, but that the data set available substantially alters the final result, being a balanced dataset and a large number of samples a critical aspect. The main complications found are divided between the two most relevant goals: the pre-processing of images and the model. The selected dataset has greatly weighed the final accuracy of the application, requiring a much more aggressive data increase than has currently been done. Nonetheless, despite not providing the best results, it offers a clear vision about the importance of the available data, as well as the functioning of the neural networks in terms of the perception of images and how they are treated. In terms of the problems that have occurred in the implementation of the model, there are large hardware demands, especially in terms of RAM memory, that is not available in the equipment used, which has made it impossible to implement a more efficient architecture.

The case study of this paper, the identification of whales by their tail pattern where there are numerous classes belonging the same species, shares similarities with facial recognition applications. Within this field, one of the techniques with the greatest impact and performance are the Siamese networks and the use of triplet loss [26].

However, as promising, the triplet loss has not been implemented due to the memory restrictions, the hardware used and the required large amount of memory and data sets (not available). Nevertheless, the use of a conventional CNN with the available data sets can be considered a valid, interesting and promising method for humpback tail recognition.

Author Contributions: Conceptualization, J.M.L. and N.G.B.; software, L.F.d.M.L. and A.A.A.; investigation, J.M.L., A.A.A. and L.F.d.M.L.; supervision, N.G.B.; all authors analyzed the data and wrote the paper. All authors have read and agree to the published version of the manuscript.

Funding: This research received no external funding.

Acknowledgments: This research has been partially supported by project Seguridad de Vehículos AUTOmóviles para un TRansporte Inteligente, Eficiente y Seguro. SEGVAUTO-TRIES-S2013/MIT-2713. (http://insia-upm.es/portfolio-items/proyecto-segvauto-tries-cm/?lang=en).

Conflicts of Interest: The authors declare no conflict of interest. This research has no funding sponsors so they had no role in the design of the study; in the collection, analyses, or interpretation of data; in the writing of the manuscript, or in the decision to publish the results.

References

1. Russell, S.J.; Norvig, P. *Artificial Intelligence: A Modern Approach*; Pearson Education Limited: Kuala Lumpur, Malaysia, 2016.
2. Goodfellow, I.; Bengio, Y.; Courville, A.; Bengio, Y. *Deep Learning*; MIT Press: Cambridge, UK, 2016; Volume 1.
3. LeCun, Y.; Bengio, Y.; Hinton, G. Deep learning. *Nature* **2015**, *521*, 436. [CrossRef] [PubMed]
4. Qin, Z.; Yu, F.; Liu, C.; Chen, X. How convolutional neural network see the world—A survey of convolutional neural network visualization methods. *arXiv* **2018**, arXiv:1804.11191.
5. Sewak, M.; Karim, M.R.; Pujari, P. *Practical Convolutional Neural Networks: Implement Advanced Deep Learning Models Using Python*; Packt Publishing: Birmingham, UK, 2018.

6. LeCun, Y.; Bottou, L.; Bengio, Y.; Haffner, P. Gradient-based learning applied to document recognition. *Proc. IEEE* **1998**, *86*, 2278–2324. [CrossRef]

7. He, K.; Zhang, X.; Ren, S.; Sun, J. Deep residual learning for image recognition. In Proceedings of the IEEE Conference on Computer Vision and Pattern Recognition, Las Vegas, NV, USA, 27–30 June 2016; pp. 770–778.

8. Alom, M.Z.; Taha, T.M.; Yakopcic, C.; Westberg, S.; Hasan, M.; Esesn, B.C.V.; Awwal, A.A.S.; Asari, V.K. The History Began from AlexNet: A Comprehensive Survey on Deep Learning Approaches. *arXiv* **2018**, arXiv:1803.01164.

9. Theckedath, D.; Sedamkar, R.R. Detecting Affect States Using VGG16, ResNet50 and SE-ResNet50 Networks. *SN Comput. Sci.* **2020**, *1*, 79. [CrossRef]

10. Yan, K.; Wang, Y.; Liang, D.; Huang, T.; Tian, Y. CNN vs. SIFT for Image Retrieval: Alternative or Complementary? In Proceedings of the 24th ACM International Conference on Multimedia; Association for Computing Machinery, Amsterdam, The Netherlands, 15–19 October 2016; pp. 407–411. [CrossRef]

11. Pouyanfar, S.; Sadiq, S.; Yan, Y.; Tian, H.; Tao, Y.; Reyes, M.P.; Shyu, M.L.; Chen, S.C.; Iyengar, S.S. A Survey on Deep Learning: Algorithms, Techniques, and Applications. *ACM Comput. Surv.* **2018**, *51*. [CrossRef]

12. Komkov, S.; Petiushko, A. AdvHat: Real-world adversarial attack on ArcFace Face ID system. *arXiv* **2019**, arXiv:1908.08705.

13. Wang, H.; Wang, Y.; Zhou, Z.; Ji, X.; Li, Z.; Gong, D.; Zhou, J.; Liu, W. CosFace: Large Margin Cosine Loss for Deep Face Recognition. *arXiv* **2018**, arXiv:1801.09414.

14. Mokhayeri, F.; Granger, E. Video Face Recognition Using Siamese Networks with Block-Sparsity Matching. *IEEE Trans. Biom. Behav. Identity Sci.* **2020**, *2*, 133–144. [CrossRef]

15. Cheng, G.; Yang, C.; Yao, X.; Guo, L.; Han, J. When Deep Learning Meets Metric Learning: Remote Sensing Image Scene Classification via Learning Discriminative CNNs. *IEEE Trans. Geosci. Remote Sens.* **2018**, *56*, 2811–2821. [CrossRef]

16. Sanakoyeu, A.; Tschernezki, V.; Buchler, U.; Ommer, B. Divide and Conquer the Embedding Space for Metric Learning. In Proceedings of the IEEE Conference on Computer Vision and Pattern Recognition (CVPR), Long Beach, CA, USA, 16–20 June 2019.

17. Feng, F.; Wang, S.; Wang, C.; Zhang, J. Learning Deep Hierarchical Spatial-Spectral Features for Hyperspectral Image Classification Based on Residual 3D-2D CNN. *Sensors* **2019**, *19*, 5276. [CrossRef] [PubMed]

18. Obara, K.; Yoshimura, H.; Nishiyama, M.; Iwai, Y. Low-resolution person recognition using image downsampling. In Proceedings of the 2017 Fifteenth IAPR International Conference on Machine Vision Applications (MVA), Nagoya, Japan, 8–12 May 2017; pp. 478–481. [CrossRef]

19. Buda, M.; Maki, A.; Mazurowski, M.A. A systematic study of the class imbalance problem in convolutional neural networks. *Neural Netw.* **2018**, *106*, 249–259. [CrossRef] [PubMed]

20. Ioffe, S.; Szegedy, C. Batch normalization: Accelerating deep network training by reducing internal covariate shift. *arXiv* **2015**, arXiv:1502.03167.

21. Theano. Categorical Cross Entropy. Available online: deeplearning.net/software/theano (accessed on 28 April 2020).

22. Zhang, Z.; Sabuncu, M.R. Generalized Cross Entropy Loss for Training Deep Neural Networks with Noisy Labels. *arXiv* **2018**, arXiv:1805.07836.

23. Wattenberg, M.; Viégas, F.; Johnson, I. How to Use t-SNE Effectively. *Distill* **2016**. [CrossRef]

24. Kingma, D.P.; Ba, J. Adam: A method for stochastic optimization. *arXiv* **2014**, arXiv:1412.6980.

25. Kaggle. Humpback Whale Identification Challenge. Available online: kaggle.com/c/humpback-whale-identification (accessed on 28 April 2020).

26. Schroff, F.; Kalenichenko, D.; Philbin, J. Facenet: A unified embedding for face recognition and clustering. In Proceedings of the IEEE conference on computer vision and pattern recognition, Boston, MA, USA, 7–12 June 2015; pp. 815–823.

Article

An Approach to Hyperparameter Optimization for the Objective Function in Machine Learning

Yonghoon Kim and Mokdong Chung *

Department of Computer Engineering, Pukyong National University, Pusan 48513, Korea;
kimyhjava@pukyong.ac.kr
*   Correspondence: mdchung@pknu.ac.kr; Tel.:+82-51-629-6253

Received: 24 September 2019; Accepted: 26 October 2019 ; Published: 1 November 2019

Abstract: In machine learning, performance is of great value. However, each learning process requires much time and effort in setting each parameter. The critical problem in machine learning is determining the hyperparameters, such as the learning rate, mini-batch size, and regularization coefficient. In particular, we focus on the learning rate, which is directly related to learning efficiency and performance. Bayesian optimization using a Gaussian Process is common for this purpose. In this paper, based on Bayesian optimization, we attempt to optimize the hyperparameters automatically by utilizing a Gamma distribution, instead of a Gaussian distribution, to improve the training performance of predicting image discrimination. As a result, our proposed method proves to be more reasonable and efficient in the estimation of learning rate when training the data, and can be useful in machine learning.

Keywords: bayesian optimization; gaussian process; learning rate; acauisition function; machine learning

---

1. Introduction

At Google's I/O 2017 conference, its CEO, Sundar Pichai, made some rather striking comments on AutoML. He said "AutoML means machine learning designed by machine learning". Since then, they have opened the Cloud AutoML site, offering automated machine learning related to sight, language, and structured data. An important question is how freely our model can train the data. The parameters in machine learning can be considered the final result after learning, which are determined by many tests. Therefore, AutoML should be able to estimate the variables in advance for machine learning, or estimate these parameters for learning.

Related research on AutoML has typically considered automated feature learning [1], architecture search [2], and hyperparameter optimization; where hyperparameter optimization includes optimizing the Learning Rate, Mini-batch Size, and Regularization Coefficient. Therefore, it must be decided which are the most appropriate values for each model's learning rate, mini-batch size, and normalization coefficient which should be set in advance for learning. However, in most cases, the default parameters of the existing researchers are used as they are.

The learning rates used in the AlexNet [3] model and for various learning models using CNN since 2011 have been defined in previous studies. Table 1 below shows the parameters of currently existing machine learning methods. However, even with this, if these values are used as they are, it will be difficult to derive optimal learning results because the data sets used in previous studies are different from actual data sets [4]. To solve this problem, we will consider optimizing the hyperparameters using grid search and random search [5]. However, there is a problem: a large number of result values must be derived and compared. In order to calculate the most appropriate learning rate with a minimum preliminary result value, a Tree-structured Parzen Estimator (TPE) [6] has been studied. Regarding optimization, techniques using Taub search [7] or other methods [8,9] have been presented.

Table 1. Examples of hyperparameters for machine learning.

Category	Learning Rate	Epochs	Batch Size	Regularization Coefficient
AlexNet	Gradient Descent (lr = 0.01)	10	128	Random Normalization
GoogleNet	SGD (lr = 0.1, decay = 1 $\times 10^{-6}$, momentum = 0.9)	Related to steps	64	Alpha = 0.001, beta = 0.75
ResNet	SGD (if batch size <8192, plr = 0.5, epochs = 5; elif batch size <16,384, plr=10.0, epochs = 5; else: (Depends on batch size)			Mean = 127, Stddev = 60

Bayesian optimization estimates the parameter distribution using prior values. The most typical Gaussian distribution is used, and this is called the Gaussian Process [10]. Each parameter has an individual problem, which contributes towards solving the multidimensional problem. In this regard, Bayesian optimization has been studied in relation to the Manifold Gaussian Process (mGP) [11] in higher dimensions, and Bayesian optimization using exponential distributions has also been actively researched. Therefore, various researchs have been conducted to automatically search configuration coefficients such as learning rate using Bayesian optimization. The most common algorithm is to estimate the hyperparameter using loss values [12].

In this paper, we attempt to optimize several parameters, based on Bayesian optimization. For this, we focus on the automation of selecting the learning rate at each epoch by utilizing a Gamma distribution. The exponential and gamma distributions are based on the Poisson distribution, and the Poisson distribution is used in the case where n is large and p is small in X to $B(n, p)$, following the binomial distribution [13]. Therefore, when the learning rate is estimated, it is judged to be the most similar. In Section 2, we show the related works on Bayesian optimization and Acquisition Functions. In Section 3, we describe an objective function called a black box. In Section 4, the results of an experiment on the MNIST data set are presented, and we propose the method of validation in Section 5, and we prove the experiment for the proposed searching technique in Section 6. Finally, we conclude in Section 7.

2. Related Work

2.1. Bayesian Optimization

The most common use of Bayesian Optimization (BO) is to solve global optimization problems, where the objective is related to a black box function [4]. In this regard, a number of approaches for this kind of global optimization have been studied in the literature [14–17]. Stochastic approximation methods, such as interval optimization and branch and bound methods, are efficient in optimizing unknown objective functions in machine learning [11]. Therefore, hyperparameter optimization [10] in machine learning refers to values set before learning, and BO can serve as an alternative to one of the optimization methods for setting these values automatically.

The general objective is to find the optimal solution x^* which maximizes the function value $f(x)$ using an unknown objective function f which receives an input value x, where the actual objective function is unknown [18,19]. However, we need two things to examine the function values sequentially for the input value candidates and find the optimal solution x^* which maximizes $f(x)$: The first is a surrogate model, which performs probabilistic estimation of the type of unknown objective function based on input values and function values investigated so far. The second is composed of an Acquisition Function, which derives the optimal input value x^* based on the probabilistic estimation results up to present. Gaussian Processes (GPs) [20] have used in probabilistic models, which have been widely used as Surrogate Models [10]. Gaussian Processes provide models for Gaussian distributions, as well as several other random variables commonly used in probabilistic statistics. The relevant model is shown in Equation (1):

$$f(x) \sim gp(m(x), k(x, x^{'})),\tag{1}$$

where the function f is distributed as a GP with mean function m and covariance function k.

2.2. Acquisition Functions for Bayesian Optimization

An Acquisition Function is based on the probability of improvement over the incumbent $f(x^+)$, where $x^+ = argmax_{x_i \in x_{1:t}} f(x_i)$; which appears as [18] Equation (2)

$$\begin{aligned} PI(X) &= P(f(x) \geq f(x^+)), \\ &= \Phi\left(\frac{\mu(x) - f(x^+)}{\sigma(x)}\right). \end{aligned}\tag{2}$$

This function is called Maximum Probablity of Improvement (MPI), or P-algorithm. However, its performance can be improved by the addition of a trade-off parameter $\xi \geq 0$, as shown in Equation (3):

$$\begin{aligned} PI(X) &= P(f(x) \geq f(x^+) + \xi), \\ &= \Phi\left(\frac{\mu(x) - f(x^+) - \xi}{\sigma(x)}\right). \end{aligned}\tag{3}$$

With regards to the theory, few researchers have studied the impact of different values of ξ in certain domains [14,21,22]. An Acquisition Function [10] is a function which recommends the next function value candidate $x(t + 1)$ to investigate, based on the probabilistic estimated results up that point. Among exploration points, exploitation is the strategy of looking near the point where the function value is maximum, and Exploration is the strategy of looking where there is the possibility that the optimal input value x^* may exist. Expected Improvement (EI) denotes the appropriate use of these two strategies. Trying many observations using these functions is more efficient than using the objective function directly. In order to find the optimum saddle, many observations are required. When observation is performed using the objective function, a lot of time and resources are required.

Therefore, faster observation can be achieved by using an Acquisition Function. In fact, it has been shown that, the higher the observation point, the more reliably the observation point can be estimated [23]. The related equation is shown in Equations (4) and (5).

$$EI(x) = \begin{cases} (\mu(x) - f(x^+) - \xi)\Phi(Z) + \sigma(x)\phi(Z) & if\ \sigma(x) > 0 \\ 0 & if\ \sigma(x) = 0 \end{cases},\tag{2}$$

where

$$Z = \begin{cases} \dfrac{(\mu(x) - f(x^+) - \xi)}{\sigma(x)} & if\ \sigma(x) > 0 \\ 0 & if\ \sigma(x) = 0 \end{cases}.\tag{3}$$

In Equation (4), $\Phi(.)$ and $\phi(.)$ are the Cumulative Distribution Function (CDF) and Probability Density Function (PDF) of a standard normal distribution, respectively.

Figure 1 shows, simply how the Bayesian Optimization (BO) operates, where it is assumed that it is predicted for a one-dimensional continuous input; the figure on the top shows the objective function and the bottom figure describes the Acquisition Function. The Objective function, which is Black Box we have to estimate, consists of two initial points and a deriviation. If we, then, continuously add the x, we can compute y. However, as shown already, this method involves a high cost in terms of time and performance. So, we typically first use the Acquisition Function. The y value (high point of the blue line) of the Acquisition Function is calculated corresponding to the selected x value, and checked to determine whether it is the optimal point. Currently, the area marked with a red area is the most likely high point candidate. Then, to decide the candidate point, it is finally computed for the real result. If it is not a saddle point, we return to the first process. This advantage of the process is that we can avoid calculating the black box directly; this mean that, by only computing the Acquisition Function, we can predict the type of distribution that the black box has.

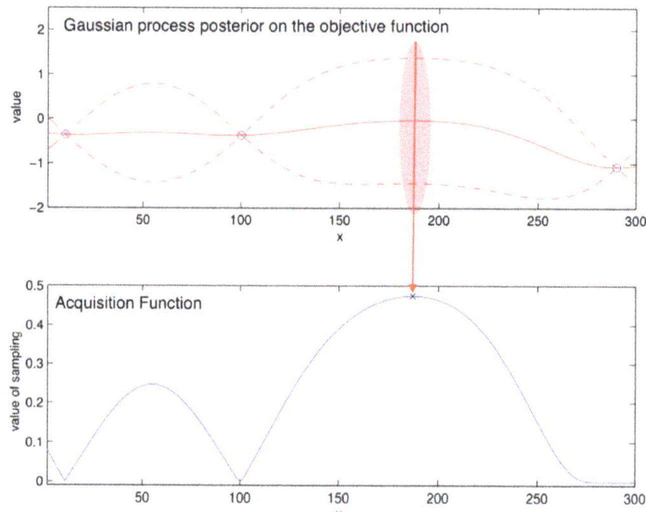

Figure 1. Illustration of Bayesian Optimization, maximizing an objective function f with a one-dimensional continuous input [20].

2.3. Gaussian Distribution and Gamma Distribution

The Gaussian distribution is used to represent continuous value distributions in discrete values in binomial distributions, such as selecting a coin's side, and a distribution with an average of 0 and a standard deviation of 1 is called the standard normal distribution.

The Gamma distribution represents the time taken for a total number of Poisson events which occur on average λ times per unit time, and is used under the normal data distribution assumption. The Poisson distribution is used when the event number N is large and the probability P is low, and is used for discrete distributions. For continuous variables, the Gamma distribution is used [13].

3. Proposed Object Function

An object function is the function to be predicted finally. In this paper, we need to predict the function of the result of the MNIST learning module. In previous papers, Bayesian Optimization (BO) has been used to predict the accuracy by returning accuracy and setting the relationship between this value and learning rate as a function, then setting this as the objective function. However, the problem in the existing studies is that the value varied greatly, even with a small change in the learning rate, due to the sensitivity of the accuracy value. This means that the graph was not stable, as a whole. Therefore, we applied the loss value, which is typically used to evaluate well-training in machine learning.

However, the result was worse than the existing accuracy. The reason for this is that the loss value is usually set to a low value. Also, low loss does not necessarily mean high accuracy. In order to solve this problem, we propose a method that considers the loss value and estimation accuracy, as follows (Equation (6)):

$$\kappa = \begin{cases} \lambda \times M_a(x) - log(M_l(x)) & if \ M_l(x) > 0 \\ 0 & if \ M_l(x) \leq 0 \end{cases}. \tag{4}$$

In Equation (6), M_a stands for accuracy, M_l stands for loss, λ was inserted as the adjusted value in the range from 0 to 1, and log was applied to the M_l function as we could not estimate the range of the loss value. This is because we hope that M_l will have the lowest value and that M_a will have the highest value. For this, we subtract M_l from M_a to determine the satisfied result; note that the result may be negative.

This study is related to the Acquisition Function, rather than the Surrogate Model of BO. In existing Expected Improvement (EI), the distribution of the Surrogate Model (SM) is referred to, and the area expected to have the next highest value is searched for, as shown in Figure 2.

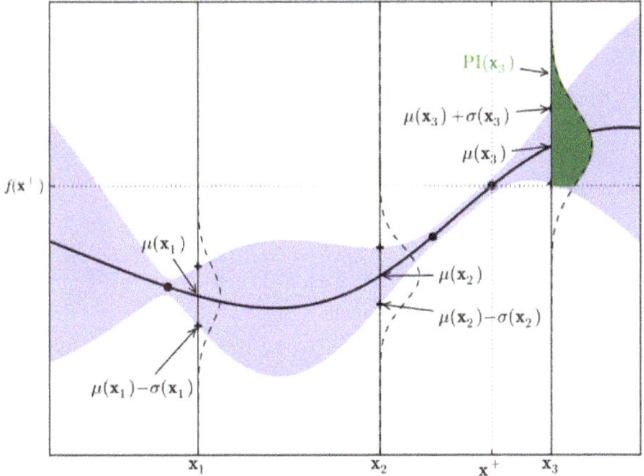

Figure 2. An example of the region of probable improvement [18].

As shown in Figure 2, the Gaussian Process (GP) in Figure 3 shows areas for EI.

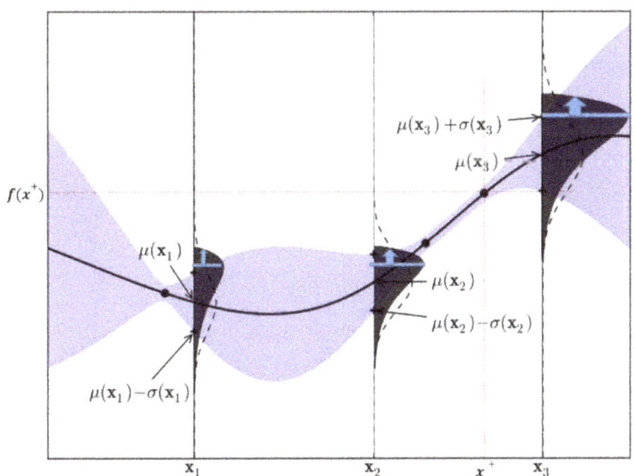

Figure 3. An example of region of probable improvement using a Gamma Distribution [24].

The maximum observation is x^+. In the overlapping Gaussian, over the dotted line, the dark shaded area can be used as a measure of improvement, $I(x)$. In this model, sampling at x_3 is more likely to be improved at $f(x^+)$, as compared to at x_1. However, using a gamma distribution rather than a Gaussian distribution results in a higher $PI(x)$ value, which results in better results in the objective function of the MNIST model. In this study, the same method was applied, as follows: Figure 4 shows the test for BO, where it can be seen that the optimal value was found 10 consecutive times. The noise-free object function used in the test was a curve with a paddle point (left) and a local point (right), and The black points in this curve describe the newly discovered location in the test. As shown in Figure 5, it can be confirmed that the sixth saddle point was found in the existing study. In Figure 4, the distance value (left) and the calculation value of EI (right) are shown.

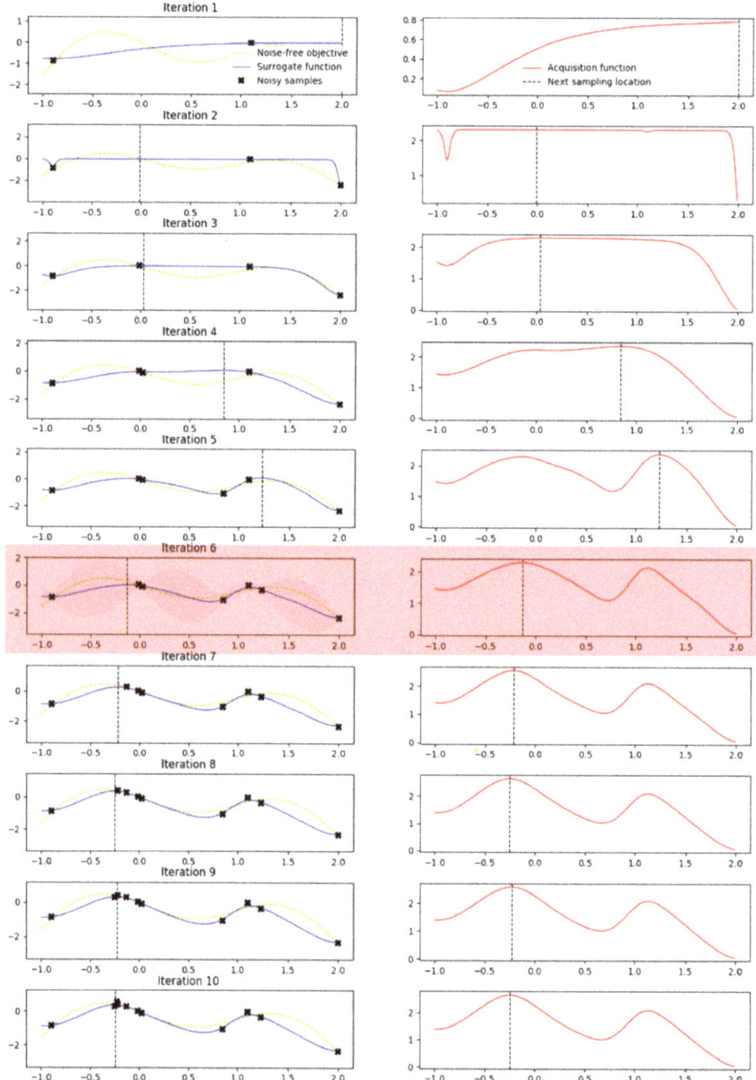

Figure 4. Tested of Gaussian Process with Gaussian distribution.

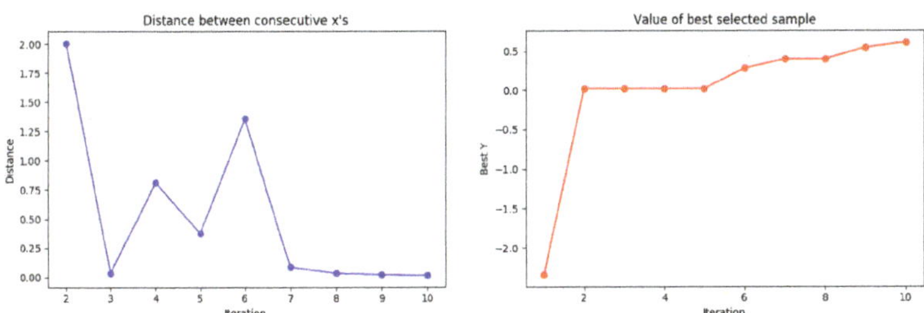

Figure 5. Tested results of distances and the best selected samples to Gaussian distribution.

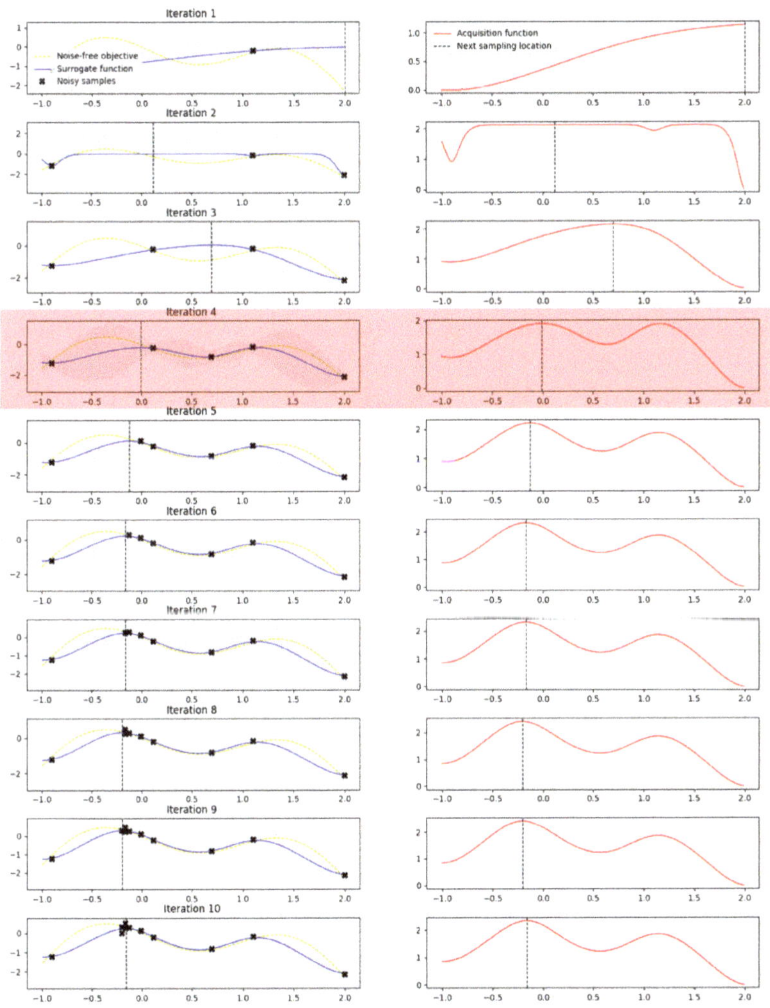

Figure 6. Tested of Gaussian process with Gamma distribution.

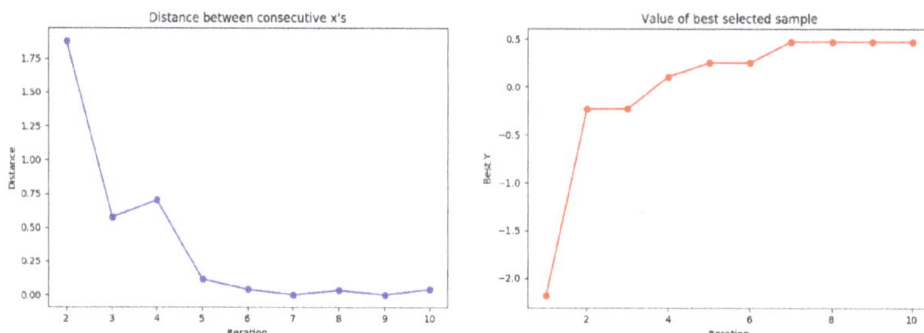

Figure 7. Tested results of distances and the best selected samples to Gamma distribution.

In Figure 6, we can see that the fourth highest distance value in the left figure was found to be higher than the local point, following which, we search around the paddle point and find the highest point. Figures 6 and 7 show the results when applying the method presented in this study, from which it can be confirmed that the paddle point was already found in the fourth iteration, unlike the conventional method. In other words, since it proceeds in the form of searching left to right within the entire range, it was easily found that there is a position higher than the local point. The distance calculation for each round can be confirmed to be decreasing flatter than when the Gaussian distribution was used. The reason for measuring the distance is that, the greater the distance from the estimated point, the greater the uncertainty will be. In this regard, it is related to the Exploration–exploitation trade-off. Additionally, in sampling, we can see that the value has been changed from the fourth to the higher value, which shows that the sub-probability between the points in the general GP was passive and, so, the local maximum may be particularly useful in many graphs [18].

As shown in Figure 7, at the third time point the actual local point was passed, but it was found that the paddle point was more stable than the existing one. In other words, the proposed method shows that the Gamma distribution can converge faster by using the existing Gaussian process as a method of determining the next measurement point.

4. Experiment on MNIST

We applied grid search, random search, and Bayesian optimization to MNIST and compared the Gaussian Process of Bayesian optimization with the Gaussian and Gamma distributions. In addition, a MNIST model objective function is proposed as a method of applying the loss value and accuracy together in the existing technique. In order to obtain the result quickly, the training data was limited to 500, and the verification data was set to 20%. In other words, the experiment was conducted with the purpose of providing the values for constructing the objective function. The graph part of some programs was detailed in Martin Krasser's Blog, and the Gaussian Process Regression (GPR) was provided by the scikit-learn package and PyOpt. The detailed conditions of MNIST, which are generally used, are shown in Table 2.

Table 2. Circumstances of test on MNIST.

Circumstances of Test	Contents
Data Set	500 images
Verification Data	Total data set of 20%
Hidden Layer	6
Loss Function	Cross Entropy Error
Optimizer	SGD
Activative Function	ReLU
Initial Value of Weight	Kaiming He

Figure 8 shows a graph of the MNIST learning module in a grid method using 100 LRs. Among them, graphs of the accuracy value of the verification data as the *y* value are shown above, and graphs of the loss value as the *y* value are shown below. The bounds of LR were set from 0–0.5.

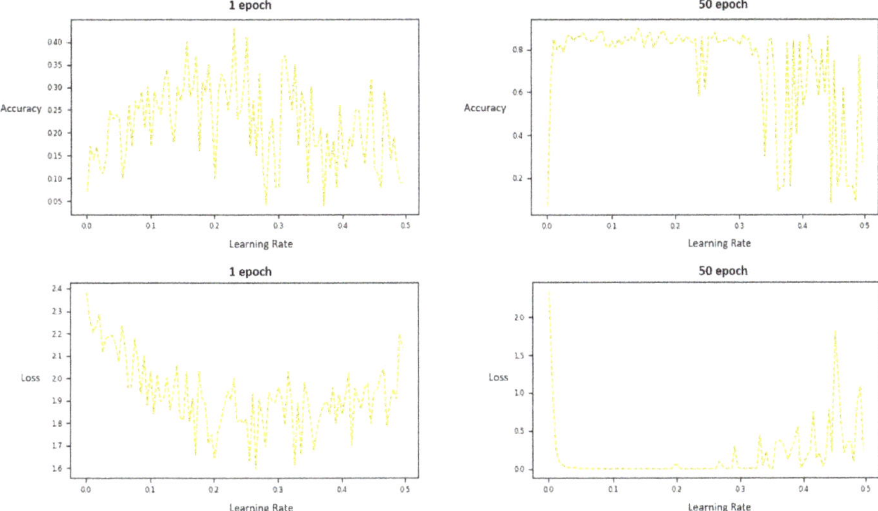

Figure 8. Estimation graph of objective function for the MNIST model. The 1 epoch and 50 epoch graphs in the top row are objective functions with accuracy as *y* values, and the two plots in the bottom row show the loss values.

As can be seen from the above Figure 8, a well-trained model indicates the highest accuracy or the lowest loss value. Importantly, there were many local maxima in the graph, and it is important to overcome these local maxima and find the saddle point. Generally, training evaluation in the machine learning module uses the loss value. However, in this study, the estimation accuracy and loss values were considered together, as shown in Figure 9. Figure 8, it shows a graph with κ as *y* values for 100 LR values. The highest accuracy is 90 and the corresponding LR value is 0.13. Compared to the case of using only the existing loss value or the accuracy only, it that can be seen that the local maxima disappeared significantly before the LR = 0.2 point. By using two values at the same time, the local maxima of the objective function can be minimized; this makes the objective function easier to estimate.

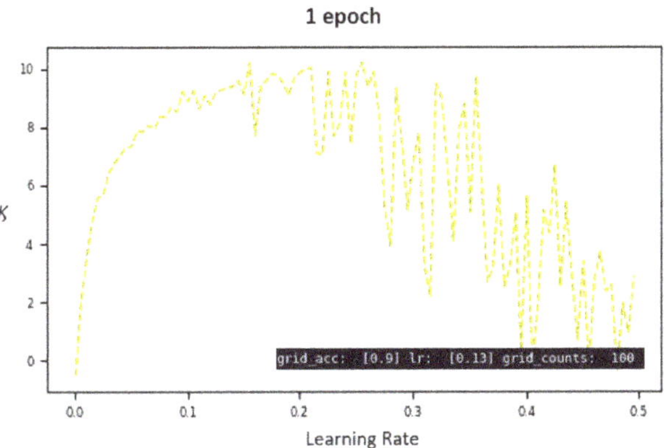

Figure 9. Estimation graph of objective function to MNIST model using κ.

Therefore, the method presented in this paper is of great importance, and the results are explained below, based on the accuracy comparison in Section 5 and the comparison using Gaussian distribution and Gamma distribution. In addition, the graphs after LR 0.2 seem problematic. However, if LR shows the highest accuracy at 0.13, it may be excluded; but further research is needed.

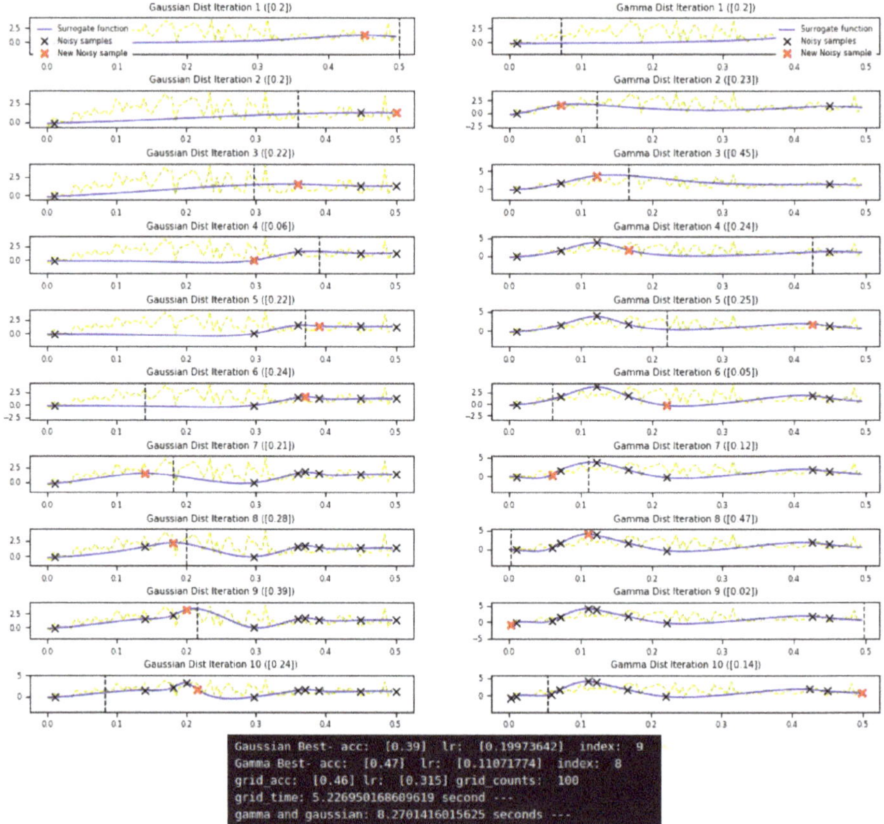

Figure 10. Comparison of Gaussian distribution and Gamma distribution at 1 epoch. The yellow dashed line shows non-free observations of the objective function f at 300 points; that is, the graph of the y value with an input of 100 x values (grid type) directly for the objective function. The blue line is an estimated graph of the objective function for the x values, which is 10 times sequentially calculated using the Acquisition Function. The red x points represent the final predicted points for each step.

Figure 10 shows the results at 1 epoch, using MNIST as the objective function. The left side shows the results of using the Gaussian distribution search and the right side shows the results of using the Gamma distribution search. The purpose of this paper is to estimate the distribution of the objective function of the MNIST module using the minimum number of estimates. Regarding the minimum number of times between the Gaussian and Gamma distribution searches, we could confirm that it had an advantage over the existing results. However, compared with the grid search, existing research has been applied only to the search times, such that a clear comparison between the searching techniques could not be carried out. In this regard, more accurate analysis would be possible if the comparison was made based on accuracy and, thus, the maximum difference value could be compared, with a comparison of when similar accuracy values were derived. In addition, the experiment was conducted by utilizing κ, as in described Section 3 of this paper. As shown in Figure 10, the accuracy of the proposed method was 47%, and the number of search iterations was the smallest in finding the highest accuracy. The method using the Gaussian distribution search was 39%, and the number of search iterations was nine. Finally, for the grid method, the accuracy was 46% and the number of iterations was 100.

In other words, the number of iterations needed for a similar level of accuracy were lower than the Gaussian distribution search, as the difference of accuracy was 7% and difference in number of iterations was one. In addition, the LR of the proposed method was about 0.11, where grid search showed a result of about 0.32. In this case, it is not clear whether the optimal position of the LR was 0.11 or around 0.32.

As described above, it was found that the Local maximum frequently occurred again after a certain LR value. Considering this, the accurate saddle point could be estimated as 0.11.

Figure 11 shows the results at 50 epochs. For the proposed technique, the accuracy was 94% and the LR was about 0.12; for the Gaussian distribution search, the accuracy was 93% and the LR was about 0.2. In the case of the grid search, the accuracy was 93% and the LR was 0.16. As a result, in the case of grid search, an estimated 100 iterations were required, but the proposed method had similar effects with about only 7 iterations.

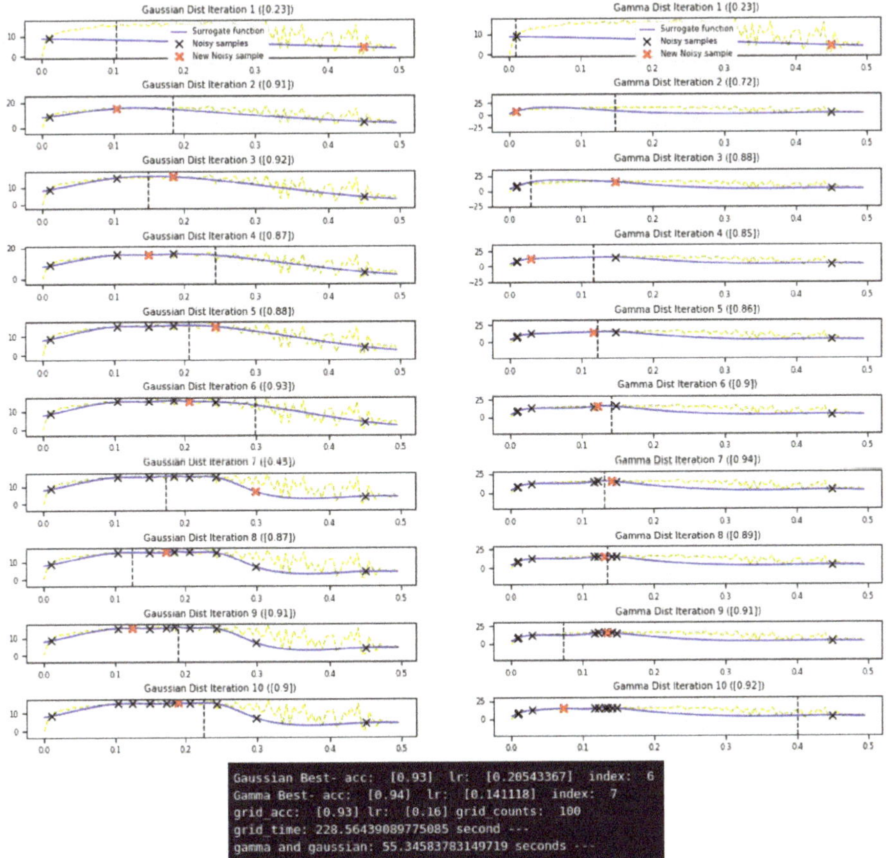

Figure 11. Comparison of Gaussian distribution search and Gamma distribution search at 50 epochs.

5. Performance Evaluation

In terms of performance evaluation, the evaluation metrics used to measure the predictive performance of the model include the sensitivity, specificity, accuracy, and MCC (mathew correlation coefficient) of the evaluation. TP, FP, TN, and FN are shown as true positive, false positive, true negative, and false negative, respectively. In addition to the following equations, this study evaluates the analysis speed based on the time required for learning to the level that the accuracy is matched based on the number of grids [25].

$$Sensitivity = \frac{TP}{TP + FN} \tag{5}$$

$$Specificity = \frac{TN}{TN + FP} \tag{6}$$

$$Accuracy = \frac{TP + TN}{TP + FP + TN + FN} \tag{7}$$

$$MCC = \frac{(TP \times TN) - (FP \times FN)}{\sqrt{(TP + FP)(TP + FN)(TN + FP)(TN + FN)}} \tag{8}$$

6. Evaluation

The grid search, Gamma distribution search, and Gaussian distribution search methods were compared to the existing methods (except for random search, as the uncertainty in the method is large). The actual random search method is excellent, in terms of learning speed, but the number of epochs or steps may increase infinitely when learning a lot of data. In particular, the final evaluation was made by comparing the accuracy with and without the added κ value in Table 3 and Figures 11 and 12.

Table 3 shows the comparative evaluation of the existing system and the proposed system. The proposed method can improve the performance by applying κ in the previous research to derive more convex output value from the objective function. Epoch did not proceed further. Because the evaluated data is the small data, further progression does not affect the accuracy (use small data). However, there is no shortage in comparing the proposed system with the existing system. In other words, it is possible to confirm the high value in terms of accuracy, sensitivity, or specificity.

Figures 12 and 13 are visual representations of the results in Table 3, with the vertical axis representing the actual label and the horizontal axis representing the value of the result inferred from learning. The darker the color, the more inconsistent.

Table 3. Comparison of random search, Gaussian distribution search, and Gamma distribution search using 400 traing data set and 100 validation data set.

Epoch	Category	Existing System				Proposed System			
		Sens	Spec	Acc	MCC	Sens	Spec	Acc	MCC
	Grid Search	43.69	93.94	47.0	35.61	46.98	94.61	52.0	38.20
1	Gaussian Distn Search	42.78	93.42	42.0	35.48	43.05	94.40	50.0	41.23
	Gamma Distn Search	44.02	94.00	46.0	38.90	47.46	94.42	50.0	39.89
	Grid Search	90.04	98.67	88.0	87.61	91.76	99.02	91.0	89.50
2	Gaussian Distn Search	89.37	98.55	87.0	86.86	88.70	98.70	88.0	85.92
	Gamma Distn Search	89.38	98.56	87.0	86.83	90.13	98.80	89.0	87.36

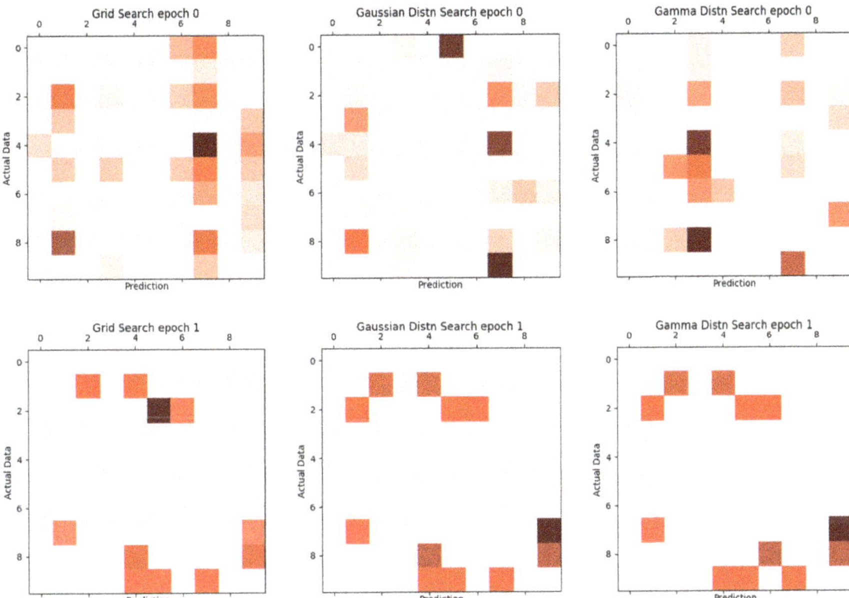

Figure 12. Existing comparison of grid search, Gaussian distribution search, and Gamma distribution search.

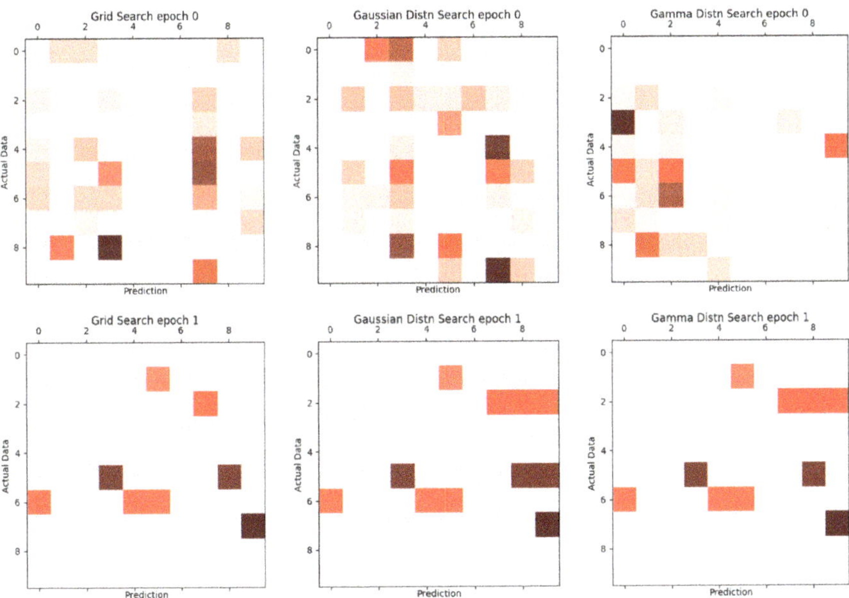

Figure 13. Proposed comparison of grid search, Gaussian distribution search, and Gamma distribution search.

Table 4 shows the training results using 60,000 training data and 10,000 verification data in units of epoch. From the learning results, it can be seen that the Gamma distn search method shows the maximum 98.36% in 2 Epoch. This table is not compared with previous studies, but the results were smaller than the current results without κ. Compared with the general CNN, it shows that the learning accuracy is high. In Table 4, 1 epoch means that is 10 epoch to the same learning model because we try to 12 optimization per 1 epoch. Thus, the above results represent accuracy for a total of 24 epochs. Figure 14 shows the result of Table 4.

Table 4. Comparison of random search, Gaussian distribution search, and Gamma distribution search using 60,000 traing data set and 10,000 validation data set. Detailed results are in Appendix A.

Epoch	Category	Sens	Spec	Acc	MCC
	Grid Search	95.59	99.51	95.62	95.10
1	Gaussian Distn Search	95.73	99.53	95.75	95.25
	Gamma Distn Search	95.61	99.52	95.64	95.12
	Grid Search	98.15	99.80	98.18	97.97
2	Gaussian Distn Search	98.11	99.79	98.13	97.91
	Gamma Distn Search	98.35	99.82	98.36	98.17
	Grid Search	97.81	99.76	97.83	97.58
3	Gaussian Distn Search	98.19	98.21	98.21	98.00
	Gamma Distn Search	98.21	99.80	98.23	98.03

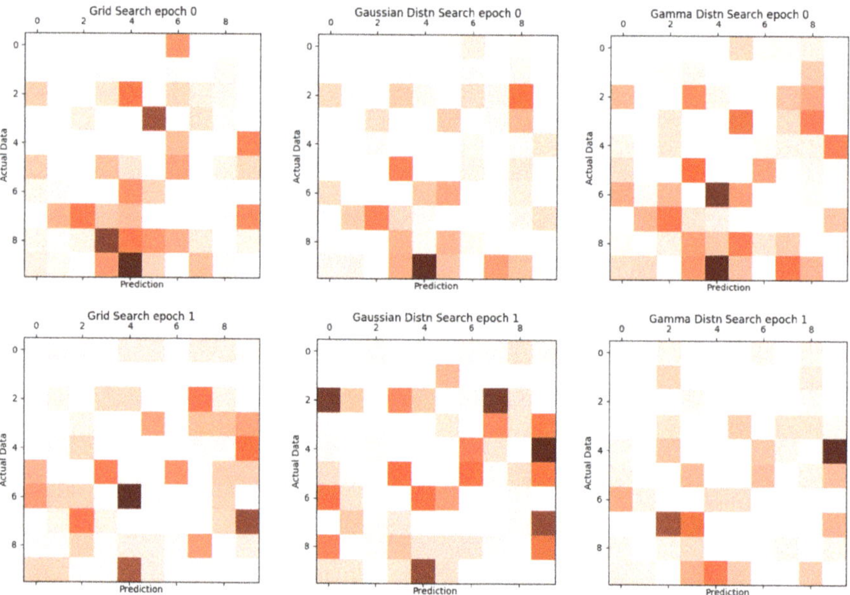

Figure 14. Proposed comparison of grid search, Gaussian distribution search, and Gamma distribution search.

Table 5 shows the results of examining the increase and decrease for each epoch using CIFAR-10. The model applied the model suggested by keras and shows a learning effect of 79% at a maximum of 50 epochs. This data set was chosen to increase the discrimination of the proposed method regardless of the model. As the result, the method presented in this paper shows that the accuracy increases regardless of the model type, and shows high values of sensitivity and specificity. Figure 15 shows the agreement for each epoch.

Table 5. Comparison of keras CNN, Gaussian distribution search of keras CNN, and Gamma distribution search of keras CNN using 50,000 traing data set of CIFAR-10 and 10,000 validation data set of CIFAR-10.

Epoch	Category	Sens	Spec	Acc	MCC	Loss
	Keras CNN	48.76	94.41	49.09	42.69	1.4160
1	Keras CNN+Gaussian Distn Search	32.37	92.37	29.95	21.83	1.8700
	Keras CNN+Gamma Distn Search	46.69	94.03	45.79	39.25	1.4728
	Keras CNN	73.36	96.80	70.72	68.19	0.8619
10	Keras CNN+Gaussian Distn Search	56.64	95.20	56.30	51.14	1.2100
	Keras CNN+Gamma Distn Search	74.35	97.06	73.07	95.12	0.7942
	Keras CNN	76.32	97.29	75.30	72.73	0.7133
20	Keras CNN+Gaussian Distn Search	65.43	96.18	65.30	95.25	0.9930
	Keras CNN+Gamma Distn Search	76.71	97.36	76.06	73.48	0.6998
	Keras CNN	76.76	97.31	75.55	73.15	0.7362
50	Keras CNN+Gaussian Distn Search	63.84	95.98	63.45	59.19	1.0278
	Keras CNN+Gamma Distn Search	83.22	98.10	82.78	80.92	0.5169

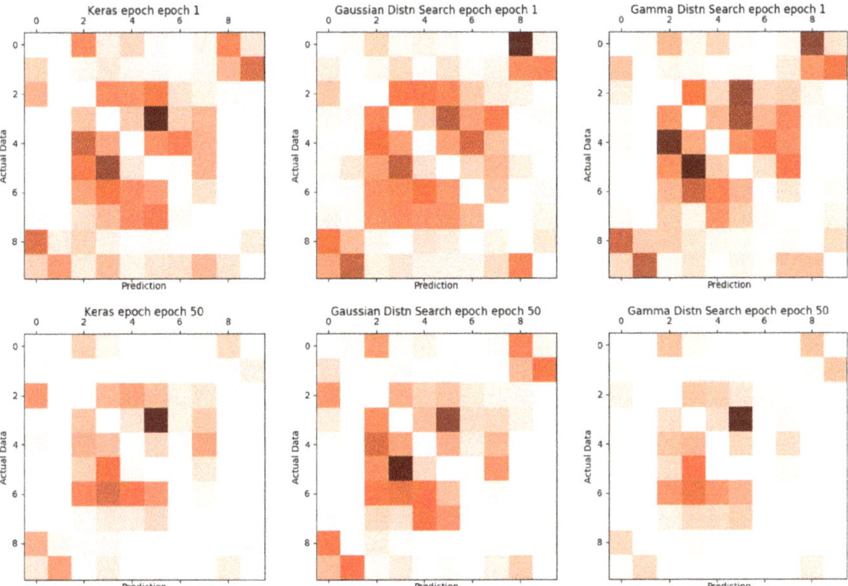

Figure 15. Proposed comparison of grid search, Gaussian distribution search, and Gamma distribution search.

Table 6 shows the comparison of the grid search, Gaussian distribution search, and Gamma distribution search methods, to which 10 optimizations and 1 epoch were applied. We can see that, the larger the grids, the more the accuracy was increased in the grid search; however, the time required increased accordingly. When comparing the averages, the time required for the Gaussian distribution search and the proposed search were almost unchanged and similar, but the accuracy was more than 7% improved, which was in agreement with that suggested in the previous studies.

Table 6. Comparison of grid search, Gaussian distribution search and Gamma distribution search, applying 10 counts of optimization and 1 epoch. LT, learning time.

Number of Grid	Existing System				Proposed System	
	Grid Search		Gaussian Distn Search		Gamma Distn Search	
	Acc (%)	LT (Sec)	Acc (%)	LT (Sec)	Acc (%)	LT (Sec)
100	47	5.3	42	4.4	52	4.3
200	49	10.4	42	4.1	48	4.2
300	50	16.6	42	4.7	47	4.7
400	52	20.1	37	5.1	41	4.9
500	49	25.6	39	4.6	51	4.3
Avg	49.4	15.6	40.4	4.58	47.8	4.48

Table 7 describes the comparison of the grid search, Gaussian distribution search, and Gamma distribution search, with 10 optimizations and 50 epochs. In particular, even though the number of grids increased, the accuracy was not affected significantly. Comparing the average for grid size from 100–500, there was a 0.7% difference between the suggested search and the grid search. However, the time of grid search was 729.9 s, and the suggested search took 29.38, which was about 24.8 times faster. Additionally, in comparison with the Gaussian distribution search, it was confirmed that the proposed method showed a 2% increase in accuracy. To confirm the improvement when comparing both the loss value and the accuracy estimate, the experiment was conducted again with 20 optimizations. The results are shown in Table 8.

Table 7. Comparison of grid search, Gaussian distribution search, and Gamma distribution search, applying 10 counts of optimization and 50 epochs. LT, learning time.

Number of Grid	Existing System				Proposed System	
	Grid Search		Gaussian Distn Search		Gamma Distn Search	
	Acc (%)	LT (Sec)	Acc (%)	LT (Sec)	Acc (%)	LT (Sec)
100	87	226.2	87	27.0	90	27.1
200	90	443.5	89	27.2	88	26.9
300	91	679.2	87	26.8	89	27.0
400	89	881.3	86	25.9	89	26.1
500	87	1419.3	81	39.2	84	39.8
Avg	88.8	729.9	86	29.22	88	29.38

As mentioned earlier, Table 8 will be compared with existing research data, and we will also study LR. It is important to see whether the actual optimized point was a saddle point or not. However, for points obtained by grid search, we can think of them as an approximations to a saddle point. In other words, with high accuracy and similar LR to a saddle point, we can expect to find a more accurate point if we increase the number of optimizations.

Comparing the average values in Table 7, κ can be used to confirm the overall improvement in learning speed, compared to previous studies. Compared with grid search, the time difference was much higher than the accuracy difference. When looking at the accuracy of time, the existing method was faster and had higher accuracy. Furthermore, the results of the proposed method were higher in accuracy than the Gaussian distribution search. In addition, in the case of LR, the proposed method had a value closer to that of LR of grid search than the Gaussian distribution search; in other words, the proposed search method was superior to the existing systems, in terms of accuracy and required time.

Table 8. Comparison of grid search, Gaussian distribution search, and Gamma distribution search applying 20 counts of optimization and 50 epochs. LT, learning time; LR, learning rate.

Number of Grid	Existing System						Proposed System		
	Grid Search			Gaussian Distn Search			Gamma Distn Search		
	Acc (%)	LT (Sec)	LR	Acc (%)	LT (Sec)	LR	Acc (%)	LT (Sec)	LR
100	93	215.9	0.085	92	26.1	0.307	93	27.7	0.265
200	92	459.1	0.070	88	28.55	0.075	91	27.8	0.113
300	94	647.58	0.156	91	27	0.117	92	26.7	0.178
400	94	899.84	0.096	93	28.1	0.138	92	27.9	0.077
500	94	1124.46	0.078	93	26.58	0.186	94	26.4	0.230
Avg	93.4	669.38		91.4	27.266		92.4	27.3	

7. Conclusions

At present, Google is indispensable for machine learning, and even more so for AutoML. In view of recent issues, it is expected that AutoML will become pivotal in future machine learning. Variables directly related to AutoML include learning rate, mini-batch size, and normalization coefficients, and these variables have always been created and distributed by the machine trainers. In this regard, when the data set is changed and needs to be retrained, the problems that need to be reviewed for the hyperparameters and the decisions about variables expected by the actual person or guessed are closely related to the time, cost, and performance. For this, unfortunately, a full understanding of the learning parameters, expert experience, and numerous experiments are required. However, if possible, the smart way to solve this problem is to let the machine obtain these parameters for you.

In this study, we attempted to find a solution for the learning rate, among these problems. Therefore, we reviewed the commonly used random search, grid search, and Bayesian optimization methods by applying the well-known MNIST data, and presented a method to apply a Gamma distribution to the acquisition function for Bayesian optimization.

In addition, we presented a method for evaluating the learning rate by using both the accuracy and loss values in the sampling, in order to analyze the distribution of the objective function of the MNIST model.

Although the results of some experiments showed higher values than the existing methods, as seen in the Section 6, it was proved that the proposed search technique presented better results, in most evaluations.

In addition, we can confirm that the result of using the κ function has a convex distribution; if we can predict a more convex distribution, we can expect to find a more reasonable saddle point. In further research, we will study the construction of the more convex models, among the studies on the values output from the black box.

Author Contributions: Conceptualization, Y.K. and M.C.; methodology, Y.K. and M.C.; software, Y.K.; validation, Y.K. and M.C.; formal analysis, Y.K. and M.C.; investigation, Y.K. and M.C.; resources, Y.K.; data curation, Y.K.; writing—original draft preparation, Y.K.; writing—review and editing, Y.K. and M.C.; visualization, Y.K.; supervision, M.C.; project administration, M.C.; funding acquisition, Y.K. and M.C.

Funding: This research was supported by Basic Science Research Program through the National Research Foundation of Korea (NRF) funded by the Ministry of Education (NRF-2017R1D1A1B03030033).

Acknowledgments: The first draft of the paper was presented in the section "Deep Learning and Application" of The 15th International Conference on Multimedia Information Technology and Applications (MITA2019) held by Korean Multimedia Society(KMMS) and University of Economics and Law(UEL).

Conflicts of Interest: The authors declare no conflict of interest.

Appendix A. Training Progress Detailed Results

Gamma Distn Search Maximum Acurracy Detail

————————epoch————————(step: 0) gamma-bounds: [[1.e-04 5.e-01]]''' gaussian-bounds: [[1.e-04 5.e-01]] grid-bounds: [[1.e-04 5.e-01]] Gamma Best- acc: [0.9564] lr: [0.15058182] index: 3 Gaussian Best- acc: [0.9575] lr: [0.20195701] index: 3 grid-acc: [0.9562] lr: [0.110078] grid-counts: 100 grid-time: 1525.9515149593353 second — gamma gaussian: 320.44497871398926 seconds — ————————Gamma———————— accuracy: 0.99128 confusion: 4545.6265772727265 precision: 0.9561963371512057 recall: 0.9560669819496841 mcc: 0.9512168890174788 sensitivity: 0.9560669819496841 specificity: 0.9951604430746006 ————————Gaussian———————— accuracy: 0.9914999999999999 confusion: 4545.906891110488 precision: 0.9573654579832793 recall: 0.9573670489406423 mcc: 0.9525345363214628 sensitivity: 0.9573670489406423 specificity: 0.9952837099770188 ————————Grid———————— accuracy: 0.99124 confusion: 4545.906947140898 precision: 0.956117721154721 recall: 0.9558912053696996 mcc: 0.9510239135134221 sensitivity: 0.9558912053696996 specificity: 0.9951402792741156 ————————epoch————————(step: 1) gamma-bounds: [[0.06023273 0.21081454]] gaussian-bounds: [[0.0807828 0.28273981]] grid-bounds: [[0.0440312 0.1541092]] Gamma Best- acc: [0.9836] lr: [0.06023273] index: 0 Gaussian Best- acc: [0.9813] lr: [0.0807828] index: 0 grid-acc: [0.9818] lr: [0.06934914] grid-counts: 100 grid-time: 3392.1185035705566 second — gamma gaussian: 308.5484368801117 seconds — ————————Gamma———————— accuracy: 0.99672 confusion: 4562.803546984866 precision: 0.9835683589709514 recall: 0.9835058725009371 mcc: 0.9817028741071395 sensitivity: 0.9835058725009371 specificity: 0.9981785294864007 ————————Gaussian———————— accuracy: 0.99626 confusion: 4545.908262967948 precision: 0.981316407524821 recall: 0.9810898959265562 mcc: 0.9791059831360185 sensitivity: 0.9810898959265562 specificity: 0.997922830382359 ————————Grid———————— accuracy: 0.9963599999999999 confusion: 4545.908146741083 precision: 0.981879115324593 recall: 0.9815012472351764 mcc: 0.9796568657792945 sensitivity: 0.9815012472351764 specificity: 0.9979768937659553 ————————epoch————————(step: 2) gamma-bounds: [[0.02409309 0.08432582]] gaussian-bounds: [[0.03231312 0.11309592]] grid-bounds: [[0.02773966 0.0970888]] Gamma Best- acc: [0.9823] lr: [0.02409309] index: 0 Gaussian Best- acc: [0.9821] lr: [0.03550456] index: 6 grid-acc: [0.9783] lr: [0.02982013] grid-counts: 100 grid-time: 5223.958828687668 second — gamma gaussian: 343.67405676841736 seconds — ————————Gamma———————— accuracy: 0.9964599999999999 confusion: 4545.632518181818 precision: 0.9823391951737168 recall: 0.982118716591801 mcc: 0.9802569569051883 sensitivity: 0.982118716591801 specificity: 0.9980324760992 ————————Gaussian———————— accuracy: 0.99642 confusion: 4545.635281241946 precision: 0.9820085701478038 recall: 0.9819438346460284 mcc: 0.9799852874924644 sensitivity: 0.9819438346460284 specificity: 0.9980115536490326 ————————Grid———————— accuracy: 0.99566 confusion: 4546.037013372861 precision: 0.9781982768860729 recall: 0.978143086587466 mcc: 0.9757536738243268 sensitivity: 0.978143086587466 specificity: 0.9975895338444459 ————————epoch————————(step: 3) gamma-bounds: [[0.00963724 0.03373033]] gaussian-bounds: [[0.01420183 0.04970639]] grid-bounds: [[0.01192805 0.04174818]] Gamma Best- acc: [0.9813] lr: [0.02185781] index: 2 Gaussian Best- acc: [0.9812] lr: [0.0207021] index: 2 grid-acc: [0.9837] lr: [0.01312086] grid-counts: 100 grid-time: 7069.49636721611 second — gamma gaussian: 350.1422553062439 seconds — ————————Gamma———————— accuracy: 0.99626 confusion: 4545.63475035308 precision: 0.9812261638722205 recall: 0.9811773662104988 mcc: 0.9791226848339004 sensitivity: 0.9811773662104988 specificity: 0.9979225414069898 ————————Gaussian———————— accuracy: 0.9962399999999999 confusion: 4545.635264046948 precision: 0.9810652055934064 recall: 0.9810538040503666 mcc: 0.9789703109590817 sensitivity: 0.9810538040503666 specificity: 0.9979120643565299 ————————Grid———————— accuracy: 0.99674 confusion: 4546.036578721596 precision: 0.9836110493048572 recall: 0.9834916677629966 mcc: 0.9817387961353237 sensitivity: 0.9834916677629966 specificity: 0.998189550424739

References

1. Zhang, Z.; Jiang, T.; Li, S.; Yang, Y. Automated feature learning for nonlinear process monitoring–An approach using stacked denoising autoencoder and k-nearest neighbor rule. *J. Process Control* **2018**, *64*, 49–61. [CrossRef]

2. Zoph, B.; Le, Q.V. Neural architecture search with reinforcement learning. *arXiv* **2016**, arXiv:1611.01578.

3. Krizhevsky, A.; Sutskever, I.; Hinton, G.E. Imagenet classification with deep convolutional neural networks. In Proceedings of the Advances in Neural Information Processing Systems 25, Neural Information Processing Systems(NIPS), Lake Tahoe, NV, USA, 3–8 December 2012; pp. 1097–1105.

4. Jones, D.R.; Schonlau, M.; Welch, W.J. Efficient global optimization of expensive black-box functions. *J. Glob. Optim.* **1998**, *13*, 455–492. [CrossRef]

5.	Bergstra, J.; Bengio, Y. Random search for hyper-parameter optimization. *J. Mach. Learn. Res.* **2012**, *13*, 281–305.

6.	Bergstra, J.S.; Bardenet, R.; Bengio, Y.; Kégl, B. Algorithms for hyper-parameter optimization. In Proceedings of the Advances in Neural Information Processing Systems 24, Neural Information Processing Systems(NIPS), Granada, Spain, 12–17 December 2011; pp. 2546–2554.

7.	Guo, B.; Hu, J.; Wu, W.; Peng, Q.; Wu, F. The Tabu Genetic Algorithm A Novel Method for Hyper-Parameter Optimization of Learning Algorithms. *Electronics* **2019**, *8*, 579. [CrossRef]

8.	Je, S.M.; Huh, J.H. An Optimized Algorithm and Test Bed for Improvement of Efficiency of ESS and Energy Use. *Electronics* **2018**, *7*, 388. [CrossRef]

9.	Monroy, J.; Ruiz-Sarmiento, J.R.; Moreno, F.A.; Melendez-Fernandez, F.; Galindo, C.; Gonzalez-Jimenez, J. A semantic-based gas source localization with a mobile robot combining vision and chemical sensing. *Sensors* **2018**, *18*, 4174. [CrossRef] [PubMed]

10.	Snoek, J.; Larochelle, H.; Adams, R.P. Practical bayesian optimization of machine learning algorithms. In Proceeding of the Advances in Neural Information Processing Systems 25, Neural Information Processing Systems(NIPS), Lake Tahoe, NV, USA, 3–8 December 2012; pp. 2951–2959.

11.	Calandra, R.; Peters, J.; Rasmussen, C.E.; Deisenroth, M.P. Manifold Gaussian processes for regression. In Proceedings of the 2016 International Joint Conference on Neural Networks, Vancouver, BC, Canada, 24–29 July 2016; pp. 3338–3345.

12.	Bergstra, J.; Yamins, D.; Cox, D.D. Making a science of model search: Hyperparameter optimization in hundreds of dimensions for vision architectures. In Proceedings of the 30th International Conference on Machine Learning, Atlanta, GA, USA, 17–19 June 2013; pp. 115–123.

13.	DeGroot, M.H.; Schervish, M.J. *Probability and Statistics*; Pearson Education Limited: London, UK, 2014; pp. 302–325.

14.	Törn, A.; Žilinskas, A. *Global Optimization*; Springer: Berlin, Germany, 1989.

15.	Mongeau, M.; Karsenty, H.; Rouzé, V.; Hiriart-Urruty, J.B. Comparison of public-domain software for black-box global optimization. *Optim. Methods Softw.* **1998**, *13*, 203–226. [CrossRef]

16.	Liberti, L.; Maculan, N. *Global Optimization: From Theory to Implementation*; Springer Optimization and Its Applications; Springer: Berlin, Germany, 2006.

17.	Zhigljavsky, A.; Žilinskas, A. *Stochastic Global Optimization*; Springer Optimization and Its Applications; Springer: Berlin, Germany, 2007.

18.	Brochu, E.; Cora, V.M.; De Freitas, N. A tutorial on Bayesian optimization of expensive cost functions, with application to active user modeling and hierarchical reinforcement learning. *arXiv* **2010**, arXiv:1012.2599.

19.	Mockus, J. Application of Bayesian approach to numerical methods of global and stochastic optimization. *J. Glob. Optim.* **1994**, *4*, 347–365. [CrossRef]

20.	Frazier, P.I. A tutorial on bayesian optimization. *arXiv* **2018**, arXiv.1807.02811.

21.	Jones, D.R. A taxonomy of global optimization methods based on response surfaces. *J. Glob. Optim.* **2001**, *21*, 345–383. [CrossRef]

22.	Lizotte, D. Practical Bayesian Optimization. Ph.D. Thesis, University of Alberta, Edmonton, AB, Canada, 2008.

23.	Pelikan, M.; Goldberg, D.E.; Cantú-Paz, E. BOA: The bayesian optimization algorithm. In Proceedings of the 1st Annual Conference on Genetic and Evolutionary Computation, Orlando, FL, USA, 13–17 July 1999; Volume 1, pp. 525–532.

24.	Kim, Y.; Chung, M. An Approch of Hyperparameter Optimization using Gamma Distribution. In Proceedings of the 15th International Conference on Multimedia Information Technology and Applications, University of Economics and Law, Ho chi minh, Vietnam, 27 June–1 July 2019; Volume 1, pp. 75–78.

25.	Le, N.Q.K.; Huynh, T.T.; Yapp, E.K.Y.; Yeh, H.Y. Identification of clathrin proteins by incorporating hyperparameter optimization in deep learning and PSSM profiles. *J. Comput. Methods Prog. Biomed.* **2019**, *177*, 81–88. [CrossRef] [PubMed]

MDPI

Article

Learning to See in Extremely Low-Light Environments with Small Data

Yifeng Xu [1,2], Huigang Wang [1,*], Garth Douglas Cooper [1], Shaowei Rong [1] and Weitao Sun [1]

[1] School of Marine Science and Technology, Northwestern Polytechnical University, Xi'an 710072, China; xuyifeng123@mail.nwpu.edu.cn (Y.X.); Coopergd1@mail.nwpu.edu.cn (G.D.C.); rsw1986@mail.nwpu.edu.cn (S.R.); Sunwt1223@gmail.com (W.S.)
[2] Jinhua Polytechnic, Jinhua 321017, China
* Correspondence: wanghg74@nwpu.edu.cn; Tel.: +86-029-88460521

Received: 25 May 2020; Accepted: 14 June 2020; Published: 17 June 2020

Abstract: Recent advances in deep learning have shown exciting promise in various artificial intelligence vision tasks, such as image classification, image noise reduction, object detection, semantic segmentation, and more. The restoration of the image captured in an extremely dark environment is one of the subtasks in computer vision. Some of the latest progress in this field depends on sophisticated algorithms and massive image pairs taken in low-light and normal-light conditions. However, it is difficult to capture pictures of the same size and the same location under two different light level environments. We propose a method named NL2LL to collect the underexposure images and the corresponding normal exposure images by adjusting camera settings in the "normal" level of light during the daytime. The normal light of the daytime provides better conditions for taking high-quality image pairs quickly and accurately. Additionally, we describe the regularized denoising autoencoder is effective for restoring a low-light image. Due to high-quality training data, the proposed restoration algorithm achieves superior results for images taken in an extremely low-light environment (about 100× underexposure). Our algorithm surpasses most contrasted methods solely relying on a small amount of training data, 20 image pairs. The experiment also shows the model adapts to different brightness environments.

Keywords: low light; image restoration; denoise; noise reduction; deep leaning; machine learning

---

1. Introduction

Image restoration is a challenging task, particularly in an extremely dark environment. For example, the recovery of images captured in a scene where the light is very dark is a difficult problem. There are two ways to solve the problem: relying on hardware and relying on software. In the aspect of hardware, the adjustment of the camera settings can partially solve the problem, but there are still difficulties: (1) a higher sensitivity value of the image sensor increases brightness, but it also increases high sensitivity noise, and, although the ISO value in the latest cameras can be set to 25,600, the images with an ISO over 400 should have more noise; (2) the larger aperture receives more light, but it also leads to worse sharpness and a shallower depth of field; (3) extending exposure time is one of the most direct solutions, but a little movement should lead to blurred imaging under the condition; (4) the larger size of the photosensitive element receives more photons, although the size of the photosensitive element is limited by camera size and cost; (5) using flash helps to capture more light, however, the flash range is limited and flash is forbidden in some situations.

In addition to the suitable hardware settings, some sophisticated algorithms have been designed to restore the images in the dark. Many denoising, deblurring, color calibration, and enhancement algorithms [1–5] are applied to low-light images. These algorithms only deal with normal low-light images but are inefficient for the extremely low-light images with brightness as low as 1 lumen.

An alternative method, burst alignment algorithms [6–8], uses multiple pictures taken continuously. However, this kind of method still loses efficacy on the pictures in extremely low-light conditions. Chen et al. [9] proposed a deep learning end-to-end method which can restore images with only 0.1 lumens. However, this method was trained with massive amounts of training data and needed a huge computational cost. It is well known the deep learning algorithms require big data. One of the reasons is that the quality of partial data is poor. By obtaining better training data, the algorithm can accomplish the same result with less data.

In the past, the low-light data were basically collected in the low-light environment, while the ground truth data were collected by long exposure. This collecting method had many difficulties and generated low-quality training data. We propose a new method to collect dark pictures and corresponding normal exposure pictures by adjusting camera settings in the daytime.

We use an end-to-end neural network to restore the extremely low-light image only by 20 high-quality image pairs based on the work of [9,10]. Our contributions in this paper are summarized as follows: (1) we propose a new low-cost method to capture high-quality image pairs, including dark pictures and corresponding normal exposure pictures; (2) we use the theory of the regularized denoising autoencoder to explain why the algorithm works effectively; (3) our proposed algorithm can restore images taken in an extremely dark environment (100× underexposure light level) by only using 20 image pairs.

The rest of this paper is organized as follows. Related work about low-light images restoration is proposed in Section 2. The image acquisition method and the framework of the neural network are shown in Section 3. The detailed experimental results are shown in Section 4. Several problems that require further research are put forward in Section 5. The paper concludes with Section 6.

2. Related Work

The restoration of low-light images has been extensively studied in the literature. In this section, we provide a short review of related work.

2.1. Low-Light Image Datasets

Well-known public image datasets, such as PASCAL VOC [11], ImageNet [12], and COCO [13] have played a significant role in traditional computer vision tasks. Because less than 2% of the images in these datasets were taken in a low-light environment, the public datasets were unsuitable for training low-light image restoration.

Many researchers proposed low-light image datasets. LLnet [14] darken the original images to simulate low-light images. The original images were used as a ground truth; the generated dark images were artificial. The authors of [5] proposed a new dataset of 3000 underexposed images, each with an expert-retouched reference. PolyU [15] collected real-world noisy images taken from 40 scenes, including indoor normal lighting, dark lighting, and outdoor normal-light scenes. However, the brightness of the images from these datasets is dusky, not extremely black. SID [9] proposed an extremely dark image dataset captured by Sony and Fuji cameras, including 5094 short exposure images and 424 long-exposure images. LOL [16] consists of 500 image pairs. The ExDARK [17] dataset is made up of images captured in real environments, containing various objects. Because the images need careful adjustment of the camera settings in dark conditions, the previously mentioned databases have two common problems: the high cost of collecting data and very few high-quality images.

2.2. Image Denoising

According noise is one of the significant obstacles in the restoration of low-light images; denoising is a notable subtask in low-light enhancement tasks. A classic traditional method of dealing with the low-light image is scaling luminosity and the followed denoising procedure. Image denoising is an often-traversed topic in low-level computer vision. Many approaches have been proposed, such as total variation [18], sparse coding [19], and 3D transform-domain filtering (BM3D) [20], etc.

These approaches are grouped into traditional denoising methods. Their effectiveness is often based on an image prior's information, such as smoothness, low rank, and self-similarity, etc. Unfortunately, most traditional methods only work more effectively for the synthetic noise data, such as salt pepper and Gaussian noise, but the performance of the methods sharply drops for noisy images taken in real-world environments.

Researchers have also widely explored the application of deep learning networks to image denoising. Important methods of deep learning include deep illumination estimation [5], end-to-end convolutional networks [9,21], autoencoders [22,23], and multi-layer perceptron [24]. Generally, most methods based on deep learning work better than traditional methods. However, the former has the requirement of huge amounts of training data.

Except for single image denoising, the alternative, multiple-image denoising [6,7,25], achieves better results since more information is collected. However, it is difficult to select the "lucky image" and the correspondence estimation between images. On some occasions, taking more than one image was infeasible.

2.3. Low-Light Image Restoration

One basic method is histogram equalization to expand the dynamic range of the images. The recent effort on low-light image restoration is the learning-based methods. For instance, the authors of [14] proposed a deep autoencoder approach. WESPE [26] proposed a weakly supervised image-to-image network based on a Generate Adversarial Network (GAN). However, WESPE was more focused on image enhancement. In addition, other methods include the approach based on dark channel prior [27], the wavelet transform [23], and illumination map estimation [28], etc. These methods mentioned above only deal with the images captured in a normal dark environment, such as dusk, morning, and shadow, etc. The end-to-end model proposed by Chen et al. [9] could restore extremely low-light images using RAW sensor data. However, its model was heavyweight. In sum, the current research suggests that either the image in the extreme dark cannot be restored or the algorithms require big data.

3. The Approach

3.1. The End-to-End Pipeline Based on Deep Learning

The pipelines based on deep learning and the traditional method can be used to restore low-light images. Two kinds of pipelines are shown in Figure 1. The deep learning model (the top sub-image) is an end-to-end method. This method generates a model from image pairs, while the traditional method cascades a sequence of low-level vision processing procedures, such as luminosity scaling, demosaicing, denoising, sharpening, and color correction, etc. In Figure 1, luminosity scaling and cBM3D are selected as the major procedures in the traditional method.

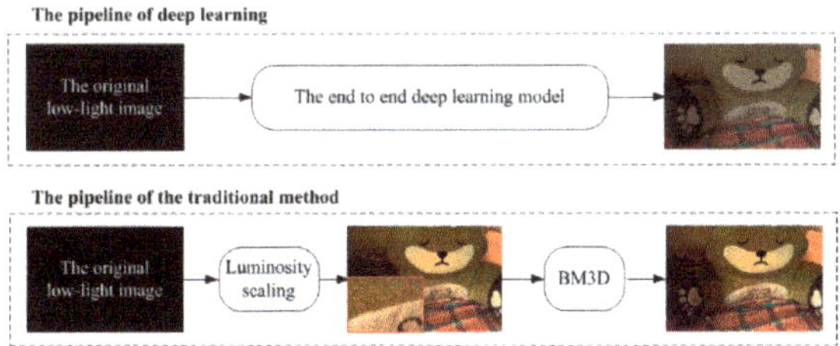

Figure 1. The two categories of pipelines concerning low-light restoration. The top sub-image shows the pipeline based on deep learning. The bottom sub-image shows a traditional image processing pipeline. The toy images on the right side are the results of the two pipelines, respectively. The sub-image surrounded by a red line box is the zoom-in image. BM3D = 3D transform-domain filtering.

In the pipeline of the traditional method, the first step is luminosity scaling. The images taken by the camera Nikon D700 are Nikon Electric Film (NEF) RAW images with 14 bits. This means that the maximum brightness value is 2^{14}, that is, 16,384. Due to the image in the extremely dark condition, the brightness values of the pixels are distributed between 1 and 50. The procedure of light scaling can be expressed as a formula: $v_x/v_{max} \times 16{,}384$. The parameter v_x in the formula represents the brightness value of a pixel, and v_{max} represents the max brightness value of all pixels. The simple luminosity scaling also amplifies the noise information in the images. The high noise is demoed in the zoom-in image surrounded by the red box after luminance scaling. The second step is noise reduction by BM3D.

In our work, the deep learning neuron network is proposed for direct single image restoration of extremely low-light images. Specifically, a convolutional neural network [29] U-net [10] is used for processing, inspired by the recent algorithms in the work of [9,10]. The structure of the network is shown in Figure 2, and the details about the structure are listed in Table 1.

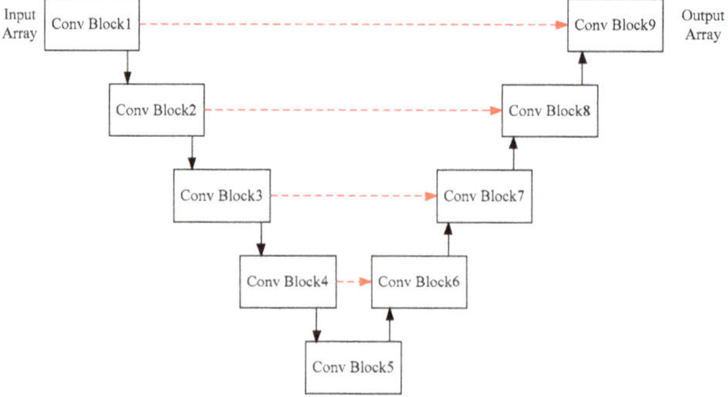

Figure 2. The structure of the network. The input array is a 4D data converted from the original RAW image. Convolutional block is abbreviated as "Conv Block", which represents a convolutional block including 2D convolutional layer and pooling layer. The red dotted arrow represents copy and crop operations.

Table 1. The parameters of the neuronal network. The parameter "32 [3,3]" represents that the output array size is 32 and the convolutional kernel size is 3×3.

ID of the Block	The First Layer	The Second Layer	The Third Layer
Block 1	Conv2d(32, [3,3])	Conv2d(32, [3,3])	Max pooling2d
Block 2	Conv2d(64, [3,3])	Conv2d(64, [3,3])	Max pooling2d
Block 3	Conv2d(128, [3,3])	Conv2d(128, [3,3])	Max pooling2d
Block 4	Conv2d(256, [3,3])	Conv2d(256, [3,3])	Max pooling2d
Block 5	Conv2d(512, [3,3])	Conv2d(512, [3,3])	none
Block 6	Conv2d(256, [3,3])	Conv2d(256, [3,3])	none
Block 7	Conv2d(128, [3,3])	Conv2d(128, [3,3])	none
Block 8	Conv2d(64, [3,3])	Conv2d(64, [3,3])	none
Block 9	Conv2d(32, [3,3])	Conv2d(32, [3,3])	Conv2d(12, [1,1])

3.2. Regularized Denoising Autoencoder

In our study, we selected an autoencoder neural network [30,31] similar to U-net to restore dark images. The autoencoder is a neural networks that is trained to attempt to map the input to the output. In other words, it is restricted in some ways to learn the useful properties of the data. It has many layers internally called the hidden layer. The network is divided into two parts: an encoder function $h = F(x)$ and a decoder function $G(h)$ which generates the reconstruction.

Regularized technology [32,33] is used to solve the invalidation of the over-complete autoencoder. The case is called over-complete when the hidden dimension is greater than the input. The over-complete autoencoders fail to learn anything useful if the encoder and decoder have a large number of parameters.

Regularized autoencoders use a loss function to learn useful information from the input. The useful information includes the sparsity of the representation, robustness to noise, and robustness to the missing input. In particular, the clear and real image data is useful information, hidden in the dark background. In one word, regularization enables the nonlinear and over-complete autoencoder to learn useful information about the data distribution.

The denoising autoencoder (DAE) [34,35] is one of the autoencoders with corrupted data as input and clear data as output by a trained model. The structure of a DAE is shown in Figure 3. It aims to learn a reconstruction distribution $p_{recontstruct}(y|x)$ by the given training pairs (x,y). The DAE minimizes the function $L(y,G(F(x)))$ to obtain the useful properties, where x is a corrupted data relative to the original data y. Specifically, in our study, data x indicates the images with dark noise and data y indicates the ground truth images.

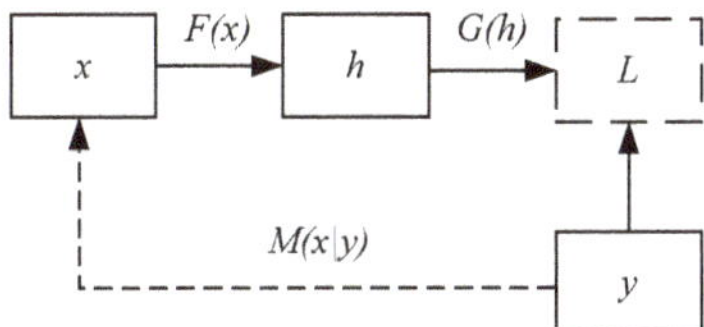

Figure 3. The structure of a denoising autoencoder (DAE). The input data tagged as x and output data tagged as y represent the noisy data and ground truth, respectably. The function $F(x)$ and $G(h)$ represent the encoder and decoder. The map $M(x|y)$ represents the procedure of generating x from y.

The first step is sampling y from the training data. The second step is sampling the corresponding data point x by $M(x|y)$. The third step is estimating the reconstruction distribution by $p_{decoder}(y|h)$ with h the output of encoder and $p_{decoder}$ defined by the decoder $G(h)$. DAE is a feedforward network and trained by the methods of any other feedforward network. We can perform gradient-based approximate minimization on the negative log-likelihood. For example, the stochastic gradient descent can be written by:

$$- E_{y \sim \hat{p}data(y)} E_{x \sim M(x|y)} \log p_{decoder}(y|h = F(x)) \tag{1}$$

In Equation (1), the $p_{decoder}$ is the distribution calculated by the decoder and the $\hat{p}data$ is the training data distribution.

3.3. The Procedure of Collecting Data

The traditional low-light enhancement methods cascade the procedures: scaling, denoising, and color-correcting. The traditional methods do not need the ground truth images during processing. On the contrary, a deep learning neural network must train the data before the testing phase. The training and testing phases are shown in Figure 4. The upper part of this figure shows that the training data consists of two parts: the low-light (dark) images and corresponding normal-light images. Every image pair in the two parts has the same size and shooting range and aligns pixel by pixel. There are only a few low-light image datasets available, an example from one of these datasets is seen in the upper left. The learned-based model can learn the fitting parameters to map the image pairs. The mapping relationship from the low-light images to normal-light images is non-linear, and thus deep learning is appropriate.

The bottom sub-image of Figure 4 shows the test phase. In the test phase, the low-light images are inputted into the trained model and the restored normal-light images are outputted. In addition, the output restored image (bottom right corner) is unknown, the other three kinds of images are known. Another significant point is that the training and the test images are independent.

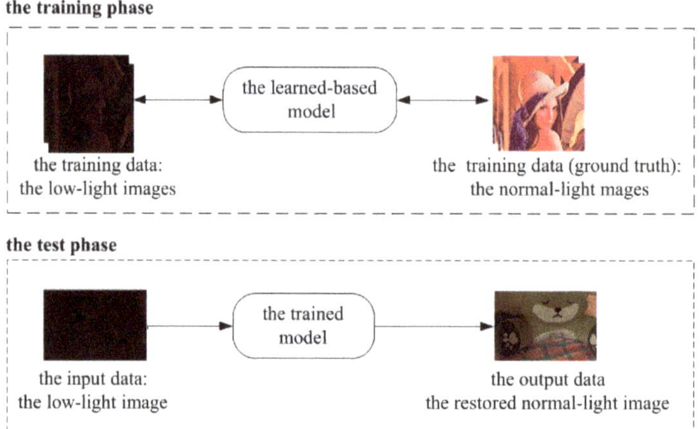

Figure 4. The schematic diagram of the train and test phase in the deep learning approach. The upper sub-image shows the training phase. The bottom sub-image shows the test phase using the trained model generated in the training phase.

In order to improve the effectiveness of the restoration, we can start from the two aspects of the algorithm and the training data. In this section, we describe the training data. Due to the large computational cost of the training data, data collection becomes an obstacle to the deep learning algorithms.

The data collecting method in other deep learning literature is shown as the upper sub-image of Figure 5. The low-light datasets based on deep learning in the computer vision community are almost collected in low-light conditions, shown as the left box of the upper of Figure 5. The corresponding ground truth images are also taken in low-light conditions, shown as the right box of the upper. For taking normal exposure images, the camera is set to a higher ISO, larger aperture, longer exposure time, larger light-sensing element, and flash. However, such settings reduce the quality of the images.

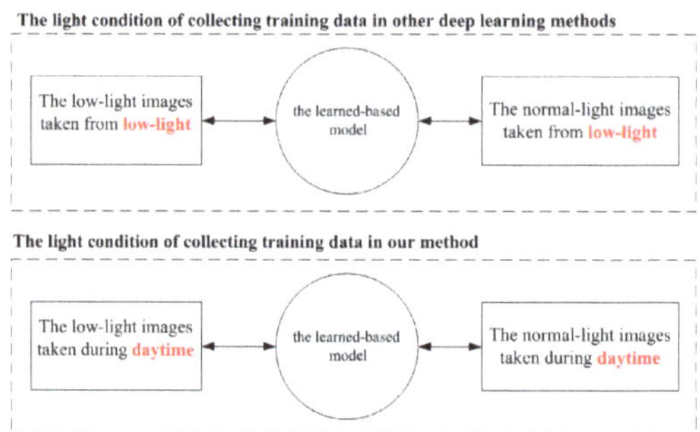

Figure 5. The different light conditions during the phase of collecting training images in deep learning. The upper sub-figure shows that the image pairs both captured in low-light conditions. In the below sub-figure, our proposed image acquisition method captures images during the daytime. In our method, the low-light images are taken in the normal-light condition.

We collected the training data during the daytime, normal-light condition, shown as the button sub-image of Figure 5. We named the collecting method the NL2LL (normal-light to low-light). The environment with enough light brings convenience to shoot high-quality data. Three pillars of the photography: shutter speed, ISO, and aperture can be set as "better" parameters during the daytime.

We set the shorter exposure speeds, lower ISO, and larger aperture values to take the dark images during the daytime.

The influence of the aperture is rarely discussed in the relevant literature. Aperture is defined as the opening in the lens through which light passes to enter the camera. Small aperture numbers represent a large aperture opening size, whereas large numbers represent small apertures. The critical effect of the aperture is the depth of field. Depth of field is the amount of the photograph that appears sharp from front to back. According to the principles of optics, a larger aperture size (smaller aperture value) leads to a shallower depth of field and therefore more defocus blur.

The effect of the aperture size on the image depth of field is shown in Figure 6. The left half of this figure has a "thin" depth of field, where the background is completely out of focus. On the contrary, the right sub-image of Figure 6 has a "deep" depth of field, where both the foreground and the background are sharp; the camera focuses on the foreground. Because the zoom-in images surrounded by the solid red box are near the focus point, both images (the first and the third image at the bottom) are clear. However, the distance between the dotted red region and the camera is far from that between the focus point and the camera. According to the principles of optics, the background area away from the focus point becomes blurred with a large aperture camera setting. The zoom-in sub-image taken in the big aperture size (the second image at the bottom) is more blurred than the respective sub-image taken in the small aperture size (the fourth image at the bottom).

Figure 6. The influence of the aperture size on the image quality. The aperture of the left and right images is set as 4 (large aperture size) and 25 (small aperture size), respectively. The center of the red circle is the location of the focal point. The area surrounded by the solid red box and dotted red box is in foreground and background, respectively.

If the images are taken in dark conditions, the aperture must be set as large as possible to receive more light. The large aperture setting in the camera inevitably results in a large amount of background blur. The training dataset used in deep learning has many image pairs. Every image pair consists of a low-light image and a corresponding normal-light image. To achieve high-quality data, each pixel pair in both images must match one-to-one. Unfortunately, the blur of some pixels impacted the quality of the training data [9,14].

The large aperture size should blur the image. It is also plain to see that the longer exposure time and the higher ISO reduces the quality of the image pair. More specifically, it is harder to align the two images pixel by pixel. When there is less light at night, the exposure time must be increased to capture more light for the images. Then, during the day it is important to decrease the exposure time; this will in turn reduce the amount of light that enters the apparatus. The exposure time of the ground truth in the literature [9] is 10 s and 30 s, while the respective exposure time in our data is between 1/10 s to 3 s.

The parameter ISO plays the same role. The value of ISO can then be adjusted to the smaller value (the better quality) in the daytime and set to 100 in our first experiment.

Moreover, we adopted Wi-Fi equipment to remotely adjust the camera settings, and the camera was fixed on the tripod. The hardware devices ensure the stability of the camera while taking image pairs.

Most researchers consider that the training data used for low-light restoration methods based on deep learning must be collected in low-light conditions. On the contrary, our experiments have shown that the training data can be collected in normal-light conditions. As opposed to previous methods to photograph in a low-light environment, our proposed method takes images in a bright environment. Figure 7 shows all the training images in our experiment. The shooting parameters are listed in Table 2. Our algorithm achieves exciting results only using 20 image pairs. The camera parameters in our method make it easier to take a high-quality image.

Figure 7. All images in our dataset. Normal-light images (ground truth) are shown at the front. The low-light images (extremely dark) are shown behind.

Table 2. The EXIF parameters of the training images. The first ID column indicates the number of image pairs. The order of ID is from left to right of every line, then the next line, and so on. The second column EXIF indicates the photos' parameters. GT = ground truth. All the parameters of the dark images are as same as that of GT except shutter time.

ID	EXIF	ID	EXIF	ID	EXIF	ID	EXIF
1	GT: F16.0, 1/8 s, ISO 100 Dark: 1/800 s	2	GT: F16.0, 1 s, ISO 100 Dark: 1/100 s	3	GT: F16.0, 1/4 s, ISO 100 Dark: 1/400 s	4	GT: F16.0, 1/8 s, ISO 100 Dark: 1/800 s
5	GT: F16.0, 1/10 s, ISO 100, +1.0 EV Dark: 1/1000 s	6	GT: F16.0, 1/10 s, ISO 100, +1.0 EV Dark: 1/1000 s	7	GT: F16.0, 1/10 s, ISO 100, +1.0 EV Dark: 1/1000 s	8	GT: F16.0, 1/8 s, ISO 100, +1.0 EV Dark: 1/800 s
9	GT: F16.0, 1/4 s, ISO 100, +1.0 EV Dark: 1/400 s	10	GT: F16.0, 1/4 s, ISO 100, +1.0 EV Dark: 1/400 s	11	GT: F16.0, 1/4 s, ISO 100, +1.0 EV Dark: 1/400 s	12	GT: F16.0, 1/2 s, ISO 100, +1.0 EV Dark: 1/200 s
13	GT: F16.0, 1/5 s, ISO 100, +1.0 EV Dark: 1/500 s	14	GT: F16.0, 1/8 s, ISO 100, +1.0 EV Dark: 1/800 s	15	GT: F16.0, 1/8 s, ISO 100, +1.0 EV Dark: 1/800 s	16	GT: F16.0, 3 s, ISO 100, +1.0 EV Dark: 1/30 s
17	GT: F16.0, 3 s, ISO 100, +1.0 EV Dark: 1/30 s	18	GT: F16.0, 2 s, ISO 100, +1.0 EV Dark: 1/50 s	19	GT: F16.0, 1/4 s, ISO 100, +1.0 EV Dark: 1/400 s	20	GT: F16.0, 1/4 s, ISO 100, +1.0 EV Dark: 1/400 s

4. Experiments

4.1. Dataset

The images in our dataset were collected from real-world scenes instead of by the artificial brightness adjustment. The images in our training dataset were taken in a cloudy environment by a Nikon D700 made in Japan. The lens was the fixed-focus lens labeled as AF-S NIKKOR 50 mm f/1.4G. Because the RAW format can save more low-light information than the sRGB format, the images were saved in the RAW format. The data collection environments of various algorithms are shown in Table 3. The dataset was divided into the training data and the test data. All the training images were taken outdoors and selected randomly from the image pairs. The training data included the 20 image pairs shown in Figure 7. To avoid overfitting, the test images were taken separately in the low-light environment and independent of the training data. The test images were taken from a variety of scenes, such as a bedroom and outdoors. They were taken from the environment about 1 lumen, approximately 100 times lower brightness than that of the training images. Some of the test images are shown in Figure 8. The exposure time of the training pictures was reduced exactly 100 times compared to the test data picture. The training images and test images were independent and identically distributed (i.i.d.) to guarantee the effectiveness of our algorithm.

Table 3. The environments of collecting images in various algorithms.

Class of Methods	The Training Data	The Test Data
Traditional methods [20]	None	Low-light environment
Methods based on deep learning [5,9,14]	Low-light environment	Low-light environment
Our methods	Normal-light environment	Low-light environment

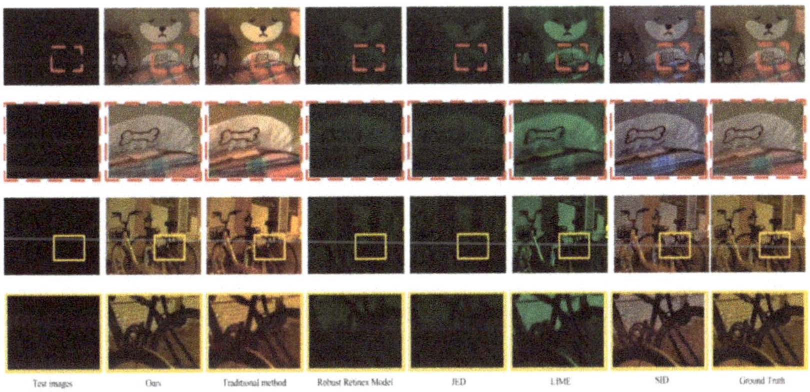

Figure 8. The results of the various algorithms. The images marked with a red dotted box on the second row are the zoom-in images of the region surrounded by the small red dotted box in line 1, respectively. The images surrounded by the orange boxes in the fourth line are also the magnified images from the solid orange box region in line 3. The indoor test image (upper left sub-image) was taken with the parameters: exposure compensation +1.0, f 6.1, 0.1 s and ISO 400. The outdoor test image (third line and first column) was taken with the parameters: f 3.5, 0.2 s and ISO 400. The first column shows the test images taken in an extremely dark condition. The second column shows our results by the end-to-end deep learning model. The third column shows the results by the traditional method which cascades luminosity scaling and BM3D [20]. The fourth to the seventh columns show the results by the Robust Retinex Model algorithm [36], by JED [37], by LIME [28] and SID [9]. The last column shows the ground truth taken with a long exposure time.

4.2. Qualitative Results and Perceptual Analysis

The methods as a comparative baseline include the traditional method and the "modern" method. The traditional method cascades a luminosity scaling algorithm and a denoising algorithm. BM3D was selected as the classic denoising algorithm in our research because it outperforms most techniques facing the images with real noise. The modern data-driven approach selects some machine learning algorithms that have been proposed in recent years.

The process of determining the level of image quality is called Image Quality Assessment (IQA). IQA is part of the quality of experience measures. Image quality can be assessed using two kinds of methods: subjective and objective. In the subjective method, the corresponding result images processed by different pipelines were presented to students, who determine which image had higher quality. The images were presented in a random order, with a random left–right order without any indication of the provenance. A total of 100 comparisons were performed by five students. The students found the results using the traditional method (the third column) still had a good aspect of light, but there were some yellow patches in the large white region and noise particles. Other "modern" algorithms [28,36,37] did not work well in the extremely dark real environments. Our results were superior in the aspects of the image contrast, color accuracy, dynamic range, and exposure accuracy. Our pipeline significantly outperformed the traditional method and the "modern" methods in the aspect of denoising and color collection, respectively. The results by the various algorithms are shown in Figure 8.

4.3. Quantitative Analysis

The objective IQA methods were used to quantitatively measure the results. The objective methods can be classified into full-reference (FR) methods, reduced-reference (RR) methods, and no-reference (NR) methods.

FR metrics try to assess an image quality by comparing it with a reference image (ground truth) that is assumed to have perfect quality. The classical FR methods, the peak signal-to-noise ratio (PSNR) and the structural similarity index (SSIM [38]) were selected in our quantitative analysis. The higher the value of the two FR methods, the better the image quality.

The ENIQA [39] and Integrated Local NIQE (IL-NIQE) [40] present the high-performance general-purpose NR IQA methods based on image entropy. IL-NIQE uses a feature-enriched completely blind image quality evaluator. NIQE [41] makes a completely blind image quality analyzer and is also one of the NR methods. SSEQ [42] evaluates image quality assessment based on spatial and spectral entropies without a reference image. The lower scores calculated by these NR methods represents better image quality.

The quantitative IQA of the experimental results is shown in Tables 4 and 5. The size of the toy image in Table 4 was set to 512 × 339 pixels. The size of the bicycle image in Table 5 was set to 512 × 340 pixels. The images in the PNG format were evaluated by the following IQA algorithms, except SID [9]. The original model in SID only accepted the 16-bit raw image taken by the Sony camera and 14-bit raw image taken by the Fuji camera. Our test images were taken with a Nikon DSLR D700 made in Japan. Therefore, we modified the test code of SID to accept our test images.

Table 4. The Image Quality Assessment (IQA) of the recovery results about the toy image. The first column shows the various IQA methods. The columns (from second to seventh) show the corresponding IQA scores of the image results by various algorithms. The last column shows the score of the ground truth. The bold font and "(1)" indicate the value that is the best score. The underline and "(2)" represent second place.

Assessment Methods	Our Method	The Traditional Method	Robust Retinex Model	JED	LIME	SID	Ground Truth
PSNR	27.4600 (2)	22.1908	12.5453	12.3313	14.3542	21.7659	**Inf (1)**
SSIM [38]	0.9479 (2)	0.8593	0.3019	0.2726	0.3442	0.5128	**1 (1)**
ENIQA [39]	0.0762 (2)	0.2004	0.5808	0.5499	0.3101	0.1900	**0.0709 (1)**
IL-NIQE [40]	34.3996 (3)	37.4977	71.1242	75.1293	70.9793	34.3299 (2)	**30.9900 (1)**
NIQE [41]	**4.6251 (1)**	5.5973	8.5458	7.6560	7.9087	5.1727	4.7387 (2)
SSEQ [42]	**5.3654 (1)**	29.0665	37.1942	19.3989	35.3465	30.2516	15.8142 (2)

Similar to the qualitative assessment, our method also exceeds most methods in quantitative analysis. The Robust Retinex Model, JED, and LIME can achieve better effect in this degree of light at dusk, but these methods do not work in extremely dark environments.

Table 5. The IQA of the recovery results about the bicycle image. The first column shows the various IQA methods. The columns (from second to seventh) show the corresponding IQA scores of the image results by various algorithms. The last column shows the score of the ground truth. The bold font and "(1)" indicate the value that is the best score. The underline and "(2)" represent second place.

Assessment Methods	Our Method	The Traditional Method	Robust Retinex Model	JED	LIME	SID	Ground Truth
PSNR	30.8136 (2)	21.1950	11.9917	11.9579	13.7948	23,4298	**Inf (1)**
SSIM [38]	0.9535 (2)	0.8106	0.1963	0.1832	0.2072	0.8374	**1 (1)**
ENIQA [39]	0.1320 (2)	0.2860	0.1476	0.2346	0.2048	0.1658	**0.0995 (1)**
IL-NIQE [40]	25.1430(3)	25.5302	50.0900	47.2871	46.7496	**23.6752 (1)**	24.7926 (2)
NIQE [41]	**3.6368 (1)**	4.1941	7.0819	5.7749	7.0317	3.7191 (2)	3.7534
SSEQ [42]	**15.6676(1)**	38.2989	35.0894	28.4950	42.2645	24.2338	19.7217 (2)

Due to the lack of reference images, the NR methods do not know which image is the ground truth. By NIQE and SSEQ (Line 5 and Line 6), the scores of our result even are a little better than that of ground truth. The close scores of ours and ground truth calculated by NIQE mean that our restored images are well.

Next, we analyzed the generalization performance of the network model. The generalization performance of a learning algorithm refers to the performance on out-of-sample data of the models learned by the algorithm. The test images in Figure 8 were shot in an environment that is as bright as that of the training images. In this experiment, the test images were taken with different camera parameters in an extremely low-light environment by different parameters. The camera metering system indicated the setting (F5, 50 s, ISO 400) can achieve normal exposure. The test images with different levels of darkness were collected by only adjusting the shutter time. For example, the exposure time was set to 0.5 second to 100 × less exposure, as shown in the second column of Figure 7. Similarly, the exposure times were set to 1 s, 1/4 s, 1/8 s, and 1/15 s, which means 50×, 200×, 400×, and 750× less exposure, respectively.

The curves are smooth in the second line of Figure 9. However, some line deformation can be found in the fourth line of Figure 9. The details of the line illustrate that our method has more powerful capabilities to restore details than the traditional method. The sub-images from the first to third column show that our algorithm adapts to the environments with varying degrees of darkness. Even in extreme exposure condition (750× less exposure), our method can restore the acceptable details.

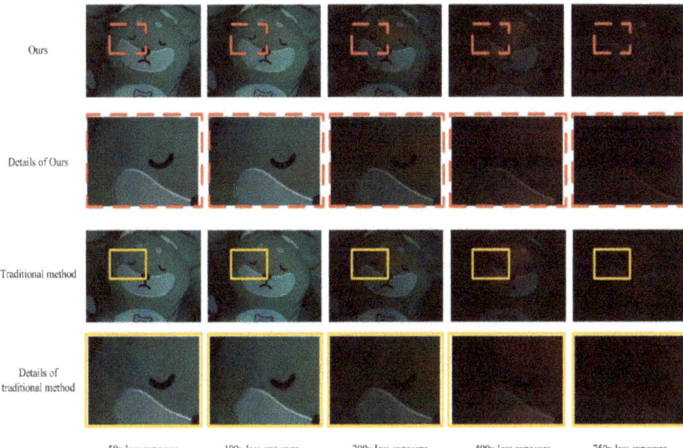

Figure 9. The restoration results in different brightness environments. The first line shows the restoration results of the test images with different black levels by our algorithm. The second line shows the details of the results. The bottom two lines show the results handled by traditional methods. The columns represent the different original brightness levels.

In addition to subjective analysis, several objective IQA methods were adopted to evaluate image quality. The quantitative analysis of the effect of different exposures is shown in Table 6. It is obvious to see our method surpasses the traditional approach.

Table 6. The IQA of the recovery results about the image with different exposures. In line 1, the word "n"× ("n" indicates a number) indicates n times reduction of exposure amount on the bases of normal exposure. The first column indicates the assessment methods and restoration algorithm. The traditional method is abbreviated as TM. Bold numbers represent the best results in this line. Underlined numbers indicate better results in different algorithms.

Assessment Methods (Restoration Algorithm)	50×	100×	200×	400×	750×
ENIQA [39] (Ours)	**0.0694**	0.0701	0.0800	0.1369	0.1856
ENIQA (TM)	0.2583	0.2318	0.2793	0.4141	0.5599
IL-NIQE [40] (Ours)	**36.5769**	38.7293	47.2164	58.6588	71.5074
IL-NIQE (TM)	50.1786	50.9244	62.1660	80.4942	96.9210
NIQE [41] (Ours)	6.1894	**5.9605**	6.0204	6.7881	7.0984
NIQE (TM)	7.5423	7.5152	8.4179	8.6997	9.1467
SSEQ [42] (Ours)	**2.4008**	4.0506	4.6036	9.4805	13.3380
SSEQ (TM)	21.6355	22.9971	19.5719	17.6767	15.8840

4.4. Implementation Details

In our study, the input images are the RAW images with the size H × W × 1. H and W are the abbreviations of the height and the width and are equal to 2832 and 4256, respectively. The input images were packed into four channels and correspondingly cut in half in each dimension. Thus, the size of the channels was 0.5 H × 0.5 W × 4. The packed data was fed into a neural network. Because the U-net network has a residual connection [43,44] and supports the full-resolution image in GPU memory, we selected a network with the architecture similar to U-net [10]. The output of the deep learning network was a 12-channel image with the size 0.5 H × 0.5 W × 12. Lastly, the 12-channel image was processed by a sub-pixel layer to recover the RGB image with the size H × W × 3.

Our implementation was based on TensorFlow and python. In all of our experiments, we used L1 loss, the Adam optimizer [45], and the Leaky ReLU (LRelu) [46] activation function. We trained the network for the Nikon D700 camera images. The initial learning rate was set to 0.0001, weight decay to 0.00001, and dampening to 0. The initial learning rate decreased according to the cosine function. According to the practical effect of the experiment, we set the training epoch as 4000.

5. Discussion

In this work, we shared a new collecting image method that can be used for future research on machine learning. Our algorithm was only trained with 20 image pairs and achieved the inspiring result in the restoration of the extremely low-light images with the help of a high-quality dataset. The method can be used in most supervised learning tasks.

In the future, we will try to improve our work in the following points. (1) It is known that Restricted Boltzmann Machines (RBMs) [47] can be used to preprocess the data and help the machine learning process become more efficient. The NL2LL model based on RMB might provide even better results. (2) Improved U-net networks can be used to improve performance. (3) The generalization performance of the method still needs to be studied. In our study, the recovered images were poor when the brightness of test images was magnified more than 400 times. Prior information about the environment light can be used for the algorithm. (4) The shooting environment can be extended to more complex scenarios, such as the condition with dark and blur and the condition with dark and scatting, etc. (5) The method can be used for low movement data (spatiotemporal data) and 3D objects (3D geometry).

Our research is of great significance in areas such as underwater robots and surveillance.

6. Conclusions

To see in the extreme dark, we have proposed a new method, NL2LL (collecting a low-light dataset in normal-light condition) to collect image pairs. The method has many potential implementations in convolutional neural network, dilated convolutional NN, regression, and graphical models. Our end-to-end approach is simple and highly effective. We have demonstrated its efficacy in low-light image restoration. The experiment shows that our approach can achieve inspiring results by only using 20 image pairs.

Author Contributions: All authors contributed to the paper. H.W. performed project administration; Y.X. conceived, designed and performed the experiments; Y.X. and G.D.C. wrote and reviewed the paper; S.R. performed the experiments; W.S. analyzed the data. All authors have read and agreed to the published version of the manuscript.

Funding: This research was funded by National Science Foundation of China (NSFC) grant No. 61571369. It was also funded by Zhejiang Provincial Natural Science Foundation (ZJNSF) grant No.LY18F010018. It was also supported by the 111 Project under Grant No. B18041.

Conflicts of Interest: The authors declare no conflict of interest.

References

1. Hu, Z.; Cho, S.; Wang, J.; Yang, M.H. Deblurring low-light images with light streaks. In Proceedings of the IEEE Conference on Computer Vision and Pattern Recognition, Columbus, OH, USA, 24–27 June 2014; pp. 3382–3389.
2. Remez, T.; Litany, O.; Giryes, R.; Bronstein, A.M. Deep Convolutional Denoising of Low-Light Images. *arXiv* **2017**, arXiv:1701.01687.
3. Zhang, X.; Shen, P.; Luo, L.; Zhang, L.; Song, J. Enhancement and noise reduction of very low light level images. In Proceedings of the 21st International Conference on Pattern Recognition (ICPR2012), Tsukuba, Japan, 11–15 November 2012.
4. Plotz, T.; Roth, S. Benchmarking denoising algorithms with real photographs. In Proceedings of the IEEE Conference on Computer Vision and Pattern Recognition, Honolulu, HI, USA, 21–26 July 2017; pp. 1586–1595.
5. Wang, R.; Zhang, Q.; Fu, C.W.; Shen, X.; Zheng, W.S.; Jia, J. Underexposed Photo Enhancement using Deep Illumination Estimation. In Proceedings of the IEEE Conference on Computer Vision and Pattern Recognition, Long Beach, CA, USA, 16–20 June 2019.
6. Hasinoff, S.W.; Sharlet, D.; Geiss, R.; Adams, A.; Barron, J.T.; Kainz, F.; Chen, J.; Levoy, M. Burst photography for high dynamic range and low-light imaging on mobile cameras. *ACM Trans. Graph.* **2016**. [CrossRef]
7. Liu, Z.; Yuan, L.; Tang, X.; Uyttendaele, M.; Suny, J. Fast burst images denoising. *ACM Trans. Graph.* **2014**. [CrossRef]
8. Mildenhall, B.; Barron, J.T.; Chen, J.; Sharlet, D.; Ng, R.; Carroll, R. Burst Denoising with Kernel Prediction Networks. In Proceedings of the IEEE Conference on Computer Vision and Pattern Recognition, Salt Lake City, UT, USA, 18–22 June 2018. [CrossRef]
9. Chen, C.; Chen, Q.; Xu, J.; Koltun, V. Learning to See in the Dark. In Proceedings of the IEEE Conference on Computer Vision and Pattern Recognition, Salt Lake City, UT, USA, 18–22 June 2018. [CrossRef]
10. Ronneberger, O.; Fischer, P.; Brox, T. U-Net: Convolutional Networks for Biomedical Image Segmentation. In *International Conference on Medical Image Computing and Computer-Assisted Intervention*; Springer: Cham, Switzerland, 2015; pp. 234–241. [CrossRef]
11. Everingham, M.; van Gool, L.; Williams, C.K.I.; Winn, J.; Zisserman, A. The pascal visual object classes (VOC) challenge. *Int. J. Comput. Vis.* **2010**. [CrossRef]
12. Russakovsky, O.; Deng, J.; Su, H.; Krause, J.; Satheesh, S.; Ma, S.; Huang, Z.; Karpathy, A.; Khosla, A.; Bernstein, M.; et al. ImageNet Large Scale Visual Recognition Challenge. *Int. J. Comput. Vis.* **2015**. [CrossRef]
13. Cheng, M.M.; Zhang, Z.; Lin, W.Y.; Torr, P. BING: Binarized normed gradients for objectness estimation at 300fps. In Proceedings of the IEEE Conference on Computer Vision and Pattern Recognition, Columbus, OH, USA, 23–28 June 2014. [CrossRef]
14. Lore, K.G.; Akintayo, A.; Sarkar, S. LLNet: A deep autoencoder approach to natural low-light image enhancement. *Pattern Recognit.* **2017**. [CrossRef]

15. Xu, J.; Li, H.; Liang, Z.; Zhang, D.; Zhang, L. Real-World Noisy Image Denoising: A New Benchmark. *arXiv* **2018**, arXiv:1804.02603.
16. Wei, C.; Wang, W.; Yang, W.; Liu, J. Deep Retinex Decomposition for Low-Light Enhancement. *arXiv* **2018**, arXiv:1808.04560.
17. Loh, Y.P.; Chan, C.S. Getting to know low-light images with the Exclusively Dark dataset. *Comput. Vis. Image Underst.* **2019**. [CrossRef]
18. Rudin, L.I.; Osher, S.; Fatemi, E. Nonlinear total variation based noise removal algorithms. *Phys. D Nonlinear Phenom.* **1992**. [CrossRef]
19. Mairal, J.; Bach, F.; Ponce, J.; Sapiro, G.; Zisserman, A. Non-local sparse models for image restoration. In Proceedings of the 2009 IEEE 12th International Conference on Computer Vision, Kyoto, Japan, 29 September–2 October 2009. [CrossRef]
20. Dabov, K.; Foi, A.; Katkovnik, V.; Egiazarian, K. Image denoising by sparse 3-D transform-domain collaborative filtering. *IEEE Trans. Image Process.* **2007**. [CrossRef]
21. Zhang, K.; Zuo, W.; Chen, Y.; Meng, D.; Zhang, L. Beyond a Gaussian denoiser: Residual learning of deep CNN for image denoising. *IEEE Trans. Image Process.* **2017**. [CrossRef] [PubMed]
22. Xie, J.; Xu, L.; Chen, E. Image denoising and inpainting with deep neural networks. *Adv. Neural Inf. Process. Syst.* **2012**. [CrossRef]
23. Łoza, A.; Bull, D.; Achim, A. Automatic contrast enhancement of low-light images based on local statistics of wavelet coefficients. In Proceedings of the 2010 IEEE International Conference on Image Processing, Hong Kong, China, 26–29 September 2010. [CrossRef]
24. Burger, H.C.; Schuler, C.J.; Harmeling, S. Image denoising: Can plain neural networks compete with BM3D? In Proceedings of the 2012 IEEE Conference on Computer Vision and Pattern Recognition, Providence, RI, USA, 16–21 June 2012. [CrossRef]
25. Joshi, N.; Cohen, M.F.; Seeing, M.T. Rainier: Lucky imaging for multi-image denoising, sharpening, and haze removal. In Proceedings of the 2010 IEEE International Conference onComputational Photography (ICCP), Cambridge, MA, USA, 29–30 March 2010. [CrossRef]
26. Ignatov, A.; Kobyshev, N.; Timofte, R.; Vanhoey, K.; van Gool, L. WESPE: Weakly supervised photo enhancer for digital cameras. In Proceedings of the IEEE Conference on Computer Vision and Pattern Recognition Workshops, Salt Lake City, UT, USA, 18–22 June 2018. [CrossRef]
27. Dong, X.; Wang, G.; Pang, Y.; Li, W.; Wen, J.; Meng, W.; Lu, Y. Fast efficient algorithm for enhancement of low lighting video. In Proceedings of the 2011 IEEE International Conference on Multimedia and Expo, Barcelona, Spain, 18–15 July 2011. [CrossRef]
28. Guo, X.; Li, Y.; Ling, H. LIME: Low-light image enhancement via illumination map estimation. *IEEE Trans. Image Process.* **2017**. [CrossRef] [PubMed]
29. LeCun, Y.; Boser, B.; Denker, J.S.; Henderson, D.; Howard, R.E.; Hubbard, W.; Jackel, L.D. Backpropagation Applied to Handwritten Zip Code Recognition. *Neural Comput.* **1989**, *1*, 541–551. [CrossRef]
30. le Cun, Y.; Fogelman-Soulié, F. Modèles connexionnistes de l'apprentissage, Intellectica. *Rev. l'Association Pour Rech. Cogn.* **1987**. [CrossRef]
31. Hinton, G.E.; Zemel, R.S. Autoencoders Minimum Description Length and Helmholtz free Energy. *Adv. Neural Inf. Process. Syst.* **1994**. [CrossRef]
32. Poggio, T.; Torre, V.; Koch, C. Computational vision and regularization theory. *Nature* **1985**. [CrossRef]
33. Friedman, J.H. Regularized discriminant analysis. *J. Am. Stat. Assoc.* **1989**. [CrossRef]
34. Vincent, P.; Larochelle, H.; Bengio, Y.; Manzagol, P.A. Extracting and composing robust features with denoising autoencoders. In Proceedings of the 25th International Conference on Machine Learning, Helsinki, Finland, 5–9 July 2008. [CrossRef]
35. Goodfellow, A.C.I. Yoshua Bengio, Deep Learning Book. *Deep Learn.* **2015**. [CrossRef]
36. Li, M.; Liu, J.; Yang, W.; Sun, X.; Guo, Z. Structure-Revealing Low-Light Image Enhancement Via Robust Retinex Model. *IEEE Trans. Image Process.* **2018**. [CrossRef]
37. Ren, X.; Li, M.; Cheng, W.H.; Liu, J. Joint enhancement and denoising method via sequential decomposition. In Proceedings of the 2018 IEEE International Symposium on Circuits and Systems (ISCAS), Florence, Italy, 27–30 May 2018; pp. 1–5.
38. Wang, Z.; Bovik, A.C.; Sheikh, H.R.; Simoncelli, E.P. Image quality assessment: From error visibility to structural similarity. *IEEE Trans. Image Process.* **2004**. [CrossRef] [PubMed]

39. Chen, X.; Zhang, Q.; Lin, M.; Yang, G.; He, C. No-reference color image quality assessment: From entropy to perceptual quality, Eurasip. *J. Image Video Process.* **2019**. [CrossRef]
40. Zhang, L.; Zhang, L.; Bovik, A.C. A feature-enriched completely blind image quality evaluator. *IEEE Trans. Image Process.* **2015**. [CrossRef] [PubMed]
41. Mittal, A.; Soundararajan, R.; Bovik, A.C. Making a "completely blind" image quality analyzer. *IEEE Signal Process. Lett.* **2013**. [CrossRef]
42. Liu, L.; Liu, B.; Huang, H.; Bovik, A.C. No-reference image quality assessment based on spatial and spectral entropies. *Signal Process. Image Commun.* **2014**. [CrossRef]
43. He, K.; Zhang, X.; Ren, S.; Sun, J. Deep Residual Learning for Image Recognition. In Proceedings of the 2016 IEEE Conference on Computer Vision and Pattern Recognition, Las Vegas, NV, USA, 27–30 June 2016. [CrossRef]
44. Xu, Y.; Wang, H.; Liu, X. An improved multi-branch residual network based on random multiplier and adaptive cosine learning rate method. *J. Vis. Commun. Image Represent.* **2019**, *59*, 363–370. [CrossRef]
45. Kingma, D.P.; Ba, J.L. Adam: A method for stochastic optimization. *arXiv* **2015**, arXiv:1412.6980.
46. Xu, B.; Wang, N.; Chen, T.; Li, M. Empirical Evaluation of Rectified Activations in Convolutional Network. *arXiv* **2015**, arXiv:1505.00853.
47. Salakhutdinov, R.; Mnih, A.; Hinton, G. Restricted Boltzmann machines for collaborative filtering. In Proceedings of the 24th International Conference on Machine Learning, Corvalis, OR, USA, 20 June 2007; pp. 791–798.

 *electronics* 

Article

WS-AM: Weakly Supervised Attention Map for Scene Recognition

Shifeng Xia [1], Jiexian Zeng [1,2], Lu Leng [1,*] and Xiang Fu [1]

[1] School of Software, Nanchang Hangkong University, Nanchang 330063, China; 1716083500006@stu.nchu.edu.cn (S.X.); zengjx58@163.com (J.Z.); fxfb163@163.com (X.F.)

[2] Science and Technology College, Nanchang Hangkong University, Gongqingcheng 332020, China

* Correspondence: leng@nchu.edu.cn

Received: 24 August 2019; Accepted: 19 September 2019; Published: 21 September 2019

Abstract: Recently, convolutional neural networks (CNNs) have achieved great success in scene recognition. Compared with traditional hand-crafted features, CNN can be used to extract more robust and generalized features for scene recognition. However, the existing scene recognition methods based on CNN do not sufficiently take into account the relationship between image regions and categories when choosing local regions, which results in many redundant local regions and degrades recognition accuracy. In this paper, we propose an effective method for exploring discriminative regions of the scene image. Our method utilizes the gradient-weighted class activation mapping (Grad-CAM) technique and weakly supervised information to generate the attention map (AM) of scene images, dubbed WS-AM—weakly supervised attention map. The regions, where the local mean and the local center value are both large in the AM, correspond to the discriminative regions helpful for scene recognition. We sampled discriminative regions on multiple scales and extracted the features of large-scale and small-scale regions with two different pre-trained CNNs, respectively. The features from two different scales were aggregated by the improved vector of locally aggregated descriptor (VLAD) coding and max pooling, respectively. Finally, the pre-trained CNN was used to extract the global feature of the image in the fully- connected (fc) layer, and the local features were combined with the global feature to obtain the image representation. We validated the effectiveness of our method on three benchmark datasets: MIT Indoor 67, Scene 15, and UIUC Sports, and obtained 85.67%, 94.80%, and 95.12% accuracy, respectively. Compared with some state-of-the-art methods, the WS-AM method requires fewer local regions, so it has a better real-time performance.

Keywords: convolution neural network; scene recognition; vector of locally aggregated descriptor; weakly supervised attention map

---

1. Introduction

Scene recognition, as a sub-problem of image recognition, has attracted increasing attention. It has important applications in robotics, intelligent security, driving assistant technique, and human-computer interaction, etc. However, scene recognition is quite different from general object recognition:

- Scene images, especially indoor scene images, commonly contain a large number of objects and a complex background;
- Human ability in scene recognition is much lower than that in object recognition;
- The number of datasets of scene recognition is much less than that of object recognition.

There are also several difficulties in scene recognition, such as variances of illumination, scale, and so on. The variability and difference of scene content lead to inter-class similarity and intra-class variation. Figure 1 shows some difficulties in scene recognition.

The focus of scene recognition is to extract more robust and generalized features, including hand-crafted features and learning-based features. Traditional scene recognition methods generally use hand-crafted features, e.g., oriented texture curves (OTC) [1], census transform histogram (CENTRIST) [2], histogram of oriented gradient (HOG) [3], and scale-invariant feature transform (SIFT) [4]. Hand-crafted features are constructed based on image color, texture, structure, and other information. They have no semantic information and are difficult to use in complex scene recognition. With the wide use of deep learning in computer vision, learning-based features have been applied to scene recognition. Convolutional neural network (CNN) is a typical representative of learning-based features [5–8]. Latent feature representation containing high-level semantic information can be learnt from large-scale data without human intervention. Even though the CNN features perform well in scene recognition [9], they still use global information while ignoring local information, and cannot satisfactorily solve between-class similarity and within-class difference.

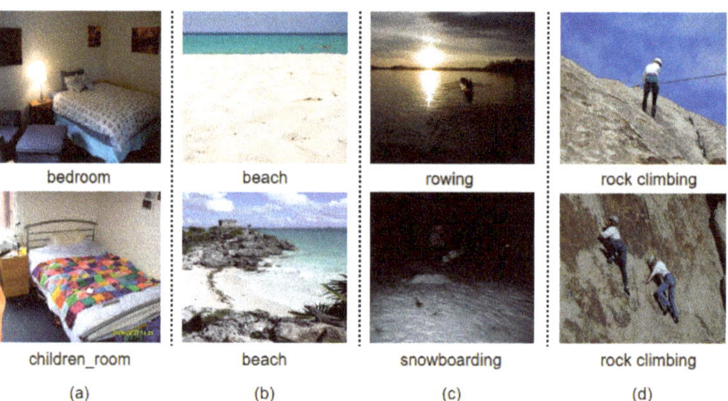

Figure 1. Some difficulties in scene recognition. (**a**) inter-class similarity. (**b**) intra-class variations. (**c**) illumination problem. (**d**) shooting angle problem.

Intuitively, the images of scene recognition are different from the general images of object recognition. Many scene images contain a large number of objects, especially indoor scenes, and have a complex background, which brings severe difficulties for feature extraction. Many CNN-based methods extract features from local regions at different scales and complement global representation; however, they do not sufficiently consider the relationship between the local region and the context of the scene. Many extracted local regions are redundant and degrade the classification results of scenes. The scene images of different categories often contain the same and similar object regions, while the scene images of the same category probably contain very different object regions. In this paper, we focus on the discriminative regions in scene images. The feature extraction of discriminative regions can effectively solve the problems of between-class similarity and within-class difference. Figure 2 shows the images of two categories ('bedroom' and 'children's room') of MIT indoor 67 dataset [10]. It can be seen from each column that the samples of different scene categories are very similar. Significant intra-class differences are remarkable in each row, which is caused by different backgrounds, objects, and angles. In order to achieve significant recognition results, a suitable way is to find discriminative region blocks that are good representations helpful to classification.

In this paper, we propose a weakly supervised attention map (WS-AM) method, which uses the gradient-weighted class activation mapping (Grad-CAM) [11] technique to obtain a small-scale attention map (AM) for each image. WS-AM uses the maximum output value information of the last fully-connected (fc) layer of CNN, but the image-level label information is absent, so it can be considered as weakly supervised. The regions with large local mean and large local center value in AM correspond to the regions of the original image that have strong discriminative power, while the others correspond to the redundant regions in the original image. The features are extracted from multi-scale discriminative regions per image. The features in small-scale regions are extracted in the softmax layer using CNN that is pre-trained on the ImageNet dataset [12] (i.e., ImageNet-CNN), and then they are coded by improved vector of locally aggregated descriptor (VLAD) [13] and normalized with L2-normalization. The features of large-scale regions are extracted in the softmax layer using CNN that is pre-trained on the Places365 dataset [9] (i.e., Places365-CNN), and then they are aggregated by max pooling. In order to obtain the global feature of the image, we use Places205-CNN (i.e., CNN pre-trained on Places205 dataset) [9] to extract the feature vector in the first fc layer (i.e., fc6 layer), and they are normalized with L2-normalization. Finally, the three feature vectors are concatenated to form the final image representation. In order to verify the effectiveness of WS-AM, the experiments were carried out on three datasets and achieved good performance.

Figure 2. Images of two categories from the MIT indoor 67 dataset. Each row shows the difference within the class. Each column shows the similarity between classes.

The remainder of this paper is organized as follows. The related works are reviewed in Section 2. Section 3 introduces our method, including the pipeline and details of the whole algorithm. Sections 4 and 5 introduce the experiments and analysis in detail. Finally, we summarize the whole work in Section 6.

2. Related Work

In this section, the related work is briefly reviewed, including scene representation, discriminative region discovery, feature coding, and scene classification.

2.1. Scene Representation

In traditional scene recognition, hand-crafted features are widely used because they are relatively simple and have low computational cost. Traditional scene recognition can be divided into the following steps: extract patches, represent patches, encode features, and pool features. In the patch representation, the features, such as SIFT, HOG, and speeded-up robust features (SURF) [14], are extracted from local regions. Effective hand-crafted features can not only depict the texture characteristics but also reflect the deep structure information. The Bag of Features (BOF) model based on SIFT feature has been widely used in scene recognition, but the lack of location information makes it difficult to use in a complex scene. Lazebnik et al. [15] improved the BOF model based on SIFT feature and proposed the spatial pyramid matching (SPM) model, which achieved good results in scene recognition. HOG feature was initially used in pedestrian detection. Later, Felzenswalb et al. [16] proposed the deformable parts model (DPM) on the basis of HOG feature. Pandey et al. [17] improved the DPM and applied it to large-scale scene image recognition. After clustering and coding of local features of scene images, pooling operations are needed. Max pooling and average pooling are commonly used in pooling operations. The experimental results of Yang et al. [18] on several benchmark databases show that the effect of max pooling is better than that of average pooling.

Recently, CNNs have made prominent progress on computer vision, especially in image recognition. AlexNet [19] won the championship in the ImageNet image recognition competition in 2012. Since then, CNNs have made breakthroughs in object detection, semantic segmentation, and image generation. Benefiting from large-scale well-labeled datasets, more CNN structures have been proposed, such as VGGNet [20], GoogLeNet [21], and ResNet [22]. CNNs are also widely used in scene recognition. Zhou et al. [9] used CNNs to train and test on a new large-scale scene dataset Places and achieved great results. Although the global features extracted by CNN have achieved remarkable results in scene recognition, they only represent the global information, and ignore the local information. Shi et al. [5] recently proposed a novel approach which utilized the visually sensitive features combining with CNN features for scene recognition. Wang et al. [23] proposed a multi- resolution CNN structure to capture visual content and structure at multiple levels of images. Javed et al. [24] proposed a deep network structure, which uses the position relations of a group of objects to infer the scene category, and then establishes the semantic context model of the scene. Many methods do not train CNNs from scratch, but directly use the CNNs, namely Places205-CNN, Places365-CNN, and ImageNet-CNN, pre-trained on the three large datasets (i.e., Places205, Places365, and ImageNet) to extract features.

2.2. Discriminative Region Discovery

Local region information is very important for scene recognition, but current methods do not sufficiently focus on the discriminative region of the scene image. Some methos densely sampled local regions in a multi-scale way for scene images [6,25,26]. Dense sampling extracts all regions of the image, but it inevitably produces many redundant regions, most of which are in the background without objects or contain similar regions in different scene categories. Dense sampling also leads to high computational cost. Uijlings et al. [27] proposed a selective search method for generating a set of regions that are likely to contain objects. Selective search is a region proposal method and widely used in object detection. Intuitively, most scenes consist of many objects, so the region proposal method can be used to generate local regions containing objects. Wu et al. [8] used multi-scale combinatorial grouping (MCG) [28] to generate high-quality local regions for scene images. Javed et al. [24] utilized edge boxes [29] to extract image candidate regions, and feature maps of the same size can be generated by region of interest (RoI) pooling for the candidate regions. However, those unsupervised region proposal approaches fail to consider the relationship between object regions and scene categories, and still produce some redundant regions. Discriminative power analysis [30,31] can help judge whether the regions are discriminative or redundant.

Zhou et al. [32] proposed a method to generate class activation mapping (CAM) using the global average pooling (GAP) in CNNs. The CAM of a specific category represents the discriminative image region for identifying this category. CAM forces the CNN structure to include GAP, but some CNN structures do not have GAP, such as AlexNet and VGGNet. In order to solve this problem, Selvaraju et al. [11] put forward the Grad-CAM technique, which uses the gradient of the interested class to propagate back to the convolutional layer to generate a coarse localization map. It highlights the discriminative regions to predict the interested category. Recently, the attention mechanism has been widely used in computer vision tasks, such as fine-grained image recognition [33–35], scene text recognition [36–38], and so on. Fu et al. [33] proposed a novel recurrent attention convolutional neural network (RA-CNN) for fine-grained image recognition. RA-CNN learns discriminative region attention and region-based feature representation in a recursive way, without the use of any bounding box annotation information. Gao et al. [37] introduced a text attention module in the text feature extraction process to focus on text regions and avoid background noise. These works utilize attention modules to capture category-specific objects and parts. Lorenzo et al. [39] proposed a new attention-based CNN for selecting bands from hyperspectral images. This method uses gating mechanisms to obtain the most informative regions of the spectrum. Attention mechanisms are also widely used in other network structures, e.g., long short-term memory (LSTM) [40] and gated recurrent (GRU) [41] neural networks. Vaswani et al. [42] proposed a new simple network architecture based on attention mechanisms, called the Transformer. The Transformer has achieved outstanding results on two machine translation tasks. Inspired by these works, we apply the attention module to scene recognition. Our Grad-CAM based method has obvious advantages:

- Our method uses pre-trained CNN as the backbone network of the attention module, instead of training from scratch or fine-tuning;
- Different from other attention modules that select a fixed number of regions per image, our method obtains an adaptive number of regions for each image, which is more conducive to scene recognition;
- Different from other attention modules, our AM does not use image-level label information.
- Compared with other methods, our method is simpler and does not require adding new components to the network structure to drive the attention mechanism.

2.3. Feature Coding

In traditional scene recognition, clustering and coding local features are needed to obtain image embedding. The feature coding methods can be mainly divided into two types: global coding and local coding. Global coding is usually used to estimate the probability density distribution of features, while local coding is used to describe each feature. Typical feature coding includes bag of visual words (BoVW) [43,44], fisher vector (FV) [45,46], VLAD, and salient coding (SC) [47]. FV coding uses the Gaussian mixture model (GMM) to estimate the distribution of features. GMM consists of weights, means, and covariance matrices of several Gaussian distributions, each of which reflects a feature pattern. As a simplification of FV, VLAD calculates the residuals between the features and the nearest neighbor visual dictionary. VLAD takes into account the value of each dimension of features and describes the local information of images in a more detailed, simple, and effective way, so it has been widely used in scene recognition.

Feature coding is also important for scene recognition based on deep learning. Many traditional feature coding methods have been improved to be more suitable for deep learning. Dixit et al. [6] proposed semantic FV for scene recognition by combining the local features extracted from traditional FV and CNNs. Khan et al. [48] proposed Deep Un-structured Convolutional Activation (DUCA), which extracts the features from middle-level regions of images through CNNs and encodes them according to their association with the codebook of representative regions of scenes.

2.4. Scene Classification

There are mainly two types of classifiers for scene classification: discriminative models and generative models. The learning of the discriminative model is a conditional probability, which mainly focuses on the classification boundary of data. The discriminative model seeks the optimal separating hyperplane between different categories and reflects the difference between the different types of data. The advantages of the discriminative model are as follows:

- It can distinguish well the differences between categories;
- It is suitable for the identification of more categories;
- It is relatively simple and easy to understand.

However, the discriminative model does not reflect well the characteristics of the data. Commonly used discriminative models include k-nearest neighbor (KNN), logistic regression (LR), and support vector machine (SVM). In particular, SVM is widely used in scene recognition [6,8,25].

Different from the discriminative model, the generative model learns the joint probability distribution, which represents the distribution of data from a statistical perspective and can reflect the similarity of similar data. The generative model gives the joint probability density, which contains more information, and its training speed is much faster than the discriminative model. However, the learning and calculation process of the generative model is complex, and the accuracy of the classification problem is lower than that of the discriminative model. The widely used generative models include the naive Bayesian model (NBM), hidden Markov model (HMM), and GMM.

3. Proposed Method

In order to distinguish one scene category from another, the most effective approach is to obtain category-specific objects or regions. Although many methods can obtain regions containing objects, many objects are not category-specific. Some regions contain objects that are common in different scenes, which introduce noise for feature extraction. To avoid common object regions, we propose a Grad-CAM based method to capture regions that only contain category-specific objects. The proposed method can be divided into two parts: WS-AM and scene representation. Figure 3 shows the main flow of our method. First, Grad-CAM is employed to generate AM for the input image, in which weakly supervised information (i.e., the maximum output value of the last fc layer) is used. The regions with large local mean and large local center value in AM correspond to the regions with strong discriminative power in the images. Second, we extract the multi-scale CNN features from these discriminative regions. Different scale regions are input to different pre-trained networks (i.e., ImageNet-CNN and Places365-CNN) and the feature vectors are extracted in the softmax layer. The features extracted from small-scale regions are aggregated by improved VLAD coding and normalized by L2-normalization. While max pooling is used for the features extracted from large-scale regions. The global feature is extracted in the first fc layer (i.e., fc6 layer) on Places205-CNN and normalized by L2-normalization. Finally, the three extracted features are concatenated to form the final image representation.

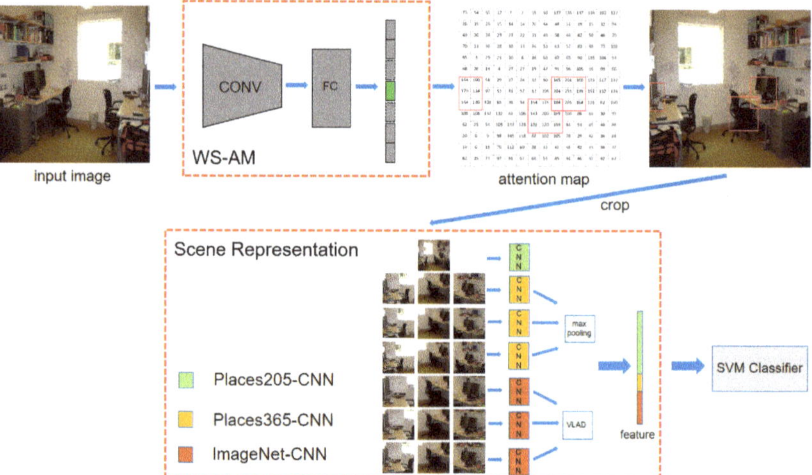

Figure 3. The framework of our method. The framework can be divided into two parts: WS-AM and scene representation.

3.1. Weakly Supervised Attention Map

WS-AM is used for discovering discriminative regions in scene images. Scene recognition is different from general object recognition, which is composed of complex background and various objects. Inspired by the work of Grad-CAM on the visual interpretation of CNNs, we use this method to generate the AM for each image. The backbone network for Grad-CAM is VGGNet pre-trained on the Places205 dataset, i.e., Places205-VGGNet. We do not use Place205-VGGNet to fine-tune the datasets, so the image-level label information is not used. Instead, the maximum output value in the last fc layer of Places205-VGGNet is used as the backpropagation information to generate AM, which can be considered as weakly supervised. The gradient information is back-propagated to the last convolution (conv) layer to calculate the importance of each neuron to the final classification.

As shown in Figure 4, the input image I is resized into the size of 224×224 and propagated forward through the CNN to obtain the output value of the last fc layer. The maximum output value S is back-propagated to calculate the gradient of the feature maps A at the last conv layer, i.e., $\partial S / \partial A$. A^k represents the k_{th} feature map of A, so the gradient of A^k is $\partial S / \partial A^k$. Then the gradients of k_{th} feature map are averaged to obtain the neuron importance weight α^k as follows:

$$\alpha^k = \frac{1}{Z} \sum_i \sum_j \frac{\partial S}{\partial A_{ij}^k} \tag{1}$$

where Z denotes the size of the k_{th} feature map, which is 14×14. The weight α^k represents the local linearization of the feature map A^k, and also indicates the importance of the k_{th} feature map to the maximum output value S. We take the sum of weighted feature maps, and by the activation function ReLU to obtain AM:

$$\text{AM} = \text{ReLU}\left(\sum_k \alpha^k A^k \right) \tag{2}$$

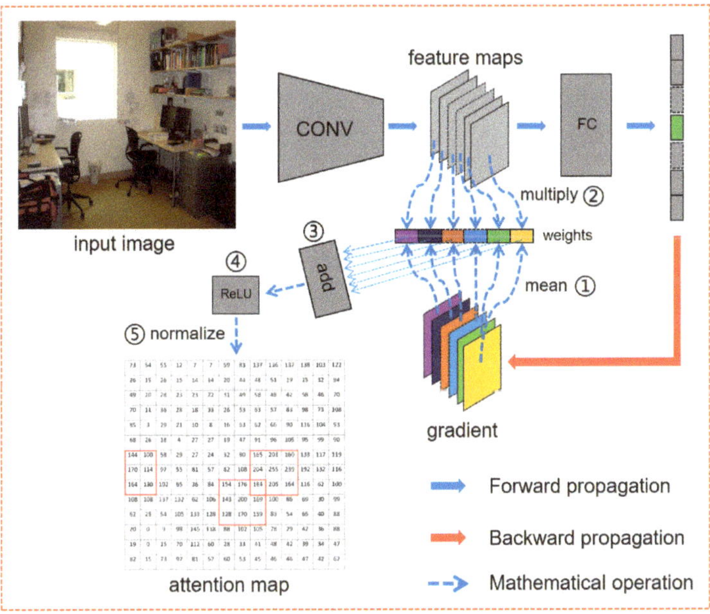

Figure 4. The pipeline of WS-AM. We use weakly supervised information to generate the attention map for each image.

ReLU is applied to linear combinations of feature maps and weights because we are only interested in those features which have a positive impact on the maximum output value, and the intensity of those feature pixels should be increased to enhance the category with the maximum output value [11]. The backbone network we used for Grad-CAM is VGGNet, so the AM size is 14×14. In order to facilitate calculation and visualization, the values in AM are normalized to the range of $(0, 255)$. If the AM is up-sampled to the input image size (i.e., 224×224), each pixel value in the AM represents the importance of the corresponding pixel in the input image to the final classification result. Figure 5 shows four Grad-CAM visualization examples of the VGGNet pre- trained in the Places205 dataset. We only use weakly supervised information, but the discriminative region for each image is consistent with a human attention mechanism.

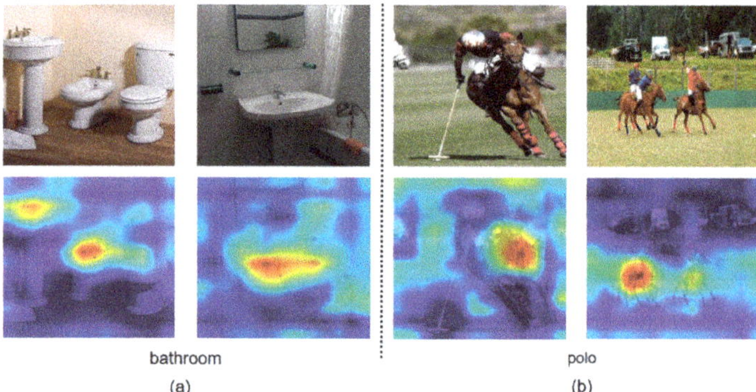

Figure 5. Some examples of Grad-CAM visualization.

A sliding window with 3×3 size and 1 stride is used to slide AM. In order to obtain the discriminative regions, two strategies are employed:

- The mean of the 9 numbers in the window is greater than the AM mean;
- The value of the window center is the maximum value of the 9 numbers and needs to be greater than the threshold value.

If both strategies are satisfied, the corresponding regions in the original image are considered as discriminative regions. The first strategies eliminate the exception window of AM in which the center value is larger than the threshold but other values are too small. Each discriminative region of the original image is cropped in the size of $s \times s$ in a multi-scale way, where $s \in \{64, 80, 96, 112, 128, 144\}$. Then, we resize each scale region into the size of 224×224 in order to adapt to the input size of VGGNet. Intuitively, small-scale regions ($s = 64, 80, 96$) contain 'object', while large-scale regions ($s = 112, 128, 144$) contain 'scene', so ImageNet-CNN and Places365-CNN are respectively used to extract local features.

3.2. Improved Vlad

In general, VLAD coding first carries out k-means cluster for local features and then calculates the accumulated residuals between the local features and their nearest neighbor cluster centers, and finally forms the image embedding as the local representation through pooling. VLAD has two shortcomings:

- It only considers the residual with the nearest neighbor cluster center;
- The encoded feature dimension is too high.

Furthermore, the number of small-scale regions is unbalanced, so the general VLAD coding cannot work well. To solve the above problems, VLAD coding is improved.

The feature vectors $l = [l_1, \ldots, l_j, \ldots, l_M]$ of the small-scale local regions in each image are non-Euclidean, so they are difficult to carry out for VLAD coding, M denotes the number of the small-scale local features of each image. Natural parameterization is used to transform these feature vectors into linear Euclidean space as follows:

$$v_j = \sqrt{l_j} \tag{3}$$

where v_j is the transformed feature vector. The conversion from non-Euclidean space to linear Euclidean space is more conducive to VLAD coding. Mini Batch k-means method clusters all small-scale local features, and obtains codebook with k cluster centers $c = [c_1, \ldots, c_i, \ldots, c_k]$. For the local features $v = [v_1, \ldots, v_j, \ldots, v_M]$ of each image, we calculate the residuals between each feature v_j and all clustering centers. Then, the residuals of each cluster center are aggregated, and the formula is as follows:

$$r_i = \sum_{j=1}^{M} w_{ji}(v_j - c_i) \tag{4}$$

$$w_{ji} = \frac{1}{1 + d_{ij}} \tag{5}$$

where w_{ij} is the weight of the residual $v_j - c_i$, which is a decreasing function of the Euclidean distance d_{ij} between v_j and c_i. VLAD embedding result is:

$$Z = [r_1 \ldots r_i \ldots r_k] \tag{6}$$

Each small-scale region is inputted into ImageNet-CNN to extract feature vectors with 1000 dimensions of the softmax layer, so each cluster center and VLAD embedding are both 1000-dimensional vectors. In this way, each image obtains a $k \times 1000$ dimensional local representation of the small-scale local regions. We do not use the vectors directly, because the dimensions are too large, so they are not very computationally friendly. Max pooling is conducted on $[r_1, \ldots, r_i, \ldots, r_k]$ to form a 1000-dimensional vector. Finally, we average the results to eliminate the impact of an unbalanced number of features in each image. The final local representation of the small-scale local regions is:

$$V_{\{64,80,96\}} = \frac{1}{M}\mathrm{max-pooling}\big([r_1, \ldots, r_i, \ldots, r_k]\big) \tag{7}$$

The numbers of feature in small-scale regions extracted for the images are different, which leads to a large difference in the residual of each cluster center. Averaging the results can eliminate this effect.

3.3. Multi-Scale Fusion Feature

Multi-scale feature fusion is widely used in scene recognition. Different scales need to be unified in order to fuse. Fusion makes features more robust and easier to learn [49–51]. WS-AM generates many discriminative regions for each image, and multi-scale ($s = 64, 80, 96, 112, 128, 144$) are taken for each discriminative region p_i. The form of local regions extracted from each image is:

$$P = [p_1, \ldots, p_i, \ldots, p_N] \tag{8}$$

where N represents the number of local regions. Small-scale regions ($s = 64, 80, 96$) can be considered to contain 'object', so they are inputted to ResNet18 pre-trained on ImageNet (i.e., ImageNet-ResNet18) to extract the 1000-dimensional feature vectors in the softmax layer. The large-scale regions ($s = 112, 128, 144$) which can be considered to contain 'scene', are inputted to ResNet18 pre-trained on Places365 (i.e., Places365-ResNet18) to extract the 365-dimensional feature vectors in the softmax layer. After improved VLAD coding and pooling, we obtain feature vectors $V_{\{64,80,96\}}$ for the small-scale region. Also, we use max pooling to aggregate the features of large-scale regions and obtain the feature vector $V_{\{112,128,144\}}$ for each image. In order to get the global information, we resize each image into the size of 224×224 and input the entire image into VGGNet pre-trained on Places205 (i.e., Places205-VGGNet) to extract the feature vector V_{GR} of the fc6 layer. We use L2-normalization on $V_{\{64,80,96\}}$ and V_{GR} to obtain $V_{\{64,80,96\}-L2}$ and V_{GR-L2}, respectively. L2-normalization is not used on $V_{\{112,128,144\}}$ because the feature vectors are extracted from the softmax layer which can play the role of normalization. Finally, three feature vectors are concatenated to form the final image representation:

$$[V_{GR-L2} \; V_{\{112,128,144\}} \; V_{\{64,80,96\}-L2}] \tag{9}$$

Table 1 shows the tensor dimensionalities in the processing pipeline.

Table 1. The tensor dimensionalities in the processing pipeline.

Tensor	Dimensionality
input image	$3 \times 224 \times 224$
feature map	$512 \times 14 \times 14$
gradient	$512 \times 14 \times 14$
weight	$512 \times 1 \times 1$
attention map	$1 \times 14 \times 14$
V_{GR} / V_{GR-L2}	1×4096
$V_{\{112,128,144\}}$	1×365
$V_{\{64,80,96\}}$ / $V_{\{64,80,96\}-L2}$	1×1000
$[V_{GR-L2}\ V_{\{112,128,144\}}\ V_{\{64,80,96\}-L2}]$	1×5461

3.4. Classification

In this paper, linear SVM classifier implementing 'one-vs-the-rest' multi-class strategy is trained on three datasets. Other kernel functions, such as the polynomial kernel, radial basis function (RBF), and sigmoid kernel, are not suitable for our task. Compared with other kernel functions, the linear kernel function has two advantages:

- The linear kernel has fewer hyperparameters and faster training speed.
- The linear function is suitable for high-dimensional features. In this paper, each image has a 5461-dimensional feature vector.

Therefore, we choose linear SVM as the classifier. Penalty parameter C, as an important parameter for the SVM model, represents the tolerance of error. Here $C = 1.0$.

4. Experiments and Results

The experiments are performed on three datasets: MIT indoor 67, Scene 15 [15] and UIUC Sports [52]. The three datasets contain different types of scene images: MIT indoor 67 mainly contains indoor scene images; Scene 15 contains both indoor and outdoor scene images; UIUC Sports contains event scene images. Then, some parameters of our method are evaluated, including the number of cluster centers, the threshold to extract discriminative regions on AM, different backbone networks of Grad- CAM, and the different scales of the discriminative region. Figure 6 shows some images in the three datasets.

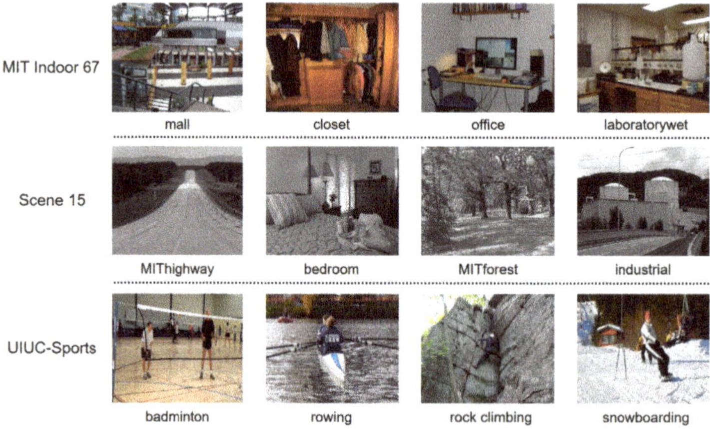

Figure 6. Some image examples of the three scene datasets.

4.1. Datasets

MIT indoor 67: This dataset contains 67 categories of indoor scene images. There are 15620 images in total, with at least 100 images in each category. We follow the division of training set and test set in ref. [10]; 80 images of each category are used for training, and 20 images are used for testing.

Scene 15: There are 15 categories in this dataset, a total of 4485 grayscale indoor and outdoor images. The dataset does not provide criteria for dividing the training set and test set. We randomly divide the dataset five times, 100 images of each category are for training, and the rest are used as test images. Finally, we calculated the average accuracy of five times of division.

UIUC Sports: This dataset contains eight sports event scene categories, including rowing, badminton, polo, bocce, snowboarding, croquet, sailing, and rock climbing. There are 1579 color images. The dataset does not provide criteria for dividing the training set and the test set. We randomly divide the dataset five times and select 70 training images and 60 test images for each category. Finally, we calculate the average accuracy of five times of division.

4.2. Comparisons with State-of-the-Art Methods

MIT indoor 67 dataset mainly verifies the performance in indoor scenes, while Scene 15 verifies the performance both in indoor and outdoor scenes. UIUC Sports verifies the performance in event scenes. The experimental parameters are the same on the three datasets.

Table 2 shows the performance of our method on MIT indoor 67 dataset and its comparison with other methods. The references [1,2,15,53–55] are traditional methods, which mainly use some low-level features and mid-level features, such as SIFT, Object Bank [53], and BOF. Because these features only consider the shape, texture, and color information without any semantic information, they do not have high recognition accuracy. The references [5,6,8,56,57] are based on CNNs, and their overall recognition accuracies are higher than those of traditional methods. The CNN features of scene image have certain semantic information, and these features are learnt from a large number of well-labeled data, while not designed artificially. Our method is remarkably superior to the compared state-of-the-art methods in Table 2, which uses both semantic information and discriminative regions. In addition, the number of local regions used by our method is less than those in other methods, so the overall running time is significantly reduced. Figure 7 shows the confusion matrix of the MIT indoor 67 dataset. We see that the probability of classification is mostly concentrated on the diagonal line, and the overall performance is great. However, some categories have lower recognition accuracy than others, such as 'museum' and 'library' categories. These categories do not work well because the images of these categories are similar to each other and have complex backgrounds.

Table 2. Accuracy comparison on MIT indoor 67 dataset.

Method	Accuracy (%)
SPM [15]	34.40
CENTRIST [2]	36.90
Object Bank [53]	37.60
Discriminative Patches [54]	38.10
OTC [1]	47.33
FV + Bag of parts [55]	63.18
Places-CNN [56]	68.24
Hybrid-CNN [56]	70.80
Semantic FV [6]	72.86
MetaObject-CNN [8]	78.90
VS-CNN [5]	80.37
LS-DHM [57]	83.75
Our WS-AM {64,80,96,112,128,144}	81.79
Our WS-AM {64,80,96,112,128,144} + fc6 features	85.67

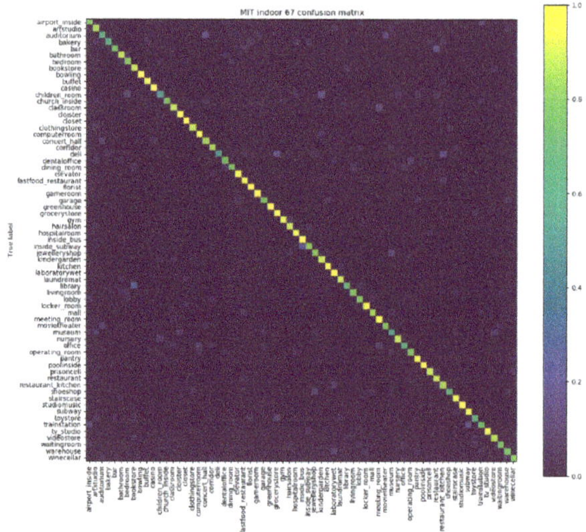

Figure 7. Confusion matrix of MIT indoor 67 dataset.

The experiments are carried out on Scene 15 dataset, which contains both outdoor and indoor scenes. Table 3 tabulates the comparison results in Scene 15 dataset. Our method achieves the recognition accuracy of 94.80% and is markedly superior to the compared state-of-the-art methods. Figure 8 shows the confusion matrix of Scene 15 dataset. The accuracy of the 'CALsuburb' class reaches 100%. The accuracies of 'MITcoast', 'MITforest', 'MIThighway', and 'MITmountain' categories is very high, and it can be concluded that our method also performs well in outdoor scenes. However, it can be clearly seen from the confusion matrix that the accuracies in outdoor scenes are relatively lower than those in indoor scenes.

Table 3. Accuracy comparison on Scene 15 dataset.

Method	Accuracy (%)
LDA [15]	59.00
BoW [15]	74.80
Object Bank [58]	80.90
SPMSM [59]	82.30
OTC [1]	84.37
Places-CNN [56]	90.19
Hybrid-CNN [56]	91.59
DGSK [60]	92.30
Our WS-AM {64,80,96,112,128,144}	92.58
Our WS-AM {64,80,96,112,128,144} + fc6 features	94.80

Table 4 tabulates the comparison results on UIUC Sports dataset. Our method achieves an accuracy of 95.12% and is superior to the compared state-of-the-art methods. UIUC Sports is a dataset of sport event scenes, which is different from the general scenes. The confusion matrix of the UIUC Sports dataset is indicated in Figure 9. We see that the recognition accuracy of the 'sailing' category reaches 100%, and the accuracies of the classes except 'bocce' and 'croquet' are good. It is because the contents of these two scene categories are similar, e.g., 'people' and 'ball'.

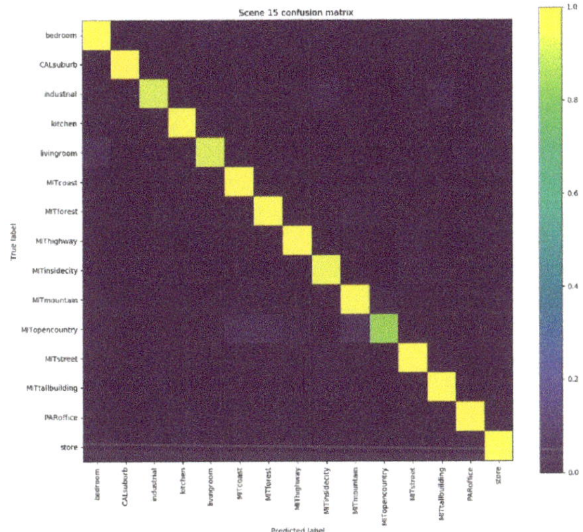

Figure 8. Confusion matrix of Scene 15 dataset.

Table 4. Accuracy comparison on UIUC Sports dataset.

Method	Accuracy (%)
GIST-color [61]	70.70
MM-Scene [62]	71.70
Object Bank [53]	76.30
CENTRIST [2]	78.25
SPMSM [59]	83.00
DF-LDA [7]	87.34
VC + VQ [63]	88.40
ISPR + IFV [64]	92.08
Our WS-AM {64,80,96,112,128,144}	93.07
Our WS-AM {64,80,96,112,128,144} + fc6 features	95.12

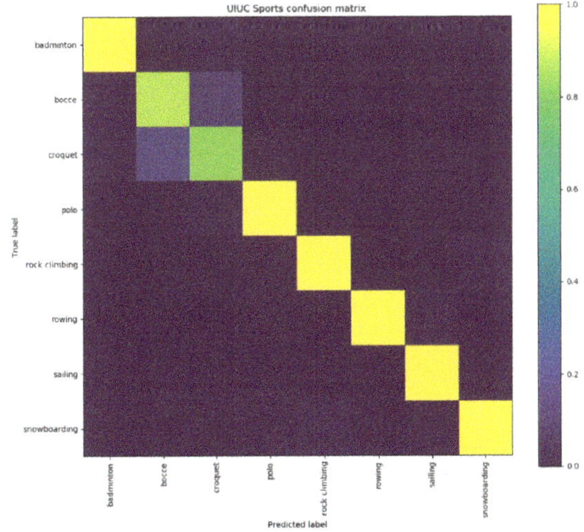

Figure 9. Confusion matrix of UIUC Sports dataset.

5. Experiments Analysis

In this section, we evaluate several important parameters of our method. First, we compare the performance of different backbone network structures for the Grad-CAM method. Second, we evaluate the impact of different scale combinations of discriminative regions on the results. Third, the effect of different thresholds is evaluated. Fourth, the number of cluster centers is very important for the aggregation of local features, so the influence of a different number of cluster centers is compared. Fifth, we prove the importance of L2-normalization. Sixth, we compare the performance of the different parameter C. Finally, in order to demonstrate the effectiveness of the WS-AM method for obtaining discriminative regions of scene images, we visualized discriminative regions of some categories. All of these evaluations are performed on MIT indoor 67 dataset.

5.1. Evaluation

Backbone network. In our WS-AM method, VGGNet pre-trained on Places205 dataset (i.e., Places205-VGGNet) from ref. [65] is used as the backbone network to obtain AM. Three pre-trained networks are evaluated including VGG11, VGG16, and VGG19. Table 5 lists the recognition results of three backbone networks on the MIT indoor 67 dataset. It shows that VGG11 performs better than the other networks, and its accuracy is 2.17% (1.72%) higher than that of VGG19 because the discriminative regions extracted from VGG11 are more representative. On the other side, the VGGNet is also used to extract the global feature in fc6 layer for each image, so the final recognition accuracy is affected by two factors: discriminative regions and global features.

Table 5. Performance of different backbone networks on MIT indoor 67 dataset.

Network	Accuracy (%) (+fc6 Features)
VGG11	81.79 (85.67)
VGG16	80.07 (84.25)
VGG19	79.62 (83.95)

Scale. Six rectangular regions of different scales ($s = 64, 80, 96, 112, 128, 144$) are cropped for each discrimination region and the performances of different scale combinations are compared on MIT indoor 67 dataset. The regions at the scales of ($s = 64, 80, 96$) contain 'object', while the regions at the scales of ($s = 112, 128, 144$) contain 'scene', so these regions with two different scales are inputted into different CNNs to extract features in the softmax layer. Table 6 indicates the influence of different scale combinations on recognition accuracy. We see that the scales of ($s = 64, 80, 96, 112, 128, 144$) perform better than other combinations of scales because the objects in the scene are basically multi-scale, and we can obtain features containing more scale information by using more scales. On the one hand, from rows 1–3, 4–6, and 7–9 in Table 6, it can be seen that the coarse local scales ($s = 112, 128, 144$) are important to extract global information. On the other hand, from rows 2, 5, 8, and 3, 6, 9 in Table 6, we can see that the fine local scales ($s = 64, 80, 96$) are significant to extract local information.

Table 6. Performance of different scales on MIT indoor 67 dataset.

Scale	Accuracy (%) (+fc6 Features)
64, 112	78.65 (83.50)
64, 112, 128	80.37 (84.32)
64, 112, 128, 144	80.74 (84.77)
64, 80, 112	79.02 (83.95)
64, 80, 112, 128	80.82 (84.25)
64, 80, 112, 128, 144	81.19 (84.92)
64, 80, 96, 112	80.00 (84.32)
64, 80, 96, 112, 128	81.26 (84.62)
64, 80, 96, 112, 128, 144	81.79 (85.67)

Threshold. Two strategies are used in Section 3 to screen the discriminative regions in AM. For the first strategy, different thresholds (0, 50, 100, 150) are experimented on MIT indoor 67 dataset and its impact evaluated on the recognition results. From the results in Figure 10, we see that the recognition accuracy is the highest when the threshold is 100 and the lowest when the threshold is 150 (without fc6 features). This indicates that more discriminative regions will improve the performance of the recognition, and fewer regions will result in a lack of local information. However, when the global features (fc6 features) are combined, the recognition accuracy is the highest when the threshold is 100, which is 85.67%. It is because the threshold only affects local representations, and when combined with global features, the overall trend will change. In this paper, 50 spacing is used to evaluate the threshold without considering smaller spacing. In future work, further optimization may lead to performance improvement.

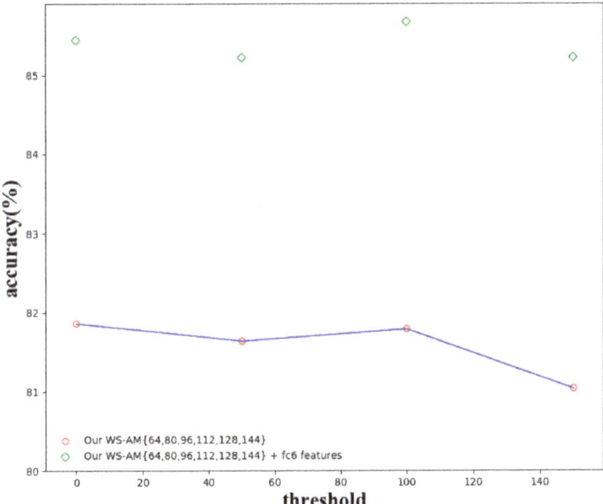

Figure 10. The recognition accuracies of different thresholds on MIT indoor 67 dataset.

Cluster center. To evaluate the impact of a different number of cluster centers, an experiment on MIT indoor 67 dataset is carried out with a various number of cluster centers. Figure 11 shows the effects of a various number of cluster centers. It can be seen that when the number of centers is 40 and 70, the recognition accuracy is the highest (82.23% without fc6 features). The unreasonable number of cluster center leads to poor generality, and further, degrades accuracy. However, combined with the global features (fc6 features) the recognition accuracy reaches 85.67% when the number of centers is 10 because the VLAD centers only affect local representations.

L2-normalization. Normalization is the process of scaling individual samples to have unit norm. After normalization, features are easier to be trained by SVM, which means it is easier to find the classification hyperplane of features. If the features are not normalized, SVM may not converge because the numerical range of each dimension is different. In this paper, features are normalized with L2-normalization. Table 7 shows the accuracy with L2-normalization or without L2-normalization on the MIT indoor 67 dataset. We can see that the feature with L2-normalization achieves better results. However, when $V_{\{112,128,144\}}$ is normalized, the accuracy is reduced by 0.97%. It is because the feature vectors are extracted from the softmax layer which can play the role of normalization.

Parameter C. Penalty parameter C is an important parameter for the SVM model. C represents the tolerance of error. When the parameter C is large, the SVM model will be over-fitting. Therefore, a suitable C will bring better results to the recognition. Table 8 shows the accuracy of the different C on MIT indoor 67 dataset. It can be seen that with the increase of C, the accuracy will decline.

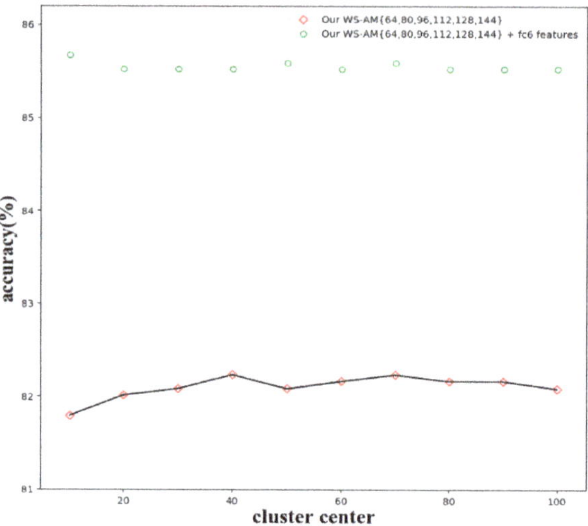

Figure 11. The recognition accuracies of a different number of cluster centers on MIT indoor 67 dataset.

Table 7. Accuracy with L2-normalization or without L2-normalization on MIT Indoor 67 dataset.

V_{GR}	$V_{\{112,128,144\}}$	$V_{\{64,80,96\}}$	Accuracy (%)
✗	✗	✗	81.86
✔	✔	✔	84.70
✔	✗	✔	85.67

[7] ✗: without L2-normalization; ✔: with L2-normalization.

Table 8. Accuracy of the different parameter C on MIT indoor 67 dataset.

Parameter C	Accuracy (%)
1	85.67
2	84.55
3	84.40
4	84.17
5	84.02

5.2. Visualization of Discriminative Regions

In order to demonstrate that the extracted regions are discriminative, we visualize some discriminative regions of four scene categories ('nursery', 'museum', 'croquet', 'industrial') from different datasets. In Figure 12, we show some examples of discriminative regions from four categories. The discriminative regions correspond to the visual mechanism of human observation scenes, for examples, a baby's cot in a nursery, a ball club on a court, and a painting of a museum. This indicates that the discovered regions contain the objects specific to the context of the scene image, and they are helpful to scene recognition.

Figure 12. Examples of discriminative regions discovered by our WS-AM method.

6. Conclusions

In this paper, we proposed a WS-AM method to discover discriminative regions in scene images. Combined with the improved VLAD coding, we could extract more robust features for scene images. Compared with existing methods, our method selects fewer local regions containing semantic information to avoid the influence of redundant regions. The improved VLAD coding is more suitable for our method than the general VLAD coding. The experiments were carried out on three benchmark datasets: MIT indoor 67, Scene 15, and UIUC Sports, and obtained better performance. Our work was inspired by fine-grained image recognition, whose main task was to find the discriminative regions within the class. In the future, we will improve our methods and apply them to other recognition tasks.

Author Contributions: Conceptualization, data curation and formal analysis, S.X.; Funding acquisition, J.Z.; Investigation, S.X and L.L.; Methodology, S.X.; Project administration, S.X and X.F.; Resources, J.Z. and L.L.; Software, S.X.; Supervision, J.Z. and X.F.; Validation, visualization and writing—original draft, S.X; Writing—review & editing, J.Z, L.L and X.F.

Funding: This work was supported in part by the National Natural Science Foundation of China under Grant 61763033, Grant 61662049, Grant 61741312, Grant 61866028, Grant 61881340421, Grant 61663031, and Grant 61866025, in part by the Key Program Project of Research and Development (Jiangxi Provincial Department of Science and Technology) under Grant 20171ACE50024 and Grant 20161BBE50085, in part by the Construction Project of Advantageous Science and Technology Innovation Team in Jiangxi Province under Grant 20165BCB19007, in part by the Application Innovation Plan (Ministry of Public Security of P. R. China) under Grant 2017YYCXJXST048, in part by the Open Foundation of Key Laboratory of Jiangxi Province for Image Processing and Pattern Recognition under Grant ET201680245 and Grant TX201604002, and in part by the Innovation Foundation for Postgraduate Student of Nanchang Hangkong University under Grant YC2018095.

Conflicts of Interest: The authors declare no conflict of interest.

References

1. Margolin, R.; Zelnik-Manor, L.; Tal, A. OTC: A novel local descriptor for scene classification. In Proceedings of the European Conference on Computer Vision (ECCV), Zurich, Switzerland, 6–12 September 2014.

2. Wu, J.; Rehg, J.M. Centrist: A visual descriptor for scene categorization. *IEEE Trans. Pattern Anal. Mach. Intell.* **2011**, *33*, 1489–1501. [PubMed]

3. Dalal, N.; Triggs, B. Histograms of oriented gradients for human detection. In Proceedings of the IEEE Computer Society Conference on Computer Vision and Pattern Recognition (CVPR), San Diego, CA, USA, 20–26 June 2005.

4. Lowe, D.G. Distinctive image features from scale-invariant keypoints. *Int. J. Comput. Vis.* **2004**, *60*, 91–110. [CrossRef]

5. Shi, J.; Zhu, H.; Yu, S.; Wu, W.; Shi, H. Scene Categorization Model Using Deep Visually Sensitive Features. *IEEE Access.* **2019**, *7*, 45230–45239. [CrossRef]

6. Dixit, M.; Chen, S.; Gao, D.; Rasiwasia, N.; Vasconcelos, N. Scene classification with semantic Fisher vectors. In Proceedings of the IEEE Conference on Computer Vision and Pattern Recognition (CVPR), Boston, MA, USA, 7–12 June 2015.

7. Feng, J.; Fu, A. Scene Semantic Recognition Based on Probability Topic Model. *Information* **2018**, *9*, 97. [CrossRef]

8. Wu, R.; Wang, B.; Wang, W.; Yu, Y. Harvesting Discriminative Meta Objects with Deep CNN Features for Scene Classification. In Proceedings of the IEEE International Conference on Computer Vision (ICCV), Santiago, Chile, 7–13 December 2015.

9. Zhou, B.; Lapedriza, A.; Khosla, A.; Oliva, A.; Torralba, A. Places: A 10 Million Image Database for Scene Recognition. *IEEE Trans. Pattern Anal. Mach. Intell.* **2018**, *40*, 1452–1464. [CrossRef] [PubMed]

10. Quattoni, A.; Torralba, A. Recognizing indoor scenes. In Proceedings of the IEEE Conference on Computer Vision and Pattern Recognition (CVPR), Miami, FL, USA, 20–25 June 2009.

11. Selvaraju, R.R.; Cogswell, M.; Das, A.; Vedantam, R.; Parikh, D.; Batra, D. Grad-CAM: Visual explanations from deep networks via gradient-based localization. In Proceedings of the IEEE International Conference on Computer Vision (ICCV), Venice, Italy, 22–29 October 2017.

12. Deng, J.; Socher, R.; Li, F.-F.; Dong, W.; Li, K.; Li, L.-J. ImageNet: A large-scale hierarchical image database. In Proceedings of the IEEE Conference on Computer Vision and Pattern Recognition (CVPR), Miami, FL, USA, 20–25 June 2009.

13. Jégou, H.; Perronnin, F.; Douze, M.; Sánchez, J.; Pérez, P.; Schmid, C. Aggregating local image descriptors into compact codes. *IEEE Trans. Pattern Anal. Mach. Intell.* **2012**, *34*, 1704–1716. [CrossRef]

14. Bay, H.; Ess, A.; Tuytelaars, T.; Gool, L.J.V. Speeded-up robust features (SURF). *Comput. Vis. Image Underst.* **2008**, *110*, 346–359. [CrossRef]

15. Lazebnik, S.; Schmid, C.; Ponce, J. Beyond Bags of Features: Spatial Pyramid Matching for Recognizing Natural Scene Categories. In Proceedings of the IEEE Computer Society Conference on Computer Vision and Pattern Recognition (CVPR), New York, NY, USA, 17–22 June 2006.

16. Felzenszwalb, P.F.; McAllester, D.A.; Ramanan, D. A Discrimin-atively Trained, Multiscale, Deformable Part Model. In Proceedings of the IEEE Computer Society Conference on Computer Vision and Pattern Recognition (CVPR), Anchorage, AL, USA, 24–26 June 2008.

17. Pandey, M.; Lazebnik, S. Scene recognition and weakly supervised object localization with deformable part-based models. In Proceedings of the IEEE 9th International Conference on Computer Vision (ICCV), Barcelona, Spain, 6–13 November 2011.

18. Yang, J.; Yu, K.; Gong, Y.; Huang, T.S. Linear spatial pyramid matching using sparse coding for image classification. In Proceedings of the IEEE Conference on Computer Vision and Pattern Recognition (CVPR), Miami, FL, USA, 20–25 June 2009.

19. Krizhevsky, A.; Sutskever, I.; Hinton, G.E. Imagenet classification with deep convolutional neural networks. In Proceedings of the Twenty-sixth Annual Conference on Neural Information Processing Systems (NIPS), Lake Tahoe, NV, USA, 3–8 December 2012.

20. Simonyan, K.; Zisserman, A. Very Deep Convolutional Networks for Large-Scale Image Recognition. In Proceedings of the International Conference on Learning Representations (ICLR), San Diego, CA, USA, 7–9 May 2015.

21. Szegedy, C.; Liu, W.; Jia, Y. Going deeper with convolutions. In Proceedings of the IEEE Conference on Computer Vision and Pattern Recognition (CVPR), Boston, MA, USA, 7–12 June 2015.

22. He, K.; Zhang, X.; Ren, S.; Sun, J. Deep residual learning for image recognition. In Proceedings of the IEEE Conference on Computer Vision and Pattern Recognition (CVPR), Las Vegas, NV, USA, 26 June–1 July 2016.

23. Wang, L.; Guo, S.; Huang, W.; Xiong, Y.; Qiao, Y. Knowledge Guided Disambiguation for Large-Scale Scene Classification with Multi-Resolution CNNs. *IEEE Trans. Image Process.* **2017**, *26*, 2055–2068. [CrossRef]

24. Javed, S.A.; Nelakanti, A.K. Object-Level Context Modeling for Scene Classification with Context-CNN. *arXiv* **2017**, arXiv:1705.04358.

25. Wang, Z.; Wang, L.; Wang, Y.; Zhang, B.; Qiao, Y. Weakly Supervised PatchNets: Describing and Aggregating Local Patches for Scene Recognition. *IEEE Trans. Image Process.* **2017**, *26*, 2028–2041. [CrossRef]

26. Herranz, L.; Jiang, S.; Li, X. Scene Recognition with CNNs: Objects, Scales and Dataset Bias. In Proceedings of the IEEE Conference on Computer Vision and Pattern Recognition (CVPR), Las Vegas, NV, USA, 26 June–1 July 2016.

27. Uijlings, J.R.R.; van de Sande, K.E.A.; Gevers, T.; Smeulders, A.W.M. Selective Search for Object Recognition. *Int. J. Comput. Vis.* **2013**, *104*, 154–171. [CrossRef]

28. Arbeláez, P.A.; Pont-Tuset, J.; Barron, J.T.; Marqués, F.; Malik, J. Multiscale combinatorial grouping. In Proceedings of the IEEE Conference on Computer Vision and Pattern Recognition (CVPR), Columbus, OH, USA, 24–27 June 2014.

29. Zitnick, C.; Dollár, P. Edge boxes: Locating object proposals from edges. In Proceedings of the European Conference on Computer Vision (ECCV), Zurich, Switzerland, 6–12 September 2014.

30. Leng, L.; Zhang, J.; Khan, M.K.; Chen, X.; Alghathbar, K. Dynamic weighted discrimination power analysis: A novel approach for face and palmprint recognition in DCT domain. *Int. J. Phys. Sci.* **2010**, *5*, 2543–2554.

31. Leng, L.; Zhang, J.; Xu, J.; Khan, M.K.; Alghathbar, K. Dynamic weighted discrimination power analysis in DCT domain for face and palmprint recognition. In Proceedings of the International Conference on Information and Communication Technology Convergence IEEE(ICTC), Jeju, Korea, 17–19 November 2010.

32. Zhou, B.; Khosla, A.; Lapedriza, A.; Oliva, A.; Torralba, A. Learning Deep Features for Discriminative Localization. In Proceedings of the IEEE Conference on Computer Vision and Pattern Recognition (CVPR), Las Vegas, NV, USA, 26 June–1 July 2016.

33. Fu, J.; Zheng, H.; Mei, T. Look Closer to See Better: Recurrent Attention Convolutional Neural Network for Fine-Grained Image Recognition. In Proceedings of the IEEE Conference on Computer Vision and Pattern Recognition (CVPR), Honolulu, HI, USA, 21–26 July 2017.

34. Xiao, T.; Xu, Y.; Yang, K.; Zhang, J.; Peng, Y.; Zhang, Z. The application of two-level attention models in deep convolutional neural network for fine-grained image classification. In Proceedings of the IEEE Conference on Computer Vision and Pattern Recognition (CVPR), Boston, MA, USA, 7–12 June 2015.

35. Liu, X.; Xia, T.; Wang, J.; Yang, Y.; Zhou, F.; Lin, Y. Fully Convolutional Attention Localization Networks: Efficient Attention Localization for Fine-Grained Recognition. *arXiv* **2016**, arXiv:1603.06765.

36. Luo, C.; Jin, L.; Sun, Z. MORAN: A Multi-Object Rectified Attention Network for scene text recognition. *Pattern Recognition.* **2019**, *90*, 109–118. [CrossRef]

37. Gao, Y.; Huang, Z.; Dai, Y. Double Supervised Network with Attention Mechanism for Scene Text Recognition. *arXiv* **2018**, arXiv:1808.00677.

38. Wang, Q.; Jia, W.; He, X.; Lu, Y.; Blumenstein, M.; Huang, Y. FACLSTM: ConvLSTM with Focused Attention for Scene Text Recognition. *arXiv* **2019**, arXiv:1904.09405.

39. Lorenzo, P.; Tulczyjew, L.; Marcinkiewicz, M.; Nalepa, J. Band Selection from Hyperspectral Images Using Attention-based Convolutional Neural Networks. *arXiv* **2018**, arXiv:1811.02667.

40. Hochreiter, S.; Schmidhuber, J. Long short-term memory. *Neural Comput.* **1997**, *9*, 1735–1780. [CrossRef]

41. Chung, J.; Gülçehre, Ç.; Cho, K.; Bengio, Y. Empirical evaluation of gated recurrent neural networks on sequence modeling. *arXiv* **2014**, arXiv:1412.3555.

42. Vaswani, A.; Shazeer, N.; Parmar, N.; Uszkoreit, J.; Jones, L.; Gomez, N.; Kaiser, L.; Polosukhin, I. Attention is All you Need. In Proceedings of the 31st Conference on Neural Information Processing Systems (NIPS), Long Beach, CA, USA, 4–9 December 2017.

43. Csurka, G.; Bray, C.; Dance, C.; Fan, L. Visual categorization with bags of keypoints. In Proceedings of the European Conference on Computer Vision Workshop (ECCV Workshop), Prague, Czech Republic, 11–14 May 2004.

44. Sivic, J.; Zisserman, A. Video google: A text retrieval approach to object matching in videos. In Proceedings of the IEEE 9th International Conference on Computer Vision (ICCV), Nice, France, 14–17 October 2003.

45. Perronnin, F.; Sánchez, J.; Mensink, T. Improving the fisher kernel for large-scale image classification. In Proceedings of the European Conference on Computer Vision (ECCV), Heraklion, Greece, 5–11 September 2010.

46. Sánchez, J.; Perronnin, F.; Mensink, T.; Verbeek, J.J. Image classification with the fisher vector: Theory and practice. *Int. J. Comput. Vis.* **2013**, *105*, 222–245. [CrossRef]

47. Huang, Y.; Huang, K.; Yu, Y.; Tan, T. Salient coding for image classification. In Proceedings of the IEEE Conference on Computer Vision and Pattern Recognition (CVPR), Colorado Springs, CO, USA, 20–25 June 2011.

48. Khan, S.H.; Hayat, M.; Bennamoun, M.; Togneri, R.; Sohel, A.F. A Discriminative Representation of Convolutional Features for Indoor Scene Recognition. *IEEE Trans. Image Process.* **2016**, *25*, 3372–3383. [CrossRef] [PubMed]

49. Leng, L.; Li, M.; Kim, C.; Bi, X. Dual-source discrimination power analysis for multi-instance contactless palmprint recognition. *Multimed. Tools Applic.* **2017**, *76*, 333–354. [CrossRef]

50. Leng, L.; Teoh, A.B.; Li, J.M.; Khan, M.K. A remote cancelable palmprint authentication protocol based on multi-directional two-dimensional PalmPhasor-fusion. *Securit. Commun. Netw.* **2014**, *7*, 1860–1871. [CrossRef]

51. Leng, L.; Zhang, J. PalmHash Code vs. PalmPhasor Code. *Neurocomputing* **2013**, *108*, 1–12. [CrossRef]

52. Li, L.-J.; Li, F.-F. What, where and who? Classifying events by scene and object recognition. In Proceedings of the IEEE 11th International Conference on Computer Vision (ICCV), Rio de Janeiro, Brazil, 14–20 October 2007.

53. Li, L.-J.; Su, H.; Xing, E.P.; Li, F.-F. Object Bank: A High-Level Image Representation for Scene Classification & Semantic Feature Sparsification. In Proceedings of the 24th Annual Conference on Neural Information Processing Systems (NIPS), Vancouver, BC, Canada, 6–11 December 2010.

54. Singh, S.; Gupta, A.; Efros, A.A. Unsupervised Discovery of Mid-Level Discriminative Patches. In Proceedings of the European Conference on Computer Vision (ECCV), Florence, Italy, 7–13 October 2012.

55. Juneja, M.; Vedaldi, A.; Jawahar, V.; Zisserman, A. Blocks that shout: Distinctive parts for scene classification. In Proceedings of the IEEE Conference on Computer Vision and Pattern Recognition (CVPR), Portland, OR, USA, 25–27 June 2013.

56. Zhou, B.; Lapedriza, A.; Xiao, J.; Torralba, A.; Oliva, A. Learning deep features for scene recognition using places database. In Proceedings of the 28th Annual Conference on Neural Information Processing Systems (NIPS), Montréal, QC, Canada, 8–13 December 2014.

57. Guo, S.; Huang, W.; Wang, L.; Qiao, Y. Locally supervised deep hybrid model for scene recognition. *IEEE Trans. Image Process.* **2017**, *26*, 808–820. [CrossRef] [PubMed]

58. Li, L.-J.; Su, H.; Lim, Y.; Li, F.-F. Object Bank: An Object-Level Image Representation for High-Level Visual Recognition. *Int. J. Comput. Vis.* **2014**, *107*, 20–39. [CrossRef]

59. Kwitt, R.; Vasconcelos, N.; Rasiwasia, N. Scene Recognition on the Semantic Manifold. In Proceedings of the European Conference on Computer Vision (ECCV), Florence, Italy, 7–13 October 2012.

60. Sun, X.; Zhang, L.; Wang, Z.; Chang, J.; Yao, Y.; Li, P.; Zimmermann, R. Scene categorization using deeply learned gaze shifting Kernel. *IEEE Trans. Cybern.* **2018**, *49*, 2156–2167. [CrossRef] [PubMed]

61. Oliva, A.; Torralba, A. Modeling the Shape of the Scene: A Holistic Representation of the Spatial Envelope. *Int. J. Comput. Vis.* **2001**, *42*, 145–175. [CrossRef]

62. Zhu, J.; Li, L.-J.; Li, F.-F.; Xing, E.P. Large Margin Learning of Upstream Scene Understanding Models. In Proceedings of the Twenty-fourth Annual Conference on Neural Information Processing Systems (NIPS), Vancouver, BC, Canada, 6–11 December 2010.

63. Li, Q.; Wu, J.; Tu, Z. Harvesting mid-level visual concepts from large-scale internet images. In Proceedings of the IEEE Conference on Computer Vision and Pattern Recognition (CVPR), Portland, OR, USA, 25–27 June 2013.

64. Lin, D.; Lu, C.; Liao, R.; Jia, J. Learning Important Spatial Pooling Regions for Scene Classification. In Proceedings of the IEEE Conference on Computer Vision and Pattern Recognition (CVPR), Columbus, OH, USA, 24–27 June 2014.

65. Wang, L.; Guo, S.; Huang, W.; Qiao, Y. Places205-VGGNet models for scene recognition. *arXiv* **2015**, arXiv:1508.01667.

 *electronics*

Article

A Novel Rate Control Algorithm Based on ρ Model for Multiview High Efficiency Video Coding

Tao Yan [1], In-Ho Ra [2,*], Qian Zhang [3], Hang Xu [1] and Linyun Huang [1]

[1] School of Information Engineering, Putian University, Putian 351100, China; yantaoshu@aliyun.com (T.Y.);
    hangxu520@hotmail.com (H.X.); huanglinyun@ptu.edu.cn (L.H.)
[2] School of Computer, Information and Communication Engineering, Kunsan National University,
    Gunsan 54150, Korea
[3] School of Information, Mechanical and Electrical Engineering, Shanghai Normal University,
    Shanghai 200234, China; qianzhang@shnu.edu.cn
* Correspondence: ihra@kunsan.ac.kr; Tel.: +82-63-469-4691

Received: 30 December 2019; Accepted: 10 January 2020; Published: 16 January 2020

Abstract: Most existing rate control algorithms are based on the rate-quantization (R-Q) model. However, with video coding schemes becoming more flexible, it is very difficult to accurately model the R-Q relationship. Therefore, in this study we propose a novel ρ domain rate control algorithm for multiview high efficiency video coding (MV-HEVC). Firstly, in order to further improve the efficiency of MV-HEVC, this paper uses our previous research algorithm to optimize the MV-HEVC prediction structure. Then, we established the ρ domain rate control model based on multi-objective optimization. Finally, it used image similarity to analyze the correlation between viewpoints, using encoded information and frame complexity to proceed in bit allocation and bit rate control of the inter-view, frame lay, and base unit. The experimental simulation results show that the algorithm can simultaneously maintain high coding efficiency, where the average error of the actual bit rate and the target bit rate is only 0.9%.

Keywords: multiview high efficiency video coding; ρ model; bit allocation; rate control; image similarity; frame complexity

1. Introduction

Recently, three-dimensional video (3DV) has become increasingly popular, because it provides real depth perception, immersive vision, and novel visual enjoyment for multimedia application. With the development and application of information technology, traditional two-dimensional video technology cannot meet the user's visual demands, and high-definition (HD), three-dimensional (3D), and wireless mobile have become the mainstream trends in video application. However, the compression efficiency of existing coding standards remains insufficient to address HD and ultra HD video applications, and more efficient coding compression schemes are still needed. The Telecommunication Standardization Sector and Moving Picture Experts Group established the Video Coding Joint group (Joint Collaborative Team on Video Coding, or JCT-VC) to solve this problem. In 2013, the first generation of the high-efficiency video coding (HEVC) standard was completed [1]. In 2015, multi-view high-efficiency video coding (MV-HEVC), as one of the new 3D standards based on HEVC, was introduced; it had a strong sense of stereoscopic and flexible interaction, which can vividly present a video scene, and showed promise of having wide application in the areas of 3DTV, video conferencing, and so on [2,3]. It has become one research focus in the field of international video coding [4,5].

Rate control plays an important role in video application, particularly in real-time communication applications. Bit rate control makes the generated bitstream conform with the needs of different channel bit rates, by controlling the encoding parameters and achieving a high quality of coding. It is one of the very important technologies for video coding. When any video compression standard lack rate control, its application will be limited. Previous video compression standards, such as MPEG-2, MPEG-4, H.263, H.264, and multi-view video coding (MVC), have provided a bit rate control model. Currently, the internationally published test model of MV-HEVC has not yet provided an effective code rate control algorithm [6].

2. Discussion of the Pros and Cons of the Various Approaches

Recently, researchers have been working on the MV-HEVC code rate control, both locally and abroad; most researchers are engaged in research on the MVC code rate control. Woo et al. studied the optimal bit allocation problem in 3D video coding, based on rate distortion theory [7]. They proposed a reasonable bit allocation algorithm, but the coding complexity is high, the computation is large, and it is difficult to meet the application requirements. Lim et al. proposed a code rate control algorithm based on the multi-view video bit rate control of the binomial model [8]. The algorithm, using motion prediction and parallax forecast spatial structure relations, places all the images into a variety of coding types. Then, it models various types of images and calculates the target bit number and frame level quantization parameters of each type of frame, according to the parameters of the model. However, in video coding of a multi-view point, the parallax prediction feature of each viewpoint has a large difference; thus, the encoding image with the same prediction relation may have different encoding characteristics. At this time, the target bit number obtained using the same model parameter will be biased. South Korea's Seanae Park and others have considered MVC using the effect of a hierarchical B frame. It performs bit allocation on MVC based on H.264 and maintains efficient coding efficiency [9]. However, its bit rate control error is relatively large, and the average bit rate control error is greater than 1%, which is not operable in practical applications. At the German Karlsruhe Institute of Technology, Bruno Boessio Vizzotto used a uniform buffer for both the right and left views in the bit rate control algorithm of stereo video coding, and then used MPEG-2's code rate control model, termed TM5, to control the code flow rate [10]. However, the accuracy of the target bit allocation based on TM5 worsens with an increase in the encoding image type in the MVC.

The aforementioned code rate control models for MVC were based on H.264. Currently, there is limited research on video coding bit rate control based on HEVC. In 2013, Shao et al. established the distortion equation of texture bit and virtual viewpoint, and the distortion equation of depth bit and virtual viewpoint [11]. They combined a texture and depth virtual viewpoint distortion function to solve the texture and depth code rate, and minimize distortion of the viewpoint. However, this method does not consider the efficiency of the bit rate of the texture and depth. The virtual viewpoint distortion caused by the texture, and the virtual viewpoint distortion caused by the depth map, are regarded as the same weight. In 2014, Pan et al. proposed a deep 3D-HEVC code rate control algorithm, with a fixed color and depth bit rate ratio of 4:1 [12], but it could not obtain the optimal rendering quality of virtual viewpoints. In 2015, Zhao Zhenjun and others proposed a joint bit allocation algorithm based on 3D-HEVC multi-view texture and depth, which is based on the statistical properties of video series [13]. This algorithm establishes a model of texture bit rate and depth map bit rate, and virtual viewpoint distortion to control the bit rate. Xiao et al. proposed the depth and texture grading bit rate control algorithm [14]. Wang et al. proposed the 3D-HEVC bit rate control algorithm based on the binomial R-D model [15]. The accuracy of the code rate control is low because of the direct use of the H.264 rate control model. In 2016, Yang et al. solved the bit rate of texture and depth, by combining the texture and depth virtual view distortion functions [16]. However, this method does not consider the efficiency of the bit rate of texture and depth, and the error of the bit rate control accuracy is lower than the average bit rate of 2.4%. Li et al. proposed the rate control algorithm for high efficiency video coding [17], but it is necessary to further study the optimal model of 3D-HEVC bit rate control. Lei J et al. proposed a novel rate control algorithm based on the region adaptive R-λ model, which can achieve considerable bjøntegaard delta peak signal-to-noise rate (BD-PSNR) gains [18].

The above studies have not taken into consideration the relationship between the MV-HEVC bit rate control model and related coding performance. Most of the studies are engaged in research on rate control for HEVC or MVC [19–27]. There have been many studies on rate control for HEVC based on scene switching [28,29], but most of the rate control algorithms for HEVC are concentrated in the single-channel video coding standard, which is not applicable to multiview video coding. Li et al. proposed the rate control algorithm for HEVC, but it is necessary to further study the optimal model of MV-HEVC bit rate control [30]. We also preliminarily explored the MVC bit rate control model, and proposed an MVC bit rate control optimization algorithm based on the binomial R-D model [31]. We believe the MV-HEVC bit rate control model in the design can also be seen as a multi-objective optimization problem. It needs to adaptively adjust the parameters of the rate control model, according to the characteristics of the video content and the requirements of the specific application, so that the accuracy of the bit rate control and the subjective quality cannot fluctuate significantly and the best balance between the two is achieved. The experimental simulation results show that the average error between the actual bit rate and the target bit rate of this rate control algorithm is only 0.90%. At the same time, efficient coding efficiency has theoretically reached the basic requirements for practical application.

This paper is structured as follows: In Section 2, we review the previous work on rate control. Section 3 addresses the ρ domain rate control model, and describes the rate control for MV-HEVC based on this model in detail. In Section 4, extensive experiments are conducted to evaluate the performance of the proposed method. Finally, conclusions are drawn in Section 5.

3. ρ Model for MV-HEVC

The reference code of the latest video coding standard, HEVC, usually adopts the rate-Lambda (R-Lambda) model for bit rate control, but the R-Lambda model allocates too many target bit rates for I-frames, causing subsequent video frames to have insufficient target bit rates. The quality of reconstruction has deteriorated severely. In High Efficient Video Coding (HEVC), the bit rate control algorithm achieves good results for both the accuracy and efficiency of the bit rate output, but the algorithm does not take into account the complexity of the actual video encoding content.

The rate control algorithm based on the ρ domain is proposed by He Zhihai [32,33], where ρ represents the percentage of the zero coefficients, after the quantization of the transform coefficients to all the coefficients. Through a large number of experiments and theoretical proofs, the paper reached the following conclusion: For video signals, ρ has a linear relationship with the texture bit encoding rate $T(\rho)$. The linear model is:

$$T(\rho) = \theta(1 - \rho) \tag{1}$$

In order to introduce the ρ model into rate control algorithm for MV-HEVC, we have done a lot of experiments to study the relationship between ρ and the encoding bit rate of textured parts. The platform used in the experiment is the MV-HEVC test model published internationally in 2016 [6]. Using the "Exit" test sequence, the frame rate is 25 frames/second, and the quantization parameter (QP) ranges from 0 to 51. Figure 1 shows the $R(\rho)$ curve. It can be seen from the figure that $R(\rho)$ is approximately a quadratic curve passing through the (1, 0) point. Our previous research has shown that ρ has the following quadratic relationship with the texture bit encoding bit rate $R(\rho)$:

$$R(\rho) = \chi \cdot (1 - \rho)^2 + \psi \cdot (1 - \rho) \tag{2}$$

where, χ, ψ can be provided by the following statistical analysis method. Let $(\rho_1, R_1(\rho)), (\rho_2, R_2(\rho)), \cdots, (\rho_n, R_n(\rho))$ be the existing n sample values, thus

$$\begin{cases} R_1(\rho) = \chi \cdot (1-\rho_1)^2 + \psi \cdot (1-\rho_1) \\ R_2(\rho) = \chi \cdot (1-\rho_2)^2 + \psi \cdot (1-\rho_2) \\ \qquad\qquad \vdots \\ R_n(\rho) = \chi \cdot (1-\rho_n)^2 + \psi \cdot (1-\rho_n) \end{cases} \tag{3}$$

Suppose that $\rho'_{1i}(\rho) = (1-\rho_i)^2$, $\rho'_{2i} = 1 - \rho_i$, and

$$\rho' = \begin{pmatrix} \rho'_{11} & \rho'_{21} \\ \rho'_{12} & \rho'_{22} \\ \vdots & \vdots \\ \rho'_{1n} & \rho'_{2n} \end{pmatrix} \qquad R = \begin{pmatrix} R_1 \\ R_2 \\ \vdots \\ R_n \end{pmatrix} \qquad X = \begin{pmatrix} \chi \\ \psi \end{pmatrix} \tag{4}$$

Using multiple regression techniques, the model parameter, N, can be calculated as follows:

$$\chi = ((\rho'^T \cdot \rho')^{-1} \cdot \rho'^T \cdot R)_{11} \tag{5}$$

$$\psi = ((\rho'^T \cdot \rho')^{-1} \cdot \rho'^T \cdot R)_{21} \tag{6}$$

where ρ'^T is the transpose matrix of ρ', and $(\rho'^T \rho')^{-1}$ is the inverse matrix of $\rho'^T \rho'$.

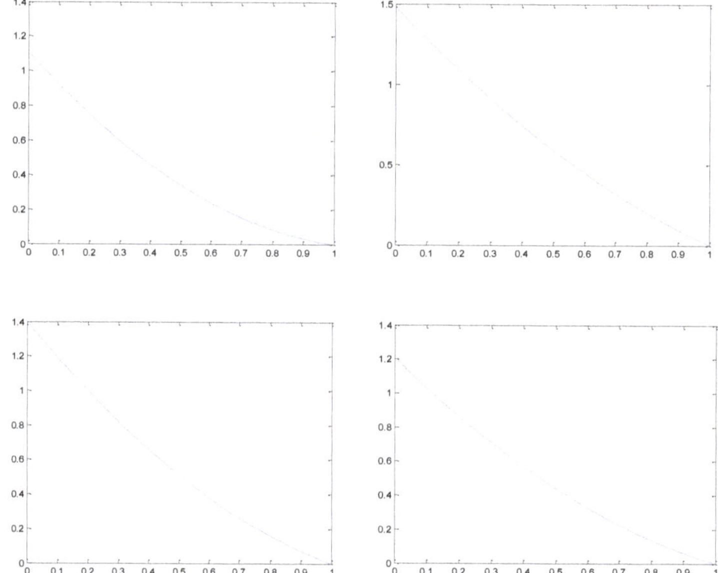

Figure 1. Experimental results for R(ρ) curve.

4. Rate Control Algorithm for MV-HEVC

To be compatible with the latest video coding standard HEVC, the bit allocation and bit rate control proposed in this study is based on the HEVC bit rate control algorithm. In order to further improve the efficiency of MV-HEVC, this paper uses our previous research algorithm to optimize the MV-HEVC prediction structure, before performing rate control for MV-HEVC [34]. The main problem of the rate control algorithm for MV-HEVC is how to perform bit allocation among viewpoints and how to use the correlation among viewpoints to perform bit allocation. The key steps of the MV-HEVC bit rate control algorithm are as follows:

4.1. View Layer Rate Control

In this study, reasonable allocation of bits to different viewpoints were based on viewpoint similarity and encoded information. The weight, w_k, was used to indicate the degree of importance of the viewpoint, k. The larger w_k was, the more important the viewpoint was. The total number of bits allocated to the Kth viewpoint, GOP_K, within each coded $GGOP$ picture group, is provided by Equation (7).

$$T_{GOP}(n_{k,0}) = T_{GGOP}(sn_{i,0}) \cdot w_k \tag{7}$$

The initial value of $w_k(k = 0, 1, 2, L, N_{view} - 1)$ is provided by Equation (8).

$$w_k = \frac{\frac{1}{N} \cdot \sum\limits_{j=0, j \neq k}^{N-1} S(V_j, V_k)}{\sum\limits_{k=0}^{N-1} \frac{1}{N-1} \cdot \sum\limits_{j=0, j \neq k}^{N-1} S(V_j, V_k)} \tag{8}$$

where N is the number of encoded viewpoints, and $S(V_j, V_k)$ is the similarity between viewpoints V_j and V_k. The bilinear similarity measurement algorithm was adopted. This algorithm has been successfully used in the field of image retrieval. The algorithm is superior to traditional distance metrics, and there are no restrictions Among them, S_j^d and S_k^d are the feature vectors of the two images, respectively.

$$S(V_j, V_k) = \frac{S_j^k \cdot S_k^d}{\left| S_j^d \right| \cdot \left| S_k^d \right|} \tag{9}$$

4.2. Frame Layer Rate Control

In the HEVC frame layer rate allocation, the bit allocation per frame is determined by the frame rate, target buffer capacity, actual buffer size, etc. The residual energy of the coded frame is not considered, which is likely to cause image quality degradation and a jump phenomenon in the frame. Previous research results in [15] have proposed the following optimal frame target bit allocation method, according to the residual energy of the coded frame:

$$T(j) - \frac{MAD_j}{MAD_a} \frac{(T - \sum\limits_{m=1}^{M} C_m)}{M} + C_j \tag{10}$$

In the aforementioned equation, T is the sum of the number of bits consumed for encoding an M frame; MAD_a represents the average of all frame MAD (mean absolute deviation difference); MAD_j represents MAD at frame j; C_j and C_m occupy bits of the header information of the j-th frame and the m-th frame, respectively. In Equation (10), it can be seen that the larger a and b were, the more target bits were allocated to image frames.

In the multi-view video code, the target bit of the j frame assignment is as follows:

$$T_r'(j-1) = \left[\frac{MAD_{j-1}}{MAD_a} \cdot \left(\frac{T_{GOP}(n_{x,0})}{N(i)} - C_a \right) + C_{j-1} \right] \tag{11}$$

In Equation (10), C_a represents the average value of the bits consumed for encoding the header information of the encoded frame in the current GOP.

In general, the smaller the active time domain of the frame, the fewer bits are needed; conversely, the larger the active time domain of the frame, the more bits are needed. To make the MVC rate control more accurate, the code rate control method in Equation (11) was further improved. The current frame target bit is calculated using Equation (12):

$$T_r(j) = T_r'(j-1) \cdot \frac{\sum\limits_{l=1}^{L} W(l) \cdot 2^n}{\sum\limits_{l=1}^{L} \frac{MAD_{l-1}}{MAD_a} \cdot \frac{FD(l-1)^2}{\frac{1}{L-1} \cdot \sum\limits_{k=1}^{L-1} FD(k)^2} \cdot W(l) + \sum\limits_{l=1}^{L} W_B(l) \cdot (2^n - 1)} + T_j \qquad (12)$$

In Equation (12), T_j is the bit consumed by the frame header information of the frame, and j and n represent the current time level. $FD(j)$ is the temporal activity for jth frame. $W(l)$ represents the weight of each frame complexity. $W_B(l)$ represents the weight of the B frame.

4.3. Macroblock Layer Rate Control

According to macroblock layer rate control algorithm for HEVC, it is known that the bits allocated in each frame are evenly distributed to each basic unit layer of the frame, so that different macroblocks in the same basic unit layer are encoded using a uniform quantization parameter (QP). However, even the macroblocks in the same basic unit have great differences in the complexity of image content, texture, and active time domain. Therefore, in order to control the MV-HEVC bit rate more accurately, different quantization values are used, according to the complexity of its image content, texture, and active time domain. ρ can be obtained from the ρ model. Our previous research has obtained the relationship between ρ and the quantization parameter (QP) [31]. Therefore, we can calculate the quantization parameter (QP) of the basic unit layer. The specific algorithm flow is shown in Figure 2.

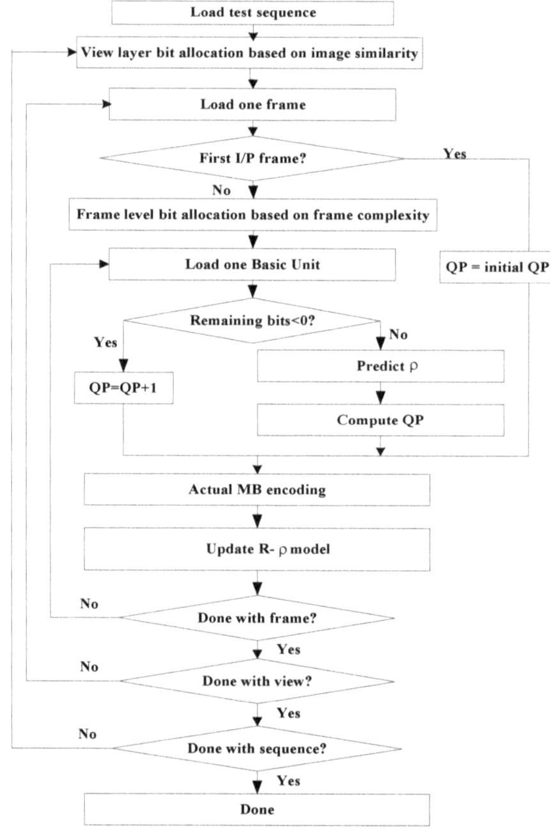

Figure 2. Rate control algorithm based on ρ model for MV-HEVC.

The main steps of the rate control algorithm based on the ρ model for MV-HEVC are as follows:

Step 1: Firstly, establish a framework for continuous encoding of multiple viewpoints, and realize continuous encoding for multiple viewpoints. Multi-view sequence is then decomposed into several *GGOPs* (the group of group of pictures), and the programming parameters are initialized.

Step 2: *GGOP* (the group of group of pictures) layer bit allocation and code rate control—Get the current target number of *GGOP* bits according to the frame rate, bandwidth, buffer, etc.

Step 3: *GOP* (group of pictures) layer bit allocation and code rate control—Firstly, calculate the weight factor (*Wk*) of the *GOP* of each viewpoint according to the correlation between viewpoints. Then obtain the target number of bits of a *GOP* for the current viewpoint.

Step 4: Frame layer bit allocation and code rate control—Obtain the number of bits allocated to the current encoding frame, according to the frame complexity.

Step 5: Macroblock layer bit allocation and code rate control—According to the number of frame bits obtained in Step 4, the number of bits allocated by the current coding basic unit is then obtained according to the complexity of the basic unit; then, ρ is calculated according to the code rate control model (ρ model), and finally the quantization parameters of the current macroblock are determined.

Step 6: Encode the current macroblock according to the quantization parameter calculated in Step 5.

Step 7: Determine whether all macroblocks in the current frame are encoded. If they are all encoded, go to Step 8; if they are not all encoded, repeat Steps 5 to 6 until they are all encoded, then go to Step 8.

Step 8: Determine whether all the frames in the current *GOP* are encoded. If they are all encoded, go to Step 9; if they are not all encoded, repeat Steps 4 to 7 until all the frames of the current *GOP* are edited.

Step 9: Determine whether all *GOPs* in the current *GGOP* are encoded. If they are all encoded, go to Step 10; if they are not all encoded, repeat Steps 3 to 8 until all *GOPs* in the current *GGOP* are edited.

Step 10: Determine whether the current *GGOP* is the last *GGOP*. If it is the last *GGOP*, the entire code rate control process ends; otherwise, repeat Steps 2 to 9.

5. Experimental Classification Results and Analysis

In order to verify the algorithm of this paper, on the platform of the MV-HEVC system provided by The Joint Collaborative Team on 3D Video Coding Extension Development (JCT-3V), this paper compares the coding performance of this bit rate control algorithm with the multi-view point bit rate control algorithm proposed in [17,18]. Due to experimental platforms and technical limitations, some algorithms in the references are just simulated data. This paper uses five standard 3DV test sequences from Poznanstreet, Akko & Kayo, Rena, Breakdancers, Uli, and Dalloons. The resolution of the sequence includes 1920×1088 pixels, 1024×768 pixels and 640×480 pixels.

Compared with [17,18], Figure 3 shows that the rate control algorithm proposed in our paper can distribute more bits consumed in the dramatic motion scene frame to several subsequent frames, thus avoiding the large fluctuation in video quality. From Figure 3, it show that the frame quality of the algorithm proposed in this paper fluctuates most smoothly after the video scene is switched.

The performance measures include the x and the x variation (σ_x), which is calculated as

$$\sigma_x = \frac{1}{N}\sum_{i=1}^{N}\left(x_i - \frac{1}{N}\sum_{i=1}^{N}x_i\right)^2 \tag{13}$$

where N denotes the number of total encoded frames.

Figure 4 shows the PSNR fluctuation for the sequences "Balloons" and "Poznanstreet." The results show that the bit rate control algorithm used in this study significantly reduces the PSNR fluctuations and improves the subjective effect.

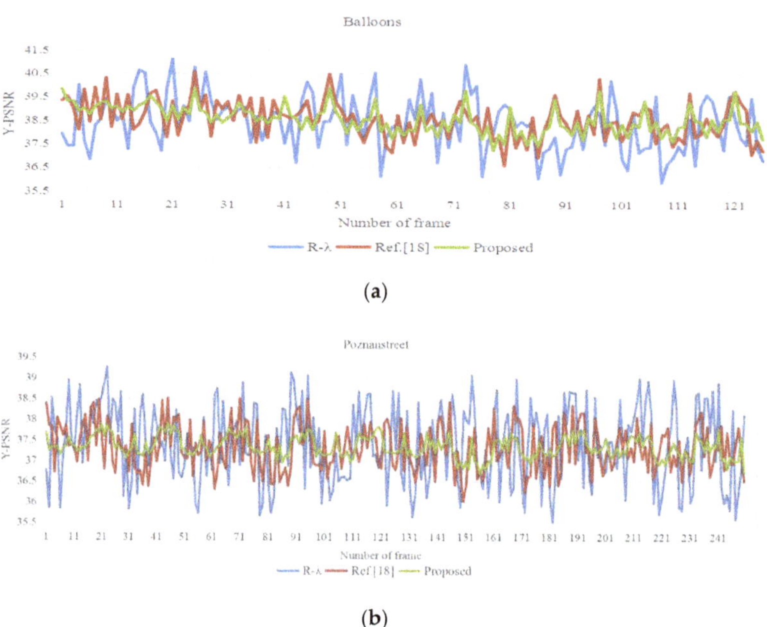

Figure 3. Experimental results of sequences: (**a**) "Balloons", and (**b**) "Poznanstreet".

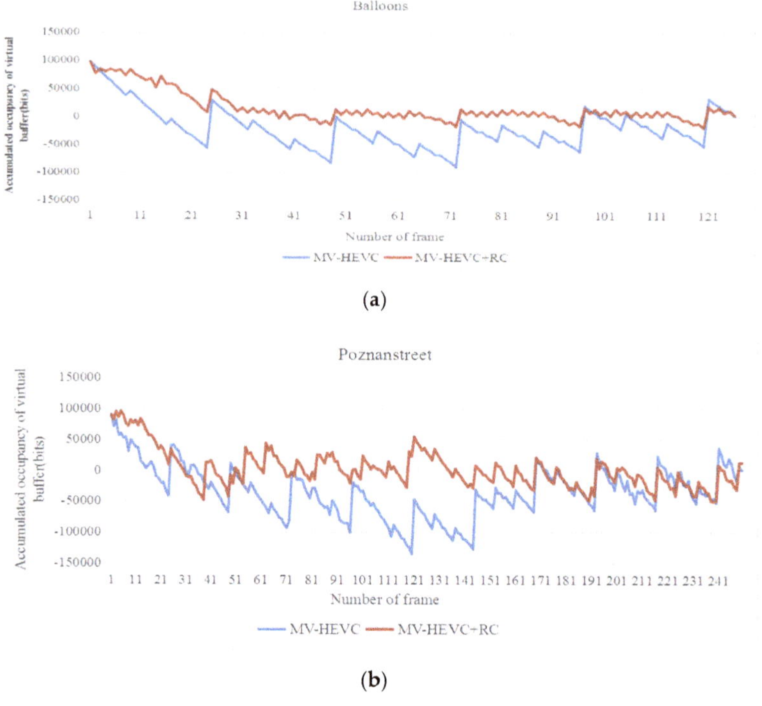

Figure 4. Experimental results of sequences: (**a**) "Balloons", and (**b**) "Poznanstreet".

Table 1 shows the simulation results of MV-HEVC rate control. Our method can accurately control the output bit rate of MV-HEVC. The synthesized virtual views algorithm is similar to [18]. In most cases, the actual bit rate and target bit rate error can be controlled to approximately 1.0% or less. Compared to [17] and [18], Table 1 shows that the rate control algorithm proposed in this study has a more accurate bit rate, smaller bit rate deviation, and a small average error rate of 1%, which can meet practical application requirements. The main reason for this is that not only is there reasonable bit rate control between viewpoints, but also that the bit rate control performs well at both the frame layer and the base unit layer. When the target bit of "Poznanstreet" sequence is 1350 kbps, the bit rate control error in [17] is relatively large (2.85%). This is mainly because "Balloons" had a relatively intense sequence of motion, and it was difficult to perform accurate bit allocation. The rate control error proposed in our study is controlled to 0.9%. The main reason is that the similarity between viewpoints deviates, resulting in inaccurate bit allocation between viewpoints. Table 1 shows that the code rate control algorithm proposed in this study also improved, compared to that of [17,18]. The experimental simulation results show that the algorithm can maintain high coding efficiency, and that the average error between the actual bit rate and the target bit rate is only 0.90%.

Table 1. Simulation results of our method.

	Sequence	Target Bit Rate (Kbps)	Actual Generated Bits (kbps)						Rate Control Error (%)					
			Fixed Ratio	[17]	[4]	[31]	[18]	Proposed	Fixed Ratio	[17]	[4]	[31]	[18]	Proposed
VGA	Akko & Kayo	250.00	256.90	255.70	254.70	254.18	253.23	251.88	2.76	2.28	1.88	1.67	1.29	0.75
		400.00	411.20	408.96	408.76	406.76	405.04	402.36	2.80	2.24	2.19	1.69	1.26	0.59
		500.00	512.75	509.60	508.60	507.85	506.55	503.35	2.55	1.92	1.72	1.57	1.31	0.67
		1000.00	1022.00	1015.40	1014.70	1013.60	1009.50	1004.90	2.20	1.54	1.47	1.36	0.95	0.49
	Rena	250.00	256.70	255.00	254.58	253.35	252.63	251.25	2.68	2.00	1.83	1.34	1.05	0.50
		400.00	411.52	409.44	408.84	407.04	405.52	403.16	2.88	2.36	2.21	1.76	1.38	0.79
		500.00	514.70	511.20	509.85	508.65	507.05	505.20	2.94	2.24	1.97	1.73	1.41	1.04
		1000.00	1027.80	1019.70	1017.80	1015.40	1013.00	1008.30	2.78	1.97	1.78	1.54	1.30	0.83
XGA	Break Dancers	500.00	515.00	510.70	509.65	510.00	507.55	505.80	3.00	2.14	1.93	2.00	1.51	1.16
		760.00	786.98	779.23	776.49	774.82	771.48	767.83	3.55	2.53	2.17	1.95	1.51	1.03
		1000.00	1030.50	1023.30	1019.80	1016.90	1013.10	1009.60	3.05	2.33	1.98	1.69	1.31	0.96
		2050.00	2118.06	2106.17	2097.15	2080.55	2077.27	2070.50	3.32	2.74	2.30	1.49	1.33	1.00
	Uli	500.00	518.45	514.90	513.15	510.95	509.20	506.50	3.69	2.98	2.63	2.19	1.84	1.30
		760.00	782.80	777.63	775.20	774.14	771.02	767.83	3.00	2.32	2.00	1.86	1.45	1.03
		1000.00	1026.80	1020.30	1017.10	1015.00	1012.10	1008.90	2.68	2.03	1.71	1.50	1.21	0.89
		2050.00	2121.34	2110.68	2100.84	2091.21	2081.98	2071.32	3.48	2.96	2.48	2.01	1.56	1.04
HD	Balloons	1530.00	1575.75	1563.66	1555.70	1555.40	1549.58	1540.10	2.99	2.20	1.68	1.66	1.28	0.66
		800.00	829.92	823.36	819.84	813.52	812.00	806.56	3.74	2.92	2.48	1.69	1.50	0.82
		450.00	466.38	462.47	460.53	460.35	458.73	455.72	3.64	2.77	2.34	2.30	1.94	1.27
		265.00	274.33	272.16	271.33	269.77	268.84	267.36	3.52	2.70	2.39	1.80	1.45	0.89
	Poznan-Street	3900.00	4029.48	4008.81	3986.19	3960.06	3954.21	3931.59	3.32	2.79	2.21	1.54	1.39	0.81
		1350.00	1398.33	1388.48	1387.13	1386.99	1380.92	1374.17	3.58	2.85	2.75	2.74	2.29	1.79
		600.00	616.80	613.32	612.90	608.76	607.68	603.60	2.80	2.22	2.15	1.46	1.28	0.60
		300.00	310.86	308.34	307.05	304.74	304.17	302.19	3.62	2.78	2.35	1.58	1.39	0.73
	Average								3.11	2.41	2.11	1.76	1.42	0.90

Figure 5 shows the experimental results of the sequences "Newspaper" and "Poznan Hall2". Compared with [17,18], the algorithm in this paper can effectively control the bit rate of MV-HEVC and maintain a high coding efficiency at the same time. Mathematical quantity analysis of the Figures 3 and 4 is shown in Table 2. For data unification, the data in Table 2 is obtained after further processing.

Table 2. Mathematical quantity analysis of the Figures 3 and 4.

Sequence	σ_{PSNR} (Figure 3)			σ_{buffer} (Figure 4)		Compared with (%)		
	R-λ	Ref. [18]	Proposed	MV-HEVC	Proposed	R-λ	Ref. [18]	MV-HEVC
Balloons	0.72	0.66	0.53	0.42	0.28	26.39	19.70	33.33
Poznanstreet	0.51	0.48	0.37	0.21	0.13	27.45	22.92	38.10

In summary, the proposed rate control algorithm is more accurate than that of [17,18]. The simulation results show that the proposed algorithm achieves up to 0.23–0.78 dB in improvement in PSNR. Meanwhile, it can efficiently control the bit rate with an average rate control error of 0.90%. The main reason is that this paper not only uses our previous research algorithm to optimize the MV-HEVC prediction structure, but also performs rate control algorithm based on ρ model for MV-HEVC.

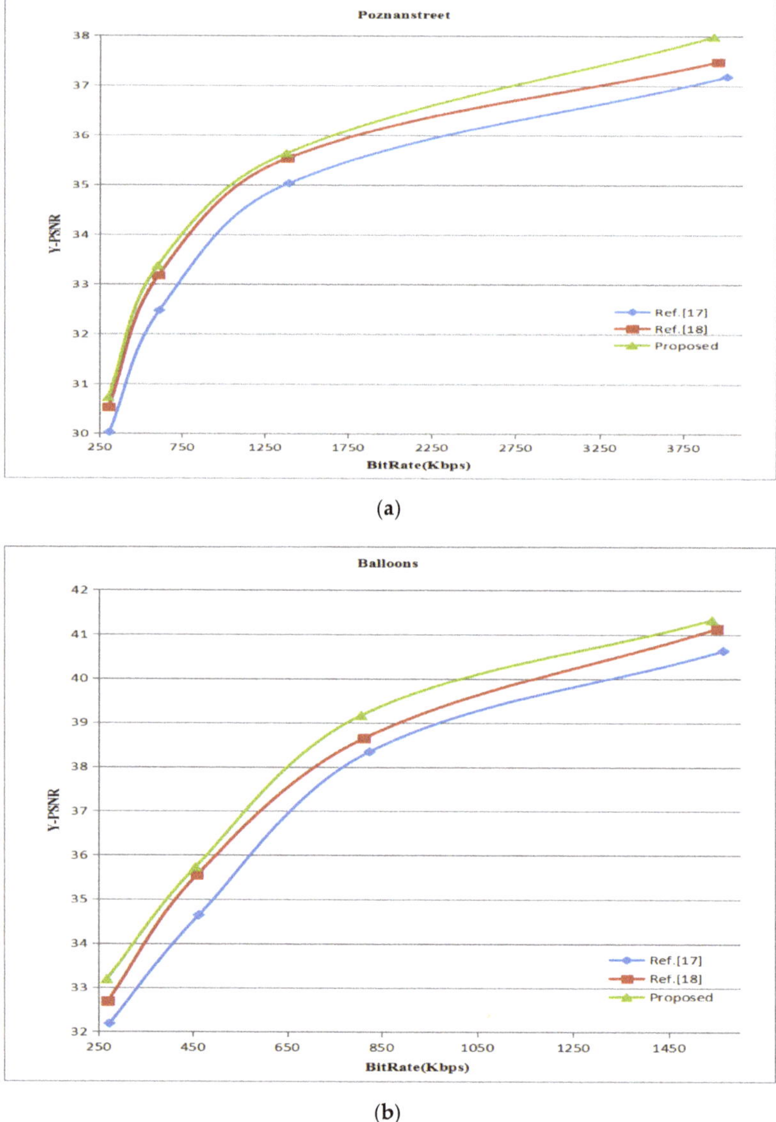

(a)

(b)

Figure 5. PSNR results. (**a**) Experimental results for the sequence "Poznanstreet", and (**b**) Experimental results for the sequence "Balloons".

6. Conclusions

The current research on multi-view video coding rate control based on the MV-HEVC has not been expanded thoroughly. In this paper, by analyzing the deficiency of the bit rate distortion model and the characteristics of multi-view video coding in current video bit rate control, a bit code rate control algorithm based on MV-HEVC multi-view video coding was proposed. The algorithm involves the entire bit rate control process, from the bit rate model design to each model's bit allocation and bit rate control, to ensure the accuracy of the bit rate control algorithm. The experimental results show that the proposed MV-HEVC bit allocation and bit rate control algorithm can effectively control the bit rate, based on the given coding parameters. It will further study the correlation between viewpoints and improve the bit rate control algorithm. In addition, this paper has not considered multi-view scene switching, which is the focus of future work research.

Author Contributions: Conceptualization, T.Y. and I.-H.R.; methodology, T.Y.; software, T.Y.; validation, T.Y., I.-H.R., and Q.Z.; formal analysis, H.X.; investigation, L.H.; data curation, T.Y. and Q.Z.; writing—original draft preparation, T.Y. and I.-H.R.; writing—review and editing, H.X. and I.-H.R.; project administration, I.-H.R.; funding acquisition, I.-H.R. and T.Y. All authors have read and agreed to the published version of the manuscript.

Funding: This research was funded by the Natural Science Foundation of China (No. 61741111). This work was Supported by the Program for New Century Excellent Talents in Fujian Province University (Tao Yan); This research was supported by Basic Science Research Program through the National Research Foundation of Korea (NRF), funded by the Ministry of Education, Science and Technology (2016R1A2B4013002) and in part by Natural Science Foundation of Fujian (No. 2019J01816), the Natural Science Foundation of Jiangxi (20181BAB202011), and the Putian University's Initiation Fee Project for Importing Talents for Scientific Research (2019003).

Acknowledgments: We thank the Interactive Visual Media group of Mitsubishi Electric Research Laboratories and Nagoya University/Tanimoto Lab for the data we used. The authors would like to thank the editors and the reviewers for their professional suggestions.

Conflicts of Interest: The authors declare no conflict of interest.

References

1. Sullivan, G.; Ohm, J.R.; Han, W.J. Overview of the high efficiency video coding (HEVC) standard. *IEEE Trans. Circuits Syst. Video Technol.* **2013**, *22*, 1649–1668. [CrossRef]
2. Yan, T.; Ra, I.H. Bit allocation algorithm based on SSIM for 3D video coding. *Int. J. Perform. Eng.* **2019**, *15*, 1813–1821. [CrossRef]
3. Lee, J.Y.; Han, J.K.; Kim, J.G. Efficient Inter-View Motion Vector Prediction in Multi-View HEVC. *IEEE Trans. Broadcast.* **2018**, *64*, 666–680. [CrossRef]
4. Lee, P.J.; Lai, Y.C. Vision perceptual based rate control algorithm for multi-view video coding. *IEEE Trans. Circuits Syst. Video Technol.* **2011**, *25*, 139–152.
5. Zhang, Q.W.; Zhang, N.; Wei, T.; Huang, K.Q.; Qi, X.L.; Gan, Y. Fast depth map mode decision based on depth–texture correlation and edge classification for 3D-HEVC. *J. Vis. Commun. Image Represent.* **2017**, *45*, 170–180. [CrossRef]
6. Ikai, T.; Kei, K.; Suzuki, T. JCT3V-MV-HEVC and 3D-HEVC Conformance Draft 4. ISO/IEC JTC 1/SC 29/WG 11, JCT3V-N0008. In Proceedings of the 14th Meeting, San Diego, CA, USA, 13–17 March 2016; pp. 22–26.
7. Woo, W.; Ortega, A. Optimal blockwise dependent quantization for stereo image coding. *IEEE Trans. Circuits Syst. Video Technol.* **2013**, *9*, 861–867.
8. Lim, J.E.; Kim, J.; Ngan, K. Advanced rate control technologies for 3D-HDTV. *IEEE Trans. Consum. Electron.* **2013**, *49*, 1–6.
9. Park, S.; Sim, D. An efficient rate-control algorithm for multi-view video coding. In Proceedings of the 13th IEEE International Symposium on consumer Electronics, Kyoto, Japan, 25–28 May 2018; pp. 115–118.
10. Vizzotto, B.B.; Zatt, B. A model predictive controller for frame-level rate control in multiview coding. In Proceedings of the 2012 IEEE International Conference on Multimedia and Expo, Melbourne, Australia, 9–13 July 2012; pp. 485–490.
11. Shao, F.; Jiang, G.Y.; Yu, M. Joint bit allocation and rate control for coding multi-view video plus depth based 3D video. *IEEE Trans. Multimed.* **2013**, *15*, 1843–1854. [CrossRef]

12. Gao, P.; Xiang, W. Rate-Distortion Optimized Mode Switching for Error-Resilient Multi-View Video Plus Depth Based 3-D Video Coding. *IEEE Trans. Multimed.* **2014**, *16*, 1797–1808. [CrossRef]

13. Zhao, Z.J.; Shen, L.Q.; Hu, Q.Q. Joint bit allocation algorithm for multi-view texture and depth based on 3D-HEVC. *J. Optoelectron. Laser* **2015**, *26*, 149–155.

14. Xiao, J.; Hannuksela, M.M.; Tillo, T. Scalable Bit Allocation Between Texture and Depth Views for 3-D Video Streaming Over Heterogeneous Networks. *IEEE Trans. Circuits Syst. Video Technol.* **2015**, *25*, 139–152. [CrossRef]

15. Wang, X.; Kwong, S.; Yuan, H.; Zhang, Y.; Pan, Z.Q. View synthesis distortion model based frame level rate control optimization for multiview depth video coding. *Signal Process.* **2015**, *112*, 189–198. [CrossRef]

16. Yang, C.; An, P.; Shen, L.; Liu, D. Adaptive Bit Allocation for 3D Video Coding. *Circuitssyst. Signal Process.* **2017**, *36*, 2102–2124. [CrossRef]

17. Li, B.; Li, H.; Li, L.; Zhang, J. λ domain rate control algorithm for high efficiency video coding. *IEEE Trans. Image Process.* **2014**, *23*, 3841–3854. [CrossRef] [PubMed]

18. Roodaki, H.; Iravani, Z.; Hashemi, M.R. A view-level rate distortion model for multi-view/3D video. *IEEE Trans. Multimed.* **2016**, *18*, 14–24. [CrossRef]

19. Gao, W.; Kwong, S.; Jiang, Q.; Fong, C.K.; Wong, P.H.W.; Yuen, W.Y.F. Data-driven rate control for rate-distortion optimization in HEVC based on simplified effective initial QP Learning. *IEEE Trans. Broadcast* **2019**, *65*, 94–108. [CrossRef]

20. Zhou, M.; Wei, X.; Wang, S.; Kwong, S.; Fong, C.K.; Wong, P.H.W.; Yuen, W.Y.F.; Gao, W. SSIM-based global optimization for CTU-level rate control in HEVC. *IEEE Trans. Multimed.* **2019**, *21*, 94–108. [CrossRef]

21. Yan, T.; An, P.; Shen, L.Q.; Li, Z.; Wang, H.; Zhang, Z.Y. Rate control algorithm based on frame complexity estimation for MVC. In Proceedings of the 2010 IEEE 10th International Conference on Visual Communications and Image Processing (IEEE VCIP 2010), Huang Shan, China, 11–14 July 2010; pp. 77–85.

22. Liu, X.G.; Li, Y.Y.; Liu, D.Y.; Wang, P.H.; Yang, L.T. An Adaptive CU Size Decision Algorithm for HEVC Intra Prediction Based on Complexity Classification Using Machine Learning. *IEEE Trans. Circuits Syst. Video Technol.* **2019**, *29*, 144–155. [CrossRef]

23. Zhu, L.; Zhang, Y.; Kwong, S.; Wang, X.; Zhao, T. Fuzzy SVM-Based Coding Unit Decision in HEVC. *IEEE Trans. Broadcast.* **2018**, *64*, 681–694. [CrossRef]

24. Guo, H.; Zhu, C.; Li, S.; Gao, Y. Optimal bit allocation at frame level for rate control in HEVC. *IEEE Trans. Broadcast.* **2019**, *65*, 270–281. [CrossRef]

25. Li, L.; Li, B.; Liu, D.; Li, H. λ-Domain rate control algorithm for HEVC scalable extension. *IEEE Trans. Multimed.* **2016**, *18*, 2023–2039. [CrossRef]

26. Corbera, J.R.; Lei, S.M. A frame-layer bit allocation for H.263+. *IEEE Trans. Circuits Syst. Video Technol.* **2000**, *10*, 1154–1158. [CrossRef]

27. Yan, T.; An, P.; Shen, L.Q.; Li, Z.; Wang, H.; Zhang, Z.Y. Frame-layer bit allocation for multi-view video coding based on frame complexity estimation. *J. Shanghai Univ.* **2010**, *14*, 50–54. [CrossRef]

28. Amer, H.; Yang, E. Scene-based low delay HEVC encoding framework based on transparent composite modeling. In Proceedings of the 2016 IEEE International Conference on Image Processing (ICIP), Phoenix, AZ, USA, 25–28 September 2016; pp. 809–813.

29. Eorn, Y.; Park, S.; Yoo, S.; Choi, J.S.; Cho, S. An analysis of scene change detection in HEVC bitstream. In Proceedings of the 2015 IEEE 9th International Conference on Semantic Computing (IEEE ICSC 2015), Anaheim, CA, USA, 7–9 February 2015; pp. 470–474.

30. Li, S.; Xu, M.; Wang, Z.; Sun, X. Optimal Bit Allocation for CTU Level Rate Control in HEVC. *IEEE Trans. Circuits Syst. Video Technol.* **2017**, *27*, 2409–2424. [CrossRef]

31. Yan, T.; An, P.; Shen, L.Q.; Zhang, Z.Y. Bit allocation and rate control algorithm for MVC. *Imaging Sci. J.* **2011**, *59*, 202–210. [CrossRef]

32. He, Z.H.; Kim, Y.K.; Mitra, S.K. Low-delay rate control for DCT video coding via p-domain source modeling. *IEEE Trans. Circuits Syst. Video Technol.* **2001**, *11*, 928–940.

33. He, Z.H.; Mitra, S.K. A linear source model and a unified rate control algorithm for DCT video coding. *IEEE Trans. Circuits Video Technol.* **2002**, *12*, 970–982.

34. Yan, T.; Ra, I.; Weng, M.; Huang, L. Efficient predictive structure algorithm for MV-HEVC. *J. Comput. (Taiwan)* **2019**, *30*, 205–212.

 *electronics*

Article

Design of Efficient Perspective Affine Motion Estimation/Compensation for Versatile Video Coding (VVC) Standard

Young-Ju Choi [1], Dong-San Jun [2], Won-Sik Cheong [3] and Byung-Gyu Kim [1,*]

[1] Department of IT Engineering, Sookmyung Women's University, Seoul 04310, Korea;
    yj.choi@ivpl.sookmyung.ac.kr
[2] Department of Information and Communication Engineering, Kyungnam University,
    Changwon 51767, Korea; dsjun9643@kyungnam.ac.kr
[3] Immersive Media Research Section, Electronics and Telecommunications Research Institute (ETRI),
    Daejeon 34129, Korea; wscheong@etri.re.kr
* Correspondence: bg.kim@sookmyung.ac.kr; Tel.: +82-2-2077-7293

Received: 14 August 2019; Accepted: 2 September 2019; Published: 5 September 2019

Abstract: The fundamental motion model of the conventional block-based motion compensation in High Efficiency Video Coding (HEVC) is a translational motion model. However, in the real world, the motion of an object exists in the form of combining many kinds of motions. In Versatile Video Coding (VVC), a block-based 4-parameter and 6-parameter affine motion compensation (AMC) is being applied. In natural videos, in the majority of cases, a rigid object moves without any regularity rather than maintains the shape or transform with a certain rate. For this reason, the AMC still has a limit to compute complex motions. Therefore, more flexible motion model is desired for new video coding tool. In this paper, we design a perspective affine motion compensation (PAMC) method which can cope with more complex motions such as shear and shape distortion. The proposed PAMC utilizes perspective and affine motion model. The perspective motion model-based method uses four control point motion vectors (CPMVs) to give degree of freedom to all four corner vertices. Besides, the proposed algorithm is integrated into the AMC structure so that the existing affine mode and the proposed perspective mode can be executed adaptively. Because the block with the perspective motion model is a rectangle without specific feature, the proposed PAMC shows effective encoding performance for the test sequence containing irregular object distortions or dynamic rapid motions in particular. Our proposed algorithm is implemented on VTM 2.0. The experimental results show that the BD-rate reduction of the proposed technique can be achieved up to 0.45% and 0.30% on Y component for random access (RA) and low delay P (LDP) configurations, respectively.

Keywords: video coding; motion estimation; motion compensation; affine motion model; perspective motion model; VVC

---

1. Introduction

Video compression standard technologies are increasingly becoming more efficient and complex. With continuous development of display resolution and type along with enormous demand for high quality video contents, video coding also plays a key role in display and content industries. After standardizing H.264/AVC [1] and H.265/HEVC [2] successfully, Versatile Video Coding (VVC) [3] is being standardized by the Joint Video Exploration Team (JVET) of ITU-T Video Coding Experts Group (VCEG) and ISO/IEC Moving Picture Experts Group (MPEG). Obviously, the HEVC is a reliable video compression standard. Nevertheless, more efficient video coding scheme is required for higher-resolution and the newest services such as UHD and VR.

To develop the video compression technologies beyond HEVC, experts in JVET have been actively conducting much research. VVC provides a reference software model called as VVC Test Model (VTM) [4]. At the 11th JVET meeting, VTM2 [5] was established with the inclusion of a group of new coding features as well as some of HEVC coding elements.

The basic framework of VVC is the same as HEVC, which consists of block partitioning, intra and inter prediction, transform, loop filter and entropy coding. Inter prediction, which aims to obtain a similar block in the reference frames in order to reduce the temporal redundancy, is an essential part in video coding. The main tools for inter prediction are motion estimation (ME) and motion compensation (MC). Finding precise correlation between consecutive frames is important to final coding performance. Block matching based ME and MC have been implemented in the reference software model of the previous video compression standards such as H.264/AVC and H.265/HEVC. The fundamental motion model of the conventional block-based MC is a translational motion model. In the early research, a translational motion model-based MC cannot address complex motions in natural videos such as rotation and zooming. Such being the case, during the development of the video coding standards, further elaborate models are required to handle non-translational motions.

Non-translational motion model-based studies have also been presented in the early research on video coding. Seferidis [6] and Lee [7] proposed deformable block based ME algorithms, in which all motion vectors (MVs) at any position inside a block can be calculated by using control points (CPs). Besides, Cheung and Siu [8] proposed to use the neighboring block's MVs to estimate the affine motion transformation parameters and added an affine mode. After those, affine motion compensation (AMC) has begun to attract attention. A local zoom motion estimation method was proposed to achieve more coding gain by Kim et al. [9]. In this method, they used to estimate some zoom-in/out cases of the object or background part. However they dealt with just zoom motion cases using the H.264/AVC standard.

Later, Narroschke and Swoboda [10] proposed an adjusted AMC to HEVC coding structure by investigates the use of an affine motion model with analyzing variable block size. Huang et al. [11] extended the work in [8] for HEVC and included the affine skip/direct mode to improve coding efficiency. Also, Heithausen and Vorwerk [12] investigated different kinds of higher order motion models. Moreover, Chen et al. [13] proposed the affine skip and merge mode. In addition, Heithausen [14] developed a block-to-block translational shift compensation (BBTSC) technique which related to the advanced motion vector prediction (AMVP) [15] and improved the BBTSC algorithm by applying the translational motion vector field (TMVF) in [16]. Li [17] proposed the six-parameter affine motion model and extended by simplifying model to four-parameter and adding gradient-based fast affine ME algorithm in [18]. Because the trade-off between the complexity and coding performance is attractive, the scheme in [18] was proposed to JVET [19] and was accepted as one of the core modules of Joint Exploration Model (JEM) [20,21]. After that, Zhang [22] proposed a multi model AMC approach. At the 11th JVET meeting in July 2018, modified AMC of JEM was integrated into VVC and Test Model 2 (VTM2) [5] based on [22].

Although AMC has significantly improved performance over the conventional translational MC, there is still a limit to finding complex motion accurately. Affine transformation is a model that maintains parallelism based on the 2D plane, and thus cannot work efficiently for some sequences containing object distortions. In actual videos, motion by a non-affine transformation appears more generally than by an affine transformation with such restriction. Figure 1 shows an example of a non-affine transformation in nature video. When a part of an object is represented by a rectangle, the four vertices must operate independently of each other to illustrate the deformation of the object most similarly. Even though different frames have the same object, if the depth or viewpoint information changes, the motion can not be completely estimated by affine transformation model. For this reason, more flexible motion model is desired for new coding tool to raise the encoding quality.

The method using basic warping transformation model results in high computational complexity and bit overhead because of the large number of parameters. Therefore, it is necessary to apply a model that is not greatly increased for bit overhead compared to the existing AMC and has flexibility enough to replace the warping transformation model.

In this paper, we propose a perspective affine motion compensation (PAMC) method which improve coding efficiency compared with the AMC method of VVC. Compared to prior-arts, this paper presents two practical contributions to AMC. First, a perspective transformation model is designed in the form of MVs so that it can be used in AMC. It is an eight parameter based motion model that requires four CPMVs. Second, we propose a multi-motion model switch approach based framework to operate adaptively with AMC. In other words, six and four parameter model-based AMC and eight parameter-based perspective ME/MC are performed to select the best coding mode adaptively.

This paper is organized as follows. In Section 2, we first present AMC in VVC briefly. The proposed perspective affine motion compensation (PAMC) is introduced in Section 3. The experimental results are shown in Section 4. Finally, Section 5 concludes this paper.

Figure 1. Example of a non-affine transformation [23].

2. Affine Motion Estimation/Compensation in VVC

HEVC standard apply translational motion model to find a corresponding prediction block. The translational motion model cannot describe complex motion such as rotation and zooming. Moreover, it cannot represent combined multiple motion. In VVC, an affine motion compensation (AMC) is implemented which supports 4-parameter and 6-parameter motion model. The motion model for the AMC prediction method in the VVC is defined for three motions: translation, rotation and zooming. Affine transformation is based on the use of a 6-parameter model. Furthermore, a simplified 4-parameter model is applied for AMC in VVC. In addition, two affine motion modes namely affine inter-mode and affine merge-mode are added to AMC module. If affine inter-mode is used for a coding unit (CU), algorithm for affine inter-mode is designed to predict the MVs at CPs. In prediction process, a gradient-based ME algorithm is used as an encoder. When a CU is applied in affine merge-mode, the MVs at CPs are derived from the spatial neighbouring CU.

2.1. 4-Parameter and 6-Parameter Affine Model

As shown in Figure 2, the affine motion vector field (MVF) of a CU is described by control point motion vectors (CPMVs): (a) two CPs (4-parameter) or (b) three CPs (6-parameter). CP_0, CP_1 and CP_2 are defined as the top-left, top-right and bottom-left corners. For 4-parameter affine motion model, MV at sample position (x, y) in a CU is derived as

$$\begin{cases} mv^h(x, y) = \frac{mv_1^h - mv_0^h}{W} x - \frac{mv_1^v - mv_0^v}{W} y + mv_0^h, \\ \\ mv^v(x, y) = \frac{mv_1^v - mv_0^v}{W} x + \frac{mv_1^h - mv_0^h}{W} y + mv_0^v. \end{cases} \tag{1}$$

For 6-parameter affine motion model, MV at sample position (x, y) in a CU is derived as

$$\begin{cases} mv^h(x,y) = \dfrac{mv_1^h - mv_0^h}{W}x + \dfrac{mv_2^h - mv_0^h}{H}y + mv_0^h, \\[3mm] mv^v(x,y) = \dfrac{mv_1^v - mv_0^v}{W}x + \dfrac{mv_2^v - mv_0^v}{H}y + mv_0^v. \end{cases} \tag{2}$$

where (mv_0^h, mv_0^v), (mv_1^h, mv_1^v) and (mv_2^h, mv_2^v) are MVs of CP_0, CP_1 and CP_2 respectively. W and H present width and height of the current CU. The $mv^h(x,y)$ and $mv^v(x,y)$ are the horizontal and vertical components of MV for the position (x,y).

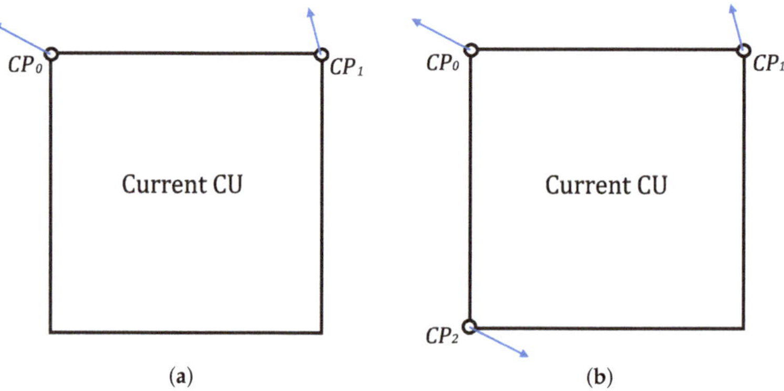

Figure 2. Affine motion vector control points: (**a**) 4-parameter motion model, (**b**) 6-parameter motion model.

To simplify the AMC, a block based AMC is applied. Figure 3 shows an example of sub block based MV derivation in a CU. The MV at the center position of each 4×4 sub block is derived from CPMVs and rounded to 1/16 fraction accuracy. Then the motion compensation interpolation filters are used to generate the prediction block of each sub-block with derived motion vector.

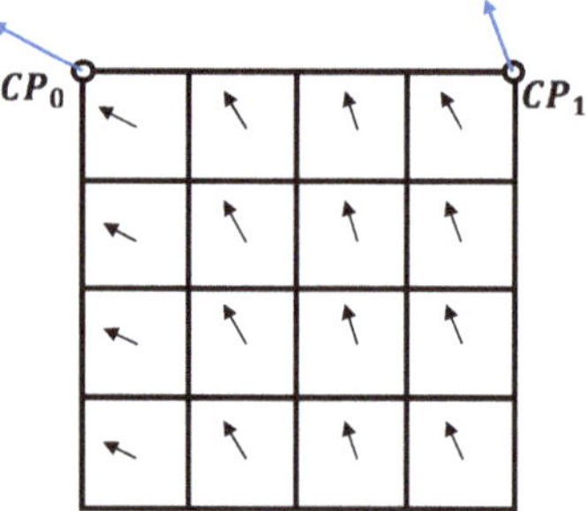

Figure 3. Affine motion vector field per sub-block.

2.2. Affine Inter Mode and Merge Mode

If a CU is coded with affine inter-mode, $\{mv_0, mv_1\}$ for 4-parameter model or $\{mv_0, mv_1, mv_2\}$ for 6-parameter model are signaled directly from the encoder to the decoder. At this moment, the difference of the CPMV of current CU and the control point motion vector prediction (CPMVP) is signaled in the bitstream. Moreover, flags for parameter type and affine mode are also signaled. Affine inter mode can be applied for CUs with both width and height larger than or equal to 16. The CPMVP candidate list size is 2 and it is derived by using the three types of CPMVP candidate generation phase in order:

1. CPMVPs extrapolated from the CPMVs of the spatial neighbour blocks

2. CPMVPs constructed using the translational MVs of the spatial neighbour blocks

3. CPMVPs generated by duplicating each of the HEVC AMVP candidates

As shown in Figure 4, neighboring blocks A, B, C, D, E, F and G are involved for generating CPMV candidate. First, if there are affine coded blocks through searching from A to G, add the CPMVs of the neighbour blocks to the CPMVP candidate list of the current CU. If the number of candidate list is smaller than 2, construct virtual CPMVP set which is composed of translational MVs $\{(mv_0, mv_1, mv_2) | mv_0 = \{mv_A, mv_B, mv_C\}, mv_1 = \{mv_D, mv_E\}, mv_2 = \{mv_F, mv_G\}, \}$. When the number of candidate list is still less than 2, finally, the list padded by the MVs composed by duplicating each of the AMVP candidates. An RD cost check process is applied to determine best CPMVP of current CU and an index indicating best CPMVP is signaled in bitstream.

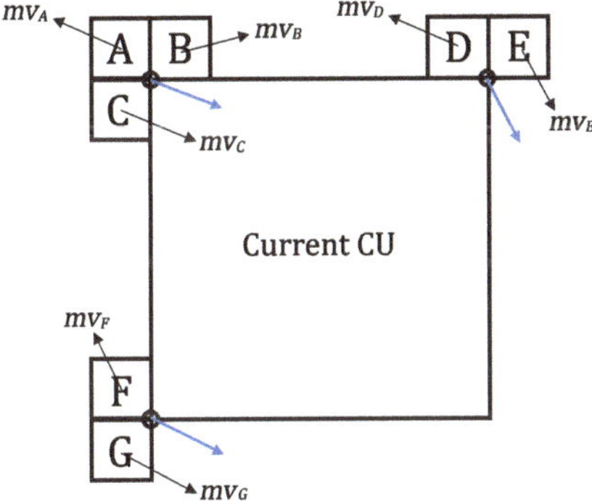

Figure 4. CPMVP candidate list for affine inter mode.

When a CU is applied in affine merge mode, the process finds the first coded block by affine mode among the neighbour candidate blocks. The selection order for the candidate block is from left, above, above right, left bottom to above as shown in Figure 5. After the CPMVs of the current CU are derived from the first neighbour block according to the affine motion model equation, the motion vector field of the current CU is generated. Like the affine inter mode, mode flag is signaled in bitstream.

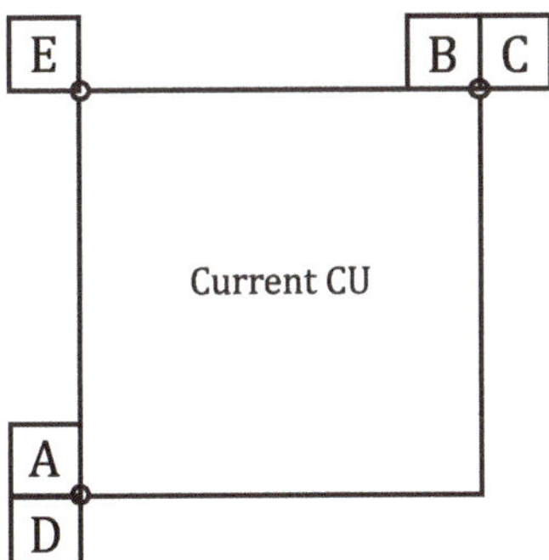

Figure 5. Candidate list for affine merge mode.

3. Proposed Perspective Affine Motion Estimation/Compensation

Affine motion estimation in the VVC is applied since it is more efficient than translational motion compensation. The coding gain can be increased by delicately estimating motion on the video sequence in which complex motion is included. However, still it has a limit to accurately find all motions in the natural video.

Affine transformation model has properties to maintain parallelism based on the 2D plane, and thus cannot work efficiently for some sequences containing object distortions or dynamic motions such as shear and 3D affine transformation. In real world, numerous moving objects have irregular motions rather than regular translational, rotation and scaling motions. So, more elaborated motion model is needed for video coding tool to estimate motion delicately.

The basic warping transformation model can estimate motion more accurately, but this method is not suitable because of its high computational complexity and bit overhead by the large number of parameters. For these reasons, we propose a perspective affine motion compensation (PAMC) method which improve coding efficiency compared with the existing AMC method of the VVC. The perspective transformation model-based algorithm adds one more CPMV, which gives degree of freedom to all four corner vertices of the block for more precise motion vector. Furthermore, the proposed algorithm is integrated while maintaining the AMC structure. Therefore, it is possible to adopt an optimal mode between the existing encoding mode and the proposed encoding mode.

3.1. Perspective Motion Model for Motion Estimation

Figure 6 shows that the proposed perspective model with four CPs (b) can estimate motion more flexible compared with the affine model with three CPs (a). Affine motion model-based MVF of a current block is described by three CPs which are matched to $\{mv_0, mv_1, mv_2\}$ in illustration. On the other hand, one more field is added for perspective motion model-based MVF. It is composed of four CPs which are matched to $\{mv_0, mv_1, mv_2, mv_3\}$. As can be seen from Figure 6, one vertex of the block can be used additionally, so that motion estimation can be performed on various types of rectangular bases. Each side of the prediction block obtained through motion estimation based on the perspective motion model has various lengths and does not has to be parallel. The typical eight-parameter perspective motion model can be described as:

$$\begin{cases} x' = \frac{p_1 x + p_2 y + p_3}{p_7 x + p_8 y + 1}, \\[2mm] y' = \frac{p_4 x + p_5 y + p_6}{p_7 x + p_8 y + 1}. \end{cases} \tag{3}$$

where p_1, p_2, p_3, p_4, p_5, p_6, p_7 and p_8 are eight perspective model parameters. Among them, parameters p_7 and p_8 serve to give the perspective to motion model. With this characteristic, as though it is a conversion in the 2D plane, it is possible to obtain an effect that the surface on which the object is projected is changed.

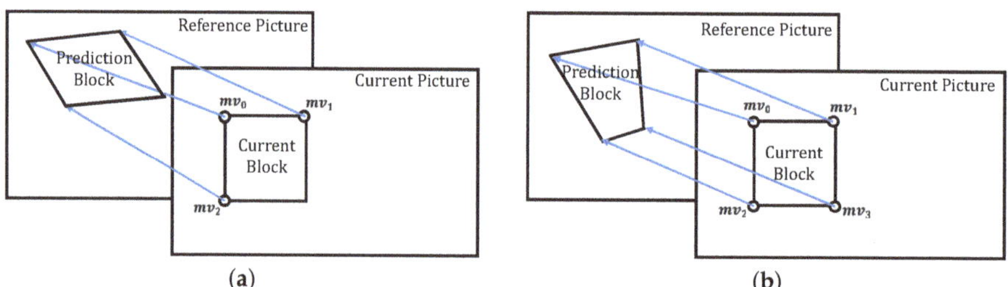

Figure 6. The motion models: (**a**) 6-parameter affine model with three CPs, (**b**) perspective model with four CPs.

Instead of these eight parameters, we used four MVs to equivalently represent the perspective transformation model like the technique applied to AMC of the existing VTM. In video codecs, using MV is more efficient in terms of coding structure and flag bits. Those four MVs can be chosen at any location of the current block. However, in this paper, we choose the points at the top left, top right, bottom left and bottom right for convenience of model definition. In a W x H block as shown in Figure 7, we denote the MVs of $(0,0)$, $(W,0)$, $(0,H)$, and (W,H) pixel as mv_0, mv_1, mv_2 and mv_3. Moreover, we replace $p_7 \cdot W + 1$ and $p_8 \cdot H + 1$ with a_1 and a_2 to simplify the formula. The six parameters p_1, p_2, p_3, p_4, p_5 and p_6 of model can solved as following Equation (4):

$$\begin{cases} p_1 = \dfrac{a_1(mv_1^h - mv_0^h)}{W}, \\[2mm] p_2 = \dfrac{a_2(mv_2^h - mv_0^h)}{H}, \\[2mm] p_3 = mv_0^h, \\[2mm] p_4 = \dfrac{a_1(mv_1^v - mv_0^v)}{W}, \\[2mm] p_5 = \dfrac{a_2(mv_2^v - mv_0^v)}{H}, \\[2mm] p_6 = mv_0^v. \end{cases} \tag{4}$$

In addition, $p_7 \cdot W$ and $p_8 \cdot H$ can solved as Equation (5):

$$\begin{cases} p_7 \cdot W = \dfrac{(mv_3^h - mv_2^h)(2mv_0^v - mv_1^v) + (mv_3^v - mv_2^v)(mv_1^h - 2mv_0^h)}{(mv_3^v - mv_2^v)(mv_3^h - mv_1^h) + (mv_3^h - mv_2^h)(mv_3^v - mv_1^v)}, \\[3mm] p_8 \cdot H = \dfrac{(mv_3^h - mv_1^h)(2mv_0^v - mv_2^v) + (mv_3^v - mv_1^v)(mv_2^h - 2mv_0^h)}{(mv_3^v - mv_1^v)(mv_3^h - mv_2^h) + (mv_3^h - mv_1^h)(mv_3^v - mv_2^v)}. \end{cases} \tag{5}$$

Based on Equations (4) and (5), we can derive MV at sample position (x, y) in a CU by following Equation (6):

$$\begin{cases} mv^h(x, y) = \dfrac{\frac{a_1\left(mv_1^h - mv_0^h\right)}{W}x + \frac{a_2\left(mv_2^h - mv_0^h\right)}{H}y + mv_0^h}{\frac{a_1 - 1}{W}x + \frac{a_2 - 1}{H}y + 1}, \\[4mm] mv^v(x, y) = \dfrac{\frac{a_1\left(mv_1^v - mv_0^v\right)}{W}x + \frac{a_2\left(mv_2^v - mv_0^v\right)}{H}y + mv_0^v}{\frac{a_1 - 1}{W}x + \frac{a_2 - 1}{H}y + 1}. \end{cases} \tag{6}$$

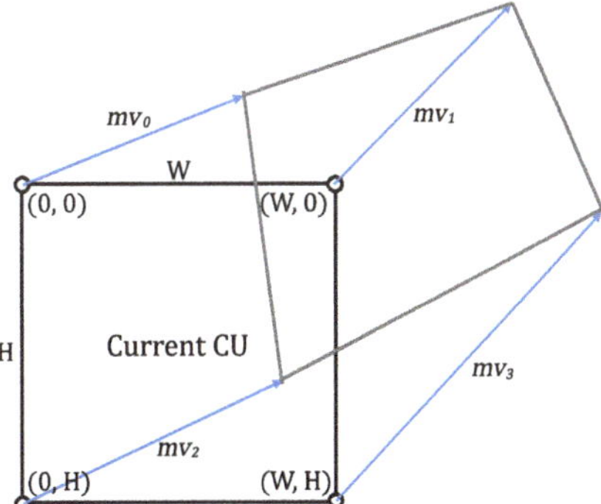

Figure 7. The representation of vertices for perspective motion model.

With the AMC, the designed perspective motion compensation also is also applied by 4×4 sub block-based MV derivation in a CU. Similarly, the motion compensation interpolation filters are used to generate the prediction block.

3.2. Perspective Affine Motion Compensation

Based on the aforementioned perspective motion model, the proposed algorithm is integrated into the existing AMC. A flowchart of the proposed algorithm is shown in Figure 8. Each motion model has its own strength. As the number of parameters increases, the precision of generating a prediction block increases. So at the same time, more bit signaling for CPMVs is required. It is effective to use the perspective motion model with four MVs is appropriate for reliability. On the other hand, if only two or three MVs are sufficient, it may be excessive to use four MVs. To take advantage of each motion model, we propose an adaptive multi-motion model-based technique.

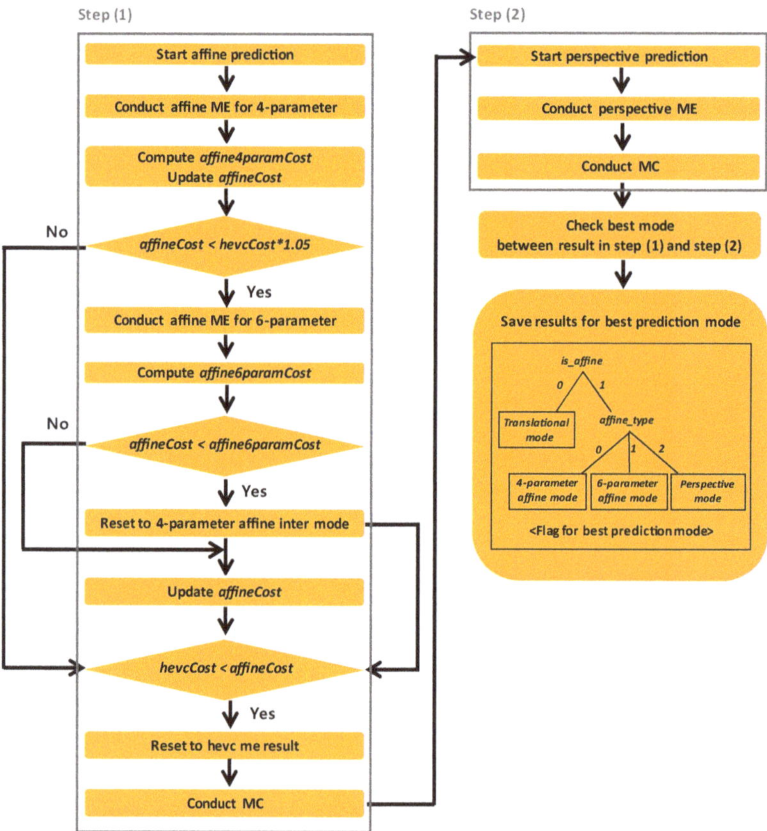

Figure 8. Flowchart of the proposed overall algorithm.

After performing fundamental translational ME and MC as in HEVC, the 4-parameter and 6-parameter affine prediction process is conducted first in step (1). Then, the proposed perspective prediction process is performed as step (2). After that, we check the best mode between result in step (1) and step (2) by RD cost check process in a current CU. Once the best mode is determined, the flag for prediction mode are signaled in the bitstream. At this time, two flags are required: affine flag and affine type flag. If the current CU is finally determined in affine mode, the affine flag is true and false otherwise. In other words, if the affine flag is false, only translational motion is used for ME. An affine type flag is signaled for a CU when its affine flag is true. When an affine type flag is 0, 4-parameter affine motion model is used for a CU. If an affine type flag is 1, 6-parameter affine motion model-based mode is used. Finally, when an affine type flag is 2, it means that the current CU is coded in the perspective mode.

4. Experimental Results

To evaluate the performance of the proposed PAMC module, the proposed algorithm was implemented on VTM 2.0 [24]. The 14 test sequences used in the experiments were from class A to class F specified in the JVET common test conditions (CTC) [25]. Experiments are conducted under random access (RA) and low delay P (LDP) configurations and four base layer quantization parameters (QP) values of 22, 27, 32 and 37. We used 50 frames in each test sequence. The objective coding performance comparison was evaluated by the Bjontegaard-Delta Rate (BD-Rate) measurement [26]. The BD-Rate was calculated by using piece-wise cubic interpolation.

Table 1 shows the overall experimental results of the proposed algorithm for each test sequence compared with VTM 2.0 baseline. Compared with the VTM anchor, we can see that proposed PAMC algorithm can bring about 0.07% and 0.12% BD-Rate gain on average Y component in RA and LDP cases respectively, and besides it can be up to 0.45% and 0.30% on Y component for random access (RA) and low delay P (LDP) configurations, respectively. Especially in LDP, shows better gain averagely. Compared to the RA, which allow bi-directional coding schemes so have two or more prediction blocks, LDP has one predication block. For the inter prediction algorithm, the coding performance depends on the number of reference frames. For these reason, when the novel algorithm is applied to the existing encoder, the coding efficiency is higher in the LDP configuration.

Table 1. BD-Rate (%) performance of the proposed algorithm, compared to VTM 2.0 Baseline.

Class	Sequence	Resolution	RA			LDP		
			Y	U	V	Y	U	V
A	Campfire	3840 × 2160	−0.09%	−0.02%	0.06%	-	-	-
	CatRobot1	3840 × 2160	−0.10%	0.41%	0.33%	-	-	-
B	RitualDance	1920 × 1080	−0.04%	0.15%	0.22%	−0.06%	0.55%	0.01%
	BasketballDrive	1920 × 1080	−0.14%	−0.05%	0.44%	−0.06%	0.28%	0.47%
	BQTerrace	1920 × 1080	−0.08%	−0.11%	−0.03%	−0.08%	0.38%	−0.16%
C	BasketballDrill	832 × 480	−0.09%	−0.12%	−0.16%	−0.02%	0.04%	−0.15%
	PartyScene	832 × 480	−0.06%	−0.21%	−0.04%	0.01%	0.46%	0.24%
D	BQSquare	416 × 240	0.03%	0.45%	−0.04%	−0.03%	−0.42%	−1.87%
	RaceHorses	416 × 240	0.07%	−0.76%	0.09%	−0.25%	0.21%	0.49%
E	FourPeople	1280 × 720	-	-	-	−0.16%	−0.58%	0.15%
	KristenAndSara	1280 × 720	-	-	-	0.07%	−0.22%	0.03%
F	BasketballDrillText	832 × 480	−0.01%	−0.18%	0.25%	−0.07%	−0.57%	−0.28%
	SlideEditing	1280 × 720	−0.07%	−0.03%	−0.03%	−0.28%	−0.22%	−0.47%
	SlideShow	1280 × 720	−0.30%	1.29%	0.19%	−0.45%	−1.39%	0.08%
Avg.			**−0.07%**	**0.07%**	**0.11%**	**−0.12%**	**−0.12%**	**−0.12%**

Although there were a small BD-rate losses of chroma components, the chroma components are usually less sensitive to the perception of human eye for recognition. So the luminance component is more important to measure the performance. The proposed algorithm achieved up to 0.45% and in the most of sequences, the coding gain was obtained in Y component (luminance component) although they were small value in some sequences such as BasketballDrill, PartyScene, and BQsquare with RA configuration. For RaceHorse, SlideEditing, and FourPeople sequences in LDP configuration, 0.25%, 0.28%, and 0.45% of DB-rate savings of Y component were observed in Table 1.

Some examples of rate-distortion (R-D) curves are shown in Figure 9. The R-D curves also verify that the proposed perspective affine MC can achieve better coding performance compared with the VTM baseline. It can be seen from Figure 9 that the proposed algorithm works more efficiently than the existing affine MC algorithm in both the LDP and RA configurations. For LDP coding condition, it is more efficient in $QP = 32$ and $QP = 37$ cases and for RA coding condition, it seems to have an effect on $QP = 22$ and $QP = 27$.

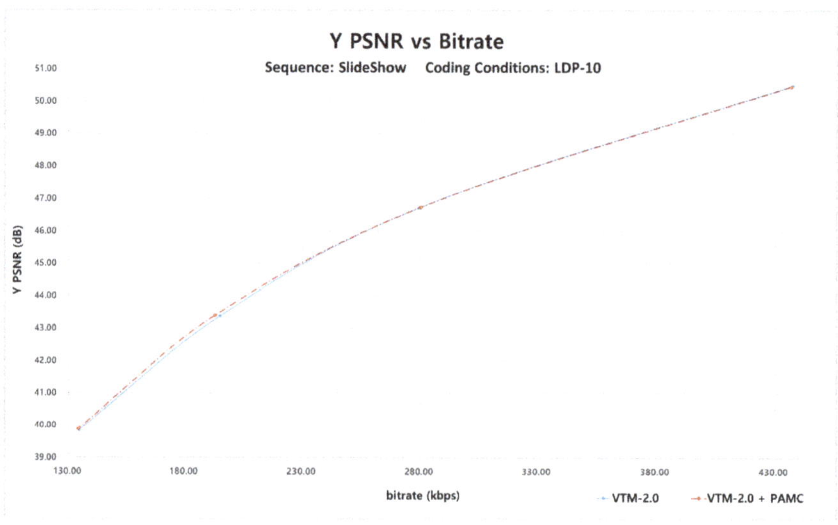

(**a**) SlideShow in LDP mdoe.

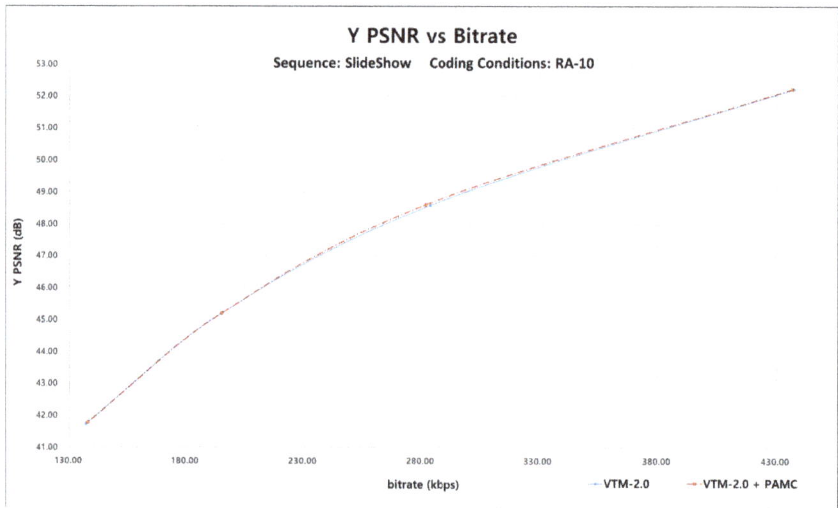

(**b**) SlideShow in RA mdoe.

Figure 9. The R-D curves of the proposed perspective affine MC framework.

To improve the coding efficiency through the proposed algorithm, the test sequence have to contain irregular object distortions or dynamic rapid motions. Figure 10 shows the examples of perspective motion area in test sequences. Figure 10a–c present the examples of sequence "Campfire", "CatRobot1" and "BQTerrace", respectively. The class A sequence "Campfire" contains a bonfire that moves inconsistently and steadily. Also "CatRobot1" contains a lot of flapping scarves and a flipped book, and class B sequence "BQTerrace" involves the ripples on the surface of the water. All of such moving objects commonly include object distortions whose shape changes. Because of this, those sequences can be compressed more efficiently by the proposed framework. The results of "Campfire" and "CatRobot1" sequences show that proposed PAMC achieves 0.09% and 0.10% BD-Rate savings respectively on Y component in RA. The result of "BQTerrace" sequence shows a coding gain of 0.08% on Y component for both RA and LDP.

Figure 11 shows an example of comparing AMC in VVC and the proposed PAMC using "CatRobot1" sequence. It is a result for POC 24 encoded by setting QP 22 in RA. Figure 11a presents the coded CU with affine mode in VVC baseline and Figure 11b shows the coded CU with affine and perspective mode in the proposed framework. If the unfilled rectangles imply the CUs coded in affine mode and the filled rectangles imply the CUs coded in perspective mode, in Figure 11b, some filled rectangles can be found on scarves and on the pages of a book. The class B sequences "RitualDance" and "BasketballDrive" and class C sequence "BasketballDrill", which contain dynamic fast movements, can also be seen to bring coding gains in both RA and LDP configurations. For the three sequences mentioned above, performance result shows that proposed PAMC results in 0.04%, 0.14%, and 0.09% BD-Rate savings respectively on Y component in RA. In LDP configuration, BD-Rate gains are 0.06%, 0.06% and 0.02% respectively on Y component.

In class F which has the screen content sequences, rapid long range motions as large as half a frame often happen like browsing and document editing. Even in this case, the proposed PAMC algorithm can bring BD-Rate gain. In particular, the "SlideShow" sequence gives the largest coding gain resulting in 0.30% and 0.45% BD-Rate savings on Y component in RA and LDP respectively. Besides, on U component in LDP, it brings 1.39% of BD-Rate gain.

When the resolution of sequence is too small such as class D, the sequence contains a small amount of textures and object content. Therefore, it is possible to estimate the motion accurately with only using further enhanced configuration. For that reason, it can be seen from the result of class D that proposed algorithm contributes to overall coding gain in LDP but not in RA. The result of "BQSquare" sequence shows that proposed PAMC achieves 0.03%, 0.42% and 1.87% BD-Rate savings on Y, U and V components respectively in LDP. For "RaceHorces" sequence, the result shows 0.25% of BD-Rate gain on Y component in LDP.

As the proposed algorithm is designed to better describe the motion with distortion of object shape, some equirectangular projection (ERP) format sequences [27] are selected to verify the performance of the proposed algorithm. Figure 12 shows an example of ERP sequence. Figure 12a presents a frame of the "Broadway" sequence and Figure 12b shows a enlarged specific area of the frame. Figure 12c shows a picture in posterior frame for the same area. The ERP produces significant deformation, especially in the pole area. It can be obviously seen from Figure 12 that the distortion of the building object occurs. Perspective motion model can take such deformation into account when compressing the planar video of panoramic content.

The R-D performance of the proposed algorithm for the ERP test sequences is illustrated in Table 2. From Table 2, we can see that the proposed framework can be up to 0.15% on Y component for low delay P (LDP) configuration. The experimental results obviously demonstrate that the proposed perspective affine motion model can well represent the motion with distortion of object shape.

As shown in some video coding research [28,29], the results show 0.07% and 0.12% BD-Rate gain on average, respectively. Furthermore, several advanced affine motion estimation algorithms in JVET meeting documents [30–32], the results show 0.09%, 0.09% and 0.13% BD-Rate gain on average Y component. Compared with these results, the performance of the proposed algorithm is also competitive. Moreover, our proposed method contributes in that the encoder can be more robust in natural videos by proposing a flexible motion model for affine ME, one of the main inter prediction tools of the existing VTM codec.

(**a**) Campfire.

(**b**) CatRobot1.

(**c**) BQTerrace.

Figure 10. An examples of perspective motion area in test sequences.

(**a**) AMC in VVC.

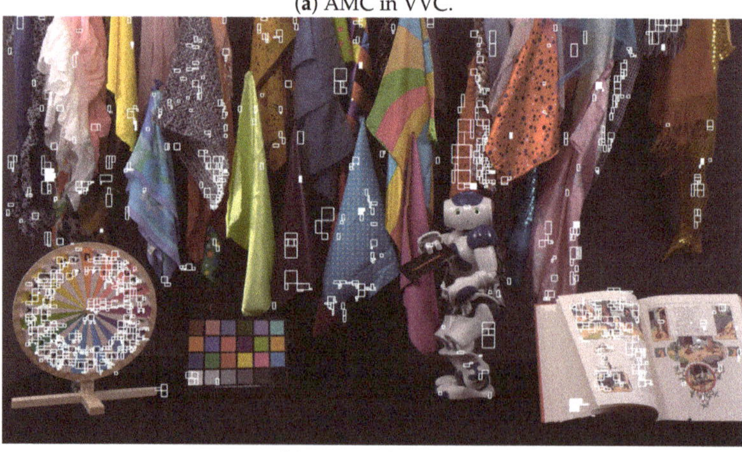

(**b**) Proposed PAMC.

Figure 11. An example of CUs with affine or perspective motion, CatRobot1, RA, QP22, POC24.

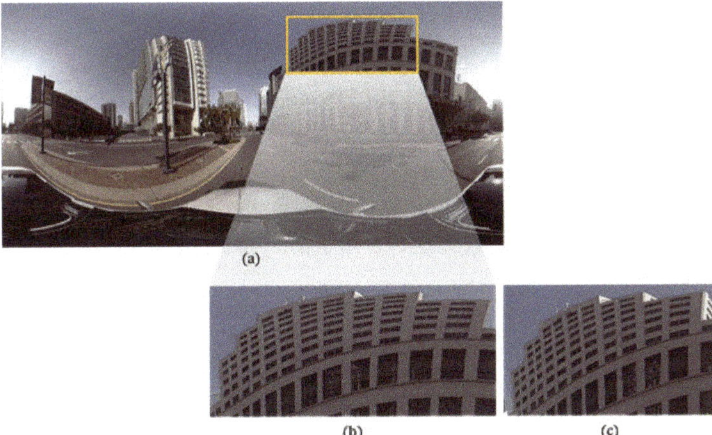

Figure 12. An example of the ERP format sequence (Broadway).

From experimental results, the designed PAMC achieved better coding gain compared to VTM 2.0 Baseline. It means that the proposed PAMC scheme can be applied for providing better video quality in terms of a limited network bandwidth environment.

Table 2. BD-Rate (%) performance of the proposed algorithm for ERP format sequences compared to VTM 2.0 Baseline.

Sequence	Resolution	LDP		
		Y	U	V
Broadway	6144 × 3072	−0.15%	0.03%	−0.19%
Freefall	6144 × 3072	−0.13%	0.12%	0.35%
BranCastle2	6144 × 3072	−0.08%	−0.11%	−0.28%
Balboa	6144 × 3072	−0.10%	−0.06%	0.17%
Avg.		**−0.12%**	**−0.01%**	**0.01%**

5. Conclusions

In this paper, an efficient perspective affine motion compensation framework was proposed to estimate further complex motions beyond the affine motion. Affine motion model has properties which maintains parallelism, and thus cannot work efficiently for some sequences containing object distortions or rapid dynamic motions. In the proposed algorithm, an eight-parameter perspective motion model was first defined and analyzed. Like the technique applied to AMC of existing VTM, we designed four MVs based motion model instead of using eight parameters. Then the perspective motion model-based motion compensation algorithm was proposed. To take advantage of each affine and perspective motion model, we proposed an adaptive multi-motion model-based technique. The proposed framework was implemented in the reference software of VVC. We experimented with two kinds of sequences. In addition to experimenting with JVET common test condition sequences, we demonstrated the effectiveness of the proposed algorithm by showing the results for the equirectangular projection format sequences. The experimental results showed that the proposed perspective affine motion compensation framework could achieve much better BD-Rate performance compared with the VVC baseline especially for sequences that contain irregular object distortions or dynamic rapid motions.

Although the proposed algorithm improved the inter-prediction of the VVC video standard technology, there is still room for further improvement. For future studies, the higher-order motion models should be investigated and applied for three-dimensional modeling of motion. Higher-order models can improve the accuracy of irregular motion, but they can result in an increase in the number of bits, as these parameters must be sent together. Considering these points, an approximation model should also be conducted to be compatible for the VVC standard.

Author Contributions: Conceptualization, B.-G.K.; methodology, Y.-J.C.; software, Y.-J.C.; validation, D.-S.J.; formal analysis, W.-S.C.; Writing—Original Draft preparation, Y.-J.C.; Writing—Review and Editing, B.-G.K.; supervision, B.-G.K.;funding acquisition, D.-S.J.

Funding: This research was supported by Basic Science Research Program through the National Research Foundation of Korea (NRF) funded by the Ministry of Education (NRF-2016R1D1A1B04934750) and partially supported by Electronics and Telecommunications Research Institute (ETRI) grant funded by ICT R&D program of MSIT/IITP [No. 2017-0-00072, Development of Audio/Video Coding and Light Field Media Fundamental Technologies for Ultra Realistic Tera-media.

Conflicts of Interest: The authors declare no conflict of interest.

References

1. Draft ITU-T Recommendation and final draft international standard of joint video specification (ITU-T Rec. H. 264 | ISO/IEC 14496-10 AVC). In Proceedings of the 5th Meeting, Joint Video Team (JVT) of ISO/IEC MPEG and ITU-T VCEG, Geneva, Switzerland, 9–17 October 2002.

2. Sze, V.; Budagavi, M.; Sullivan, G.J. High efficiency video coding (HEVC). In *Integrated Circuit and Systems, Algorithms and Architectures*; Springer: Berlin, Germany, 2014; pp. 1–375.

3. Bross, B.; Chen, J.; Liu, S. *Versatile Video Coding (Draft 2)*; JVET-K1001; JVET: Ljubljana, Slovenia, 2018.

4. Chen, J.; Alshina, E. *Algorithm Description for Versatile Video Coding and Test Model 1 (VTM 1)*; JVET-J1002; JVET: San Diego, CA, USA, 2018.

5. Chen, J.; Ye, Y.; Kim, S. *Algorithm Description for Versatile Video Coding and Test Model 2 (VTM 2)*; JVET-K1002; JVET: Ljubljana, Slovenia, 2018.

6. Seferidis, V.; Ghanbari, M. General approach to block-matching motion estimation. *Opt. Eng.* **1993**, *32*, 1464–1474. [CrossRef]

7. Lee, O.; Wang, Y. Motion compensated prediction using nodal based deformable block matching. *J. Vis. Commun. Image Represent.* **1995**, *6*, 26–34. [CrossRef]

8. Cheung, H.K.; Siu, W.C. Local affine motion prediction for H.264 without extra overhead. In Proceedings of the IEEE International Symposium on Circuits and Systems (ISCAS), Paris, France, 30 May–2 June 2010; pp. 1555–1558.

9. Kim, H.-S.; Lee, J.-H.; Kim, C.-K.; Kim, B.-G. Zoom motion estimation using block-based fast local area scaling. *IEEE Trans. Circuits Syst. Video Technol.* **2012**, *22*, 1280–1291. [CrossRef]

10. Narroschke, M.; Swoboda, R. Extending HEVC by an affine motion model. In Proceedings of the 2013 Picture Coding Symposium (PCS), San Jose, CA, USA, 8–11 December 2013; pp. 321–324.

11. Huang, H.; Woods, J.W.; Zhao, Y.; Bai, H. Affine SKIP and DIRECT modes for efficient video coding. In Proceedings of the Visual Communications and Image Processing (VCIP), San Diego, CA, USA, 27–30 November 2012; pp. 1–6.

12. Heithausen, C.; Vorwerk, J.H. Motion compensation with higher order motion models for HEVC. In Proceedings of the 2015 IEEE International Conference on Acoustics, Speech and Signal Processing (ICASSP), Brisbane, QLD, Australia, 19–24 April 2015; pp. 1438–1442.

13. Chen, H.; Liang, F.; Lin, S. Affine SKIP and MERGE modes for video coding. In Proceedings of the 2015 IEEE 17th International Workshop on Multimedia Signal Processing (MMSP), Xiamen, China, 19–21 October 2015; pp. 1–5.

14. Heithausen, C.; Bläser, M.; Wien, M.; Ohm, J.R. Improved higher order motion compensation in HEVC with block-to-block translational shift compensation. In Proceedings of the 2016 IEEE International Conference on Image Processing (ICIP), Phoenix, AZ, USA, 25–28 September 2016; pp. 2008–2012.

15. Sullivan, G.J.; Ohm, J.R.; Han, W.J.; Wiegand, T. Overview of the high efficiency video coding (HEVC) standard. *IEEE Trans. Circuits Syst. Video Technol.* **2012**, *22*, 1649–1668. [CrossRef]

16. Heithausen, C.; Meyer, M.; Bläser, M.; Ohm, J.R. Temporal prediction of motion parameters with interchangeable motion model. In Proceedings of the 2017 Data Compression Conference (DCC), Snowbird, UT, USA, 4–7 April 2017; pp. 400–409.
17. Li, L.; Li, H.; Lv, Z.; Yang, H. An affine motion compensation framework for High Efficiency Video Coding. In Proceedings of the IEEE International Symposium on Circuits and Systems (ISCAS), Lisbon, Portugal, 24–27 May 2015; pp. 525–528.
18. Li, L.; Li, H.; Liu, D.; Li, Z.; Yang, H.; Lin, S.; Chen, H.; Wu, F. An Efficient Four-Parameter Affine Motion Model for Video Coding. *IEEE Trans. Circuits Syst. Video Technol.* **2018**, *28*, 1934–1948. [CrossRef]
19. Lin, S.; Chen, H.; Zhang, H.; Maxim, S.; Yang, H.; Zhou, J. *Affine Transform Prediction for Next Generation Video Coding*; ITU-T SG16 Doc. COM16–C1016; Huawei Technologies: Shenzhen, China, 2015.
20. Chen, J.; Alshina, E.; Sullivan, G.J.; Ohm, J.R.; Boyce, J. *Algorithm Description of Joint Exploration Test Model 1*; JVET-A1001; JVET: Geneva, Switzerland, 2015.
21. Choi, Y.J.; Kim, J.H.; Lee, J.H.; Kim, B.G. Performance Analysis of Future Video Coding (FVC) Standard Technology. *J. Multimed. Inf. Syst.* **2017**, *4*, 73–78.
22. Zhang, K.; Chen, Y.W.; Zhang, L.; Chien, W.J.; Karczewicz, M. An Improved Framework of Affine Motion Compensation in Video Coding. *IEEE Trans. Image Process.* **2019**, *28*, 1456–1469. [CrossRef] [PubMed]
23. Lichtenauer, J.F.; Sirmacek, B. A semi-automatic procedure for texturing of laser scanning point clouds with google streetview images. Available online: https://repository.tudelft.nl/islandora/object/uuid% 3A8bb4d40b-0950-471f-b774-7f74449fe26e (accessed on 16 September 2019).
24. Versatile Video Coding (VVC) Test Model 2.0 (VTM 2.0). Available online: https://vcgit.hhi.fraunhofer.de/ jvet/VVCSoftware_VTM.git (accessed on 16 September 2018).
25. Bossen, F.; Boyce, J.; Suehring, K.; Li, X.; Seregin, V. *JVET Common Test Conditions and Software Reference Configurations for SDR Video*; JVET-K1010; JVET: Ljubljana, Slovenia, 2018.
26. Bjøntegaard, G. *Calculation of Average PSNR Differences between RDcurves*; ITU-T SG.16 Q.6, Document VCEG-M33; ITU-T VCEG: Austin, TX, USA, 2001.
27. Hanhart, P.; Boyce, J.; Choi, K. *JVET Common Test Conditions and Evaluation Procedures for 360 Video*; JVET-K1012: Ljubljana, Slovenia, 2018.
28. Yoon, Y.U.; Kim, H.H.; Lee, Y.J.; Kim, J.G. Methods of padding inactive regions for rotated sphere projection of 360 video. In Proceedings of the 2019 Joint International Workshop on Advanced Image Technology (IWAIT) and International Forum on Medical Imaging in Asia (IFMIA), Singapore, 6–9 January 2019.
29. Ma, S.; Lin, Y.; Zhu, C.; Zheng, J.; Yu, L.; Wang, X. Improved segment-wise DC coding for HEVC intra prediction of depth maps. In Proceedings of the Signal and Information Processing Association Annual Summit and Conference (APSIPA), Siem Reap, Cambodia, 9–12 December 2014.
30. Zhao, J.; Kim, S. H.; Li, G.; Xu, X.; Li, X.; Liu, S. *CE2: History Based Affine Motion Candidate (Test 2.2.3)*; JVET-M0125; JVET: Marrakech, Morocco, 2019.
31. Galpin, F.; Robert, A.; Le Léannec, F.; Poirier, T. *CE2.2.7: Affine Temporal Constructed Candidates*; JVET-M0256; JVET: Marrakech, Morocco, 2019.
32. Zhang, K.; Zhang, L.; Liu, H.; Xu, J.; Wang, Y.; Zhao, P.; Hong, D. *CE2-Related: History-Based Affine Merge Candidates*; JVET-M0266; JVET: Marrakech, Morocco, 2019.

 *electronics*

Article

Layer Selection in Progressive Transmission of Motion-Compensated JPEG2000 Video

José Carmelo Maturana-Espinosa [1], Juan Pablo García-Ortiz [1], Daniel Müller [2]
and Vicente González-Ruiz [1,*]

[1] University of Almería, Ctra. Sacramento, s/n, 04120 Almería, Spain;
    carmelo.maturana@gmail.com (J.C.M.-E.); jp.garcia.ortiz@gmail.com (J.P.G.-O.)
[2] European Space Agency, ESTEC, P.O. Box 299, 2200 AG Noordwijk, The Netherlands;
    Daniel.Mueller@esa.int
*   Correspondence: vruiz@ual.es

Received: 26 August 2019; Accepted: 11 September 2019; Published: 13 September 2019

Abstract: MCJ2K (Motion-Compensated JPEG2000) is a video codec based on MCTF (Motion-Compensated Temporal Filtering) and J2K (JPEG2000). MCTF analyzes a sequence of images, generating a collection of temporal sub-bands, which are compressed with J2K. The R/D (Rate-Distortion) performance in MCJ2K is better than the MJ2K (Motion JPEG2000) extension, especially if there is a high level of temporal redundancy. MCJ2K codestreams can be served by standard JPIP (J2K Interactive Protocol) servers, thanks to the use of only J2K standard file formats. In bandwidth-constrained scenarios, an important issue in MCJ2K is determining the amount of data of each temporal sub-band that must be transmitted to maximize the quality of the reconstructions at the client side. To solve this problem, we have proposed two rate-allocation algorithms which provide reconstructions that are progressive in quality. The first, OSLA (Optimized Sub-band Layers Allocation), determines the best progression of quality layers, but is computationally expensive. The second, ESLA (Estimated-Slope sub-band Layers Allocation), is sub-optimal in most cases, but much faster and more convenient for real-time streaming scenarios. An experimental comparison shows that even when a straightforward motion compensation scheme is used, the R/D performance of MCJ2K competitive is compared not only to MJ2K, but also with respect to other standard scalable video codecs.

Keywords: quantization (signal); video coding; channel allocation; scalable video coding

1. Introduction

The JPEG2000 (J2K) standard [1] is a still-image codec which also encompasses the compression of sequences of images that goes by the name Motion J2K (MJ2K). The standard relies on the J2K Interactive Protocol (JPIP) [2] to transmit J2K codestreams between client/server systems, offering a high degree of scalability (spatial, temporal and quality). These features make J2K (and its extension MJ2K) especially suitable for the management of video repositories, and for the implementation of interactive image/video streaming services [3]. In particular, JPIP has proven very effective for visualization of petabyte-scale image data of the Sun (Helioviewer Project [4,5]), allowing researchers and the general public alike to explore time-dependent image data from different space-borne observatories, interactively zoom into areas of interest and play sequences of high-resolution images at various cadences. Figure 1 shows an example of an interaction with the JHelioviewer service.

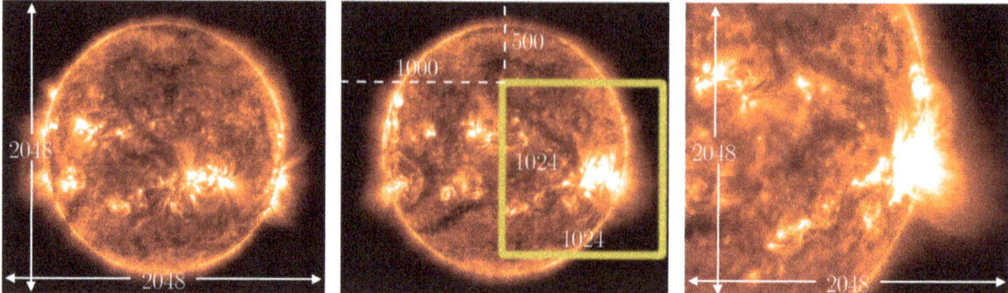

Figure 1. Different instants of a remote browsing (on a 2048 × 2048 pixels display) of a 4096 × 4096 pixels image sequence of extreme ultraviolet images of the Sun, taken by NASA's Solar Dynamics Observatory (SDO), using the JHelioviewer application. Initially, users retrieve the sequence of images with a resolution that fits in their display (left subfigure). Notice that the information of the highest spatial resolution level (4096 × 4096) is not transmitted because can not be displayed. In any moment of the transmission users can select a Window Of Interest (WOI), which in this example starts at pixel 1000 × 500 and have a (image) resolution of 1024 × 1024 (center subfigure), retrieving this WOI at the highest resolution. In the rest of the transmission (right subfigure), only the code-stream related to that WOI will be transmitted. Image credit: NASA/SDO/AIA.

MCJ2K (Motion-Compensated JPEG2000) is a combination of two fundamental stages: (1) MCTF (Motion-Compensated Temporal Filtering) and (2) J2K. Basically, MCTF is a transform that inputs a sequence of images and outputs a sequence of *MCTF-coefficients* (which will simply be called *coefs*), grouped in a collection of temporal sub-bands. Then, these coefs are compressed with J2K, resulting in a collection of J2K codestreams that can be transmitted using JPIP. The R/D performance of MCJ2K can clearly be better than that of J2K, depending on the temporal correlation among the input images. As an example, Figure 2 shows a Sun image (of a sequence) decompressed with MJ2K and MCJ2K, at similar bitrates.

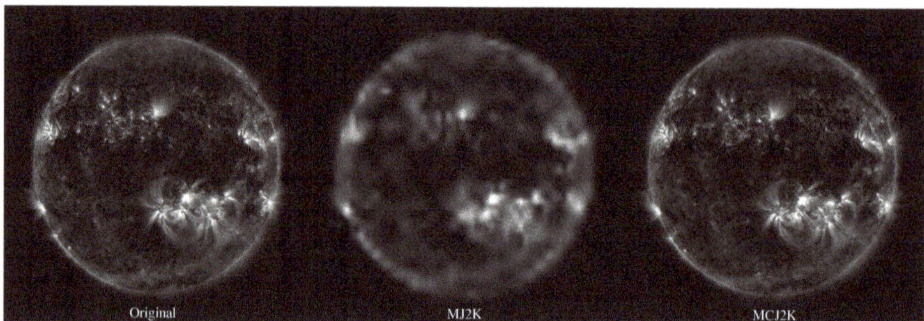

Figure 2. Reconstruction of one image of a sequence of Sun images using MJ2K and MCJ2K. Left: original Sun image with 512 × 512 pixels and a cadence of $\frac{1}{12}$ images/second. Center: same image (progressively) decompressed with MJ2K at 0.08 kbps. Right: same image (progressively) decompressed with MCJ2K at 0.04 kbps. Image credit: NASA/SDO/AIA.

MCJ2K is a straightforward extension of MJ2K, and it has been proposed previously [6]. However, the adaptation of MCJ2K to standard JPIP services, such a Helioviewer, is a novel contribution. Furthermore, two novel RA (Rate-Allocation; in this document this term refers to the action of sorting the code-stream to provide some kind of scalability, and the term "rate control" is used to decide which information is represented by the code-stream in rate-constrained scenarios) algorithms: OSLA (Optimized Sub-band Layers Allocation) and ESLA (Estimated-slope Sub-band Layers Allocation) are herein proposed and experimentally evaluated. Both algorithms are run at post-compression time to determine an efficient progression of quality layers.

The rest of this document is structured as follows. Section 2 describes the most relevant works related to MCJ2K and RA in wavelet-based video coding. MCJ2K, OSLA and ESLA are detailed in Section 3. Section 4 presents the results of an empirical performance evaluation, and Section 5 summarizes our findings and outlines future research in Section 6.

2. Background and Related Work

The combination of MCTF and J2K has been proposed in previous works. Secker et al. use these techniques to create LIMAT [7], but no RC (Rate Control) or RA algorithms are proposed. The motion information is simply placed first, followed by the texture data.

In [6], Cohen et al. propose two ME (motion estimation)-based J2K codecs. The first is a 2D-pyramid codec with an MCTF step on each spatial level, and a closed-loop coding structure, similar to H.264/SVC [8] and HEVC [9]. The second codec is open-loop, similar to MCJ2K, but the authors do not address the problem of RA among temporal texture sub-bands and motion information.

A similar approach to [7] was designed in [10] and extended in [11] by André et al. Also Ferroukhi et al. [12] have recently proposed a similar codec based on second generation J2K. In these works, using RDO (Rate-Distortion Optimization) [13], an RC algorithm is proposed to determine the contribution of each temporal sub-band. None of these works provide an RA algorithm.

In [14] Barbarien et al. provide some interesting ideas to perform optimal RC at compression time. Before using a 2D-DWT (Discrete Wavelet Transform) [15], all the residue coefficients resulting from the MCTF stage are multiplied by a scaling factor to approximate MCTF to a unitary (energy preserving) transformation. As in [11], an optimal RC among motion and texture data is proposed using RDO. An interesting alternative was proposed in [3], where, similar to the use of quantization in hybrid video coding, a set of R/D slopes can be specified to control the composition of each quality layer (by including those layers whose R/D slopes are higher than or equal to the slope of the corresponding layer). These approaches are optimal only in linear transformation scenarios, a condition which is difficult to satisfy (as will be shown in the experimental results) when ME/MC techniques are used. The compatibility with JPIP is not studied in these works.

As previously mentioned, RA can be performed at decompression time. However, in this case, it can be implemented by the sender (server), receivers (clients) or both. FAST, proposed by Aulí-Llinàs et al. in [16] and improved by Jimenez-Rodriguez et al. in [17], is a sender-driven RA algorithm for MJ2K sequences. Another interesting MJ2K/MCJ2K sender-driven RA proposal was introduced by Naman et al., which uses Conditional Replenishment (CR) [18] and Motion Compensated (MC) [19]. In Naman's proposals, a server sends those J2K packets related to the regions that the clients should refresh to optimize the quality of the video after considering bandwidth constraints. These proposals are not fully J2K compliant at the server side (a requirement in standard JPIP services) because some kind of non-J2K-standard logic must be used.

Receiver-driven RA solutions have also been proposed in previous studies. For example, in DASH [20], clients retrieve video streams, requesting (GOP by GOP) those code-stream segments that maximize the user's QoE (Quality of Experience), and the buffer fullness. In [21], Mehrotra et al. propose an improvement of the previous approach in which clients use the R/D information of the video to select (taking into account the desired startup latency, the buffer size, and the estimated network capacity) the optimal number of quality layers (in the case of H.264/SVC), or which quality-version of each GOP (in the case of H.264/simulcasting) will be transmitted.

As in [11], in [14] the authors also propose an optimal RA among motion and texture data based on Lagrangian RDO, considering that the distortions are additive (something that can be sub-optimal in those cases where the MCTF is not linear). Such optimization minimizes the distortion for a known bitrate, but not for any possible bitrate (note that when transmitting an image or a sequence of images, such bitrates established at compression time might not be met at decompression time).

3. MCJ2K

3.1. Codec Overview

MCJ2K is a two stages codec (see Figure 3): MCTF performs temporal filtering and MCJ2K compress the sequence of sub-bands. The resulting code-stream (see Figure 4) is a collection of compressed texture (each one composed by coefs) and motion sub-bands. MCJ2K is an open-loop "t+2D" structure. The "t" corresponds to a T-levels MCTF (a T-levels 1/3 linear 1D-DWT, denoted by MCTF^T) and the "2D" to a 2D-DWT, provided by the standard J2K codec. MCTF^T exploits the temporal redundancy and 2D-DWT, included as a part of the MJ2K compressors, the spatial redundancy. The set of MJ2K compressors inputs the coefs of each temporal sub-band generated by MCTF^T and perform entropy layered coding.

In Figure 3, s represents the original sequence and $[s]^Q$ a progressive approximation of s, reconstructed with MCJ2K using Q quality layers. MCTF^T transforms s into a collection of $T+1$ temporal texture sub-bands $\{L^T, \{H^t; 1 \leq t \leq T\}\}$, and T motion-"sub-bands" $\{M^t; 1 \leq t \leq T\}$. In Figure 3, the number of levels of MCTF is $T = 2$.

Compared to the MPEG/ITU standards, all the coefs (in our case, the images of index $i \times 2^T; i = 0, 1, \cdots$ of s) of L^T are I-type, and all the coefs of $\{H^t; 1 \leq t \leq T\}$) are B-type. More details about how MCTF has been implemented can be found in [22], and in our implementation published on GitHub (https://github.com/vicente-gonzalez-ruiz/MCTF-video-coding).

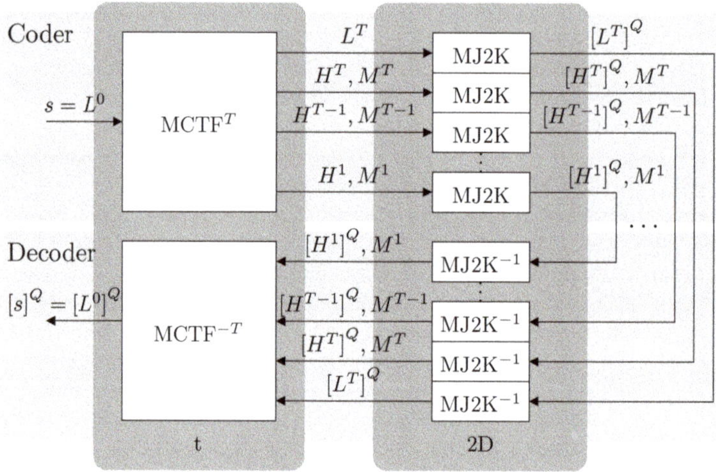

Figure 3. Codec architecture.

Figure 4 shows an example of the organization of an MCJ2K code-stream. Nine images have been compressed (although only the first six $s_0, \cdots s_5$ have been shown) using a GOP size $G = 4$ (i.e., $T = 2$ ($G = 2^T$)), except for the first GOP, which always has only one image. MCTF^2 transforms (see Figure 3) the input sequence s into 3 texture sub-bands $\{L^2, H^2, H^1\}$ and 2 motion sub-bands $\{M^2, M^1\}$. L^2 is the low-frequency texture sub-band, and represents the low-frequency temporal components of s. $\{H^2, H^1\}$ contains the high-frequency temporal components of s. $\{M^2, M^1\}$ stores a description of the motion detected in s. In Figure 4, arrows over the motion fields indicate the decoding dependencies between the coefs. When the inverse transform is applied, a succession of increasing temporal resolution levels $\{L^2, L^1, L^0\}$ are generated. By definition, $L^0 = s$.

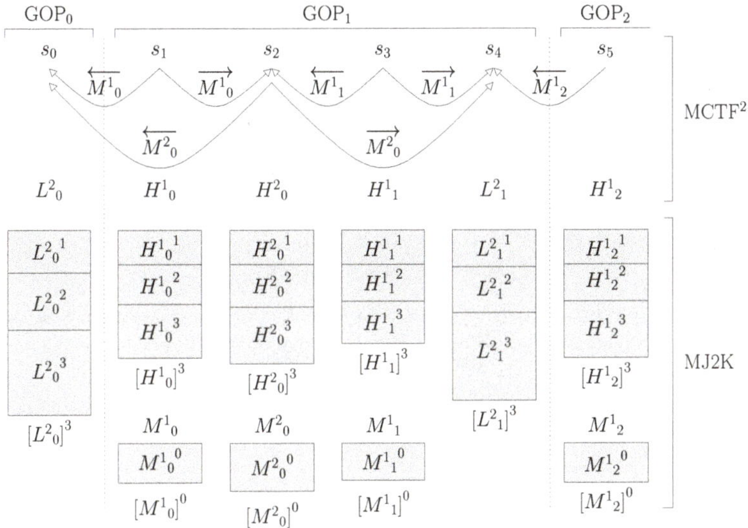

Figure 4. Example of a code-stream for $MCTF^2$.

MCJ2K implements a full/sub-pixel telescopic-search [23] bidirectional block-matching ME algorithm [24]. The block size B is constant (inside a coef), and a search area of A pixels is configurable. Exhaustive and logarithmic searches [25] are available. ME/MC operations are performed at the maximum spatial resolution of the sequence. This design decision, which is convenient for a progressive-quality visualization of the full-resolution video, implies that the inverse motion compensation process must be performed at the maximum resolution to avoid a drift error [26], when a reduced resolution of the images is decoded. Obviously, in this case, the decoder increases the computing requirements, but this does not significantly increase the memory usage if all the blocks are not processed in parallel. As an advantage, the quality of the reconstructions is higher than for the case in which the ME/MC state is performed at a lower resolution because the motion information is always used with the accuracy used at the compression, which can be sub-pixel.

Motion data is temporally and spatially decorrelated, and lossless MJ2K-compressed as a sequence of 4-component (2 vectors per macro-block) single-layer ($Q = 1$) images. (Usually, the use of approximate motion information generates severe artifacts in the reconstructed images and increases the non-linearity of the codec. Therefore, only one quality layer and lossless coding was used for the motion sub-bands.) The decorrelation process uses an algorithm in which, when no motion data is received, the inverse MCTF process supposes that all the motion vectors are zero. Thus (in the transmission process), when the decoder knows M^T, then it is supposed that the motion vectors of M^{T-1} are half of the value of M^T, and this linear prediction is used for the remaining temporal resolution levels [27].

3.2. Bitrate Control

RC is performed at compression time. In the MJ2K stage, each coef of each texture sub-band is J2K-compressed, producing a layered variable-length code-stream (in the Figure 4, $Q = 3$ quality layers). Let S_i^q be the q-th quality layer of the compressed representation of the coef S_i of sub-band S, and $(S_i)^q$ the quality (i.e., the decrease in distortion) provided by S_i^q in the progressive reconstruction of S_i. Assuming that the distortion metric is additive, we define

$$[S_i]^q = \sum_{j=1}^{q} (S_i)^j, \tag{1}$$

which is the quality of the reconstruction of the coef S_i using q layers. (In this notation, the first quality layer, in the layers decoding order, has the index 1.) We define the q-th R/D slope of coef S_i as

$$\lambda_{S_i{}^q} = \frac{[S_i]^q - [S_i]^{q-1}}{l(S_i{}^q)} = \frac{(S_i)^q}{l(S_i^q)}, \tag{2}$$

where $l(S_i{}^q)$ represents the length of $S_i{}^q$.

Owing to how the R/D slopes are chosen in the MJ2K stage, it holds that for any two different coefs i and j of sub-band S

$$\lambda_{S_i{}^q} = \lambda_{S_j{}^q} \forall q \in \{1, \cdots, Q\}. \tag{3}$$

We define a *sub-band layer* (SL) S^q (of motion (In the case of the motion, the definition is identical, but there is only one quality layer.) or texture) as the collection of quality layers

$$S^q = \{S_i{}^q, i = 0, \cdots, 2^T - 1\}. \tag{4}$$

For example, in Figure 4, SL $L^{2^1} = \{L^2{}_0{}^1, L^2{}_1{}^1\}$ and SL $M^{1^0} = \{M^1{}_0{}^0, M^1{}_1{}^0\}$.

Equation (3) has two implications: (1), in general, the total length of the code-stream of each coef will be different (depending on its content), and (2) the bitrate allocation is optimal for each sub-band layer [28].

The q-th R/D slope of SL S^q is defined as

$$\lambda_{S^q} = \frac{[S]^q - [S]^{q-1}}{l(S^q)} = \frac{(S)^q}{l(S^q)}, \tag{5}$$

where $l(S^q)$ represents the length of SL S^q, $[S]^q$ the quality of the GOP obtained after decompressing q layers, and $(S)^q$ the quality provided by the SL S^q.

3.3. Post-Compression R/D Allocation

RA is typically performed at decompression time. In accordance with Part 9, Section C.4.10 of the J2K standard [2], JPIP clients can request J2K images by quality layers. Moreover, as previously shown in [29], it is also possible to perform JPIP request for a range of images. Therefore, by extension, the JPIP standard can also be used for retrieving complete sub-band layers using a single JPIP request. For example (see Figure 4), if $T = 2$, we decompose a sequence in 3 temporal sub-bands, and the sub-band layer H^{2^1} has, for each GOP, only one coef $H^2{}_0$ with two quality layers $\{H^2{}_0{}^1, H^2{}_1{}^1\}$, which would be that which is requested by a client to retrieve the sub-band layer H^{2^1}.

It is easy to see that the SLs in a MCJ2K code-stream are

$$\begin{array}{ccccc}
L^{T^1}, & H^{T^1}, & H^{T-1^1}, & \cdots, & H^{1^1}, \\
L^{T^2}, & H^{T^2}, & H^{T-2^2}, & \cdots, & H^{1^2}, \\
\vdots & \vdots & \vdots & & \vdots \\
L^{T^Q}, & H^{T^Q}, & H^{T-1^Q}, & \cdots, & H^{1^Q}, \\
& M^T, & M^{T-1}, & \cdots, & M^1,
\end{array} \tag{6}$$

and that there are $Q(T+1)$ SLs in this set, which is also the number of optimal truncation points of a MCJ2K code-stream.

At decompression time, the order in which the SLs are retrieved from the JPIP server should minimize the R/D curve, for any bitrate. For this task, we propose the following two approaches.

3.3.1. Optimized SL Allocation (OSLA)

Starting at L^{T^1}, the optimal order of the remaining SLs of a GOP can be determined by applying Equation (5) to each feasible SL, and sorting them by slope. Thus, after retrieving L^{T^1} (which always contributes to the quality of the reconstruction more than any other SL), several alternatives $\{M^T, M^{T-1}, \cdots, M^1, H^{T^1}, H^{T-1^1}, \cdots, H^{1^1}\}$ should be checked to determine the next SL with the highest possible contribution. Considering that $\lambda_{M^T} > \lambda_{M^t}, \forall t \in \{T-1, \cdots, 1\}$, for example, if $\lambda_{M^T} > \lambda_{H^{t^1}}, \forall t \in \{T, \cdots, 1\}$, the following SL to decode should be M^T and the next set of alternatives would be $\{M^{T-1}, H^{T^1}, H^{T-1^1}, \cdots, H^{1^1}\}$. Otherwise, if for example, $\lambda_{H^{T^1}} > \lambda_{M^t}$ and $\lambda_{H^T} > \lambda_{H^t}, \forall t = T-1, \cdots, 1$, after L^{T^1} the next SL to decode should be H^{T^1}, and the current set of feasible SLs of would be $\{M^T, H^{T^2}, H^{T-1^1}, \cdots, H^{1^1}\}$. Notice also that other SLs could follow L^{T^1}, such as L^{T^2}.

This idea was implemented in the OSLA algorithm (see Algorithm 1). For each GOP, the input sequence of $Q(T+1)$ SLs is S sorted in descending order by their R/D slopes to reconstruct the GOP (see Equation (5)). The output list Λ of sorted-by-slope SLs can be stored in a COM segment of the header of the coef of L^T (the SL L^{T^1} is always the first in Λ). Next, JPIP clients retrieve the quality layers of each coef of the GOP in the order specified in Λ.

Algorithm 1: OSLA algorithm.

```
1.   for each GOP:
2.       Λ = []; i = 0
3.       Λ[i++] = input L^(T^1)
4.       S = input {L^(T^2), M^T, H^(T^1), ⋯, H^(1^1)}
5.       while S ≠ ∅:
6.           s = arg max (λ_s ≥ λ_s' ∀s' ∈ S)
                  s∈S
7.           Λ[i++] = s
8.           S = S \ {s}
9.           if s = M_i:
10.              S = S ∪ {M^(i-1)}
11.          else if s = L^(T^i):
12.              S = S ∪ {L^(T^(i-1))}
13.          else /* s = H^(t^i) */:
14.              S = S ∪ {H^(t^(i-1))}
15.      output Λ
```

3.3.2. Estimated-slope SLs Allocation (ESLA)

Ignoring any possible effect of the non-linear behavior of the ME/MC stage, our implementation of MCTF approximates to a biorthogonal transform and, therefore, each sub-band $\{L^T, H^T, \cdots, H^1\}$ contributes with a different amount of energy to the reconstruction of the sequence. This can easily be verified by comparing the energy that the different coefs of each temporal sub-band contribute to reconstruction of the sequence [30]. How much energy a coef must contribute to the code-stream to approximate MCTF to an orthonormal (energy preserving) transform is represented by attenuation values (see Table 1)

$$\alpha_{H^t} = \frac{E(L^T)}{E(H^t)},\tag{7}$$

where $E(\cdot)$ represents the signal energy. These attenuations are empirical, specifically determined for the 1/3 ME-driven DWT implemented in our codec (for a different transform, other values would be obtained).

Table 1. L_2-norm (energy) of the MCTF basis functions for the temporal sub-bands, expressed as attenuation values.

MCTF^1		MCTF^2		MCTF^3		MCTF^4		MCTF^5		MCTF^6		MCTF^7	
H^t	a_t	H^t	a_t	H^t	a_t	H^t	a_t	H^t	a_t	H^t	a_t	H^t	a_t
H^1	1.246	H^2	1.250	H^3	1.160	H^4	1.088	H^5	1.046	H^6	1.023	H^7	1.012
		H^1	1.865	H^2	2.122	H^3	2.130	H^4	2.079	H^5	2.043	H^6	2.023
				H^1	3.167	H^2	3.888	H^3	4.061	H^4	4.063	H^5	4.039
						H^1	5.802	H^2	7.431	H^3	7.936	H^4	8.031
								H^1	11.089	H^2	14.522	H^3	15.688
										H^1	21.669	H^2	28.707
												H^1	42.835

The ESLA algorithm incorporates these attenuations to scale the R/D slopes of each SL of each GOP, when these slopes have been determined taking into consideration only the reconstruction of the corresponding coef (not the reconstruction of the full GOP, as OSLA does (notice that for this reason, OSLA does not need to use such attenuations). Thus, for example, an R/D slope for a quality layer of a coef of the sub-band H^3 resulting from an MCTF^5 is divided by 4.061. In cases where there is more than one coef in a temporal sub-band, as in this example, the average of all the scaled slopes is used to determine the contribution of the corresponding SL.

This idea was implemented in ESLA (see Algorithm 2). As in OSLA, for each GOP, the input sequence of $Q(T+1)$ SLs S is sorted in descending order by their estimated R/D slope, but now the slopes of the SLs are computed directly as a weighted average of the R/D slopes of the quality layers of the corresponding coefs. If these slopes are predefined (the compression of the coefs uses the same slopes set of Q slopes for all the coefs), ESLA can be run at the receiver side without sending any R/D information. This means that the JPIP client can determine the order of SLs Λ for all the GOPs of the sequence after receiving only T, Q, and knowing the sub-band attenuations (Table 1), which does not depend on the sequence. For this reason, ESLA is more suitable than OSLA for real-time streaming scenarios.

Algorithm 2: ESLA algorithm.

```
1.   for each GOP:
2.       Λ = [];i = 0
3.       for each q ∈ {1,··· ,Q}:
4.           Λ[i++] = input{λ_{H^Tq},··· ,λ_{H^1q}}
5.       for each λ_k ∈ Λ:
6.           λ_k = λ_k/α_k
7.       Λ[i++] = input{λ_{L^T1},··· ,λ_{L^1Q}}
8.       sort_in_descending_order Λ
9.       output Λ
```

---

4. Evaluation

The performance of MCJ2K was evaluated for different working configurations and compared to previous proposals.

4.1. Materials and Methods

Several test videos were used for our evaluation:

1. `Mobile` (http://trace.eas.asu.edu/yuv/mobile/mobile_cif.7z) (YUV 4:2:0, 352×288 pixels, 30 Hz), a low-resolution video with complex movement.
2. `Container` (http://trace.eas.asu.edu/yuv/container/container_cif.7z) (YUV 4:2:0, 352×288 pixels, 30 Hz), a low-resolution video with simple movement.

3. **Crew** (ftp://ftp.tnt.uni-hannover.de/pub/svc/testsequences/CREW_704x576_60_orig_01_yuv.zip) (YUV 4:2:0 704 × 576 pixels, 60 Hz), a medium-resolution video with complex movement.

4. **CrowdRun** (ftp://vqeg.its.bldrdoc.gov/HDTV/SVT_MultiFormat/) (YUV 4:2:0, 1920 × 1080 pixels, 50 Hz), a high-resolution video with a high degree of movement.

5. **ReadySetGo** (http://ultravideo.cs.tut.fi/video/ReadySetGo_3840x2160_120fps_420_8bit_YUV_RAW.7z) (YUV 4:2:0 3840 × 2160 pixels, 120 Hz), a high-resolution high degree of movement.

6. Sun (http://helioviewer.org/jp2/AIA/2015/06/01/131/) (monochromatic, due to represent only one frequency of the spectrum radiated by the Sun, 4096 × 4096 pixels, 1/30 Hz) a sequence of images of the Sun with only small-scale frame-to-frame motion (which is, however, complex to predict due to the random motions of magnetic flux concentrations in the Sun's photosphere).

In all experiments, 129 images were compressed, and the search range for ME was 4 pixels using full-pixel accuracy of ($A = 0$). The block size (B) was 32 × 32 for Mobile, Container and Crew, 64 × 64 for CrowdRun and ReadySetGo, and 128 × 128 for Sun. The parameters used for compressing the coefs and the images were 5 levels for the DWT, no precinct partition and code-blocks of 64 × 64 coefficients. The number of quality layers (Q) was 8, which provides a good tradeoff between the compression performance and the granularity for the rate-allocation. In the case of the motion data, $Q = 1$ and no DWT were used.

4.2. Impact of Motion Compensation

Figure 5 shows the performance of MCJ2K compared to MJ2K for different GOP sizes. Each video was compressed once and decompressed progressively, sorting the subband layers using OSLA. MCJ2K was in most of cases superior to MJ2K, depending on the temporal correlation found in each video. For example, MCTF is very efficient in Container, in which it can be seen that, for example, at 300 Kbps, MCJ2K is about 10 dB better than MJ2K. However, in the case of ReadySetGo, in which MCTF is not able to generate accurate predictions, the use of a GOP size larger than 4 does not increase the quality of the reconstructions. Therefore, the GOP size has a high impact on the performance of MCJ2K and is a parameter that should be optimized for every video sequence. Nevertheless, it can be expected that GOP sizes of 4 and 8 should work well for most sequences. We would like to highlight here that the MC model used in MCJ2K is very basic. More advanced predictors, such as those used in the last video coding standards cited earlier, would facilitate the use of larger GOP sizes and, therefore, higher compression ratios.

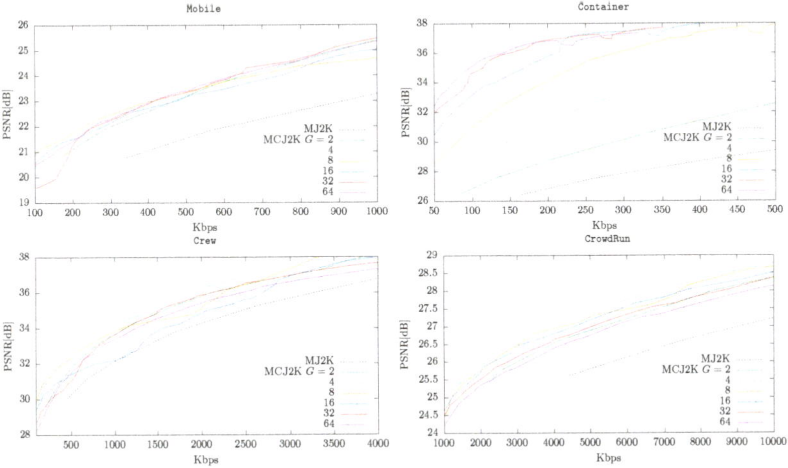

Figure 5. *Cont.*

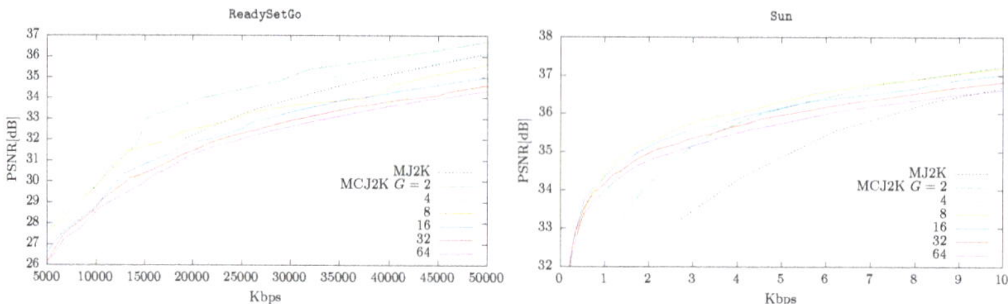

Figure 5. MCJ2K (OSLA) vs. MJ2K for different GOP sizes and sequences.

4.3. MCJ2K (Using OSLA or ESLA) vs. MJ2K

Using the information provided by the previous experiments, we selected a suitable GOP size for each sequence and compared the performance of OSLA and ESLA, respect to MJ2K. The results are shown in Figure 6. As can be seen, the performance of both RA algorithms is similar, which means that although the MCTF process used by MCJ2K is not linear, a reasonable prediction of the impact of the SLs can be made in ESLA, which runs much faster than OSLA. For this reason, in the following experiments only ESLA will be used.

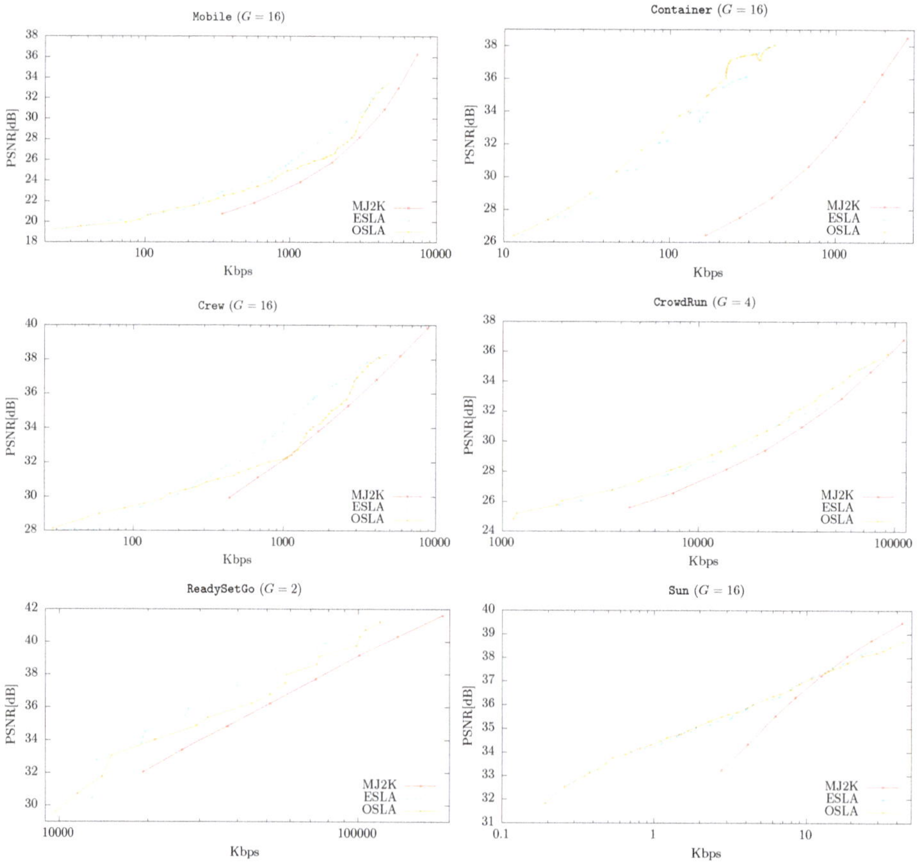

Figure 6. MCJ2K (using OSLA or ESLA) vs. MJ2K for different sequences.

4.4. MCJ2K vs. Other Video Codecs

Figure 7 shows the compression performance of MCJ2K (using ESLA and optimized compression parameters found in previous experiments) and other standard video codecs. Dashed lines represent a non-embedded decoding, while solid lines, a progressive decoding provided by scalable codecs. As can be seen, compared with non-scalable video codecs (which generally produce videos with a better R/D ratio than scalable video codecs), such as HEVC (https://hevc. hhi.fraunhofer.de/svn/svn_HEVCSoftware using `trunk/cfg/encode_randomaccess_main.cfg`) or AVC (http://www.videolan.org/developers/x264.html using `-profile high-preset placebo-tune psnr`), MCJ2K needs approximately 50% more data to achieve the same quality, but this difference is much smaller when it is compared with SHVC (https://hevc.hhi.fraunhofer.de/svn/svn_SHVCSoftware/ using `branches/SHM-dev/cfg/encoder_randomaccess_scalable.cfg` and `branches/SHM-dev/cfg/misc/layers8.cfg`) where MCJ2K produces better results for some of the test videos (even using a very basic MCTF scheme). In the case of MPEG-2 (http://linux.die.net/man/1/mpeg2enc, a codec that implements an MCTF scheme similar to the used in MCJ2K), MCJ2K outperforms it consistently. These results are consistent with the ME prediction model used in MCJ2K, which is not the focus of this research work.

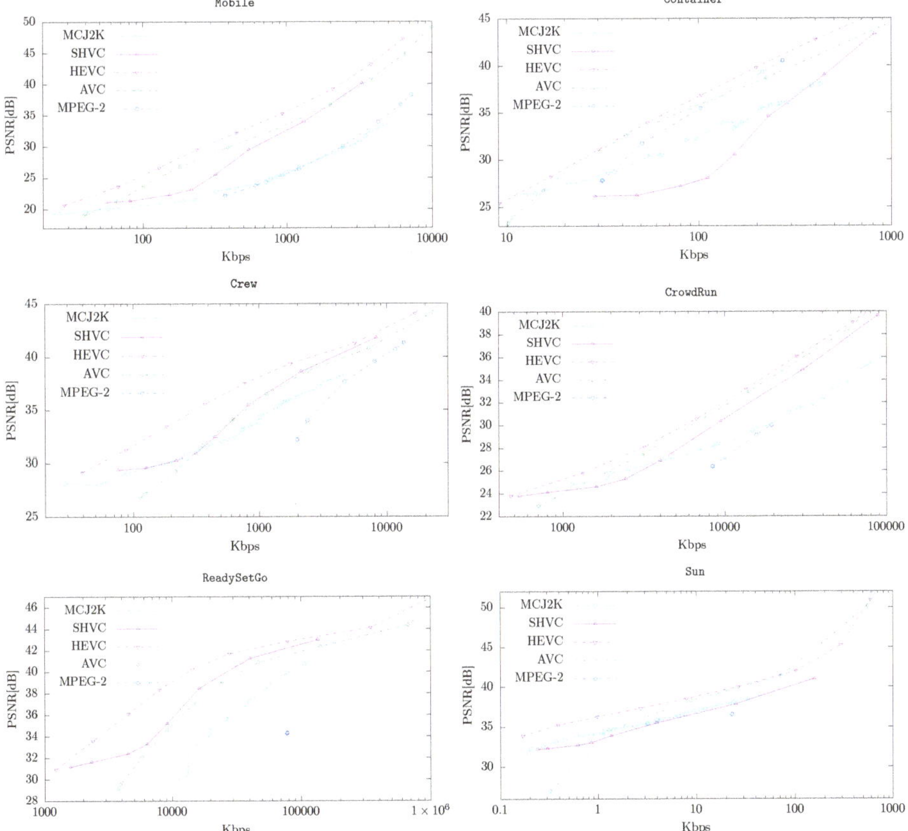

Figure 7. MCJ2K (using ESLA) vs. other codecs for different sequences.

5. Conclusions

This work presents MCJ2K, a straightforward extension (JPIP compatible) of the MJ2K standard that can be used to exploit the temporal redundancy of the sequences of images. Two different rate-allocation algorithms (OSLA and ESLA) are proposed to be used in a streaming scenario where quality scalability is used, generating a reconstruction at the receiver proportional to the amount of decoded data. After analyzing MCJ2K and the proposed rate-allocation algorithms, the following can be concluded:

1. The compression ratio obtained by MCJ2K is superior to MJ2K if enough time redundancy can be exploited in the MCTF stage. Our experiments show that the quality of the reconstructions can be up to 10 dB in terms of PSNR.
2. The increment in the compression ratio provided by OSLA compared to ESLA is small. Considering that the RD performance of OSLA is better than ESLA when the MCTF process is not linear, we conclude (1) that MCDWT is almost linear, and (2) that the position of the motion information in the progression generated by ESLA is near optimal.
3. Considering that ESLA requires less computational resources than OSLA, and that ESLA needs to be run only at the receiver, ESLA should be the rate-allocation algorithm used by default in MCJ2K.
4. MCJ2K implies recompressing the sequences of images but not modifying the JPIP servers at all. Only the JPIP clients need to implement the logic needed by MCJ2K.
5. The compression ratio obtained by MCJ2K is comparable to SHVC, when the movement of the video can be modeled using our ME proposal. However, the quality granularity and the range of decoding bitrates is higher in MCJ2K, which makes MCJ2K more suitable than SHVC for streaming scenarios.
6. In MCJ2K the GOP size G significantly affects the RD performance. G should be high if the temporal correlation of the video can be removed by the MCTF stage, and vice versa.
7. Compared to the state-of-the-art non-scalable video compressors (such as HEVC), MCJ2K require more bitrate because HEVC use more effective ME schemes than MCJ2K (an aspect out of the scope of this paper). However, at very low bitrates this gap is usually small.

6. Future Research

Future lines of work should be focused on:

1. Like the rest of video codecs based on MC, MCJ2K has a cost in terms of temporal scalability. A study on the number of bytes required for obtaining the same quality in both codecs, MJ2K and MCJ2K, when only one image of the sequence is decoded could prove worthwhile, especially in interactive browsing systems such as Helioviewer.
2. Find a quality scalable representation of the motion data. Such a contribution should reduce the minimal number of bytes required for rendering the image sequence.
3. The use of more accurate MCTF schemes should increase the compression ratios.
4. The use of encoding schemes where the motion information can be estimated at the decoder (to avoid being sent as a part of the code-stream). This can be carried out in those contexts where the large-scale motion is predictable, such as image sequences of the Sun, whose rotation rate is stable and well known.
5. How MCJ2K affects the spatial/WOI scalability provided by the J2K standard.

Author Contributions: conceptualization, J.C.M.-E., V.G.-R., J.P.G.-O. and D.M.; methodology, V.G.-R. and J.P.G.-O.; software, J.C.M.-E. and V.G.-R.; validation, J.C.M.-E., V.G.-R., J.P.G.-O. and D.M.; investigation, J.C.M.-E.; resources, D.M.; data curation, J.C.M.-E. and V.G.-R.; writing–original draft preparation, J.C.M.-E. and V.G.-R.; writing–review and editing, J.C.M.-E. and V.G.-R.; supervision, V.G.-R.; project administration, V.G.-R.; funding acquisition, V.G.-R.

Funding: Work supported by the Spanish Ministry of Economy and Competitiveness (RTI2018-095993-B-100) and Junta de Andalucía (P10-TIC-6548), in part financed by the European Regional Development Fund (ERDF) and Campus de Excelencia Internacional Agroalimentario (ceiA3).

Conflicts of Interest: The authors declare no conflict of interest.

References

1. ISO. *Information Technology-JPEG 2000 Image Coding System-Core Coding System*; ISO/IEC 15444-1:2004; ISO: Geneva, Switzerland, 2004.
2. ITU. Information Technology-JPEG 2000 Image Coding System: Interactivity Tools, APIs and Protocols. Available online: http://www.itu.int/rec/T-REC-T.808-200501-I (accessed on 26 August 2019).
3. Bilgin, A.; Marcellin, M. JPEG2000 for digital cinema. In Proceedings of the IEEE International Symposium on Circuits and Systems (ISCAS), Island of Kos, Greece, 21–24 May 2006; pp. 4–3881, doi:10.1109/ISCAS.2006.1693475. [CrossRef]
4. Müller, D.; Dimitoglou, G.; Caplins, B.; Ortiz, J.P.; Wamsler, B.; Hughitt, K.; Alexanderian, A.; Ireland, J.; Amadigwe, D.; Fleck, B. JHelioviewer: Visualizing large sets of solar images using JPEG 2000. *Comput. Sci. Eng.* **2009**, *11*, 38–47. [CrossRef]
5. Müller, D.; Nicula, B.; Felix, S.; Verstringe, F.; Bourgoignie, B.; Csillaghy, A.; Berghmans, D.; Jiggens, P.; García-Ortiz, J.; Ireland, J.; Zahniy, S.; Fleck, B. JHelioviewer-Time-dependent 3D visualisation of solar and heliospheric data. *Astron. Astrophys.* **2017**, *606*, A10. [CrossRef]
6. Cohen, R.; Woods, J. Resolution scalable motion-compensated JPEG 2000. In Proceedings of the 2007 15th International Conference on Digital Signal Processing, Cardiff, UK, 1–4 July 2007; pp. 459–462.
7. Secker, A.; Taubman, D. Lifting-based Invertible Motion Adaptive Transform (LIMAT) framework for highly scalable video compression. *IEEE Trans. Image Process.* **2003**, *12*, 1530–1542, doi:10.1109/TIP.2003.819433. [CrossRef] [PubMed]
8. Schwarz, H.; Marpe, D.; Wiegand, T. Overview of the scalable video coding extension of the H. 264/AVC standard. *IEEE Trans. Circuits Syst. Video Technol.* **2007**, *17*, 1103–1120. [CrossRef]
9. Sullivan, G.; Ohm, J.; Han, W.J.; Wiegand, T. Overview of the High Efficiency Video Coding (HEVC) standard. *IEEE Trans. Circuits Syst. Video Technol.* **2012**, *22*, 1649–1668, doi:10.1109/TCSVT.2012.2221191. [CrossRef]
10. Andre, T.; Cagnazzo, M.; Antonini, M.; Barlaud, M. JPEG2000-compatible scalable scheme for wavelet-based video coding. *EURASIP J. Image Video Process.* **2007**, *2007*, 1–11, doi:10.1155/2007/30852. [CrossRef]
11. Cagnazzo, M.; Castaldo, F.; Andre, T.; Antonini, M.; Barlaud, M. Optimal motion estimation for wavelet motion compensated video coding. *IEEE Trans. Circuits Syst. Video Technol.* **2007**, *17*, 907–911, doi:10.1109/TCSVT.2007.897110. [CrossRef]
12. Ferroukhi, M.; Ouahabi, A.; Attari, M.; Habchi, Y.; Taleb-Ahmed, A. Medical video coding based on 2nd-generation wavelets: Performance evaluation. *Electronics* **2019**, *8*, 88. [CrossRef]
13. Sullivan, G.J.; Wiegand, T. Rate-distortion optimization for video compression. *IEEE Signal Process. Mag.* **1998**, *15*, 74–90. [CrossRef]
14. Barbarien, J.; Munteanu, A.; Verdicchio, F.; Andreopoulos, Y.; Cornelis, J.; Schelkens, P. Motion and texture rate-allocation for prediction-based scalable motion-vector coding. *Signal Process. Image Commun.* **2005**, *20*, 315–342. [CrossRef]
15. Ouahabi, A. *Signal and Image Multiresolution Analysis*; Wiley Online Library: Hoboken, NJ, USA, 2012.
16. Aulí-Llinàs, F.; Bilgin, A.; Marcellin, M. FAST Rate Allocation Through Steepest Descent for JPEG2000 video transmission. *IEEE Trans. Image Process.* **2011**, *20*, 1166–1173, doi:10.1109/TIP.2010.2077304. [CrossRef] [PubMed]
17. Jiménez-Rodríguez, L.; Aulí-Llinàs, F.; Marcellin, M. FAST rate allocation for JPEG2000 video transmission over time-varying channels. *IEEE Trans. Multimed.* **2013**, *15*, 15–26, doi:10.1109/TMM.2012.2199973. [CrossRef]
18. Naman, A.; Taubman, D. JPEG2000-based Scalable Interactive Video (JSIV). *IEEE Trans. Image Process.* **2011**, *20*, 1435–1449, doi:10.1109/TIP.2010.2093905. [CrossRef] [PubMed]
19. Naman, A.; Taubman, D. JPEG2000-Based Scalable Interactive Video (JSIV) with motion compensation. *IEEE Trans. Image Process.* **2011**, *20*, 2650–2663, doi:10.1109/TIP.2011.2126588. [CrossRef] [PubMed]

20. ISO/IEC 23009-1:2012 Information Technology—Dynamic Adaptive Streaming over HTTP (DASH)—Part 1: Media Presentation Description and Segment Formats. Available online: http://www.iso.org/iso/iso_catalogue/catalogue_tc/catalogue_detail.htm?csnumber=57623 (accessed on 26 August 2019).

21. Mehrotra, S.; Zhao, W. Rate-distortion optimized client side rate control for adaptive media streaming. In Proceedings of the IEEE International Workshop on Multimedia Signal Processing (MMSP), Rio de Janeiro, Brazil, 5–7 October 2009; pp. 1–6, doi:10.1109/MMSP.2009.5293246. [CrossRef]

22. Secker, A.; Taubman, D. Motion-compensated highly scalable video compression using an adaptive 3D wavelet transform based on lifting. In Proceedings of the IEEE International Conference on Image Processing, Thessaloniki, Greece, 7–10 October 2001; Volume 2, pp. 1029–1032.

23. Suguri, K.; Minami, T.; Matsuda, H.; Kusaba, R.; Kondo, T.; Kasai, R.; Watanabe, T.; Sato, H.; Shibata, N.; Tashiro, Y.; others. A real-time motion estimation and compensation LSI with wide search range for MPEG2 video encoding. *IEEE J. Solid-State Circuits* **1996**, *31*, 1733–1741. [CrossRef]

24. Wu, S.; Gersho, A. Joint estimation of forward/backward motion vectors for MPEG interpolative prediction. *IEEE Trans. Image Process.* **1994**, *3*, 684–687. [CrossRef] [PubMed]

25. Hsieh, C.; Liu, Y. Fast Search Algorihtms for Vector Quantization of Images Using Multiple Triangle Inequalities and Wavelet Transform. *IEEE Trans. Image Proc.* **2000**, *9*, 321–328. [CrossRef] [PubMed]

26. Mokry, R.; Anastassiou, D. Minimal error drift in frequency scalability for motion-compensated DCT coding. *IEEE Trans. Circuits Syst. Video Technol.* **1994**, *4*, 392–406. [CrossRef]

27. Andreopoulos, Y.; van der Schaar, M.; Munteanu, A.; Barbarien, J.; Schelkens, P.; Cornelis, J. Fully-scalable wavelet video coding using in-band motion compensated temporal filtering. In Proceedings of the 2003 IEEE International Conference on Acoustics, Speech, and Signal Processing, (ICASSP'03), Hong Kong, China, 6–10 April 2003; Volume 3, pp. 417–420.

28. Dagher, J.; Bilgin, A.; Marcellin, M. Resource-constrained rate control for Motion JPEG2000. *IEEE Trans. Image Process.* **2003**, *12*, 1522–1529, doi:10.1109/TIP.2003.819228. [CrossRef] [PubMed]

29. Sánchez-Hernández, J.; García-Ortiz, J.; González-Ruiz, V.; Müller, D. Interactive streaming of sequences of high resolution JPEG2000 images. *IEEE Trans. Multimed.* **2015**, *17*, 1829–1838. [CrossRef]

30. Xiong, R.; Xu, J.; Wu, F.; Li, S. Adaptive MCTF based on correlation noise model for SNR scalable video coding. In Proceedings of the IEEE International Conference on Multimedia and Expo (ICME), Toronto, ON, Canada, 9–12 July 2006; pp. 1865–1868.

Article

Multiscale Image Matting Based Multi-Focus Image Fusion Technique

Sarmad Maqsood [1,2], Umer Javed [1], Muhammad Mohsin Riaz [3], Muhammad Muzammil [1], Fazal Muhammad [2] and Sunghwan Kim [4,*]

[1] Faculty of Engineering and Technology, International Islamic University, Islamabad 44000, Pakistan; sarmad.maqsood@cusit.edu.pk (S.M.); umer.javed@iiu.edu.pk (U.J.); m.muzammil@iiu.edu.pk (M.M.)
[2] Department of Electrical Engineering, City University of Science and Information Technology, Peshawar 25000, Pakistan; fazal.muhammad@cusit.edu.pk
[3] Center for Advanced Studies in Telecommunication, COMSATS University, Islamabad 44000, Pakistan; mohsin.riaz@comsats.edu.pk
[4] School of Electrical Engineering, University of Ulsan, Ulsan 44610, Korea
*   Correspondence: sungkim@ulsan.ac.kr

Received: 2 February 2020; Accepted: 8 March 2020; Published: 12 March 2020

Abstract: Multi-focus image fusion is a very essential method of obtaining an all focus image from multiple source images. The fused image eliminates the out of focus regions, and the resultant image contains sharp and focused regions. A novel multiscale image fusion system based on contrast enhancement, spatial gradient information and multiscale image matting is proposed to extract the focused region information from multiple source images. In the proposed image fusion approach, the multi-focus source images are firstly refined over an image enhancement algorithm so that the intensity distribution is enhanced for superior visualization. The edge detection method based on a spatial gradient is employed for obtaining the edge information from the contrast stretched images. This improved edge information is further utilized by a multiscale window technique to produce local and global activity maps. Furthermore, a trimap and decision maps are obtained based upon the information provided by these near and far focus activity maps. Finally, the fused image is achieved by using an enhanced decision maps and fusion rule. The proposed multiscale image matting (MSIM) makes full use of the spatial consistency and the correlation among source images and, therefore, obtains superior performance at object boundaries compared to region-based methods. The achievement of the proposed method is compared with some of the latest techniques by performing qualitative and quantitative evaluation.

Keywords: image fusion; multi-focus; trimaps; focus maps

1. Introduction

During image acquisition, one of the most important objectives is obtaining a focused region of interest. However, because of the limited field depth, the focused region contains sharp edges, whereas the other regions get blurred. Recently, multi-focus image fusion (combine images with different focused objects) has received tremendous attention amongst the researchers. This fused image offers high quality containing more detailed information [1,2]. Several methods are developed to fuse multiple images, which are broadly grouped into transform and spatial domains [3,4].

Transform domain methods fuse the corresponding transform coefficients and employ inverse transformation to construct the fused image. Spatial domain methods are further classified into pixel [5,6] and region based methods [7,8]. The spatial domain methods form the fuse image by choosing the pixels/regions/blocks that are focused. Transform domain-based methods in dynamic scenes merge these coefficients without considering the spatial properties, resulting in artifacts in the fused image. Furthermore, pixel and region-based methods are unable to produce the best fusion results for images with complicated texture patterns [1].

Zhang et al. [9] used morphological operations to extract focus regions. However, this technique suffers from block artifacts. De et al. [10] utilized morphological processes to detect the focused region and suggested a technique for calculating an optimized block size. The fused result still suffers from blocking effects. Later on, Bai et al. [11] presented a novel quadtree decomposition and a weighted focus based image fusion technique. However, this technique also provides inaccurate segmentation and low visual effects because of the smooth regions. Yin et al. [12] proposed a method based on joint dictionary and singular value decomposition (SVD) methods. Still, this method is not effective computationally because of the individual training for sub dictionaries and SVD computation.

Li et al. [13] explored guided filtering (GFF) and spatial information to improve the fusion results by mitigating the block effects. Zhang et al. [14] proposed a multifocus image scheme based upon a visual saliency method. Recently, image matting has been used for effectively differentiating the focused and out-of-focus regions. These methods can be broadly categorized as supervised matting and unsupervised matting techniques. Supervised methods require a user specified foreground and background regions known as trimap. Therefore, such techniques require human experts, are time consuming, and produce inconsistent results for images with high-textured backgrounds. However, unsupervised methods are better than supervised ones because user interaction is not required for achieving a good matting result. Chen et al. [15] used a parametric edge based method. However, these methods do not consider the artifacts among the smooth regions, and much of it depends on the performance of hand crafted features, which require much expert knowledge. Li et al. [16] proposed multifocus matting (MFM) based image fusion by combining together the focus region and its neighboring pixels. This method marginally improves the fusion results and also overcomes some shortcomings of spatial domain methods.

Xiao et al. [17] used depth information to segment an image into focus and blur regions. Zhang et al. [18] made use of log spectrum, Fourier transform, and Bayesian techniques. In [19], a definite focus region is detected by using a novel multi scale gradient information. Liu et al. [20] proposed a transform (which is scale invariant) to detect focus regions. However, this technique fails to offer sharp edges of the focus regions. Furthermore, in [21], the focus information was extracted by using texture features. Baohua et al. [22] performed the near and far focus region detection by using a sparse representation and guided filter techniques. In [23], a structure tensor was used for the detection of high and low frequency components. However, this technique fails to provide a visible difference between focus and defocus regions in many cases. Yu at.al. [24] presented a convolutional neural network (CNN) based multifocus image fusion technique. However, in this method, the precision of recognizing the focus block is very low.

In this paper, a novel multi-focus image fusion method is presented using contrast stretching and spatial gradients to enhance the edges from the source images. A multiscale sliding window method is used for detecting the local and global intensity variations to generate initial activity maps. These multiple activity maps are further processed to generate a trimap. An enhanced image matting technique is used for generating the decision maps. Finally, the fused image is obtained after processing the source images, enhanced decision maps, and employing the fusion rule.

2. Proposed Fusion Technique

The schematic diagram of the proposed algorithm is shown in Figure 1. It can be observed that in the first step, a contrast enhancement scheme is applied on the source images. In the second step, the outcome of the intensity transformed image is processed through an edge detection method. In a multi-focus image fusion scheme, the selection of near focus and far focus region plays a vital role. The region that is in focus during image acquisition tends to have sharp edges as compared to the out-of-focus region. Therefore, these sharp edges can be detected easily by applying an appropriate edge detection method.

The edge detection schemes rely heavily on the intensity distribution of an image. A poor intensity distribution can lead to an oversaturated, undersaturated, dark, or bright image. In either of those images, the edge detection algorithm cannot perform well. In order to improve the intensity distribution of an image, an intensity transformation can be performed. In Figure 2, the improvement in edge information is shown by comparing the images before and after applying the contrast enhancement scheme. In the next step, a sliding window technique with two different scales is applied on both edges of the detected images to generate activity maps. In this step, both local and global intensity variations are analyzed. The fine details are more prominent under a small sliding window scale. These masks are further fused together and processed to generate a trimap. Next, the trimap undergoes an image matting transformation to produce refined decision maps, which produces the final fused image. The proposed fusion scheme, along with the equation references, is also elaborated in Algorithm 1.

Let I_i be the source color images with $M \times N$ dimensions where, $m = 1, 2, \ldots, M, n = 1, 2, \ldots, N$ and $i \in [1, 2]$ represents near and far focus images, respectively.

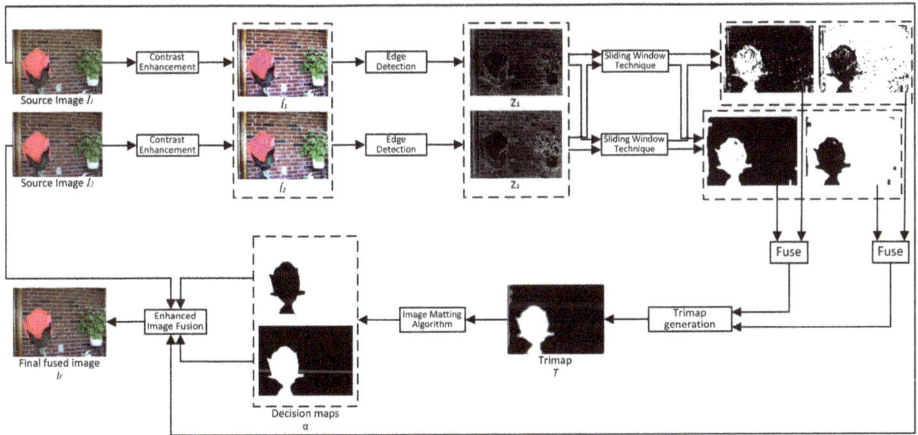

Figure 1. Schematic diagram of the proposed approach for the image fusion algorithm.

Algorithm 1: Proposed MSIM based Fusion Technique.

Require: $I_i, i \in [1, 2]$.

Step 1. Apply contrast enhancement on I_i using Equation (1).

Step 2. Apply edge map on $\acute{I}_i$ using Equation (2) to Equation (5).

Step 3. Compute activity maps, $\acute{G}_{i,\sigma} \xleftarrow{\sigma = 1} w, Z_i$ using Equation (6).

Step 4. Compute Smooth Activity maps, $G_{i,\sigma} \xleftarrow{\text{Sum Filter}, \sigma = 1} \acute{G}_{i,\sigma}$ using Equation (7).

Step 5. Compute Score maps, $\zeta_{i,\sigma} \xleftarrow{\sigma = 1} G_{i,\sigma}$ using Equations (8) and (9).

Step 6. Repeat Steps 3, 4, and 5 with $\sigma = 3$.

Step 7. Compute Near focus (D_1) and Far focus (D_2) using Equation (10).

Step 8. Generate Trimap T using Equation (11).

Step 9. Generate Alpha Matte α using Equation (14).

Step 10. Generate Fused Image A_F using Equation (15).

2.1. Contrast Enhancement

Improving the enhancement of the low contrast image, the histogram equalization seems an effective method. Non-parametric modified histogram equalization (NMHE) [25] is integrated to enhance the contrast and preserves the mean brightness of the source image I_i, i.e.,

$$\acute{I}_i \xleftarrow{\text{NMHE}} I_i \tag{1}$$

Image development in contrast centrally improves and concentrates pixel details. Figure 2 shows the enhancement in edge information. Figure 2a,b shows the far and near focus source image, respectively, and their gradients are shown in Figure 2c,d. Contrast enhanced of near and far focus images are displayed in Figure 2e,f, and their respective edge maps are shown in Figure 2g,h, respectively. From the images, it can be clearly seen that after the enhancement algorithm, the gradients of the source image were greatly improved.

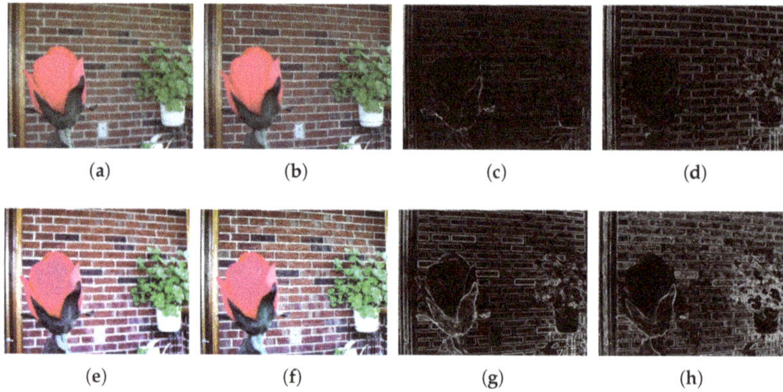

Figure 2. Results of edge detection after contrast enhancement. (**a**) Near focus image, (**b**) far focus image, (**c**,**d**) gradients of (**a**,**b**) achieved by a spatial stimuli sketch model (SSGSM) [26], (**e**,**f**) contrast enhancement using non-parametric modified histogram equalization (NMHE) [25], (**g**,**h**) gradients of (**e**,**f**) achieved by SSGSM [26].

2.2. Edge Detection

The edges of the images after contrast enhancement is done by a spatial stimuli sketch model (SSGSM) [26] technique, which principally focuses on focal intensity points and edges in an image, and then the unknown region is calculated in the coarse decision maps by implementing the concentrated information in both the activity level maps. The weight of the local stimuli is deliberated by detecting the local variation in the perceived brightness at the respective positions. The discerned brightness, P_i of a specific image is given in Equation (2) as,

$$P_i = \vartheta \log_{10}(\acute{I}_i) \tag{2}$$

where, $\acute{I}_i$ represents the source images, and ϑ denotes the scaling factor.

Gradients illustrate the sharp intensity variations in the image. Mathematically, the weight is computed as the total difference of the perceived brightness on x and y directions. The intensity variations of P_i on the x and y axis are represented by ϱ_i^x and ϱ_i^y, respectively. These variations are calculated by using their respective gradients B_i^x and B_i^y, given as in Equations (3) and (4):

$$[B_i^x, B_i^y] \xleftarrow{\text{gradient}} P_i \tag{3}$$

$$\varrho_i^x = B_i^x\left(e^{-|B_i^x|}\right); \varrho_i^y = B_i^y\left(e^{-|B_i^y|}\right) \tag{4}$$

The weight of local stimuli Z_i is expressed by using Equation (5):

$$Z_i = \sqrt{(\varrho_i^x)^2 + (\varrho_i^y)^2} \tag{5}$$

2.3. Focus Maps

A multiscale sliding window technique is applied to acquire diverse focus maps from activity maps Z_i. Two sliding windows are selected for the generation of focus maps. Firstly, a 9×9 window is initialized by setting $k = 9$, $l = 9$ and $\sigma = 1$ in Equation (6). The activity maps are divided into blocks of 9×9 pixels by using spatial domain filters, as in Equations (6) and (7):

$$G'_{i,\sigma}(m,n) = \sum_{q_1=-\sigma\times k}^{\sigma\times k} \sum_{q_2=-\sigma\times l}^{\sigma\times l} w(q_1,q_2)Z_i(m+q_1,n+q_2) \tag{6}$$

$$G_{i,\sigma}(s,t) = \sum_{(m,n)\,\in\,\Omega} G'_{i,\sigma}(m,n) \tag{7}$$

The activity of each block is stored in the form of map scores. Furthermore, the sum of intensity levels in each near ($G_{1,\sigma=1}(s,t)$) and far focus block ($G_{2,\sigma=1}(s,t)$) are calculated and compared with one another to update the score maps ($\zeta_{i,\sigma=1}$), as given in Equations (8) and (9).

$$\zeta_{1,\sigma}(m,n) = \begin{cases} 1, & \text{if } G_{1,\sigma}(s,t) > G_{2,\sigma}(s,t) \\ 0, & \text{Otherwise} \end{cases} \tag{8}$$

$$\zeta_{2,\sigma}(m,n) = 1 - \zeta_{1,\sigma}(m,n) \tag{9}$$

Similarly, 27×27 block of pixels are generated by setting $k = 9$, $l = 9$ and $\sigma = 3$ in Equation (6). The activity maps in each near ($G_{1,\sigma=3}(s,t)$) and far focus block ($G_{2,\sigma=3}(s,t)$) are calculated and compared with one another to update the score maps ($\zeta_{i,\sigma=3}$), as in Equations (8) and (9).

These multiple sliding windows result in multiple near and far focus maps. This multiscale sliding window technique reduces the blocking artifacts in the coarse decision maps. Each map offers different characteristic information, which plays a key role in improving the focus maps and the fused image. These multiscale windows extract the information from original images at different scales. It is noted that this approach has demonstrated better visual quality than the existing methods. Each scale offers different information for image fusion, for example, a small window size focuses on local intensity variations, whereas a large size window size extracts global variations in an image. The information from these multiscale near-focus ($\zeta_{1,\sigma=1}$ and $\zeta_{1,\sigma=3}$) and far-focus maps ($\zeta_{2,\sigma=1}$ and $\zeta_{2,\sigma=3}$) are combined together to form a single near-focus (D_1) and far-focus (D_2) map, respectively, carrying the attributes of both scales, as in Equation (10).

$$D_i(m,n) \xleftarrow{\text{AND}} \zeta_{i,\sigma=1}, \zeta_{i,\sigma=3} \tag{10}$$

After obtaining the focus maps, the next step is to generate a trimap that segments the given images into the three different regions, i.e., focused, definite defocused, and unknown. Pixels from the focused region have greater focus value than pixels in the defocused region [27]. The trimap T of A_1 is processed by using D_1, D_2 as in Equation (11).

$$T \xleftarrow{\text{Tri Map}} D_i \tag{11}$$

In a given image I, the image matting considers it a composite of foreground I^{Fore} and background I^{back}. Each pixel is assumed to be a linear combination of I^{Fore} and I^{back}. Let α denote the pixel foreground opacities then an image I can be represented as,

$$I_i = \alpha_i I_i^{Fore} + (1 - \alpha_i) I_i^{back} \tag{12}$$

In [28], the quadratic cost function for α is derived as,

$$J(\alpha) = \alpha^T L \alpha \tag{13}$$

where, L is defined as a matting laplacian matrix of $N \times N$ dimension.

The L is a symmetric positive definite matrix and is defined in [28] as $L = H$ - W, where, H is a diagonal matrix and W is a symmetric matrix. The neighborhood W_M is given as,

$$W_M(i,j) = \sum_{k/(i,j)\epsilon w_k=0}^{K} \frac{1}{|w_k|} \left(1 + (\chi_i - \phi_k)(\nu_k + \frac{\epsilon}{|w_k|}\Gamma)^{-1}(\chi_j - \phi_k)\right) \tag{14}$$

where, $|w_k|$ denotes the number of pixels in the window, ϕ_k and ν_k represents mean and variance of intensities in the window w_k, respectively. χ represents the pixel color, ϵ is a regularization parameter and Γ is an identity matrix.

Finally, the obtained alpha matte α from the source images and trimap is same as the focused region of I_i is constructed as in Equation (15).

$$I_F(m, n) = \alpha(m, n)I_1(m, n) + (1 - \alpha(m, n))I_2(m, n) \tag{15}$$

3. Results and Discussion

To show the superiority of the proposed MSIM, a comparison was performed with discrete wavelet transform (DWT) [29], guided filtering based fusion (GFF) [13], discrete cosine transform (DCT) [30], dense sift (DSIFT) [20], multi-scale morphological focus-measure (MSMFM) [9], and convolutional neural network (CNN) [24] on a multifocus image dataset [31]. The proposed method was evaluated by performing both subjective and objective assessments. These algorithms were tested on a Acer laptop Intel(R) Core™ i7 2.6GHz processor with 12GB RAM under a Matlab R2018b environment. All the algorithms were executed by using the original codes made available by the authors.

3.1. Comparison of Image Matting Result

Generally, an unsupervised trimap produces better results than the supervised ones. Hence, in practice, user specified trimaps are often necessary to achieve the high quality matting results; however, the making of a user supervised trimap takes time, skills, and is not available for all kind of images. In this paper, two image matting techniques have been proposed, i.e., focus maps matting and feature based matting. The results of the proposed method are compared with feature based matting and the closed form matting [28]. It is clearly observed that the proposed matting produces better results compared to the existing technique (Figure 3).

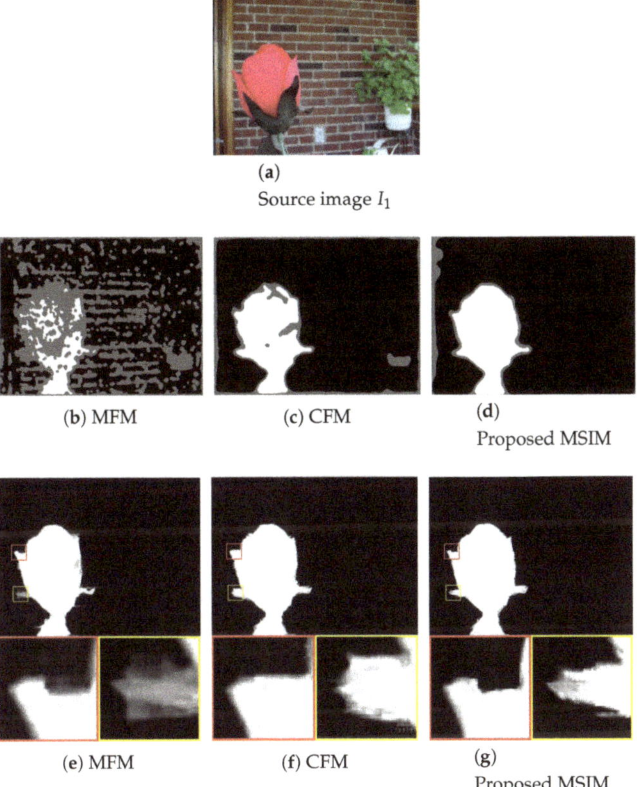

(a)

Source image I_1

(b) MFM

(c) CFM

(d)
Proposed MSIM

(e) MFM

(f) CFM

(g)
Proposed MSIM

Figure 3. Results of Trimap and Alpha matte on flower image (**b–d**) trimaps of (**a**), (**e–g**) alpha mattes of Figure 3a.

3.2. Comparison of Image Fusion with Other Methods

The proposed technique is tested on gray scale, color, and dynamic images. Figure 4 shows the results of the proposed MSIM for "Lab" image. The source near and far focus inputs are presented in Figure 4a,b, respectively. The fused results produced by other methods and the proposed technique are given in Figure 4c–i. To further investigate the effectiveness, the difference of the near-focus image with the fused images is shown in Figure 4j–p. The close up views enclosed by red and yellow boxes are also shown at the bottom of their respective difference image. It is noted that the DWT, DCT, and DSIFT methods produce poor edge information and contain artifacts (as shown in the close-ups). Furthermore, GFF, MSMFM, and CNN methods also provide limited information of the focused regions as compared to the proposed MSIM technique. Similarly, Figure 5 illustrates the results produced by several existing and proposed algorithms for "Globe" images. To further analyze the results, close-up views of important regions are placed at the bottom of each difference image. In this image, the boundary region of the hand is difficult to detect since it lies on the focus transition point. The results of fusion by other techniques in Figure 5j–o show the distorted regions and lack of sharpness in the highlighted region. However, the proposed MSIM method has successfully fused the complementary information from both the images, as shown in Figure 5p. It is very important to evaluate the results of different algorithms on the color dataset shown in Figures 6a,b and 7a,b. The outcomes of the existing techniques and the proposed method on "Flower" and "Boy" are shown in Figures 6c–i and 7c–i, respectively. The difference between the fused and out of focus source images is illustrated in Figures 6j–p and 7j–p, respectively. It is noted that in both the flower and boy images, the existing techniques are unable to mitigate the artifacts and blur in the focus transition area (as noted in the close-ups of difference images). The proposed MSIM is able to preserve contrast and details using the edge feature and multi scale image matting technique.

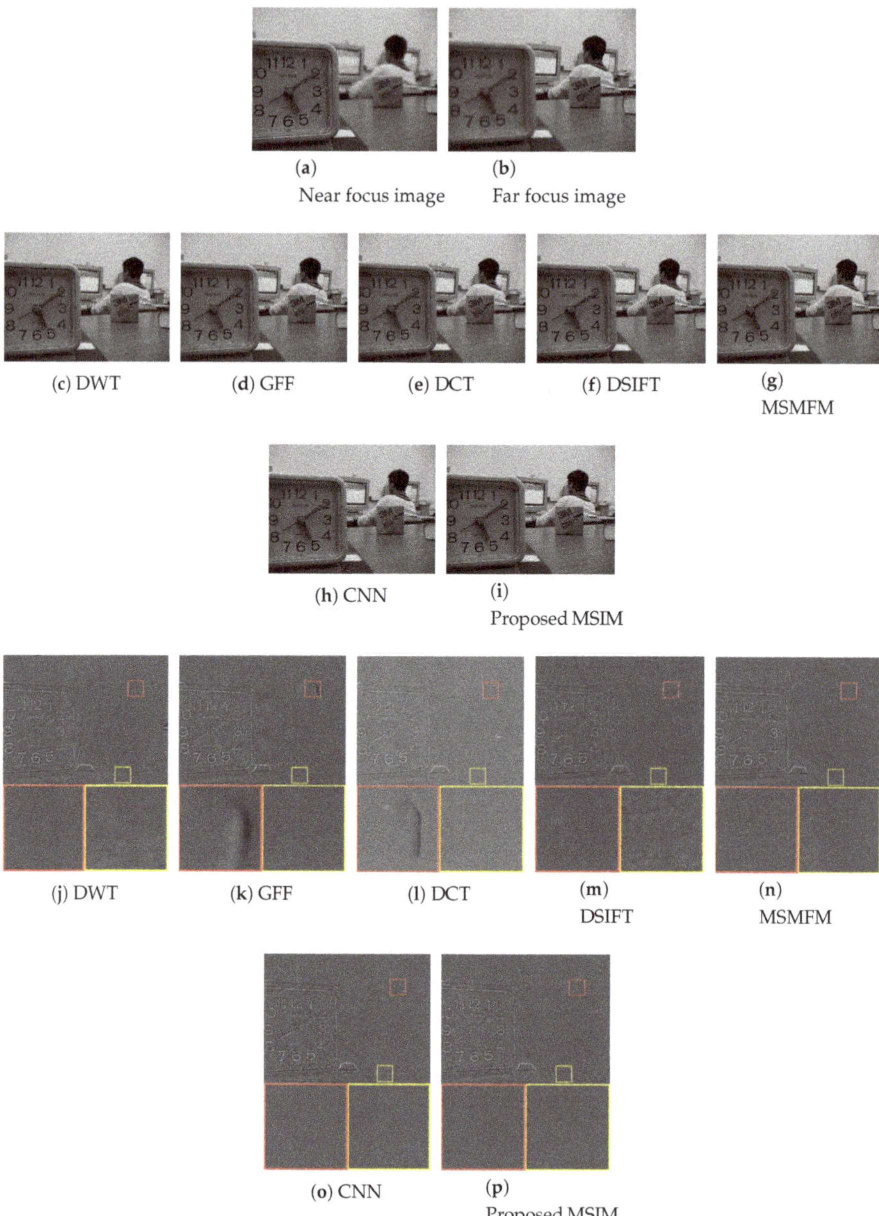

Figure 4. Results of image fusion and their difference images on the "Lab" source images. (c–i) Fused images obtained through fusion schemes. (j–p) Difference images obtained from the fusion results and Figure 4b.

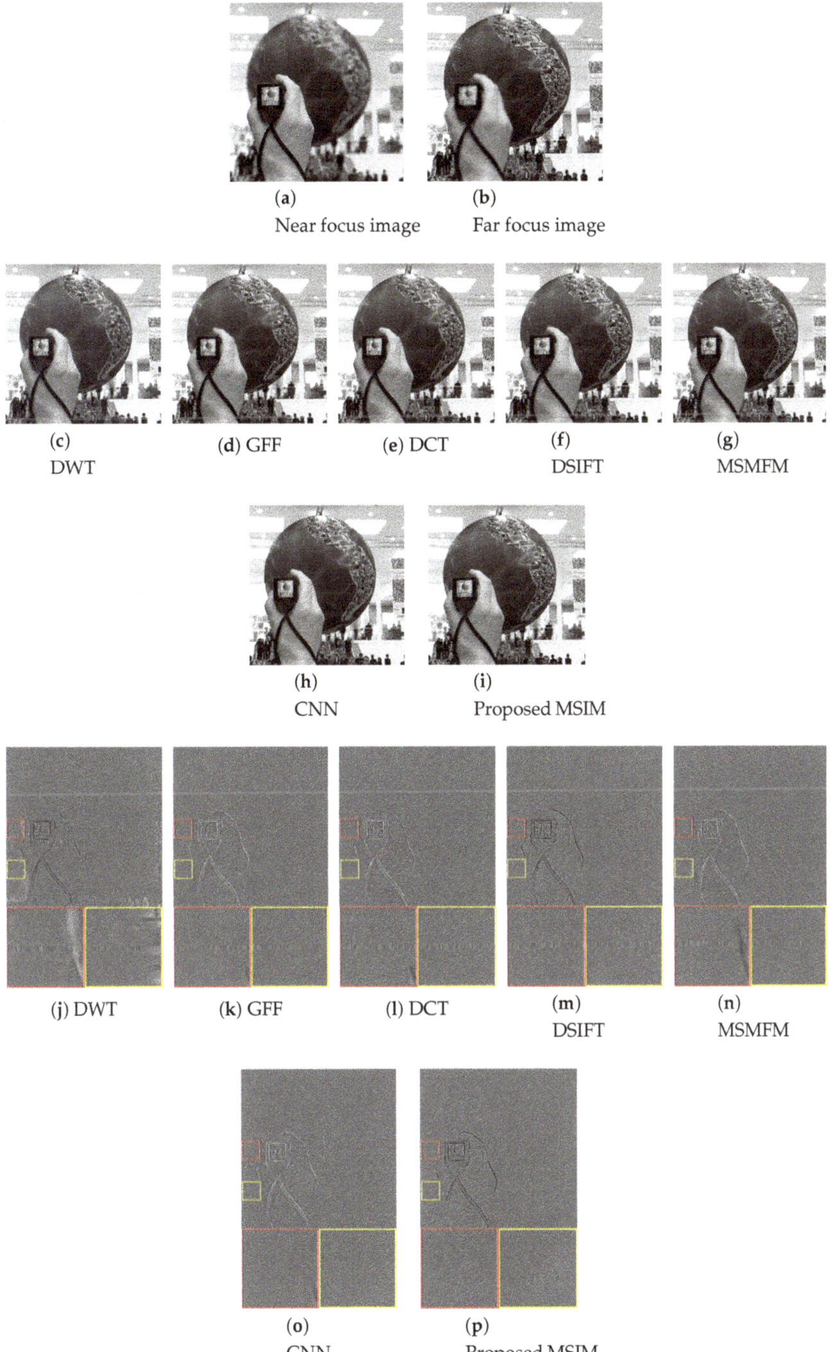

Figure 5. Results of image fusion and their difference images on the "Globe" source images. (**c–i**) Fused images obtained through fusion schemes. (**j–p**) Difference images obtained from the fusion results and Figure 5b.

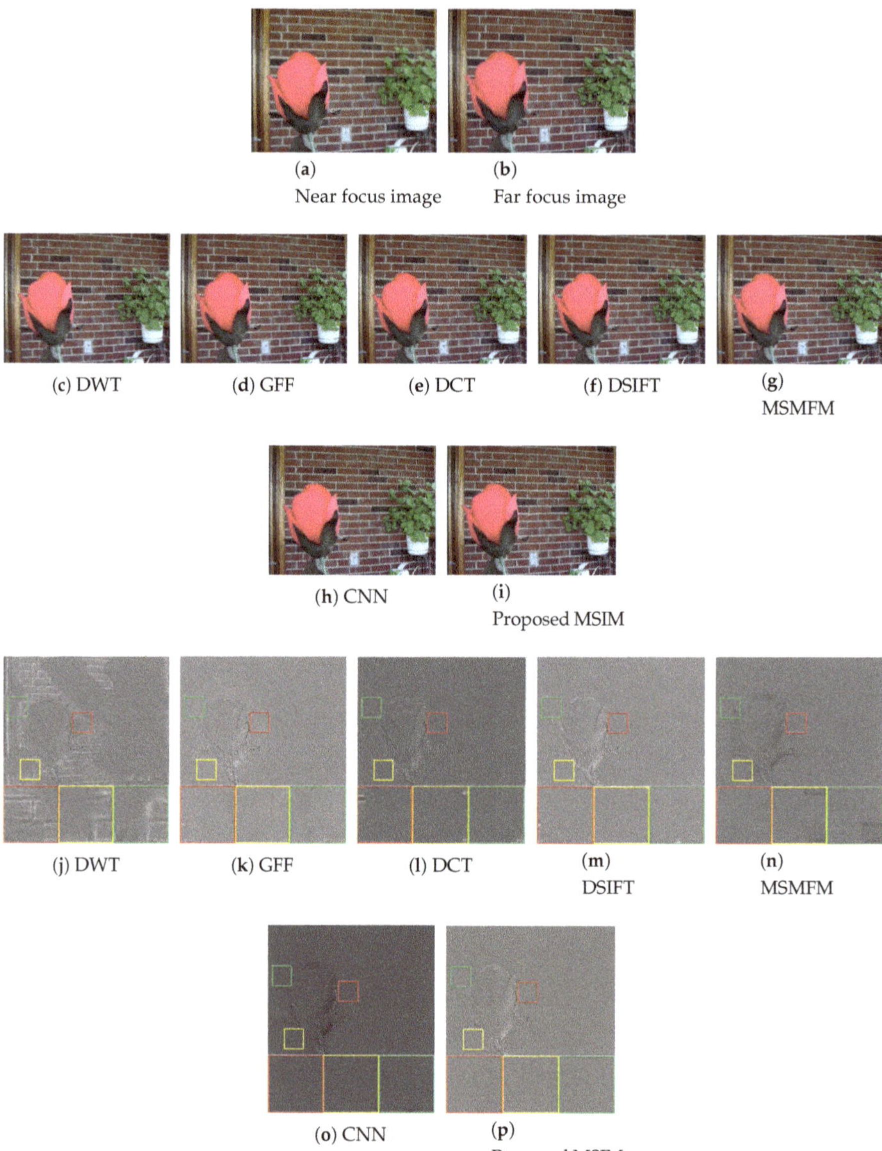

Figure 6. Results of image fusion and their difference images on the "Flower" source images. (**c–i**) Fused images obtained through fusion schemes. (**j–p**) Difference images obtained from the fusion results and Figure 6b.

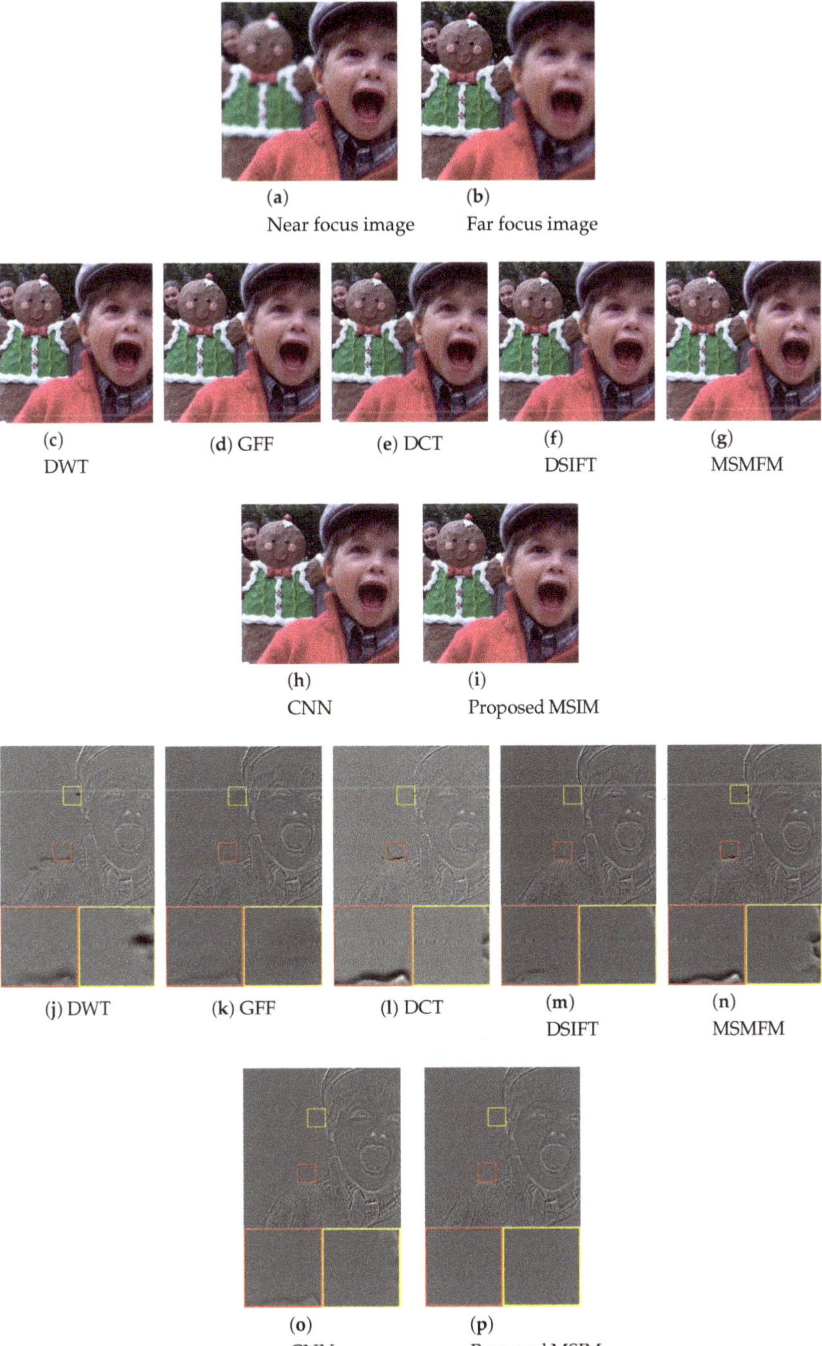

Figure 7. Results of image fusion and their difference images on the "Boy" source images. (**c–i**) Fused images obtained through fusion schemes. (**j–p**) Difference images obtained from the fusion results and Figure 7a.

Another challenge for multi-focus fusion includes the performance in dynamic scenes. The scenario occurs either due to the movement of the camera or the motion of the object. So it is important to verify the effectiveness of the MSIM result with the existing ones on such scenes. Figure 8a,b shows near and far focus "Girl" images, respectively. The results of MSIM and existing techniques are shown in Figure 8c–i, while Figure 8j–p shows difference images. As shown in the red and yellow boxes, the DWT, GFF, DCT, DSIFT, and MSMFM methods are unable to completely fuse the focus regions. Moreover, the CNN has produced erosion in the fused image, whereas the proposed MSIM has successfully mitigated the inconsistencies and limitations of the existing techniques.

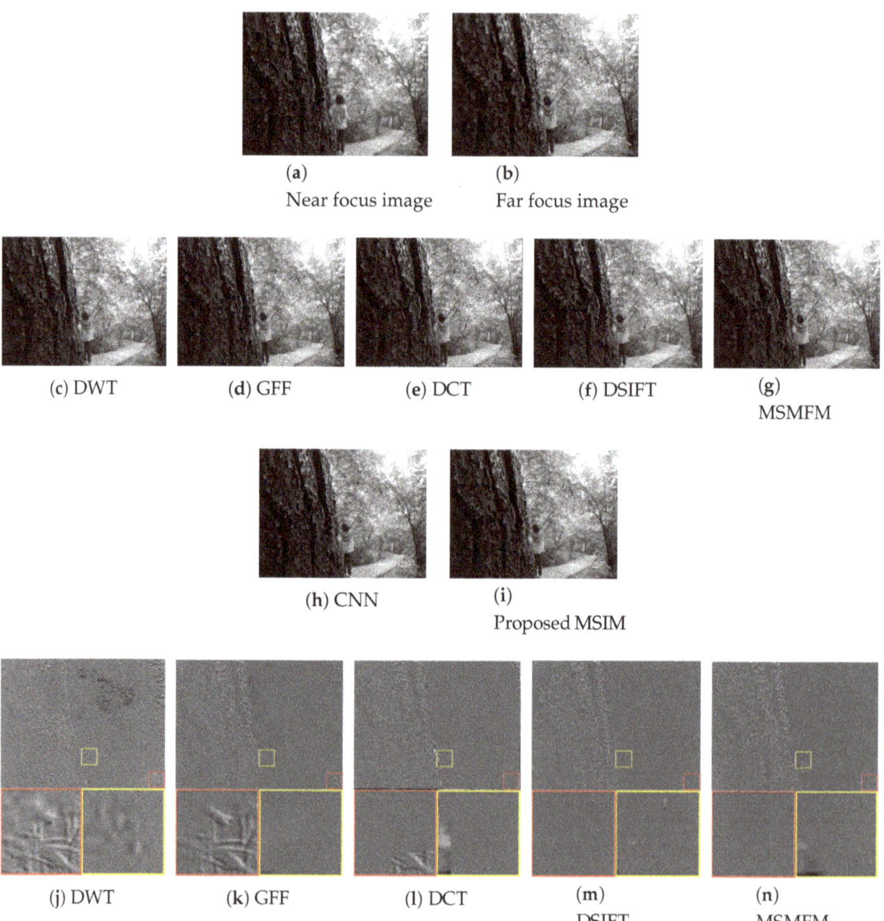

Figure 8. *Cont.*

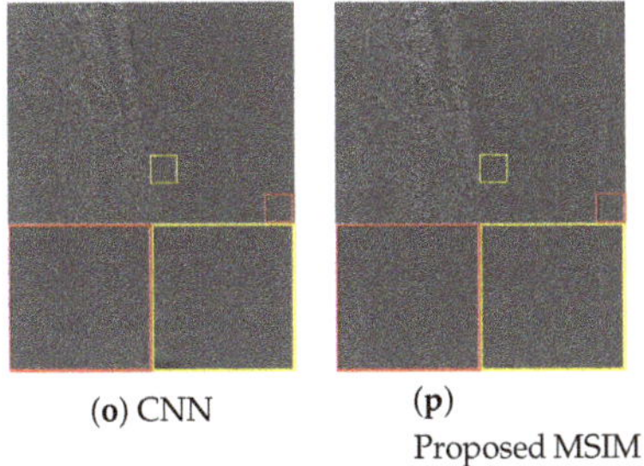

Figure 8. Results of image fusion and their difference images on the "Girl" source images. (**c–i**) Fused images obtained through fusion schemes. (**j–p**) Difference images obtained from the fusion results and Figure 8b.

It is observed from these visualizations that the existing methods produce artifacts, erosion, halo effects and are unable to produce sharp boundaries of the near and far focus images. Note that the MSIM technique not only perfectly identifies the near and far focus regions but also fuses the complementary information in an effective manner.

3.3. Objective Evaluation Metrics

After evaluating the visual quality and quantitative assessment of different methods, it can be clearly observed that MSIM produces a visually pleasant and high quality fusion result in almost all cases and outperformed the existing fusion methods for multi-focus images. Five most commonly used metrics are evaluated, i.e., Mutual Information (MI) [32], Spatial Structural Similarity (SSS) $Q^{AB/F}$ [33], Feature Mutual Information (FMI) [34], Entropy (EN) [35], and Visual Information Fidelity (VIF) [36] to verify the superiority and effectiveness of the proposed MSIM method. Table 1 shows that the proposed MSIM gives better objective assessment results than the existing methods. Although, the results of existing techniques are comparable in some cases (Flower and Boy); however, the metric values obtained using the proposed MSIM generally outperforms the existing techniques.

3.4. Comparison of Computational Efficiency

In this section, the computational efficiency of different fusion methods is compared. The execution time of these schemes for different images is shown in Table 1. The results show that the proposed MSIM, DSIFT, and GFF consume less time as compared to the other algorithms DCT, DWT, MSMFM, and CNN. The MSMFM algorithm uses a multi-scale morphological gradient based feature, therefore taking longer processing time than DSIFT. Whereas, GFF integrates the source images by using a global weight based scheme; however, it still takes less computation time and produces satisfactory results.

The proposed method utilizes the contrast enhancement, SSGSM based edge extraction, sliding window based local and global operations to create activity maps and trimap. The sliding window method, activity maps generation, their comparison, and a trimap generation are time consuming tasks. Although, the proposed algorithm consumes more processing time as compared to the existing ones; however, it produces the best unsupervised image matting and image fusion results.

Table 1. The quantitative assessment of different fusion methods.

Images	Fusion Methods	MI [32]	$Q^{AB/F}$ [33]	FMI [34]	EN [35]	VIF [36]	Time/s
Lab	DWT [29]	8.2152	0.7239	0.8190	7.0474	0.9138	4.02
	GFF [13]	7.9114	0.7279	0.8191	7.0602	0.9149	3.11
	DCT [30]	8.5263	0.7460	0.9197	6.9819	0.9143	11.56
	DSIFT [20]	8.5212	**0.7478**	0.9097	7.0759	0.9171	6.65
	MSMFM [9]	8.7995	0.6864	0.9196	6.9885	0.9161	5.79
	CNN [24]	8.6812	0.7471	0.9196	6.9974	0.9159	7.88
	Proposed	**8.8322**	0.7474	**0.9386**	**7.1759**	**0.9980**	5.08
Globe	DWT [29]	8.1910	0.7246	0.8892	7.7037	0.9240	11.09
	GFF [13]	8.7664	0.7726	0.8935	7.7412	0.9476	9.78
	DCT [30]	9.1845	0.7731	0.8939	7.6990	0.9374	10.98
	DSIFT [20]	9.1435	**0.7746**	0.8938	7.6989	0.9437	5.51
	MSMFM [9]	9.3739	0.7711	0.8940	7.7389	0.9439	6.55
	CNN [24]	9.2397	0.7701	0.8927	7.7458	0.9472	8.01
	Proposed	**9.4316**	0.7733	**0.8943**	**7.7479**	**0.9480**	7.06
Flower	DWT [29]	5.6452	0.6536	0.8773	7.1701	0.9064	16.93
	GFF [13]	7.3290	0.6944	0.8908	7.1915	0.9200	10.13
	DCT [30]	7.8561	0.6785	0.8861	7.4331	0.9263	12.03
	DSIFT [20]	8.0057	0.6947	0.8857	7.4316	0.9304	4.61
	MSMFM [9]	7.9233	0.6930	0.8915	7.1873	0.9140	5.98
	CNN [24]	3.0773	0.6951	0.8912	7.1872	0.9177	7.32
	Proposed	**8.1458**	**0.7940**	**0.8936**	**7.5897**	**0.9367**	6.91
Boy	DWT [29]	7.5321	0.7206	0.8814	7.5371	0.8935	9.95
	GFF [13]	7.6316	0.7448	0.8717	7.5310	0.8097	5.97
	DCT [30]	8.0852	0.7409	0.8714	7.5669	0.9035	11.08
	DSIFT [20]	8.1765	0.7437	0.8721	7.5388	0.9048	3.80
	MSMFM [9]	8.2081	0.7418	0.8717	7.2402	0.9026	7.55
	CNN [24]	2.9966	0.7466	0.8819	7.5386	0.9072	8.01
	Proposed	**8.2961**	**0.7487**	**0.8826**	**7.5673**	**0.9077**	7.75
Girl	DWT [29]	5.6226	0.5939	0.8189	7.8477	0.6486	5.39
	GFF [13]	8.0820	0.6902	0.8213	7.8475	0.6448	5.88
	DCT [30]	8.7222	0.6788	0.8168	7.8226	0.7332	6.44
	DSIFT [20]	8.8774	0.6834	0.8215	7.8429	0.7394	4.04
	MSMFM [9]	9.0368	0.5427	0.8219	7.6549	0.7397	7.11
	CNN [24]	8.8580	0.6919	0.8208	7.8607	0.7417	10.21
	Proposed	**9.0970**	**0.6936**	**0.8415**	**7.8749**	**0.7468**	5.65

4. Conclusions

A multiscale image fusion technique is presented for accurate construction of tri-maps, decision maps, and fused images. Firstly, the source images are pre-processed using a NMHE histogram equalization method and their gradients are computed using SSGSM. A multiscale sliding window technique calculates the focus maps from source images. Furthermore, the focus information is processed so that an accurate focused region is extracted. The proposed MSIM is robust to noise interference and is flexible to combine various fusion strategies and provides better fusion performance both visually and quantitatively when compared with other state of the art methods for multi-focus images datasets. In the future, the proposed scheme will be further considered for other application areas of image processing.

Author Contributions: Conceptualization: S.M. and U.J.; methodology: S.M. and U.J.; software: S.M., U.J. and M.M.; validation: S.M., U.J. and M.M.R.; formal analysis: S.M., U.J. and M.M.R.; investigation: S.M., U.J., M.M.R., and F.M.; data curation: S.M. and U.J.; writing—original draft preparation: S.M., U.J. and M.M.R.; supervision: U.J., and S.K.; project administration: F.M. and S.K.; funding acquisition: S.K.

Funding: This work was supported by the Research Program through the National Research Foundation of Korea (NRF-2019R1A2C1005920).

Conflicts of Interest: The authors declare no conflict of interest.

References

1. Li, S.; Kang, X.; Fang, L.; Hu J.; Yin, H. Pixel-level image fusion. A survey of the state of the art. *Inf. Fus.* **2017**, *33*, 100–112. [CrossRef]
2. Maqsood, S.; Javed, U. Multi-modal Medical Image Fusion based on Two-scale Image Decomposition and Sparse Representation. *Biomed. Signal Process. Control* **2020**, *57*, 101810. [CrossRef]
3. Thang, C.; Anh, D.; Khan, A.W.; Karim, P.; Sally, V. Multi-Focus Fusion Technique on Low-Cost Camera Images for Canola Phenotyping. *Sensors* **2018**, *18*, 1887.
4. Goshtasby, A.A.; Nikolov, S. Image fusion: Advances in the state of the art. *Inf. Fus.* **2007**, *8*, 114–118. [CrossRef]
5. Yang, Y.; Yang, M.; Huang, S.; Ding, M.; Sun, J. Robust sparse representation combined with adaptive PCNN for multifocus image fusion. *IEEE Access* **2018**, *6*, 20138–20151. [CrossRef]
6. Eltoukhy, H.A.; Kavusi, S. A computationally efficient algorithm for multi-focus image reconstruction. *SPIE Electr. Imaging Proc.* **2003**, 332–341.
7. Zribi, M. Non-parametric and region-based image fusion with Bootstrap sampling. *Inf. Fus.* **2010**, *11*, 85–94. [CrossRef]
8. Qilei, L.; Xiaomin, Y.; Wei, W.; Kai, L.; Gwanggil, J. Multi-Focus Image Fusion Method for Vision Sensor Systems via Dictionary Learning with Guided Filter. *Sensors* **2018**, *18*, 2143.
9. Zhang, Y.; Bai, X.; Wang, T. Boundary finding based multi-focus image fusion through multi-scale morphological focus-measure. *Inf. Fus.* **2017**, *35*, 81–101. [CrossRef]
10. De, I.; Chanda, B. Multi-focus image fusion using a morphology-based focus measure in a quad-tree structure. *Inf. Fus.* **2013**, *14*, 136–146. [CrossRef]
11. Bai, X.; Zhang, Y.; Zhou, F.; Xue, B. Quadtree-based multi-focus image fusion using a weighted focus measure. *Inf. Fus.* **2015**, *22*, 105–118. [CrossRef]
12. Yin, H.; Li, Y.; Chai, Y.; Liu, Z.; Zhu, Z. A novel sparse-representation based multi-focus image fusion approach. *Neurocomputing* **2016**, *216*, 216–229. [CrossRef]
13. Li, S.; Kang, X.; Hu, J. Image fusion with guided filtering. *IEEE Trans. Image Process.* **2013**, *22*, 2864–2875. [PubMed]
14. Zhang, B.; Lu, X.; Peo, H.; Liu, H. Multi-focus image fusion algorithm based on focused region extraction. *Neurocomputing* **2016**, *174*, 733–748. [CrossRef]
15. Chen, Y.; Guan, J.; Cham, W.K. Robust Multi-Focus Image Fusion Using Edge Model and Multi-Matting. *IEEE Trans. Image Process.* **2017**, *27*, 1526–1541. [CrossRef]
16. Li, S.; Kang, X.; Hu, J.; Yang, B. Image matting for fusion of multi-focus images in dynamic scenes. *Inf. Fus.* **2013**, *14*, 147–162. [CrossRef]
17. Xiao, J.; Liu, T.; Zhang, Y.; Zou, B.; Lei, J.; Li, Q. Multi-focus image fusion based on depth extraction with inhomogeneous diffusion equation. *Signal Process.* **2016**, *125*, 171–186. [CrossRef]
18. Zhang, X.; Li, X.; Feng, Y. A new multifocus image fusion based on spectrum comparison. *Signal Process.* **2016**, *123*, 127–142. [CrossRef]
19. Zhou, Z.; Li, S.; Wang, B. Multi-scale weighted gradient-based fusion for multifocus images. *Inf. Fus.* **2014**, *20*, 60–72. [CrossRef]
20. Liu, Y.; Liu, S.; Wang, Z. Multi-focus image fusion with dense sift. *Inf. Fus.* **2015**, *23*, 139–155. [CrossRef]

21. Liu, Z.; Chai, Y.; Yin, H.; Zhou, J.; Zhu, Z. A novel multi-focus image fusion approach based on image decomposition. *Inf. Fus.* **2017**, *35*, 102–116. [CrossRef]

22. Baohua, Z.; Xiaoqi, L.; Haiquan, P.; Yanxian, L.; Wentao, Z. Multi-focus image fusion based on sparse decomposition and background detection. *Dig. Signal Process.* **2015**, *58*, 50–63.

23. Li, H.; Li, X.; Yu, Z.; Mao, C. Multifocus image fusion by combining with mixed-order structure tensors and multiscale neighborhood. *Inf. Sci.* **2016**, *349–350*, 25–40. [CrossRef]

24. Liu, Y.; Chen, X.; Peng, H.; Wang, Z. Multi-focus image fusion with a deep convolutional neural network. *Inf. Fus.* **2017**, *36*, 191–207. [CrossRef]

25. Poddar, S.; Tewary, S.; Sharma, D.; Karar, V.; Ghosh, A.; Pal, S.K. Non-parametric modified histogram equalisation for contrast enhancement. *IET Image Process.* **2013**, *7*, 641–652. [CrossRef]

26. Mathew, J.J.; James, A.P. Spatial stimuli gradient sketch model. *IEEE Signal Process. Lett.* **2015**, *22*, 1336–1339. [CrossRef]

27. Gonzalez, R.C.; Woods, R.E.; Eddins, S. *Digital Image Processing Using MATLAB*; Prentice Hall: New York, NY, USA, 2004.

28. Levin, A.; Lischinski, D.; Weiss, Y. A closed-form solution to natural image matting. *IEEE Trans. Pattern Anal. Mach. Intell.* **2008**, *30*, 228–242. [CrossRef]

29. Liu, Y.; Wang, Z. Multi-focus image fusion based on wavelet transform and adaptive block. *J. Image Graph.* **2013**, *18*, 10.

30. Phamila, Y.A.V.; Amutha, R. Discrete Cosine Transform based fusion of multi-focus images for visual sensor networks. *Signal Process.* **2014**, *95*, 161–170. [CrossRef]

31. Hong, R.; Yang, Y.; Wang, M.; Hua, X. Learning Visual Semantic Relationships for Efficient Visual Retrieval. *IEEE Trans. Big Data* **2015**, *1*, 152–161. [CrossRef]

32. Hossny, M.; Nahavandi, S.; Vreighton, D. Comments on information measure for performance of image fusion. *Electron. Lett.* **2008**, *44*, 1066–1067. [CrossRef]

33. Petrovi, V.S.; Xydeas, C.S. Sensor noise effects on signal-level image fusion performance. *Inf. Fus.* **2003**, *4*, 167–183. [CrossRef]

34. Haghighat, M.B.A.; Aghagolzadeh, A.; Seyedarabi, H. A non-reference image fusion metric based on mutual information of image features. *Comput. Electr. Eng.* **2011**, *37*, 744–756. [CrossRef]

35. Liu, Y.; Liu, S.; Wang, Z. A general framework for image fusion based on multiscale transform and sparse representation. *Inf. Fus.* **2015**, *24*, 147–164. [CrossRef]

36. Han, Y.; Cai, Y.; Cao, Y.; Xu, X. A new image fusion performance metric based on visual information fidelity. *Inf. Fus.* **2013**, *14*, 127–135. [CrossRef]

Article

MID Filter: An Orientation-Based Nonlinear Filter For Reducing Multiplicative Noise

Ibrahim Furkan Ince [1,*], Omer Faruk Ince [2] and Faruk Bulut [3]

[1]   Department of Electronics Engineering, Kyungsung University, Busan 48434, Korea
[2]   Center for Intelligent & Interactive Robotics, Korea Institute of Science and Technology, Seoul 02792, Korea
[3]   Department of Computer Engineering, Istanbul Rumeli University, Istanbul 34570, Turkey
*   Correspondence: furkanince@ks.ac.kr; Tel.: +82-51-663-4114

Received: 13 July 2019; Accepted: 21 August 2019; Published: 26 August 2019

Abstract: In this study, an edge-preserving nonlinear filter is proposed to reduce multiplicative noise by using a filter structure based on mathematical morphology. This method is called the minimum index of dispersion (MID) filter. MID is an improved and extended version of MCV (minimum coefficient of variation) and MLV (mean least variance) filters. Different from these filters, this paper proposes an extra-layer for the value-and-criterion function in which orientation information is employed in addition to the intensity information. Furthermore, the selection function is re-modeled by performing low-pass filtering (mean filtering) to reduce multiplicative noise. MID outputs are benchmarked with the outputs of MCV and MLV filters in terms of structural similarity index (SSIM), peak signal-to-noise ratio (PSNR), mean squared error (MSE), standard deviation, and contrast value metrics. Additionally, *F* Score, which is a hybrid metric that is the combination of all five of those metrics, is presented in order to evaluate all the filters. Experimental results and extensive benchmarking studies show that the proposed method achieves promising results better than conventional MCV and MLV filters in terms of robustness in both edge preservation and noise removal. Noise filter methods normally cannot give better results in noise removal and edge-preserving at the same time. However, this study proves a great contribution that MID filter produces better results in both noise cleaning and edge preservation.

Keywords: non-linear filters; MCV and MLV filters; de-noising; noise removal; edge preserving

1. Introduction

Edge-preserving smoothing is an image processing method that smooths away textures while preserving sharp edges. Most smoothing methods are generally linear low-pass filters that effectively reduce noise at the same time wipe out edges. Since the edges might concern important image information, they have to be protected in smoothing. Non-linear filters are employed for this purpose; however, most of these techniques focus on the problem of reducing additive noise from images, since it is by far the most popular type of corrupting multiplicative noise.

In the literature, there is various research on edge-preserving noise reduction algorithms. Chinrungrueng et al. have presented a study based on edge-preserving noise reduction on ultrasound images. They have introduced a modified 2D weighted Savitzky Golay filter based on the least-squares fitting in a polynomial function to image intensities [1]. Petryniak has described a dynamic image filter using both linear and non-linear image smoothing, based on the Gaussian function. Their filter removes noises in the graphic while preserving information on edges [2]. Yuan and Wang have suggested an edge-preserving and signal-preserving noise removing method based on a Bayesian framework. This filter reduces the number of noises and also adaptively protects edges on signals [3]. Hofheinz et al. have introduced a novel study, which is suitable for bilateral filtering for noise reduction and edge-preserving in the PET image dataset. Bilateral filtering exhibits a successful increase in the smoothing of the PET images while preserving spatial resolution at edges in order to maintain the quantitative accuracy and obtain an acceptable signal-to-noise ratio (SNR) [4]. Pal et al. have presented a survey of benchmark edge-preserving smoothing methods, presented in the literature for computational photography. In their study, they have discussed various effects of the edge-preserving filters also within their optimized modifications and extensions according to their mathematical analysis [5]. Wang et al. have presented a study about a smoothing method with edge preservation for single-image de-hazing (removing haze from image). A novel variational model (VM) that optimizes the transmission in the dark channel has been proposed. This model has an effective linear time complexity in performing transmissions [6]. Storath et al. have introduced a reconstruction framework of edge-preserving and noise reducing for emerging medical imaging, magnetic particle imaging (MPI). Tikhonov regularization, a basic image reconstruction method, is used for MPIs to handle efficiently because of the high temporal resolution of 3D volumes. In their study, they improved an efficient noise removing and edge-preserving reconstruction technique for MPI, giving higher quality in reconstruction for the prototypical medical application of angioplasty [7]. A book chapter for edge-preserving smoothing filters has been written by Burger and Burge. In this detailed and extended study, they have presented noise reduction methods, adaptive smoothing filters for both color and grayscale images. They have especially stressed three conventional types of edge-preserving filters based on different strategies. These are the Kuwahara-type filters, the bilateral filters, and the anisotropic diffusion filters [8]. Additionally, Muhammad et al. have proposed a Bayesian method in which there is a hybrid filtering framework for images having more noises with an unknown variance. The framework, including an automatic parameter selection mechanism, removes noises by enabling an appropriate smoothing and feasible sharpening [9]. In another study, proposed by González-Hidalgo et al., a salt and pepper noise removal system is implemented by a special filter based on a fuzzy mathematical morphology [10]. Luengo et al. have studied noise removal differently by using a supervised learning approach. Specifically, their filter, named CNC-NOS (class noise cleaner with noise scoring), is designed on a noise scoring basis by using ensemble classifiers [11]. A noise-cleaning method for colorful images has been introduced by Pérez-Benito et al. A graph structure is constructed for each of the image pixels in the image by considering some constraints and criterions in order to characterize the pixels as the link cardinality of their connected components [12]. Tang et al. have prepared a detailed study of a smoothing method for edge-aware image manipulations by using a minimization formula of a convex objective function in order to regularize edge and texture pixels in the image [13]. Furthermore, Huang et al. have proposed a technique using an NP-hard method, rank minimization with matrix ranks for regularization in order to remove white Gaussian additive and Gamma multiplicative noises in an image [14].

Apart from those mentioned above, non-linear MCV (minimum coefficient of variation) and MLV (mean least variance) filters are proposed by Schulze et. al. [15] in which multiplicative noise is reduced while preserving the edge contours by employing sliding windows around the central pixel and selecting the pixel that has the minimum amount of coefficient of variation (MCV) and variance (MLV) in terms of intensity within its surrounding window to be low-pass filtered. By this approach, the varying contours, edge lines, and textures are preserved while multiplicative noise is reduced.

In this paper, an extended version of MCV and MLV filters are proposed by modifying its value, criterion, and selection functions to be better than MCV and MLV filters in terms of robustness in noise reduction and edge preservation.

This paper has five sections. There is an introduction with a literature review related to this proposed method in the first section above. Explanations and details of the suggested technique are placed in the second chapter. There are validations of experimental results using statistical metrics and discussion in the third section. Availability of the study is demonstrated in the fourth section. In the last summary section, future work is presented and the contributions are summarized.

2. Methods

In the literature, MCV and MLV filters are well-known filters, which eliminate noises in an image while preserving edges. All these methods similarly use a certain kernel (mask) size, which represents the size and shape of the neighborhood to be sampled while computing the corresponding value. The kernel is a $q \times q$ square matrix where q is a small odd number, generally 3, 5, or 7.

2.1. MCV and MLV Filters

The MCV and MLV filters are edge-preserving noise removing filters based on the concepts of mathematical morphology [16]. They are value-and-criterion filters that are aimed to filter an image only over regions that are generally homogeneous, have low contrast and contain less amount of edges or textures [16]. The difference between MCV and MLV filters is that MCV filter uses the coefficient of variation as the criterion function whereas MLV employs the variance to perform better on multiplicative noise [16].

The idea is basically sliding windows around each central pixel and finding the sub-window, which has the minimum criterion function output, and apply the value function (mean value) of the window belonging to the central pixel [17]. The coefficient of variation over a sliding kernel is calculated by the ratio of the standard deviation to the mean over the sliding kernel. If the image is uniform within the kernel, the variation coefficient becomes very low [17].

On the other hand, if the image has high amount of edge and texture within the kernel, both the coefficient of variation and the standard deviation will return high values [18]. The selection function of MCV and MLV filters is designed as the minimum so that the filtering operation can completely be done over the kernel, which has the smallest output of the criterion function [18]. In other words, the noise smoothing function only acts over these kernels with the smallest coefficient of variation for MCV filter or the variance for the MLV filter [18].

As the value function, mean value is employed over the regions that have the minimum amount of criterion function: The coefficient of variation for the MCV and variance for the MLV filters [18]. The filter structure and detailed explanation of filter design are explained in Section 2.2 within the proposed method.

2.2. Proposed Method: Minimum Index of Dispersion (MID) Filter

The MID filter uses the same morphological structure with MCV and MLV filters to direct low-pass filtering operation to only execute over regions decided to be most nearly constant by calculating the index of dispersion as the criterion function. As previously explained in Section 2.1, MCV and MLV filters employ the coefficient of variation and variance as the criterion function, MID filter employs the index of dispersion as the criterion function. The index of dispersion is the ratio of the variance of a random process to its mean and defined as in Equation (1).

$$D = \begin{cases} \frac{\sigma^2}{\mu} \ if \ \mu \neq 0 \\ 0 \ otherwise \end{cases} \tag{1}$$

where σ is the standard deviation and μ is the mean value of given elements of a set. For an image that is corrupted only by stationary multiplicative noise, the index of dispersion in terms of intensity and orientation in theory is constant at every point. Estimates of the index of dispersion show whether a region is nearly constant under the multiplicative noise or it includes important features. Regions that contain edges or other features generate higher estimates of the index of dispersion in terms of intensity and orientation than areas that are approximately constant. Value, criterion, and selection functions are defined as follows:

$$\omega(x) = \varphi\{f(x); N\} \tag{2}$$

$$\gamma(x) = \delta\{f(x); N\} \tag{3}$$

$$\varphi(x) = \omega(\{x' : x' \in N'_x; \, \gamma(x') = \beta\{\gamma(x); N'\}\}) \tag{4}$$

where ω is the value function, γ is the criterion function, and φ is the selection function. Additionally, $f(x)$ denotes the intensity function that gets the intensity values of each pixel in each window that has N number of elements (pixels) within the window. β is another value function that gets the value of intensity with the minimum index of dispersion. x' denotes the pixels within each sub-window around the central pixel of x and N' is the number of pixels in each sub-window. This filter structure can be described as having a set of sub-windows within an overall filter window. The value-and-criterion filter operation at a point is equivalent to examining each sub-window within the overall window centered at that point and finding which sub-window has the output of optimal criterion function as described by the selection function. Then, the value function output over this sub-window becomes the final filter output for the current point. Value functions are interpreted in Equations (5) and (6) as follows:

$$\omega(x) = \frac{1}{|N|} \sum_{y \in N_x} f(y) \tag{5}$$

$$\theta(x) = \frac{1}{|N|} \sum_{y \in N_x} g(y) \tag{6}$$

where ω represents the mean value of intensity and θ denotes the mean value of orientation. As the second metric, in addition to intensity value, normalized gradient orientation values are employed within their magnitudes defined as in Equation (7).

$$g(x) = \arctan\left(\frac{G_y}{G_x}\right) \times \sqrt{G_y{}^2 + G_x{}^2} \tag{7}$$

where G_y is the vertical gradient vector's normalized magnitude and G_x is the horizontal one. A 3×3 Sobel operator is employed to find the gradient vectors for each pixel of the input image. The regions that have a chaotic distribution of orientations are also defined as noise since regular patterns of orientation distributions land on the regions without noise. Therefore, orientation dispersion is also employed as the second metric in the proposed filter. Less dispersion in the orientation will have more impact on the intensity distribution. Thus, the criterion function is re-modeled as follows:

$$\gamma(x) = \frac{\frac{1}{|N|} \sum_{y \in N_x} [f(y) - \omega(x)]^2}{\omega(x)} \times \left(1 - \frac{\frac{1}{|N|} \sum_{y \in N_x} [g(y) - \theta(x)]^2}{\theta(x)}\right) \tag{8}$$

Elimination of multiplicative noise from images is commonly more difficult than additive noise since the noise intensity varies with the signal intensity. In order to avoid this, selection function is re-modeled by adding an alpha parameter, which adds a low-pass value to the output of selection function. Alpha is a normalized parameter and it transforms the filter into a fully mean filter when it is set to 1. The proposed selection function is defined as follows:

$$\vee(x) = \omega(\{x' : x' \in N'_x; \, \gamma(x') = \min[\gamma(y); \, y \in N'_x]\}) \tag{9}$$

$$MID\{f(x); N\} = \vee(x) \times (1 - \alpha) + \omega(x) \times \alpha \tag{10}$$

The sample mean is accepted as the value function for the MID filter. Thus, the MID filter uses the sample mean for a value function, the index of dispersion as a criterion function, and the minimum as a selection function. This value-and-criterion filter is particularly designed to remove multiplicative noise. Theoretically, index of dispersion in the images corrupted by noises is minimum in structuring elements where there is the constant signal. The MID filter in images both preserves sharp edges between flat areas and enhances the edges, which are not perfect step edges.

This study especially indicates a structural design of a special filter, which is the extended version of MCV, and MLV architectures. This filtering method is an edge-preserving noise reduction technique designed for reducing multiplicative noise by using the structure of the value-and-criterion filter. The theoretical mechanism of this method stands on the well-known fundamentals of geometrical properties. The advantage of this method is that it successfully preserves edges in the regions corrupted by the multiplicative noise and enhances them while preserving the morphological structures of the image.

3. Experimental Results and Analysis

Mainly, all of the experiments have been performed on the Computational and Subjective Image Quality (CSIQ) benchmark dataset [19] by using a normal computer. MID filtering is implemented using Processing in Java and testing is performed in the MATLAB environment.

3.1. CSIQ Image Quality Database Specifications

The CSIQ database is a popular image quality-benchmark test set in order to evaluate algorithms. The database includes 30 original images at the resolution of 512×512 pixels. The set is distorted using one of six distortions with four to five different distortion levels. CSIQ images have been tested based on linear image displacements on four calibrated LCD screens placed side by side with equal viewing distance. This database contains 5000 subjective evaluations from 35 different observers and the assessment are presented in the form of difference mean opinion scores (DMOS) in which a larger one indicates greater visual impairments compared to the corresponding reference image.

3.2. Performance Measurement Criterions

De-noising a picture requires a successful method providing that edges are to be preserved. In order to evaluate the performances of the methods, some quality metrics are preferred. Evaluation of de-noising quality is performed five fundamental metrics. These are mean squared error (MSE), peak signal-to-noise ratio (PSNR), the structural similarity index (SSIM), contrast, and standard deviation [20]. Additionally, Γ Score, which is an original hybrid metric for comparison, is proposed in this study by using the combinations of the basic measurements.

3.2.1. Mean Squared Error (MSE) and Peak Signal-to-Noise Ratio (PSNR)

Both MSE and PSNR are used to evaluate the performance for image manipulation algorithms. They are similar to each other and derived from signal processing. Implementation and calculation are straightforward, but the results are not always considered reliable as they show aspects in various situations. Nevertheless, they have a great role in the performance evaluation domain.

The MSE between the two signals is described as seen in Equation (11).

$$MSE = \frac{1}{N \times M} \sum_{i=0}^{N-1} \sum_{j=0}^{M-1} [X(i, j) - Y(i, j)]^2 \tag{11}$$

where X and Y are two arrays of size $N \times M$. The closer Y is to X, the smaller MSE will be. When the MSE is equal to zero, apparently, the maximum similarity is achieved.

The PSNR (in dB) accordingly is defined as follows:

$$PSNR = 10 \log_{10} \frac{L^2}{MSE} \tag{12}$$

In Equation (12) above, L is the maximum fluctuation in the data type of the input image. For instance, if the input image has a double-precision floating-point data type, then L is defined as 1. Similarly, if the input image has an 8-bit unsigned integer data type, L is defined as 255. Logarithm transforms the ratio into a decibel (dB) scale, which is a common scale operation in signal processing. PSNR in decibels units calculates the PSNR between original and filtered images. The lower the PSNR value, the worse the quality of de-noised image. MSE and PSNR are the two-error measurement metrics used to compare the image de-noising quality.

MSE shows the cumulative squared error between filtered and original images. PSNR displays the measure of the peak error. In a little while, the higher the MSE value, the higher the error. If there are two identical images (in the absence of artificial noise), the MSE value becomes 0 and the PSNR value becomes infinite [21].

3.2.2. The Structural Similarity Index Measurement (SSIM)

The SSIM measurement is a common and well-known quality criterion to determine the similarity between two images. The SSIM index gives a similarity percentage in the interval of [0, 1].

This measurement style compares two images in the same size, the de-noised picture, and the original picture. The original picture is assumed as it has perfect quality. The de-noised one is for test and the original is for verification. SSIM index is defined as follows:

$$SSIM(x, y) = \frac{\left(2\mu_x\mu_y + C_1\right)\left(2\sigma_{xy} + C_2\right)}{\left(\mu_x + \mu_y + C_1\right)\left(\sigma_x + \sigma_y + C_2\right)} \tag{13}$$

where x and y are the two different images with μ_x and μ_y mean values of intensity and standard deviations of σ_x and σ_y with contrast values C_1 and C_2 for the two images separately. When comparing two images, MSE does not indicate highly perceived similarity while implementation is simple. Structural similarity is aimed at addressing this hardship.

3.2.3. Contrast

Contrast of an image might be simply explained as the difference between the minimum and maximum pixel intensity. Shortly, it is the difference in color or luminance for a group of objects. In this project, edge-based contrast measure (EBCM) for image quality evaluation is selected as a performance metric [22]. This metric is based on the fact that an enhanced image normally has more edge pixels than the original image. The EBCM metric calculates the intensity of edge pixels in small windows of the image.

3.2.4. Standard Deviation

The standard deviation of the pixel intensity values is used to quantify the amount of variation or dispersion of a grayscale image. It is calculated by Equation (14).

$$\sigma = \sqrt{\frac{1}{N \times M} \sum_{i=0}^{N-1} \sum_{j=0}^{M-1} \left(x_{ij} - \mu\right)^2} \tag{14}$$

where σ is the standard deviation of matrix elements. N and M are the vertical and horizontal sizes of the image. x_{ij} is the pixel of the ith line and jth column. μ is the arithmetic mean. A low standard deviation value displays that the pixels tend to be close to the mean of the image, while a high value shows that the pixels are spread out over a wider range of values.

3.2.5. A Hybrid Assessment Metric: F Score

Handling each of the metrics separately in image quality assessment might be difficult. A hybrid approach is proposed to evaluate each of the filter methods as shown in Equation (15).

$$F\,Score = 100 \times \frac{PSNR \times SSIM \times Contrast}{Std.Dev \times MSE} \tag{15}$$

An optimal edge-preserving and noise-reducing filter should increase the PSNR, SSIM, and contrast values while reducing the standard deviation and MSE values. Therefore, a compact formula of F is generated in order to benchmark the filters. Higher values of PSNR, SSIM, and Contrast values indicate that there is a successful smoothing operation.

In contrast, higher values of standard deviation and MSE shows poor smoothing results. In other words, PSNR, SSIM, and contrast have a positive effect on image quality, whereas the others have a negative impact.

In the experiments, F scores result in very tiny values, even very close to zero. Hence, the F score results are multiplied by a constant value of 100 so as to optimize the outputs. A regularly higher F score rate indicates the successful filtering performance.

3.3. Comparison Steps of Experimental Outputs

In the experimental, the overall procedural steps are illustrated in Figure 1.

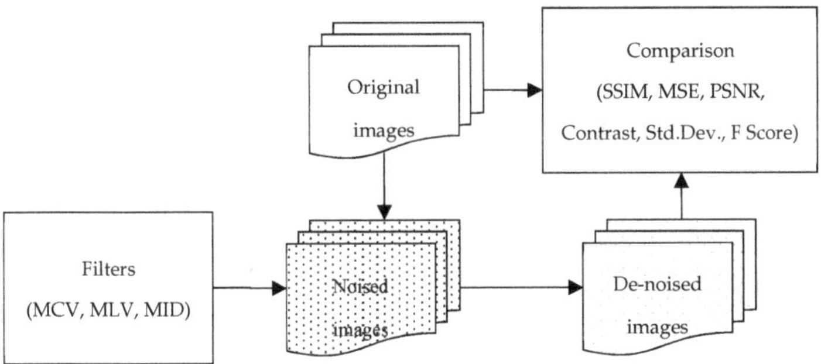

Figure 1. Procedural steps for overall comparisons.

Firstly, the original images are synthetically noised with the multiplicative noise option, using Equation (16) for an image I.

$$J = I + n \times I \tag{16}$$

where n is evenly distributed random noise with mean 0 and variance v. The default value for v is set to 0.04. Then, each of the filtering methods (MCV, MLV, and MID) is employed to de-noise the noised images. In other words, the noised images are filtered by the 3 filtering methods in order to clean the noises. Each method produces individual outputs. Lastly, the outputs are compared with the original images using the metrics of PSNR, MSE, SSIM, contrast, standard deviation, and F score.

As the experimental setup, artificial multiplicative noise is added to the 30 CSIQ images in order to quantify the performance of filters in terms of robustness to noise and edge preservation. Filtered images are compared with the original images with respect to five main metrics: PSNR, SSIM, MSE, standard deviation, and contrast. Table 1 illustrates an original image and a multiplicative noise added image.

In Table 1, the first image is the original gray-scale form of the CSIQ "1600.png" image, which has a contrast value of 74.19 and standard deviation of 66.21. The second one is the multiplicative noise-added gray-scale form of the CSIQ "1600.png" image, which has a contrast value of 67.25 and standard deviation of 75.83. After adding noise, standard deviation is increased and contrast is decreased since the noise factor increases the deviation from the mean while it wipes out the edges, which lowers the contrast accordingly.

Table 1. Original and artificially noised "1600.png" image (the v variance parameter for multiplicative noise is set to 0.04).

Original Picture	Noisy Picture
Std.Dev.: 66.21 Contrast: 74.19	Std.Dev.: 75.83 Contrast: 67.25

3.4. Numerical Outputs and Discussion

As the experimental setup, pictures are compared with the original gray-scale pictures so that selected metrics can be examined. For this purpose, pictures filtered out by MCV, MLV, and MID filters are compared with respect to selected six metrics: PSNR, MSE, SSIM, standard deviation, contrast, and *F* Score. In the performance assessments, it is observed that PSNR and SSIM values increase while MSE decreases. Standard deviation decreases if the amount of noise is decreased. Also, the contrast is increased if the edge contours are enhanced. *F* Score demonstrates the overall success rate. Table 2 demonstrates a set of sample experimental results for the selected gray-scale form of the "1600.png" image as follows.

Table 2. Sample experimental results for a gray-scale picture taken from filters with 5×5 kernel size.

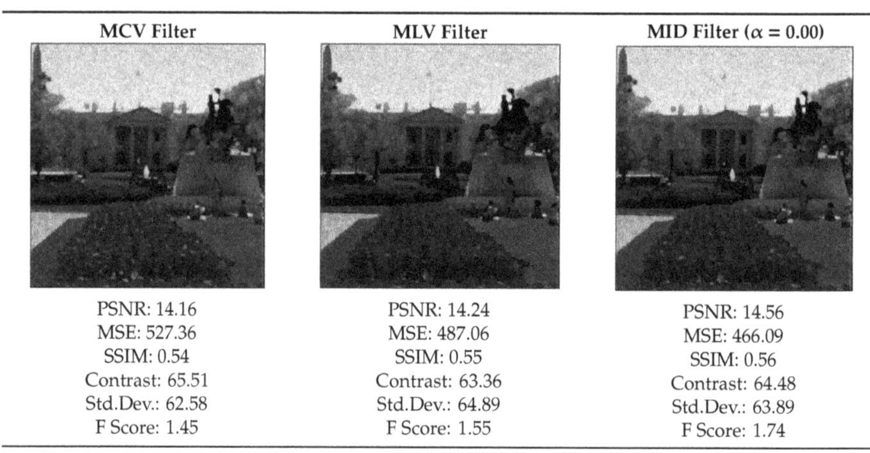

MCV Filter	MLV Filter	MID Filter ($\alpha = 0.00$)
PSNR: 14.16 MSE: 527.36 SSIM: 0.54 Contrast: 65.51 Std.Dev.: 62.58 F Score: 1.45	PSNR: 14.24 MSE: 487.06 SSIM: 0.55 Contrast: 63.36 Std.Dev.: 64.89 F Score: 1.55	PSNR: 14.56 MSE: 466.09 SSIM: 0.56 Contrast: 64.48 Std.Dev.: 63.89 F Score: 1.74

According to the experimental outputs shown in Table 2, the highest *F* Score is obtained in the MID Filter with the parameter (alpha = 0.0) when the comparison is performed among all filters. It indicates that the highest amount of noise reduction occurs with the MID Filter.

Figure 2 demonstrates the MSE, PSNR, SSIM, contrast, standard deviation, and *F* score bar charts of each filter as follows:

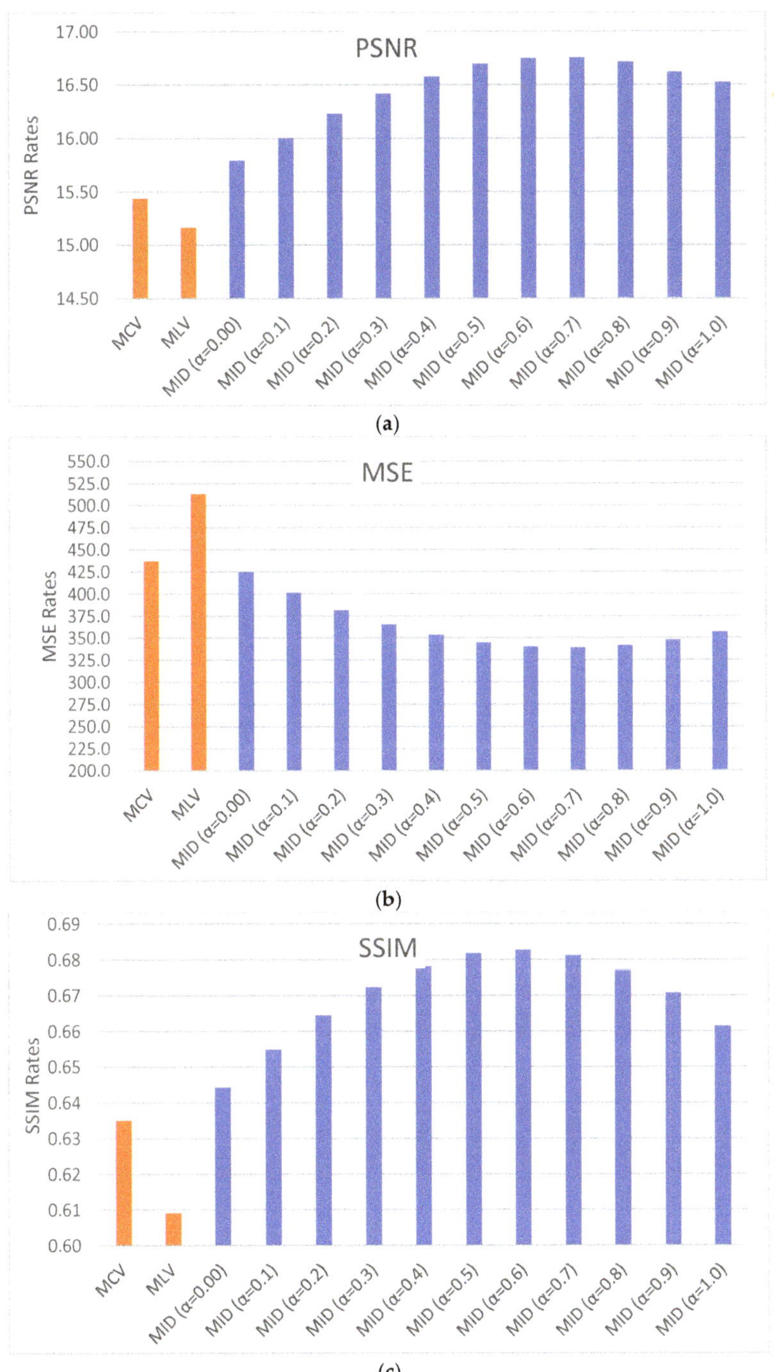

(a)

(b)

(c)

Figure 2. *Cont.*

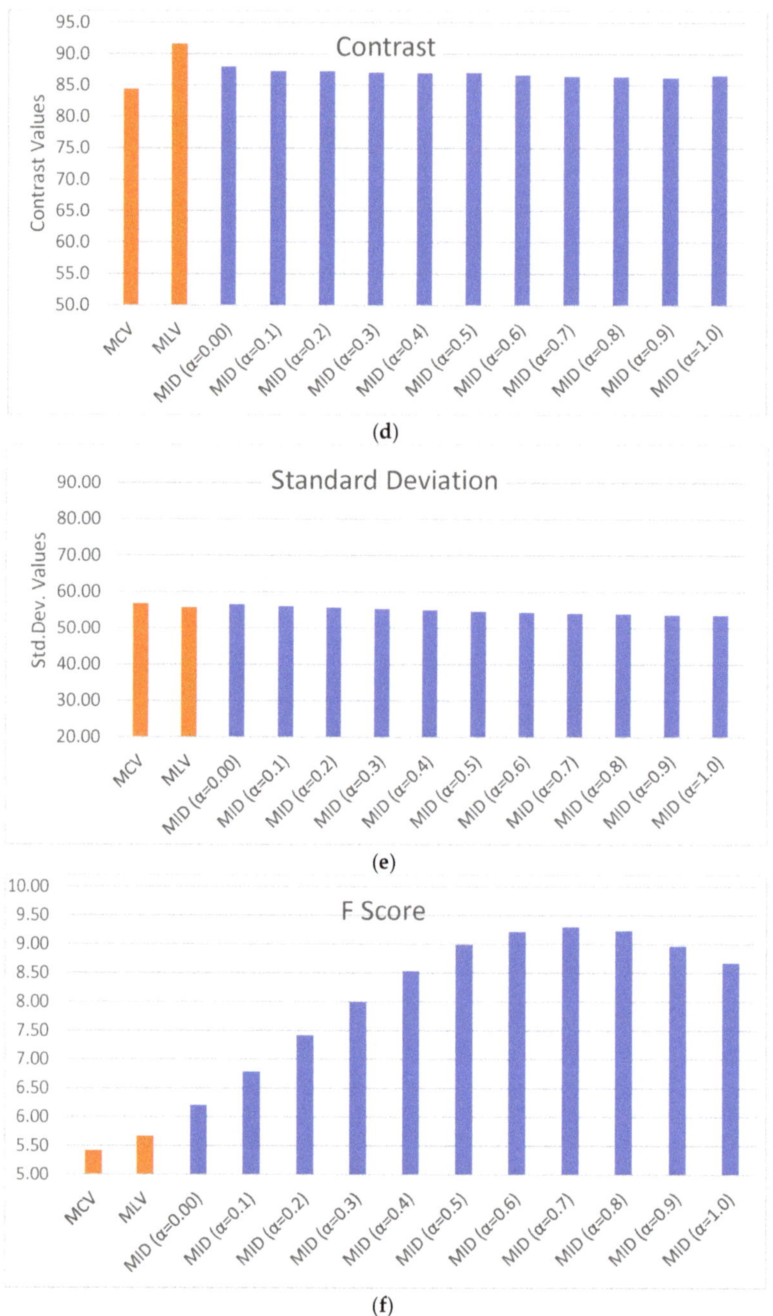

Figure 2. Average peak signal-to-noise ratio (PSNR) rates in decibel (dB) (**a**), average mean squared error (MSE) values (**b**), average structural similarity index (SSIM) rates (**c**), average contrast values (**d**), average standard deviations (**e**), and average F scores (**f**) for the gray-scale computational and subjective image quality (CSIQ) dataset.

According to Figure 2, when MCV, MLV, and MID filters are compared in terms of PSNR, MSE, and SSIM, MID filter leads in the group. It has the highest rate of SSIM, and the lowest amount of MSE and the highest rate of PSNR.

Furthermore, the success rate of MID filter will increase when the alpha parameter is increased. However, too much increment in alpha will ruin the structural similarity and reduce the contrast; therefore, the optimal value of the alpha should be determined which will balance the ratio between SSIM and contrast. According to experimental tests, optimal alpha value, which yields the best results, is discovered as 0.30.

Observations through experiments are performed with respect to six metrics: PSNR, MSE, SSIM, Standard Deviation, Contrast, and *F* Score. The most determinant metrics appear as PSNR and SSIM, which indicates the percentage of noise reduction and similarity with the original pictures. Table 3 indicates the overall results with alternative values of alpha.

Table 3. Total average results of filtering experiments when the kernel size is set to 5.

	PSNR	**MSE**	**SSIM**	**Std. Dev.**	**Contrast**	**F Score**
MCV	15.44	437.0	0.635	56.81	84.4	5.42
MLV	15.16	512.6	0.609	55.71	91.6	5.67
MID ($\alpha = 0.0$)	15.79	424.6	0.644	56.60	88.0	6.21
MID ($\alpha = 0.1$)	16.00	401.5	0.655	56.12	87.3	6.78
MID ($\alpha = 0.2$)	16.23	381.7	0.664	55.72	87.2	7.42
MID ($\alpha = 0.3$)	16.42	365.7	0.672	55.34	87.0	8.00
MID ($\alpha = 0.4$)	16.57	353.4	0.678	55.00	86.9	8.53
MID ($\alpha = 0.5$)	16.70	344.5	0.682	54.70	86.9	8.99
MID ($\alpha = 0.6$)	16.75	339.9	0.683	54.40	86.6	9.22
MID ($\alpha = 0.7$)	16.75	338.9	0.681	54.14	86.4	9.30
MID ($\alpha = 0.8$)	16.72	341.4	0.677	53.94	86.3	9.23
MID ($\alpha = 0.9$)	16.62	347.8	0.671	53.75	86.2	8.97
MID ($\alpha = 1.0$)	16.52	356.6	0.661	53.63	86.5	8.68

In Table 3, the average PSNR value with MID filter is obtained as 16.42 while MCV and MLV filters attain 15.16 and 15.44, respectively. This proves that the MID filter is superior to the MCV and MLV filters in terms of robustness to noise and the SSIM value is calculated as 0.672 while the MCV and MLV filters reach 0.609 and 0.635, respectively. This also proves that the MID filter is better than the MCV and MLV filters in terms of similarity with the original pictures, which means MID filter cleans the noise while preserving the structural similarity with the original pictures. Mean squared error (MSE) is also observed as lower than the MCV and MLV filters where the MSE of MCV is observed as 512.6 and MSE of MLV is observed as 437.0, which is much higher than the MSE of the MID filter obtained as 365.7 through the experiments. This also proves that the MID filter's outputs are much more similar to the original pictures and cleans the noise better than the MCV and MLV filters. As an overall evaluation, the MID filter is better than the MCV and MLV filters in terms of robustness to noise while preserving the edges.

Additionally, the rows of Table 4 contain some cropped sections from original and filtered images of CSIQ "1600.png", "family.png", "turtle.png", and "trolley.png", respectively. These gray-scale pictures are from the original picture, MCV, MLV, and MID filters using a kernel size of 5×5 matrix. Table 4 demonstrates the visual performances of filters on edge preservation.

According to Table 4, in the first image in the first row, there are iron fences behind the people. The fence is a good representation (sample) of edges. As it is seen, the fence in MCV is thinner than in MLV. Edge contours of the fence are not well preserved by both MCV and MLV filters, whereby MCV makes edges thinner and MLV makes thicker than normal. Furthermore, in the first and second images in the first and second rows, the heads of people in MCV almost disappear since the mean method shrinks the edges. On the other hand, the heads of people in MLV are oversized since the method expands the edges [23]. However, the heads of people in the MID filtered image looks neither oversized nor shrunken since the proposed method employs orientation information that optimizes the size of contours. In the third row, the head of the turtle loses its texture when MCV is applied, and contours become thicker when MLV is applied. However, both edge contours and textures become normal when the MID filter is applied. Additionally, in the fourth row, humans on the trolley almost disappear when MCV is applied and contours become extremely thicker when MLV is applied. On the other hand, both contours and texture look normal when the MID filter is executed.

Table 4. Some small cropped image sections from the outputs of filters.

Original Section	MCV	MLV	MID

1600.png	PSNR: 10.60 SSIM: 0.570	PSNR: 10.65 SSIM: 0.638	PSNR: 11.49 SSIM: 0.630

family.png	PSNR: 9.44 SSIM: 0.430	PSNR: 11.29 SSIM: 0.600	PSNR: 11.82 SSIM: 0.617
	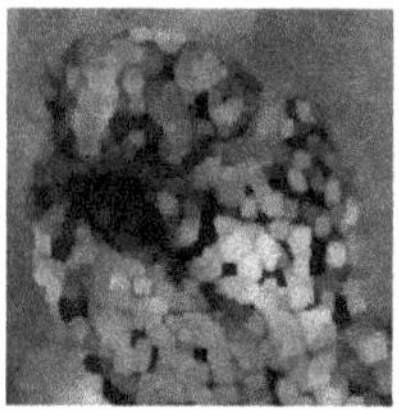	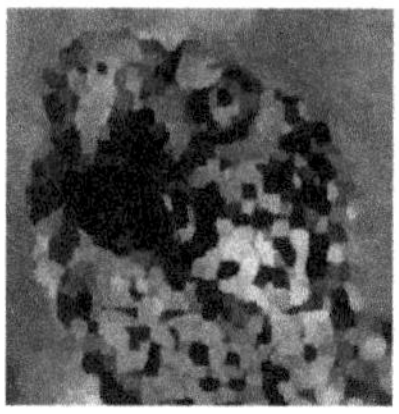	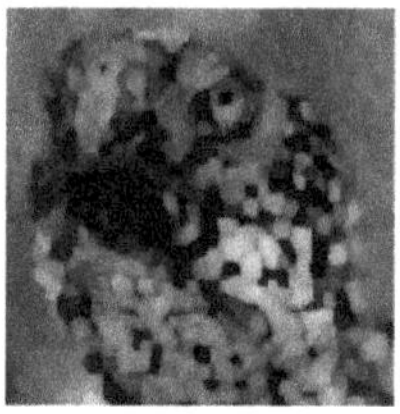
turtle.png	PSNR: 11.37 SSIM: 0.531	PSNR: 13.03 SSIM: 0.640	PSNR: 13.82 SSIM: 0.670
	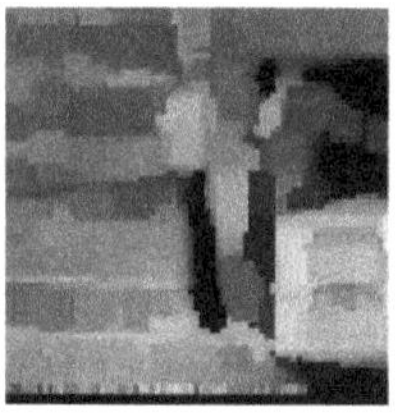	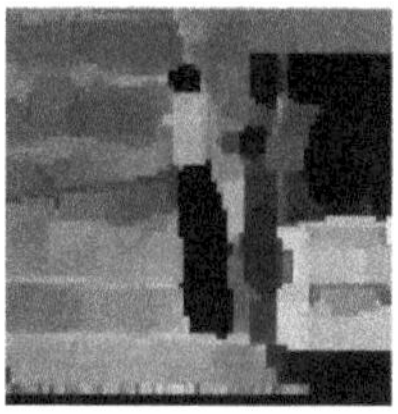	
trolley.png	PSNR: 10.78 SSIM: 0.507	PSNR: 11.37 SSIM: 0.582	PSNR: 12.24 SSIM: 0.605

As it is widely accepted, preserving edges is a great issue in noise reduction operations. The primary orientation of this study stands on two main principles, edge preservation and noise reduction. Measuring the quality of edge preservation might be performed by the SSIM index. The performance of noise-cleaning might be also assessed by the F score.

As it is seen in the experimental results above, the MID filter gives better results than the MCV and MLV filters starting from when the alpha is set to 0.30 according to SSIM index. Since the SSIM index indicates structural similarity of objects in the pictures, it also gives a sign about the rate of edge preservation. The more alpha is increased, the more the filter behaves like a mean filter, which ruins the edge preservation. Therefore, a minimum optimal value of alpha is necessary to get better results in terms of both edge preservation and noise reduction. For this reason, 0.30 might be determined as the optimum value of alpha. Even though the highest SSIM is gained when the alpha is set to 0.70, the edges partly disappear since MID behaves like a mean filter. As the main purpose of the study is to protect edges from deformations, the alpha parameter should be lessened as much as possible.

As a result, the shape of the objects changes with respect to type of filters. While MCV filters ruin the object boundaries, the MLV filter over-blurs the edge contours, which results in thick borderlines of the objects. However, the MID filter preserves the original contours of objects since the MID filter employs orientation information as the criterion function. This is the most prominent contribution of this study. This improvement can be recognized with the SSIM metric, which indicates the structural similarity of objects within the image pairs. Additionally, F score is presented as a novel comparison metric, which separates the filters in terms of edge preservation and robustness to noise.

4. Availability

This presented MID filtering model has been implemented in the Java Processing and tested in MATLAB platforms. For examinations, further studies, and citations, all of the written original codes, benchmark datasets, test images, outputs, and total experimental results including SSIM, MSE, PSNR, contrast, standard deviations, and F scores for all cases can be publicly reachable at the website: https://sites.google.com/site/bulutfaruk/study-of-mid-filtering.

5. Conclusions

In this paper, an extended version of MCV (minimum coefficient of variation) and MLV (mean least variance) filters are proposed. The proposed approach is the MID (minimum index of dispersion) filter, which employs orientation information of pixels in order to support value-criterion structure of the MCV and MLV filters. The dispersion of orientations is employed as the criterion function, which yields better results against multiplicative noise. Moreover, the value function is modified by adding an alpha parameter, which acts as low-pass filtering by the amount of alpha. Experimental results show that the proposed approach produces better results than MCV and MLV filters against multiplicative noise and eliminates the weaknesses of MCV and MLV filters. As the metric for measuring the robustness to noise, SSIM (structural similarity index), MSE (mean squared error), PSNR (peak signal-to-noise ratio), standard deviation, and contrast values are employed. Additionally, F Score, a hybrid metric that is the combination of five metrics is introduced in order to compare the filters. Benchmarking study indicates the MID filter is superior to the MCV and MLV filters. By the increment of the alpha parameter, the noise is blurred but the contrast is decreased, which acts by blurring the edges as well. Therefore, a balanced alpha parameter value is necessary, which will enhance the edges and at the same time blur the multiplicative noise. As the optimal value of the alpha parameter, 0.30 is determined according to experimental tests. This study might be an innovative guide for those who are interested in MCV and MLV filters and able to output different studies on the topic in the future.

Author Contributions: Conceptualization, I.F.I. and F.B.; methodology, I.F.I.; software, I.F.I.; validation, F.B.; formal analysis, F.B.; investigation, O.F.I.; resources, O.F.I.; data curation, F.B.; writing—original draft preparation, O.F.I.; writing—review and editing, I.F.I.; visualization, F.B.; supervision, I.F.I.; project administration, I.F.I.

Funding: This research received no external funding.

Conflicts of Interest: The authors declare no conflict of interest.

References

1. Chinrungrueng, C.; Suvichakorn, A. Fast edge-preserving noise reduction for ultrasound images. *IEEE Trans. Nucl. Sci.* **2001**, *48*, 849–854. [CrossRef]
2. Rafał, P. Edge preserving techniques in image noise removal process. *Czas. Tech.* **2014**, *5*, 301–307.
3. Yuan, S.; Wang, S. Edge-preserving noise reduction based on Bayesian inversion with directional difference constraints. *J. Geophys. Eng.* **2013**, *10*, 025001. [CrossRef]
4. Hofheinz, F.; Langner, J.; Beuthien-Baumann, B.; Oehme, L.; Steinbach, J.; Kotzerke, J.; van den Hoff, J. Suitability of bilateral filtering for edge-preserving noise reduction in PET. *EJNMMI Res.* **2011**, *1*, 23. [CrossRef] [PubMed]
5. Pal, C.; Chakrabarti, A.; Ghosh, R. A brief survey of recent edge-preserving smoothing algorithms on digital images. *arXiv* **2015**, arXiv:1503.07297.
6. Wang, D.; Zhu, J. Fast smoothing technique with edge preservation for single image dehazing. *IET Comput. Vis.* **2015**, *9*, 950–959. [CrossRef]
7. Storath, M.; Brandt, C.; Hofmann, M.; Knopp, T.; Salamon, J.; Weber, A.; Weinmann, A. Edge Preserving and Noise Reducing Reconstruction for Magnetic Particle Imaging. *IEEE Trans. Med Imaging* **2017**, *36*, 74–85. [CrossRef]
8. Burger, W.; Burge, M.J. Edge-Preserving Smoothing Filters. In *Digital Image Processing: An Algorithmic Introduction Using Java*; Springer: London, UK, 2016; Chapter 17.
9. Muhammad, N.; Bibi, N.; Wahab, A.; Mahmood, Z.; Akram, T.; Naqvi, S.R.; Oh, H.S.; Kim, D.G. Image de-noising with subband replacement and fusion process using bayes estimators. *Comput. Electr. Eng.* **2017**, *70*, 413–427. [CrossRef]
10. Gonzalez-Hidalgo, M.; Massanet, S.; Mir, A.; Ruiz-Aguilera, D. Improving salt and pepper noise removal using a fuzzy mathematical morphology-based filter. *Appl. Soft Comput.* **2018**, *63*, 167–180. [CrossRef]
11. Luengo, J.; Shim, S.O.; Alshomrani, S.; Altalhi, A.; Herrera, F. CNC-NOS: Class noise cleaning by ensemble filtering and noise scoring. *Knowl. Based Syst.* **2018**, *140*, 27–49. [CrossRef]
12. Pérez-Benito, C.; Morillas, S.; Jordán, C.; Conejero, J.A. A model based on local graphs for colour images and its application for Gaussian noise smoothing. *J. Comput. Appl. Math.* **2017**, *330*, 955–964. [CrossRef]
13. Tang, C.; Hou, C.; Hou, Y.; Wang, P.; Li, W. An effective edge-preserving smoothing method for image manipulation. *Digit. Signal Process.* **2017**, *63*, 10–24. [CrossRef]
14. Huang, Y.M.; Yan, H.Y.; Wen, Y.W.; Yang, X. Rank minimization with applications to image noise removal. *Inf. Sci.* **2018**, *429*, 147–163. [CrossRef]
15. Schulze, M.A.; Pearce, J.A. Value-and-criterion filters: A new filter structure based upon morphological opening and closing. In *Nonlinear Image Processing IV*; Dougherty, E.R., Astola, J., Longbotham, H., Eds.; SPIE: San Jose, CA, USA, 1993; pp. 106–115.
16. Schulze, M.A.; Pearce, J.A. A morphology-based filter structure for edge-enhancing smoothing. In Proceedings of the 1st International Conference on Image Processing, Austin, TX, USA, 13–16 November 1994; pp. 530–534.
17. Schulze, M.A. Biomedical Image Processing with Morphology-Based Nonlinear Filters. Ph.D. Thesis, The University of Texas at Austin, Austin, TX, USA, 1994.
18. Schulze, M.A.; Wu, Q.X. Nonlinear filtering for edge-preserving smoothing of synthetic aperture radar imagery. In Proceedings of the New Zealand Image and Vision Computing '95 Workshop, Christchurch, New Zealand, 28–29 August 1995; pp. 65–70.
19. Larson, E.C.; Chandler, D.M. Most apparent distortion: full-reference image quality assessment and the role of strategy. *J. Electron. Imaging* **2010**, *19*, 011006.
20. Kipli, K.; Krishnan, S.; Zamhari, N.; Muhammad, M.S.; Masra, S.M.; Chin, K.L.; Lias, K. Full reference image quality metrics and their performance. In Proceedings of the 2011 IEEE 7th International Colloquium on Signal Processing and its Applications (CSPA), Penang, Malaysia, 4–6 March 2011.
21. Salomon, D. *Data Compression: The Complete Reference*, 4th ed.; Springer: New York, NY, USA, 2007; p. 281.
22. Beghdadi, A.; Negrate, A.L. Contrast enhancement technique based on local detection of edges. *Comput. Vis. Graph. Image Process.* **1989**, *46*, 162–174. [CrossRef]

23. Moulick, H.N.; Ghosh, M. Biomedical image processing with nonlinear filters. *Int. J. Comput. Eng. Res.* **2013**, *3*, 7–15.

 *electronics*

Article

Adaptive Algorithm on Block-Compressive Sensing and Noisy Data Estimation

Yongjun Zhu [1,2,*]**, Wenbo Liu** [1] **and Qian Shen** [1]

[1] College of Automation Engineering, Nanjing University of Aeronautics and Astronautics,
     Nanjing 211106, China
[2] School of Electronic and Information, Suzhou University of Science and Technology, Suzhou 215009, China
* Correspondence: zyj@mail.usts.edu.cn; Tel.: +86-025-848-92-766

Received: 14 May 2019; Accepted: 29 June 2019; Published: 3 July 2019

Abstract: In this paper, an altered adaptive algorithm on block-compressive sensing (BCS) is developed by using saliency and error analysis. A phenomenon has been observed that the performance of BCS can be improved by means of rational block and uneven sampling ratio as well as adopting error analysis in the process of reconstruction. The weighted mean information entropy is adopted as the basis for partitioning of BCS which results in a flexible block group. Furthermore, the synthetic feature (SF) based on local saliency and variance is introduced to step-less adaptive sampling that works well in distinguishing and sampling between smooth blocks and detail blocks. The error analysis method is used to estimate the optimal number of iterations in sparse reconstruction. Based on the above points, an altered adaptive block-compressive sensing algorithm with flexible partitioning and error analysis is proposed in the article. On the one hand, it provides a feasible solution for the partitioning and sampling of an image, on the other hand, it also changes the iteration stop condition of reconstruction, and then improves the quality of the reconstructed image. The experimental results verify the effectiveness of the proposed algorithm and illustrate a good improvement in the indexes of the Peak Signal to Noise Ratio (PSNR), Structural Similarity (SSIM), Gradient Magnitude Similarity Deviation (GMSD), and Block Effect Index (BEI).

Keywords: block-compressive sensing (BCS); saliency; error analysis; flexible partitioning; step-less adaptive sampling

---

1. Introduction

The traditional Nyquist sampling theorem states that the sampling frequency of a signal must be more than twice its highest frequency to ensure that the original signal is completely reconstructed from the sampled value, while the compressive sensing (CS) theory breaks through the traditional limitation of the Nyquist sampling theorem in signal acquisition and can achieve reconstructing a high-dimensional sparse signal or compressible signal from the lower-dimensional measurement [1]. As an alternative to the Nyquist sampling theorem, CS theory is being widely studied, especially in the current image processing. The research of CS theory mainly focuses on several important aspects such as sparse representation, measurement matrix construction, and the reconstruction algorithm [2,3]. The main research hotspot of sparse representation is how to construct a sparse dictionary of the orthogonal system and an over-complete dictionary for suboptimal approximation [4,5]. The construction of the measurement matrix mainly includes the universal random measurement matrix and the improved deterministic measurement matrix [6]. The research of the reconstruction algorithm mainly focuses on the suboptimal solution problem and a training algorithm based on self-learning [7,8]. With the advancement of research and application about CS theory, especially in 2D or 3D image processing, the CS technology faces several challenges, including computational dimensional disaster and the spatial storage problem with the increase of the images geometric scale. To solve these challenges, the researchers proposed many fast-compressive sensing algorithms to solve the computation cost and the block-compressive sensing (BCS) algorithm to solve the space storage problem [9–12]. This article is based on the analysis of the above two points.

The CS recovery algorithm of images can mainly be divided into convex optimization recovery algorithms, non-convex recovery algorithms, and hybrid algorithms. The convex optimization algorithms include basis pursuit (BP), greedy basis pursuit (GBP), iterative hard threshold (IHT), etc. Non-convex algorithms include orthogonal matching pursuit (OMP), subspace matching basis pursuit (SP), and iteratively reweighted least square (IRLS), etc. The hybrid algorithms include sparse Fourier description (SF), chain pursuit (CP), and heavy hitters on steroids pursuit (HHSP) and other mixed algorithms [13–15]. The convex optimization algorithms based on l_1 minimization have benefits on the reconstruction effect, but with large computational complexity and high time complexity. Compared with convex optimization algorithms, the non-convex algorithms, such as the greedy pursuit algorithm, operate quickly, with a slightly poor accuracy based on l_0 minimization, and can also meet the general requirements of practical applications. In addition, the iterative threshold method has also been widely used in both of them with excellent performance. However, the iterative threshold method is sensitive to the selection of the threshold and the initial value of the iteration that affects the efficiency and accuracy of the algorithm [16,17]. The selection of thresholds in this process often uses simple error values (including absolute or relative values) or quantitative iterations as stopping criterion of the algorithm, which does not guarantee algorithm optimization [18,19].

The focus of this paper is on three aspects, namely, the block partitioning under weighted information entropy, the adaptive sampling based on synthetic features, and the iterative reconstruction through error analysis. The mean information entropy (MIE) and texture saliency (TS) are introduced in the block partitioning to provide a basis for promoting the algorithm. This part of adaptive sampling mainly improves the overall image quality through designing the block sampling rate by means of variance and local saliency (LS). The iterative reconstruction part mainly uses the relationship of three errors to provide the number of iterations required for the best reconstructed image in different noise backgrounds. Based on the above points, this paper proposes an altered adaptive block-compression sensing algorithm with flexible partitioning and error analysis, which is called FE-ABCS.

The remainder of this paper is organized as follows. In Section 2, we focus on the preliminaries of BCS. Section 3 includes the problem formulation and important factors. Then, the structure of the proposed FE-ABCS algorithm is presented in Section 4. In Section 5, the experiments and results analysis are listed to show the benefit of the FE-ABCS. The paper concludes with Section 6.

2. Preliminaries

2.1. Compressive Sensing

The algorithm theory of compressive sensing is derived from the sparse characteristic of natural signals that can be sparsely represented under a certain sparse transform basis, enabling direct sampling of sparse signals (sampling and compressing simultaneously). Set the sparse representation s of an original digital signal x which can be obtained by the transformation of sparse basis Ψ with K sparse coefficients and the signal x is observed by a measurement matrix Φ, then the observation signal y can be expressed as:

$$y = \Phi x = \Phi \Psi s = \Omega s \tag{1}$$

where, $x \in R^N$, $s \in R^N$, and $y \in R^M$. Consequently, $\Omega \in R^{M \times N}$ is the product of the matrix $\Phi \in R^{M \times N}$ and $\Psi \in R^{N \times N}$, named the sensing matrix, and the value of M is much less than N because of the compressive sensing theory.

The reconstruction process is an NP-hard problem which restores the N-dimensional original signal from the M-dimensional measurement value through nonlinear projection and cannot be solved directly. Candès et al. pointed out that the number M must meet the condition $M = O(K \log(N))$ in order to reconstruct the N-dimensional signal x accurately, and the sensing matrix Ω must satisfy the Restricted

Isometry Property (RIP) [20]. Furthermore, the former theories proved that the original signal x can be accurately reconstructed from the measured value y by solving the l_0 norm optimization problem:

$$\hat{x} = \Psi\hat{s}, \quad \hat{s} = \arg\min\|s\|_0 \; s.t. \; y = \Phi x = \Phi\Psi s \tag{2}$$

In the above formula, $\|*\|_0$ is the l_0 norm of a vector, which represents the number of non-zero elements in the vector.

With the wide application of CS technology, especially for 2D/3D image signal processing, it inevitably leads to a dimensional computing disaster problem (because the amount of calculation increases with the square/cube of dimensions), which is not directly overcome by CS technology itself. Here, it is necessary to introduce block partitioning and parallel processing to improve the algorithm, that is, the BCS algorithm improves its universality.

2.2. Block-Compressive Sensing (BCS)

The traditional method of BCS used in image signal processing is to segment the image and process the sub-images in parallel for reducing the cost of storage and calculation. Suppose the original image ($I_W \times I_H$) with $N = W \times H$ pixels in total, the observation with M-dimension and the definition of total sampling rate ($TSR = M/N$), in the normal processing of BCS, the image is divided into small blocks with a size of $B \times B$, each of which is sampled with the same operator. Let x_i represent the vectorized signal of the i-th block through raster scanning, and the output vector y_i of BCS measurement can be written as:

$$y_i = \Phi_B x_i \tag{3}$$

where, Φ_B is an $m \times n$ matrix with $n = B^2$ and $m = \lfloor n \cdot TSR \rfloor$. The matrix Φ_B is usually taken as an orthonormalized i.i.d Gaussian matrix. For the whole image, the equivalent sampling operator Φ in (1) is thus a block diagonal matrix taking the following form:

$$\Phi = \begin{bmatrix} \Phi_B & \cdots & 0 \\ \vdots & \ddots & \vdots \\ 0 & \cdots & \Phi_B \end{bmatrix} \tag{4}$$

2.3. Problems of BCS

The mentioned BCS algorithm for solving the storage space, dividing image into multiple sub-images, reduces the scale of the measurement matrix on the one hand, and on the other hand could be conducive to the parallel processing of the sub-images. However, BCS still has the following problems that need to be investigated and solved:

- Most existing research papers of BCS do not perform useful analysis on image partitioning and then segment according to the analysis result [21,22]. The common partitioning method ($n = B \times B$) of BCS only considers reducing the computational complexity and storage space problem without considering the integrity of the algorithm and other potential effects, such as providing a better foundation for subsequent sampling and reconstructing by combining the structural features and the information entropy of the image.
- The basic sampling method used in BCS is to sample each sub-block uniformly according to the total sampling rate (TSR), while the adaptive sampling method selects different sampling rates according to the sampling feature of each sub-block [23]. Specifically, the detail block allocates a larger sampling rate, and the smooth block matches a smaller sampling rate, thereby improving the overall quality of the reconstructed image at the same TSR. But the crux is that the studies of criteria (feature) used to assign adaptive sampling rates are rarely seen in recent articles.

- Although there are many studies on the improvement of the BCS iterative construction algorithm [24], few articles focus on optimizing the performance of the algorithm from the aspect of iteration stop criterion in the image reconstruction process, especially in the noise background.

In addition, the improvement on BCS also includes blockiness elimination and engineering implementation of the algorithm. Finally, although BCS technology still has some aspects to be solved, due to its advantages, the technology has been widely applied to optical/remote sensing imaging, medical imaging, wireless sensor networks, and so on [25].

3. Problem Formulation and Important Factors

3.1. Flexible Partitioning by Mean Information Entropy (MIE) and Texture Structure (TS)

Reasonable block partitioning reduces the information entropy (IE) of each sub-block to improve the performance of the BCS algorithm at the same total sampling rate (TSR), and ultimately improves the quality of the entire reconstructed image. In our paper, we adopt flexible partitioning with image sub-block shape $n = row \times column = l \times h$ to remove the blindness of image partitioning with the help of texture structure (TS) and mean information entropy (MIE) instead of the primary shape $n = B \times B$. The expression of TS is based on the gray-tone spatial-dependence matrices and the angular second moment (ASM) [26,27]. The value of TS is defined as follows using ASM:

$$g_{TS} = \sum_{i=0}^{255} \sum_{j=0}^{255} \{p(i,j,d,a)\}^2$$

$$p(i,j,d,a) = P(i,j,d,a)/R \tag{5}$$

where, $P(i,j,d,a)$ is the (i,j)-th entry in a gray-tone spatial-dependence matrix, $p(i,j,d,a)$ is the normalized form of $P(i,j,d,a)$, (i,j,d,a) is the neighboring pixel pair with distance d, orientation a, and gray value (i,j) in the image, and R denotes the number of neighboring resolution cell pairs. The definition of MIE of the whole image is as follows:

$$g_{MIE} = \frac{1}{N/n} \sum_{i=1}^{N/n} \left(-\sum_{j=0}^{255} e_{i,j} \log_2 e_{i,j}\right) = \frac{1}{T_1} \sum_{i=1}^{T_1} \left(-\sum_{j=0}^{255} e_{i,j} \log_2 e_{i,j}\right) \tag{6}$$

where, $e_{i,j}$ is the proportion of pixels with gray value j in the i-th sub-image, and T_1 is the number of sub-images.

Suppose the flexible partitioning of BCS is reasonable, increasing the similarity between pixels in each sub-block and reducing the MIE of the whole image sub-blocks will inevitably bring about a decrease in the difficulty of image sampling and recovery, which means that the flexible partitioning itself is a process of reducing order and rank. Figure 1 shows the effect on the MIE of four 256 × 256-pixel-testing gray images with 256 gray levels by different partitioning methods when the number of pixels per sub-image is limited to 256. The abscissa represents different 2-base partitioning modes, the ordinate represents the MIE of the whole image in different partitioning modes. Figure 1 indicates that images with different structures reach minimum MIE on different partitioning points which will be used in flexible partitioning to provide a basis for block segmentation.

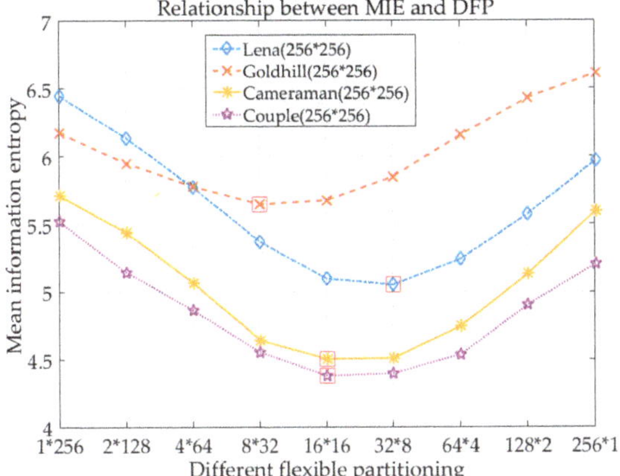

Figure 1. Effect of different partitioning methods on mean information entropy (MIE) of images (the abscissa represents flexible partitioning with shape $n = l \times h = 2^{i-1} \times 2^{9-i} = 256$).

Furthermore, the MIE guiding the partitioning of the image only considers the pixel level of the image, i.e., gray scale distribution, without considering the image optimization of the spatial level, i.e., texture structure. In fact, TS information is also very important for image restoration algorithms. Therefore, this paper uses the method of g_{MIE} combined with g_{TS} to provide the basis for flexible partitioning, that is, weighted MIE (WM) which is defined as follows:

$$g_{FB} = c_{TS} \times g_{MIE} = f(g_{TS}) \times g_{MIE} \tag{7}$$

where, c_{TS} is the weighting coefficient, $f(*)$ is the weighting coefficient function, and its value is related to the TS information g_{TS}.

3.2. Adaptive Sampling with Variance and Local Salient Factor

The feature selection for distinguishing the detail block and the smooth block is very important on the process of adaptive sampling. Information entropy, variance, and local standard deviation are often used as criteria for features, respectively. The shortcomings are found in using the above features individually as criteria for adaptive sampling, such as information entropy only reflects the probability of gray distribution, the variance is also only related to the degree of dispersion of the pixels, and the local standard deviation only focuses on the spatial distribution of the pixels. Secondly, the adaptive sampling rate is mostly set using segment adaptive sampling instead of step-less adaptive sampling in the previous literature [28], which leads to the discontinuity of sampling rate and the inadequacy utilization of the distinguishing feature.

In order to overcome the shortcomings of individual features, this paper uses the synthetic feature to distinguish between smooth blocks and detail blocks. The synthetic feature for adaptive sampling is defined as:

$$J(x_i) = L(x_i)^{\lambda_1} \times D(x_i)^{\lambda_2} \tag{8}$$

where, $D(x_i)$ and $L(x_i)$ denote the variance and local salient factor in the i-th sub-image, and λ_1 and λ_2 are the corresponding weighting coefficients. The expressions of variance and local salient factor for the sub-block are as follows:

$$D(x_i) = \frac{1}{n} \sum_{j=1}^{n} (x_{ij} - \mu_i)^2$$

$$L(x_i) = \frac{1}{n} \sum_{j=1}^{n} \frac{\sum_{k=1}^{q} \left| x_{ij}^k - x_{ij} \right|}{x_{ij}} \tag{9}$$

where, x_{ij} is the gray value of the j-th pixel in the i-th sub-image, μ_i is the gray mean of the i-th sub-block image, x_{ij}^k is the gray value of the k-th pixel in the salient operator domain around the center pixel x_{ij}, and q represents the number of pixels in the salient operator domain. The synthetic feature $J(x_i)$ can not only reflect the degree of dispersion and relative difference of sub-image pixels, but also combines the relationship between sensory amount and physical quantity of Weber's Law [29].

In order to avoid the disadvantage of segmented adaptive sampling, step-less adaptive sampling is adopted in this literature [30–32]. The key point of step-less adaptive sampling is how to select a continuous sampling rate accurately based on the synthetic feature. The selection of continuous sampling rates is achieved by setting the sampling rate factor (η_{SR}) based on the relationship between the sensory amount and the physical quantity in Weber's Law. The sampling rate factor (η_{SR}) and the step-less adaptive sampling rate (c_{SR}) are defined as follows:

$$\eta_{SR}(x_i) = \frac{\log_2 J(x_i)}{\frac{1}{T_1} \sum\limits_{j=1}^{T_1} \log_2 J(x_j)} \tag{10}$$

$$c_{SR}(x_i) = \eta_{SR}(x_i) \times TSR \tag{11}$$

where, TSR is the total sampling rate of the whole image.

3.3. Error Estimation and Iterative Stop Criterion in Reconstruction Process

The goal of the reconstruction process is to provide a good representative of the original signal:

$$x^* = \left[x_1^*, x_2^*, \cdots, x_N^* \right]^T, \quad x^* \in R^N. \tag{12}$$

Given the noisy observed output ($\widetilde{y}$) and finite-length sparsity (K), the performance of reconstruction is usually measured by the similarity or the error function between x^* and x. In addition, the reconstruction method, whether it is a convex optimization algorithm or a non-convex optimization algorithm, needs to solve the NP-hard problem by linear programming (LP), wherein the number of the correlation vectors is crucial. Therefore, the error estimation and the selection of the number of correlation vectors are two important factors of reconstruction. Especially in some non-convex optimization restoration algorithms, such as the OMP algorithm, the selection of the number of correlation vectors is linearly related to the number (v) of iterations of the algorithm. So, the two points (error estimation and optimal iteration) need to be discussed below.

3.3.1. Reconstruction Error Estimation in Noisy Background

In the second section, Equation (1) was used to describe the relationship model between the original signal and the noiseless observed signal, but the actual observation is always in the noise background, so the observed signal in this noisy environment is as shown in the following equation:

$$\widetilde{y} = \Phi x + w = \Phi \Psi s + w = \Omega s + w \tag{13}$$

where, $\widetilde{y}$ is the observed output in the noisy environment, and w is an additive white Gaussian noise (AWGN) with zero-mean and standard deviation σ_w. The M-dimension AWGN w is independent of the signal x. Here, we discuss the reconstruction error in two steps, the first step confirms the entire reconstruction model, and the second step derives the relevant reconstruction error.

Since the original signal (x) itself is not sparse, it is K-sparse under sparse basis (Ψ), so we have:

$$s = \Psi^{-1} x, \quad s = [s_1, s_2, \cdots, s_k, \cdots, s_N]^T \tag{14}$$

where, $[s_1, s_2, \cdots, s_k, \cdots, s_N]^T$ is a vector of length N which only has K non-zero elements, i.e., the remaining N-K micro elements are zero or much smaller than any of the K non-zero elements. Assuming that the first K elements of the sparse representation s are just non-zero elements without loss of generality, we can have:

$$ s = \begin{bmatrix} s_K \\ s_{N-K} \end{bmatrix} \tag{15} $$

where, s_K is a K dimensional vector and s_{N-K} is a vector of length N-K. The actual observed signal obtained by Equations (13) and (15) can be described as follows:

$$ \widetilde{y} = y + w = \Omega s + w = \begin{bmatrix} \Omega_K & \Omega_{N-K} \end{bmatrix} \begin{bmatrix} s_K \\ s_{N-K} \end{bmatrix} + w = \Omega_K s_K + \Omega_{N-K} s_{N-K} + w \tag{16} $$

where, $\Omega = \begin{bmatrix} \omega_1 & \cdots & \omega_K & \omega_{K+1} & \cdots & \omega_N \end{bmatrix}$ is an $M \times N$ matrix generated of N vectors with M-dimension.

In order to estimate the error of the recovery algorithm accurately, we define three error functions using the l_2 norm:

$$ \text{Original data error}: \ e_x = \frac{1}{N}\|x - x^*\|_2^2 \tag{17} $$

$$ \text{Observed data error}: \ e_y = \frac{1}{M}\|y - y^*\|_2^2 \tag{18} $$

$$ \text{Sparse data error}: \ e_s = \frac{1}{N}\|s - s^*\|_2^2 \tag{19} $$

where, x^*, y^*, s^* represent the reconstructed values of x, y, s, respectively. The three reconstructed values are obtained by maximum likelihood (ML) estimation using l_0 minimization. The number of iterations in the restoration algorithm is v times, that is, the number of correlation vectors. In addition, in the process of solving s^* by using pseudo-inverse, which is based on the least-squares algorithm, the value of v is smaller than M. Using Equations (13) and (15), the expressions of x^*, y^*, s^* are listed as follows:

$$ s^* = \begin{bmatrix} s_v^* \\ s_{N-v}^* \end{bmatrix} = \begin{bmatrix} s_v^* \\ 0_{N-v}^* \end{bmatrix} = \begin{bmatrix} \Omega_v^+ \widetilde{y} \\ 0_{N-v}^* \end{bmatrix} = \begin{bmatrix} \Omega_v^+ (\Omega_v s_v + \Omega_{N-v} s_{N-v} + w) \\ 0_{N-v}^* \end{bmatrix} = \begin{bmatrix} s_v + \Omega_v^+ (\Omega_{N-v} s_{N-v} + w) \\ 0_{N-v}^* \end{bmatrix} \tag{20} $$

$$ x^* = \Psi s^* \tag{21} $$

$$ y^* = \Omega s^* \tag{22} $$

where, Ω_v^+ is the pseudo inverse of Ω_v, and its expression is $\Omega_v^+ = \left(\Omega_v^T \Omega_v\right)^{-1} \Omega_v^T$.

Using Equations (20–22), the three error functions are rewritten as follows:

$$ e_x = \frac{1}{N}\left\|\left[-\Psi_v \Omega_v^+ (\Omega_{N-v} s_{N-v} + w) + \Psi_{N-v} s_{N-v}\right]\right\|_2^2 \tag{23} $$

$$ e_y = \frac{1}{M}\left\|\left[-\Omega_v \Omega_v^+ (\Omega_{N-v} s_{N-v} + w) + \Omega_{N-v} s_{N-v}\right]\right\|_2^2 \tag{24} $$

$$ e_s = \frac{1}{N}\left\|\begin{bmatrix} -\Omega_v^+ (\Omega_{N-v} s_{N-v} + w) \\ s_{N-v} \end{bmatrix}\right\|_2^2. \tag{25} $$

According to the definition of Ψ, Ω and the RIP, we know:

$$ e_s = e_x \tag{26} $$

$$ (1 - \delta_K)e_s \leq e_y \leq (1 + \delta_K)e_s \tag{27} $$

where, $\delta_K \in (0,1)$ represents a coefficient associated with Ω and K. According to Gershgorin circle theorem [33], $\delta_K = (K-1)\mu(\Omega)$ for all $K < \mu(\Omega)^{-1}$, where $\mu(\Omega)$ denotes the coherency of Ω:

$$\mu(\Omega) = \max_{1 \leq i < j \leq N} \frac{\left|\langle \omega_i, \omega_j \rangle\right|}{\|\omega_i\|_2 \|\omega_j\|_2}. \tag{28}$$

Using Equations (26) and (27), the boundaries of the original data error are as follows:

$$\frac{1}{(1+\delta_K)} e_y \leq e_x \leq \frac{1}{(1-\delta_K)} e_y. \tag{29}$$

Therefore, from the above analysis, we can conclude that the three errors are consistent, and the minimizing of the three errors is equivalent. Due to the complexity and reliability of the calculation (e_x-too complicated, e_s-insufficient dimensions), e_y is used as the target in the optimization function of the recovery algorithm.

3.3.2. Optimal Iterative Recovery of Image in Noisy Background

The optimal iterative recovery of image discussed in this paper refers to the case where the error function of the image is the smallest, as can be seen from the minimization of e_y in the form of l_2 norm:

$$v_{opt} = \left\{ v \middle| \arg\min_v e_y \right\} \tag{30}$$

$$\arg\min e_y = \arg\min \frac{1}{M} \|G_v \Omega_{N-v} s_{N-v} - C_v w\|_2^2 \tag{31}$$

$$\begin{cases} G_v = I - \Omega_v \Omega_v^+ \\ C_v = \Omega_v \Omega_v^+ \end{cases} \tag{32}$$

while G_v is a projection matrix of rank $M - v$, C_v is a projection matrix of rank v. Since the projection matrices G_v and C_v in Equation (30) are orthogonal, the inner product of the two vectors $G_v \Omega_{N-v} s_{N-v}$ and $C_v w$ is zero and therefore:

$$e_y = \frac{1}{M} \|C_v w\|_2^2 + \frac{1}{M} \|G_v \Omega_{N-v} s_{N-v}\|_2^2 = e_y^w + e_y^s \tag{33}$$

According to [34], the observed data error e_y is a Chi-square random variable with degree of freedom v, and the expected value and the variance of e_y are as follows:

$$\frac{M}{\sigma_w^2} e_y \sim \chi_v^2 \tag{34}$$

$$E(e_y) = \frac{v}{M} \sigma_w^2 + \frac{1}{M} \|G_v \Omega_{N-v} s_{N-v}\|_2^2 \tag{35}$$

$$\mathrm{var}(e_y) = \frac{2v}{M^2} \left(\sigma_w^2\right)^2 \tag{36}$$

The expected value of e_y has two parts. The first part $\frac{v}{M}\sigma_w^2$ is the noise-related part, which is a function that is positively related to the number v. The second part $\frac{1}{M}\|G_v \Omega_{N-v} s_{N-v}\|_2^2$ is a function of the unstable micro element s_{N-v}, which is decreased as the number v increases. Therefore, the observed data error e_y is normally called a bias-variance tradeoff.

Due to the existence of the uncertain part e_y^s, this results in an impossible-to-seek optimal number of iterations v_{opt} by solving the minimum value of e_y directly. As a result, another bias-variance tradeoff

e_y^* is introduced to provide probabilistic bounds on the e_y^s by using the noisy output $\widetilde{y}$ instead of noiseless output y:

$$e_y^* = \frac{1}{M}\|\widetilde{y} - y^*\|_2^2 = \frac{1}{M}\|G_v\Omega_{N-v}s_{N-v} + G_v w\|_2^2. \tag{37}$$

According to [35], the second observed data error e_y^* is a Chi-square random variable of order $M - v$, and the expected value and the variance of e_y^* are as follows:

$$\frac{M}{\sigma_w^2}e_y^* \sim \chi_{M-v}^2 \tag{38}$$

$$E\!\left(e_y^*\right) = \frac{M-v}{M}\sigma_w^2 + \frac{1}{M}\|G_v\Omega_{N-v}s_{N-v}\|_2^2 = \frac{M-v}{M}\sigma_w^2 + e_y^s \tag{39}$$

$$\mathrm{var}\!\left(e_y^*\right) = \frac{2(M-v)}{M^2}\left(\sigma_w^2\right)^2 + \frac{4\sigma_w^2}{M^2}\|G_v\Omega_{N-v}s_{N-v}\|_2^2 \tag{40}$$

So, we can derive probabilistic bounds for the observed data error e_y using probability distribution of the two Chi-square random variables:

$$\underline{e_y(p_1,p_2)} \le e_y \le \overline{e_y(p_1,p_2)} \tag{41}$$

where, p_1 is the confidence probability on a random variable of the observed data error e_y, and p_2 is the validation probability on a random variable of the second observed data error e_y^*. As both of the two errors satisfy the Chi-square distribution, Gaussian distribution can be used to estimate them. Therefore, confidence probability p_1 and validation probability p_2 can be calculated as:

$$p_1 = Q(\alpha) = \int_{-\alpha}^{\alpha} \frac{1}{\sqrt{2\pi}} e^{-\frac{x^2}{2}}\, dx \tag{42}$$

$$p_2 = Q(\beta) = \int_{-\beta}^{\beta} \frac{1}{\sqrt{2\pi}} e^{-\frac{x^2}{2}}\, dx \tag{43}$$

where, α and β denote the tuning parameters of confidence and validation probabilities, respectively. Furthermore, the worst case is considered when calculating the minimum value of e_y, that is, by calculating the minimum value of the upper bound of e_y:

$$\begin{aligned}
v_{opt} &= \left\{v\,\middle|\,\mathrm{argmin}_v \overline{e_y}\right\} = \left\{v\,\middle|\,\mathrm{argmin}_v \overline{e_y(p_1,p_2)}\right\} = \left\{v\,\middle|\,\mathrm{argmin}_v \overline{e_y(\alpha,\beta)}\right\} \\
&= \left\{v\,\middle|\,\mathrm{argmin}_v \left(\tfrac{2v-M}{M}\sigma_w^2 + e_y^* + \alpha\tfrac{\sqrt{2v}}{M}\sigma_w^2 + \beta\,\mathrm{var}\!\left(e_y^*\right)\right)\right\}
\end{aligned} \tag{44}$$

Normally, based on Akaike information criterion (AIC) or Bayesian information criterion (BIC), the optimum number of iterations can be chosen as follows:

AIC: Set $\alpha = \beta = 0$

$$\overline{e_y} = \overline{e_y(\alpha = 0, \beta = 0)} = \left(\frac{2v}{M} - 1\right)\sigma_w^2 + e_y^* \tag{45}$$

BIC: Set $\alpha = \sqrt{v}\log M$ and $\beta = 0$.

$$\overline{e_y} = \overline{e_y(\alpha = \sqrt{v}\log M, \beta = 0)} = \left(\frac{(2+\sqrt{2}\log M)v}{M} - 1\right)\sigma_w^2 + e_y^* \tag{46}$$

where, e_y^* can be calculated based on the noisy observation data and the reconstruction algorithm.

3.3.3. Application of Error Estimation on BCS

The proposed algorithm (FE-ABCS) is based on block-compressive sensing, so the optimal number of iterations (v_{opt}) derived in the above section also requires a variant to be applied to the above algorithm:

$$
\begin{aligned}
v^i_{opt} &= \left\{ v^i \,\middle|\, \underset{v^i}{\text{argmin}}\,\overline{e_{y_i}} \right\} = \left\{ v^i \,\middle|\, \underset{v^i}{\text{argmin}}\,\overline{e_{y_i}(p^i_1, p^i_2)} \right\} = \left\{ v^i \,\middle|\, \underset{v^i}{\text{argmin}}\,\overline{e_{y_i}(\alpha_i, \beta_i)} \right\} \\
&= \left\{ v^i \,\middle|\, \underset{v^i}{\text{argmin}}\, \left(\frac{2v^i - m^i + \alpha_i \sqrt{2v^i}}{m^i} \sigma^2_{w^i} + \beta_i \text{var}\!\left(e^*_{y_i}\right) + e^*_{y_i} \right) \right\}
\end{aligned}
\tag{47}
$$

where, $i = 1, 2, \cdots, T_1$ represents the serial number of all sub-images. Similarly, the values of α_i and β_i can be valued according to the AIC and BIC criteria.

4. The Proposed Algorithm (FE-ABCS)

With the knowledge presented in the previous section, the altered ABCS (FE-ABCS) is proposed for the recovery of block sparse signals in noiseless or noise backgrounds. The workflow of the proposed algorithm is presented in Section 4.1 while the specific parameter settings of the proposed algorithm is introduced in Section 4.2.

4.1. The Workflow and Pseudocode of FE-ABCS

In order to better express the idea of the proposed algorithm, the workflow of the typical BCS algorithm and the FE-ABCS algorithm are presented, as shown in Figure 2.

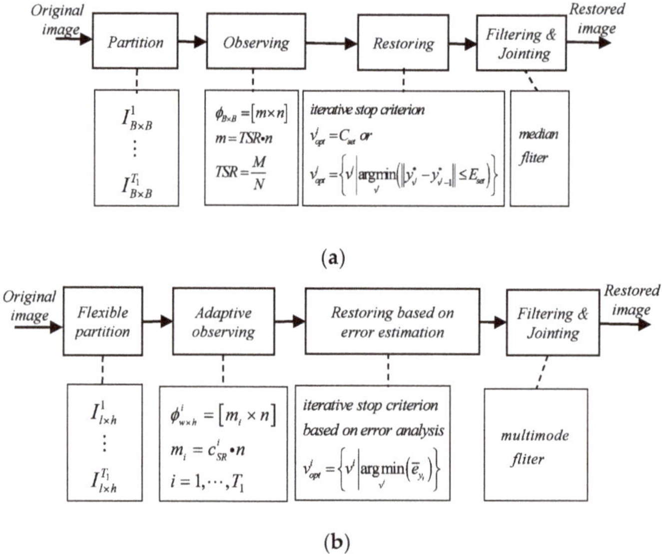

Figure 2. The workflow of two block-compressive sensing (BCS) algorithms. (**a**) Typical BCS algorithm, (**b**) FE-ABCS algorithm.

According to Figure 2, compared with the traditional BCS algorithm, the main innovations of this paper can be reflected in the following points:

- Flexible partitioning: using the weighted MIE as the block basis to reduce the average complexity of the sub-images from the pixel domain and the spatial domain;

- Adaptive sampling: adopting synthetic feature and step-less sampling to ensure a reasonable sample rate for each subgraph;
- Optimal number of iterations: using the error estimate method to ensure the minimum error output of the reconstructed image in the noisy background.

Furthermore, since the FE-ABCS algorithm is based on iterative recovery algorithm, especially the non-convex optimization iterative recovery algorithm, this paper uses the OMP algorithm as the basic comparison algorithm without loss of generality. The full pseudocode of the proposed algorithm is presented as follows.

FE-ABCS Algorithm based on OMP (Orthogonal Matching Pursuit)

1: Input: Original image I, total sampling rate TSR, sub-image dimension n ($n = 2^b, b = 2, 4, 6, \cdots$), sparse matrix $\Psi \in R^{n \times n}$, initialized measurement matrix $\Phi \in R^{n \times n}$

2: Initialization: $\{x \mid x \leftarrow I \text{ and } x \in R^N\}$;

$\quad T_1 = {N}/{n}$; // quantity of subimages

$\quad T_2 = 1 + \log_2 n$; // type of flexible partitioning

step1: flexible partitioning (FP)

3: for j = 1,…,T_2 do

4: $\quad l_j \times h_j = 2^{j-1} \times 2^{T_2 - j}; \{I_i, i = 1, \cdots, T_1\}_j \leftarrow I$;

$\quad \{x_i, i = 1, \cdots, T_1\}_j \leftarrow x$;

5: $\quad g_{MIE}^j = MIE\left(\{x_i, i = 1, \cdots, T_1\}_j\right)$;

$\quad c_{TS}^j = (f(g_{TS}))_j = (f([g_{TS}^H, g_{TS}^V]))_j$;

6: $\quad g_{FB}^j = c_{TS}^j \cdot g_{MIE}^j$;

$\quad$ // Weighted MIE--Base of FP

7: end for

8: $i_{opt} = \arg\min_j \left(\{g_{FB}^j, j = 1, \cdots, T_2\}\right)$;

9: $l \times h = 2^{J_{opt} - 1} \times 2^{T_2 - J_{opt}}; \{I_i, i = 1, \cdots, T_1\} \leftarrow I$;

$\quad \{x_i, i = 1, \cdots, T_1\} \leftarrow x$;

step2: adaptive sampling (AS)

10: for i = 1,…,T_1 do

11: $\quad D(x_i) \leftarrow x_i; L(x_i) \leftarrow I_i; J(x_i) = L(x_i)^{\lambda_1} \cdot D(x_i)^{\lambda_2}$;

$\quad$ // synthetic feature (J)--Base of AS

12: end for

13: $\eta(\{x_i, i = 1, \cdots, T_1\}) = \dfrac{\log_2 J(\{x_i, i = 1, \cdots, T_1\})}{\dfrac{1}{T_1}\sum\limits_{i=1}^{T_1} \log_2 J(x_i)}$;

14: $c_{SR}(\{x_i, i = 1, \cdots, T_1\}) = \eta(\{x_i, i = 1, \cdots, T_1\}) \cdot TSR$;

$\quad$ // c_{SR} -- AS ratio of sub-images

$\quad \{m_i, i = 1, \cdots, T_1\} = c_{SR}(\{x_i, i = 1, \cdots, T_1\}) \cdot n$;

15: $\Phi = (\phi_1, \cdots, \phi_n)^T, \chi_i = randperm(n), \Phi_{\chi_i} = \Phi(\chi_i, :)$;

16: $\{\Phi_i, i = 1, \cdots, T_1\} = \{\Phi_{\chi_i}([1, \cdots, m_i], :), i = 1, \cdots, T_1\}$;

17: $\{y_i, i = 1, \cdots, T_1\} = \{\Phi_i \cdot x_i, i = 1, \cdots, T_1\}$;

step3: restoring based on error estimation

18: $\{\tilde{y}_i, i = 1, \cdots, T_1\} = \{y_i + w_i, i = 1, \cdots, T_1\}$;

$\quad$ // $w_i : m_i - dimension\ AWGN, w_i = 0 : noiseless$

19: $\{\Omega_i, i = 1, \cdots, T_1\} = \{\Phi_i \cdot \Psi, i = 1, \cdots, T_1\}$;

20: for i = 1,…,T_1 do

21: $\quad \Omega_i = \{\omega_{i1}, \cdots, \omega_{in}\}, r = \tilde{y}_i, A = \varnothing, s^* = 0^n$;

$\quad$ // $\{\omega_{ij}, j = 1, \ldots, n\}$ -- column vector of Ω_i

22: $\quad v_{opt}^i = \left\{v^i \mid \arg\min_{v^i} \overline{e_{y_i}}\right\}$;

$\quad$ // calculate optimal iterative of sub-images

23: $\quad$ for $j = 1, \cdots, v_{opt}^i$ do

24: $\quad\quad \wedge = \arg\min_j |\langle r, w_{ij}\rangle|$;

25: $\quad\quad A = A \cup \{\wedge\}$;

26: $\quad\quad r = \tilde{y}_i - \Omega_i(:, A) \cdot [\Omega_i(:, A)]^+ \cdot \tilde{y}_i$;

27: $\quad$ end for

28: $\quad s_i^* = [\Omega_i(:, A)]^+ \cdot \tilde{y}_i$;

$\quad$ // s_i^* --reconstructed sparse representation

29: $\quad x_i^* = \Psi \cdot s_i^*$;

$\quad$ // x_i^* --reconstructed original signal

30: end for

31: $x^* = \{x_i^*, i = 1, \cdots, T_1\}, I_r^* = \{x^*, l, h\}$;

$\quad$ // I_r^* --reconstructed image without filter

step4: multimode filtering (MF)

32: *if* $\left(BEI \geq BEI^*\right)$; // BEI^* – *Threshold of block effect*

33: $I_r^* = deblock(I_r^*)$;

34: *end if*

35: *if* $\left(TSR \geq TSR^*\right)$; // TSR^* – *Threshold of TSR*

36: $I_F^* = wienerfilter(I_r^*)$;

37: *else* $I_F^* = medfilter(I_r^*)$;

38: *end if*

39: $I^* = I_F^*$

$\quad$ // I^* --reconstructed image with MF

4.2. Specific Parameter Setting of FE-ABCS

4.2.1. Setting of the Weighting Coefficient c_{TS}

The most important step in achieving flexible partitioning is to the minimum of the weighted MIE, where the design of the weighting coefficient function is the most critical point. Therefore, this section focuses on the specific design of the function which ensures optimal partitioning of the image:

$$
c_{TS} = f(g_{TS}) = \left\{ (f(g_{TS}))_j, 1, \cdots, T_2 \right\} = \left\{ (f([g_{TS}^H, g_{TS}^V]))_j, 1, \cdots, T_2 \right\}
$$
$$
= \begin{cases}
ones(1, T_2) & g_{TS}^H \leq G_{TS} \& g_{TS}^V \leq G_{TS} \\
[0, 1, \cdots, b]/(b/2) & g_{TS}^H \leq G_{TS} \& g_{TS}^V > G_{TS} \\
[b, \cdots, 1, 0]/(b/2) & g_{TS}^H > G_{TS} \& g_{TS}^V \leq G_{TS} \\
\dfrac{[b, \log_2^{(n/2+2)}, \log_2^{(n/4+4)}, \cdots, \log_2^{(4+n/4)}, \log_2^{(2+n/2)}, b]}{b/2+1} & g_{TS}^H > G_{TS} \& g_{TS}^V > G_{TS}
\end{cases}
\tag{48}
$$

where, g_{TS}^H and g_{TS}^V represent the value of horizontal and vertical TS by using ASM, and G_{TS} represents the threshold at which the TS feature value reaches a significant degree.

4.2.2. Setting of the Adaptive Sampling Rate c_{SR}

In the process of adaptive observation, the most important thing is to design a reasonable random observation matrix, and the dimension of this matrix needs to be constrained by the adaptive sampling rate, so as to assign the different sampling of each sub-image with a different complexity. Therefore, the setting of $c_{SR} = \{c_{SR}(x_i), i = 1, \cdots, T_1\}$ is crucial, and its basic form is mainly determined by the synthetic feature ($J = \{J(x_i), i = 1, \cdots, T_1\}$) and the sampling rate factor ($\eta_{SR} = \{\eta_{SR}(x_i), i = 1, \cdots, T_1\}$).

The definition of $J(x_i)$ can be implemented by setting the corresponding weighting coefficients λ_1 and λ_2. This article obtains the optimization values for λ_1 and λ_2 through analysis and partial verification experiments: $\lambda_1 = 1$ and $\lambda_2 = 2$.

The purpose of setting $\eta_{SR}(x_i)$ is to establish the mapping function relationship between $J(x_i)$ and c_{SR} by Equations (10) and (11). However, the mapping relationship established by Equation (10) does not consider the minimum sampling rate. In fact, the minimum sampling rate factor (MSRF) is considered in the proposed algorithm to improve performance, that is, the function between $\eta_{SR}(x_i)$ and $J(x_i)$ should be modified as follows.

- Initial value calculation of $\eta_{SR}(x_i)$: get the initial value of the sampling factor by Equation (10).
- Judgment of $\eta_{SR}(x_i)$ through MSRF ($\eta_{\min}$): if the corresponding sampling rate factor of all image sub-blocks meets the minimum threshold requirement ($\eta_{SR}(x_i) > \eta_{\min}, i \in \{1, 2, \cdots, T_1\}$), there is no need for modification, however, if it is not satisfied, modify it.
- Modifying of $\eta_{SR}(x_i)$: if $\eta_{SR}(x_i) \leq \eta_{\min}$, then $\eta_{SR}(x_i) = \eta_{\min}$; if $\eta_{SR}(x_i) > \eta_{\min}$, then use the following equation to modify the value:

$$
\eta_{SR}(x_i) == (1 + (1 - \eta_{\min})\frac{T_1 - T_1'}{T_1'}) \frac{\log_2 J(x_i)}{\frac{1}{T_1'} \sum\limits_{j=1}^{T_1'} \log_2 J(x_j)}
\tag{49}
$$

where, T_1' is the number of sub-images that can meet the requirement of the minimum threshold.

4.2.3. Setting of the Iteration Stop Condition v_{opt}

The focus of the proposed algorithm in the iterative reconstruction part is to make the best effect of the rebuilt image by choosing v_{opt} in the actual noisy background. This paper combines BIC and BCS to propose the calculation formula of the optimal iteration number of the proposed algorithm:

$$v_{opt} = \left\{ v_{opt}^i, i = 1, \cdots, T_1 \right\} = \left\{ v^i \middle| \underset{v^i}{\arg\min} \left(\frac{(2 + \sqrt{2}\log m^i)v^i - m^i}{m^i} \sigma_{w^i}^2 + e_{y_i}^* \right), i = 1, \cdots, T_1 \right\}. \tag{50}$$

5. Experiments and Results Analysis

In order to evaluate the FE-ABCS algorithm, experimental verification is performed in three scenarios. This paper first discusses the performance of the improved algorithm by flexible partitioning and adaptive sampling in the absence of noise, and secondly discusses how to combine the number of optimal iterations to eliminate the noise effect and achieve the best quality (comprehensive indicator) under noisy conditions. Finally, the differences between this proposed algorithm and other non-CS image compression algorithms is analyzed. The experiments were carried out in the matlab2016b software environment, and 20 typical grayscale images with 256 × 256 resolution were used for testing, which were selected from the LIVE Image Quality Assessment Database, the SIPI Image Database, the BSDS500 Database, and other digital image processing standard test Databases. The performance indicators mainly adopt Peak Signal to Noise Ratio (PSNR), Structural Similarity (SSIM), Gradient Magnitude Similarity Deviation (GMSD), Block Effect Index (BEI), and Computational Cost (CC). The above five performance indicators are defined as follows:

The PSNR indicator is an index that shows the amplitude error between the reconstructed image and the original image, which is the most common and widely used objective measure of image quality:

$$PSNR = 20 \times \log_{10}\left(\frac{255}{\sqrt{\frac{1}{N}\sum\limits_{i=1}^{N}(x_i - x_i^*)^2}} \right) \tag{51}$$

where, x_i and x_i^* are the gray value of i-th sub-image of the reconstructed image and the original image.

The SSIM indicator is adopted to indicate similarity between the reconstructed image and the original image:

$$SSIM = \frac{(2\mu_x\mu_{x^*} + c_1)(2\sigma_{xx^*} + c_2)}{(\mu_x^2 + \mu_{x^*}^2 + c_1)(\sigma_x^2 + \sigma_{x^*}^2 + c_2)} \tag{52}$$

where, μ_x and μ_{x^*} are the mean of x and x^*, σ_x and σ_{x^*} are the standard deviation of x and x^*, σ_{xx^*} represents the covariance of x and x^*, constant $c_1 = (0.01L)^2$ and $c_2 = (0.03L)^2$, and L is the range of pixel values.

The GMSD indicator is mainly used to characterize the degree of distortion of the reconstructed image. The larger the value, the worse the quality of the reconstructed image:

$$GMSD = std(\{GMS(i)|i = 1, \cdots, N\}) = std\left(\left\{ \frac{2m_x(i)m_{x^*}(i) + c_3}{m_x^2(i) + m_{x^*}^2(i) + c_3} |i = 1, \cdots, N \right\} \right) \tag{53}$$

where, $std(*)$ is the standard deviation operator, GMS is the gradient magnitude similarity between x and x^*, $m_x(i) = \sqrt{(h_H \otimes x(i))^2 + (h_V \otimes x(i))^2}$ and $m_y(i) = \sqrt{(h_H \otimes x^*(i))^2 + (h_V \otimes x^*(i))^2}$ are the gradient magnitude of $x(i)$ and $x^*(i)$, h_H and h_V represent the Prewitt operator of horizontal and vertical direction, and c_3 is an adjustment constant.

The main purpose of introducing BEI is to evaluate the blockiness of the algorithm in a noisy condition, which means that the larger the value, the more obvious the block effect:

$$BEI = \log_2 \left[\frac{sum(edge(x^*)) - sum(edge(x)) + sum(|edge(x^*) - edge(x)|)}{2} \right] \tag{54}$$

where, $edge(*)$ denotes the edge acquisition function of the image, $sum(*)$ represents the function of finding the number of all edge points of the image, and $|*|$ is an absolute value operator.

The Computational Cost is introduced to measure the efficiency of the algorithm, which is usually represented by Computation Time (CT). The smaller the value of CT, the higher the efficiency of the algorithm:

$$CT = t_{end} - t_{start} \tag{55}$$

where, t_{start} and t_{end} indicate the start time and end time, respectively.

In addition, the sparse basis and the random measurement matrices use discrete cosine orthogonal basis and orthogonal symmetric Toeplitz matrices [36,37], respectively.

5.1. Experiment and Analysis without Noise

5.1.1. Performance Comparison of Various Algorithms

In order to verify the superiority of the proposed ABCS algorithm, this paper mainly uses the OMP algorithm as the basic reconstruction algorithm. Based on the OMP reconstruction algorithm, eight BCS algorithms (including the proposed algorithm with the idea of flexible partitioning and adaptive sampling) are listed, and the performance of these algorithms under different overall sampling rates is compared, which is shown in Table 1. In this experiment, four normal grayscale standard images are used for performance testing, the dimension of the subgraph and the iterative number of reconstructions are limited to 256 and one quarter of the measurement's dimension, respectively.

These 8 BCS algorithms are named as M-B_C, M-B_S, M-FB_MIE, M-FB_WM, M-B_C-A_S, M-FB_WM-A_I, M-FB_WM-A_V, and M-FB_WM-A_S respectively, which in turn represent BCS with a fixed column block, BCS with a fixed square block, BCS with flexible partitioning by MIE, BCS with flexible partitioning by WM, BCS with a fixed column block and IE-adaptive sampling, BCS with WM-flexible partitioning and IE-adaptive sampling, BCS with WM-flexible partitioning and variance-adaptive sampling, and BCS with WM-flexible partitioning and SF-adaptive sampling (A form of FE-ABCS algorithm in the absence of noise). Comparing the data in Table 1, there are the following consensuses:

- Analysis of the performance indicators of the first four algorithms shows that for the BCS algorithm, BCS with a fixed column block is inferior to BCS with a fixed square block because square partitioning makes good use of the correlation of intra-block regions. MIE-based partitioning minimizes the average amount of information entropy of the sub-images. However, when the overall image has obvious texture characteristics, simply using MIE as the partitioning basis may not necessarily achieve a good effect, and the BCS algorithm based on weighted MIE combined with the overall texture feature can achieve better performance indicators.
- Comparing the adaptive BCS algorithms under different features in Table 1, variance has obvious superiority to IE among the single features, because the variance not only contains the dispersion of gray distribution but also the relative difference of the individual gray distribution of sub-images. In addition, the synthetic feature (combined local saliency) has a better effect than a single feature. The main reason for this is that the synthetic feature not only considers the overall difference of the subgraphs, but also the inner local-difference of the subgraphs.
- Combining experimental results of the eight BCS algorithms in Table 1 reveals that using adaptive sampling or flexible partitioning alone does not provide the best results, but the proposed algorithm combining the two steps will have a significant effect on both PSNR and SSIM.

Table 1. The Peak Signal to Noise Ratio (PSNR) and Structural Similarity (SSIM) of reconstructed images with eight BCS algorithms based on OMP. (TSR = total sampling rate).

Images	Algorithms	TSR = 0.2	TSR = 0.3	TSR = 0.4	TSR = 0.5	TSR = 0.6
		PSNR/SSIM	PSNR/SSIM	PSNR/SSIM	PSNR/SSIM	PSNR/SSIM
Lena	M-B_C	29.0945/0.7684	29.8095/0.8720	30.6667/0.9281	31.8944/0.9572	32.9854/0.9719
	M-B_S	31.1390/0.8866	31.7478/0.9227	32.4328/0.9480	33.2460/0.9651	34.2614/0.9763
	M-FB_MIE	30.7091/0.8613	31.2850/0.9115	32.0093/0.9413	32.9032/0.9600	33.8147/0.9737
	M-FB_WM	31.1636/0.8880	31.7524/0.9236	32.4623/0.9479	33.2906/0.9645	34.2691/0.9763
	M-B_C-A_I	29.1187/0.7838	29.8433/0.8803	30.8898/0.9305	32.0023/0.9577	33.2193/0.9732
	M-FB_WM-A_I	31.1763/0.8967	31.8872/0.9344	32.7542/0.9584	33.7353/0.9732	34.8647/0.9827
	M-FB_WM-A_V	31.2286/0.9087	32.0579/0.9433	33.0168/0.9643	34.1341/0.9775	35.4341/0.9856
	M-FB_WM-A_S	**31.3609/0.9138**	**32.0943/0.9487**	**33.1958/0.9681**	**34.3334/0.9807**	**35.8423/0.9878**
Goldhill	M-B_C	28.4533/0.7747	28.9144/0.8718	29.3894/0.9080	29.7706/0.9315	30.2421/0.9495
	M-B_S	29.5494/0.8785	29.9517/0.9089	30.3330/0.9341	30.8857/0.9514	31.4439/0.9640
	M-FB_MIE	29.7012/0.8882	29.9811/0.9154	30.4465/0.9364	30.9347/0.9516	31.5153/0.9642
	M-FB_WM	29.7029/0.8867	30.0277/0.9151	30.4827/0.9361	30.9555/0.9516	31.5333/0.9649
	M-B_C-A_I	28.4436/0.7809	28.8691/0.8693	29.3048/0.9089	29.7046/0.9321	30.2355/0.9499
	M-FB_WM-A_I	29.6708/0.8918	30.0833/0.9215	30.5120/0.9424	31.0667/0.9574	31.6899/0.9697
	M-FB_WM-A_V	29.5370/0.8957	30.0891/0.9253	30.5379/0.9456	31.0922/0.9607	31.8011/0.9724
	M-FB_WM-A_S	**29.7786/0.8975**	**30.1482/0.9272**	**30.5689/0.9472**	**31.1310/0.9622**	**31.8379/0.9736**
Cameraman	M-B_C	28.5347/0.7787	29.0078/0.8559	29.3971/0.9051	29.9417/0.9379	30.6612/0.9592
	M-B_S	31.1796/0.8763	31.4929/0.9121	31.9203/0.9391	32.3009/0.9581	32.7879/0.9704
	M-FB_MIE	31.1487/0.8782	31.5067/0.9123	31.8644/0.9403	32.3170/0.9577	32.7946/0.9703
	M-FB_WM	31.2118/0.8675	31.4645/0.9072	31.8130/0.9365	32.2050/0.9559	32.6811/0.9686
	M-B_C-A_I	28.5669/0.7852	28.8807/0.8612	29.3928/0.9164	29.9924/0.9461	30.6130/0.9639
	M-FB_WM-A_I	31.2554/0.8901	31.5975/0.9296	32.0955/0.9533	32.6859/0.9701	33.4007/0.9802
	M-FB_WM-A_V	31.2869/0.9085	31.8762/0.9550	32.5052/0.9746	33.3531/0.9848	34.4449/0.9904
	M-FB_WM-A_S	**31.3916/0.9287**	**31.9731/0.9621**	**32.6508/0.9790**	**33.6779/0.9877**	**34.8958/0.9918**
Couple	M-B_C	28.6592/0.7582	29.0162/0.8557	29.5471/0.9109	30.2260/0.9440	30.9136/0.9640
	M-B_S	30.1529/0.8912	30.6910/0.9289	31.2853/0.9541	31.9693/0.9695	32.7464/0.9796
	M-FB_MIE	30.1920/0.8895	30.7257/0.9282	31.2948/0.9531	31.9509/0.9692	32.7424/0.9794
	M-FB_WM	30.1357/0.8917	30.6890/0.9259	31.3185/0.9539	31.9520/0.9691	32.7622/0.9793
	M-B_C-A_I	28.5694/0.7428	29.0442/0.8589	29.5828/0.9088	30.2127/0.9444	30.9839/0.9642
	M-FB_WM-A_I	30.2105/0.9027	30.7783/0.9413	31.4680/0.9630	32.3143/0.9759	33.2604/0.9840
	M-FB_WM-A_V	30.1896/0.9099	30.8541/0.9454	31.4990/0.9670	32.3769/0.9792	33.3260/0.9864
	M-FB_WM-A_S	**30.3340/0.9117**	**30.9047/0.9475**	**31.5496/0.9686**	**32.3788/0.9798**	**33.3561/0.9869**

Figure 3 shows the reconstructed images of Cameraman using the above eight BCS algorithms and multimode filter at the overall sampling rate of 0.5. Compared with other algorithms, the graph (i) reconstructed by the proposed algorithm has good quality both in performance indicators and subjective vision. Adding multimode filtering has improved the performance of the above eight BCS algorithms. While comparing the corresponding data (SR = 0.5) in Figure 3 and Table 1, it was found that PSNR has a certain improvement after adding multimode filtering, and so as to SSIM under the first six BCS algorithms except the latter two algorithms. The reason is that the adaptive sampling rates of the latter two algorithms are both related to the variance (the more variance, the more sampling rate), and SSIM is related to both the variance and the covariance. In addition, to filter out high-frequency noise, the filtering process will also lose some high-frequency components (a contribution to the improvement of SSIM) of signal itself. Therefore, the latter two algorithms will reduce the value of SSIM for images with a lot of high-frequency components (SSIM value of graph (h) and (i) of Figure 3 is a little smaller than the corresponding value in Table 1), but for most images without lots of high-frequency information, the value of SSIM is improved.

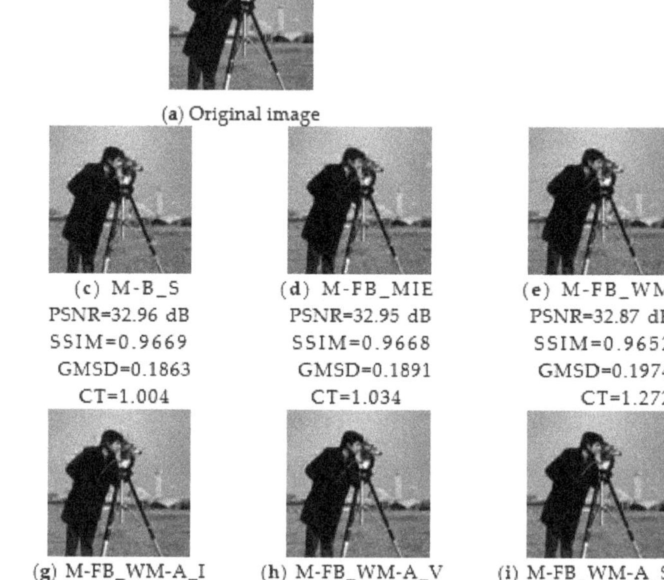

(b) M-B_C
PSNR=30.97 dB
SSIM=0.9550
GMSD=0.2096
CT=0.8980

(c) M-B_S
PSNR=32.96 dB
SSIM=0.9669
GMSD=0.1863
CT=1.004

(d) M-FB_MIE
PSNR=32.95 dB
SSIM=0.9668
GMSD=0.1891
CT=1.034

(e) M-FB_WM
PSNR=32.87 dB
SSIM=0.9652
GMSD=0.1974
CT=1.272

(f) M-B_C-A_I
PSNR=30.83 dB
SSIM=0.9559
GMSD=0.1998
CT=1.115

(g) M-FB_WM-A_I
PSNR=33.19 dB
SSIM=0.9727
GMSD=0.1845
CT=1.466

(h) M-FB_WM-A_V
PSNR=33.86 dB
SSIM=0.9793
GMSD=0.1290
CT=1.482

(i) M-FB_WM-A_S
PSNR=**34.21** dB
SSIM=**0.9865**
GMSD=**0.1225**
CT=1.671

Figure 3. Reconstructed images of Cameraman and performance indicators with different BCS algorithms (TSR = 0.5).

Secondly, in order to evaluate the effectiveness and universality of the proposed algorithm, in addition to the OMP reconstruction algorithm as the basic comparison algorithm, the IRLS and BP reconstruction algorithms were also adopted and combined with the proposed method to generate eight BCS algorithms, respectively. Table 2 shows the experimental data records of the above two types of algorithms tested with the standard image Lena. From the resulting data, the proposed method has a certain improvement for the BCS algorithm based on IRLS and BP, although it will bring a slightly higher cost in computation time due to the increase of the proposed algorithm's complexity itself.

Table 2. The PSNR and SSIM of reconstructed images with eight BCS algorithms based on iteratively reweighted least square (IRLS) and basis pursuit (BP).

Restoring Method	Algorithms	TSR = 0.4				TSR = 0.6			
		PSNR	SSIM	GMSD	CT	PSNR	SSIM	GMSD	CT
IRLS	R-B_C	32.38	0.9361	0.1943	5.729	33.42	0.9790	0.1348	13.97
	R-B_S	32.67	0.9634	0.1468	5.928	35.08	0.9843	0.0987	14.94
	R-FB_MIE	32.14	0.9593	0.1658	5.986	34.44	0.9825	0.1094	13.79
	R-FB_WM	32.46	0.9631	0.1441	6.071	34.80	0.9841	0.0992	14.25
	R-B_C-A_I	30.55	0.9460	0.1882	6.011	34.08	0.9825	0.1238	14.46
	R-FB_WM-A_I	33.05	0.9714	0.1346	6.507	36.01	0.9894	0.0863	14.98
	R-FB_WM-A_V	33.25	0.9751	0.1216	6.994	36.71	0.9914	0.0691	17.34
	R-FB_WM-A_S	**33.59**	**0.9787**	**0.1188**	7.456	**37.23**	**0.9927**	**0.0661**	19.38
BP	P-B_C	30.56	0.9380	0.1984	33.47	33.35	0.9787	0.1378	69.67
	P-B_S	32.72	0.9638	0.1484	34.22	34.61	0.9823	0.1072	71.00
	P-FB_MIE	32.00	0.9531	0.1627	34.04	34.14	0.9810	0.1149	68.78
	P-FB_WM	32.82	0.9635	0.1512	35.08	34.57	0.9832	0.1070	69.48
	P-B_C-A_I	30.72	0.9428	0.1973	33.84	33.57	0.9795	0.1335	70.27
	P-FB_WM-A_I	33.01	0.9705	0.1442	35.63	35.70	0.9884	0.0888	70.73
	P-FB_WM-A_V	33.32	0.9750	0.1277	36.98	36.37	0.9909	0.0742	72.59
	P-FB_WM-A_S	**33.49**	**0.9773**	**0.1210**	37.85	**37.45**	**0.9932**	**0.0662**	74.41

Furthermore, comparative experiments of the proposed algorithm combined with different image reconstruction algorithms (OMP, IRLS, BP, and SP) have also been carried out. Figure 4 is the data record of the above experiments tested with the standard image Lena under the conditions of TSR = 0.4 and TSR = 0.6, respectively. The experimental data shows that the proposal using these four algorithms has little difference between the PSNR and SSIM performance index. However, in terms of the GMSD index, the IRLS and BP algorithms are obviously better than the OMP and SP. In terms of calculation time, BP is based on the l_1 norm, its performance is significantly worse than the other three, which is also consistent with the content of Section 1.

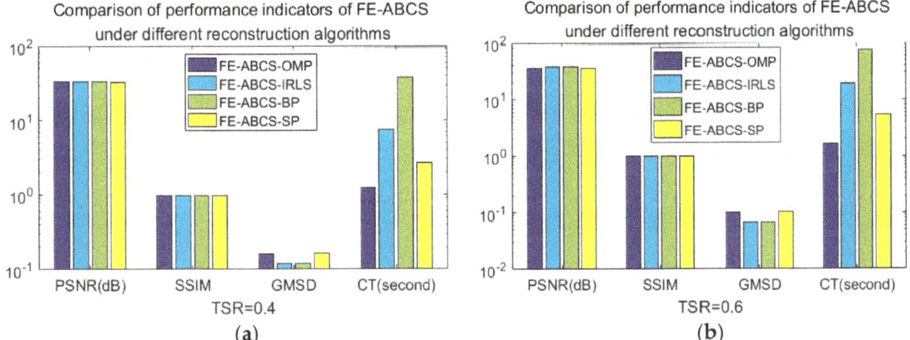

Figure 4. The comparison of the proposed algorithm with 4 reconstruction algorithms (OMP, IRLS, BP, SP): (a) TSR = 0.4, (b) TSR = 0.6.

5.1.2. Parametric Analysis of the Proposed Algorithm

The main points of this proposed algorithm involves the design and verification of weighting coefficients (c_{TS}, λ_1, λ_2) and minimum sampling rate factor (η_{min}). The design of the three weighting coefficients of the algorithm in this paper was introduced in Section 4.2, and its effect on performance was reflected in the comparison of the eight algorithms in Section 5.1.1. Here, only the selection and effect of η_{min} need to be researched, and the influence of η_{min} on the PSNR under different TSR is analyzed.

Figure 5 shows the analysis results of the test image Lena on the correlation between PSNR, TSR, and MSRF (η_{min}). It can be seen from Figure 5 that the optimal value (maximizing the PSNR of Lena's recovery image) of minimum sampling rate factor (OMSRF) decreases as the TSR increases. In addition, the gray value in Figure 5 means the revised PSNR of the recovery image ($PSNR^* = PSNR - \overline{PSNR}$).

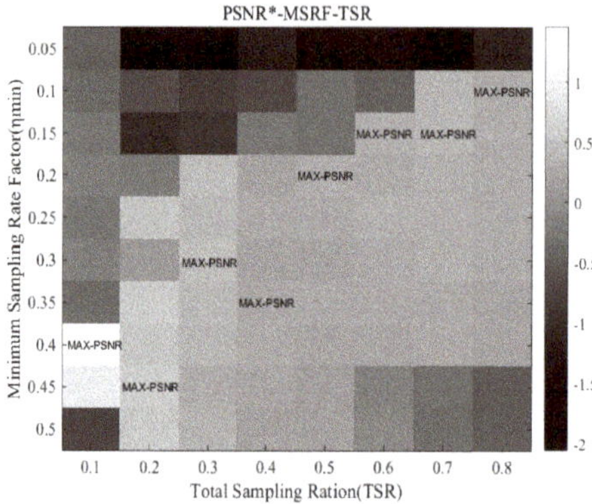

Figure 5. Correlation between PSNR*, MSRF, and TSR of test image Lena.

Then, many other test images were analyzed in this paper to verify the relationship between TSR and OMSRF ($\eta_{opt} = \left\{ \eta_{\min} \middle| \underset{\eta_{\min}}{\mathrm{argmax}}(PSNR(x, \eta_{\min})) \right\}$), and the experimental results of eight typical test images are shown in Figure 6. According to the data, the reasonable setting of MSRF (η_{opt}) in the algorithm can be obtained by the curve fitting method. The baseline fitting method (a simple curve method) is used in the proposed algorithm of this article ($\eta_{opt} = 0.1 + 6 \times (0.8 - TSR)/7$).

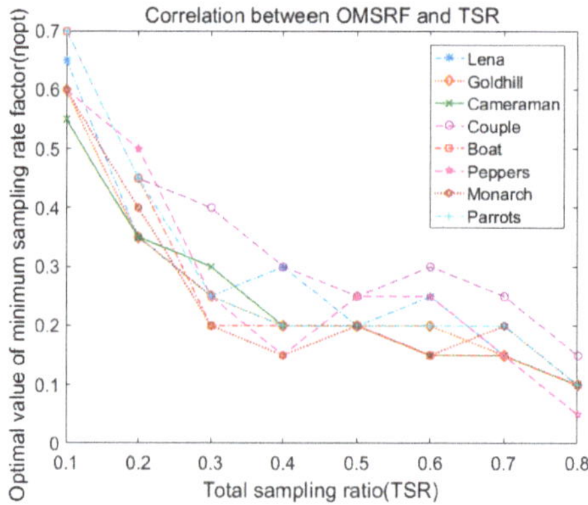

Figure 6. Correlation between OMSRF (η_{opt}) and TSR.

5.2. Experiment and Analysis Under Noisy Conditions

5.2.1. Effect Analysis of Different Iteration Stop Conditions on Performance

In the case of noiseless, the larger the iterative number (v) of the reconstruction algorithm, the better the effect of the reconstructed image. But in the noisy condition, the quality of the reconstructed image does not become monotonous with the increase of v, which has been carefully analyzed in Section 3.3. The usual iteration stop conditions are: (1) using the sparsity (ς) of signal as the stopping condition, i.e., fixed number of iterations ($v_{stop1} = \varsigma \cdot m$), and (2) using the certain differential threshold (γ) of the recovery value as the stopping condition, i.e., the difference between the adjacent two results of the iterative output less than the threshold ($v_{stop2} == \left\{ v \middle| \underset{v}{\mathrm{argmin}}\left(\|y_{v-1}^* - y_v^*\| \leq \gamma \right) \right\}$). Since the above two methods could not guarantee the optimal recovery of the original signal in the noisy background, the innovation of the FE-ABCS algorithm is to make up for the above deficiency and propose a constraint (v_{opt}) based on error analysis to ensure the best iterative reconstruction. Then the rationality of the proposed scheme would be verified through experiments in this section, and without loss of generality, OMP is used as the basic reconstruction algorithm, just like what was done in Section 5.1.

The specific experimental results of test image Lena for selecting different iteration stop conditions under different noise backgrounds are recorded in Table 3. The value of Noise-std represents the standard deviation of additional Gaussian noise signal. From the overall trend of Table 3, selecting v_{opt} has better performance than selecting v_{stop1} as the stop condition for iterative reconstruction. This advantage is especially pronounced as the Noise-std increases.

Table 3. The experimental results of Lena at different stop conditions and noise background (TSR = 0.4).

Noise-std	Sparsity	$v_{\varsigma=0.1}$	$v_{\varsigma=0.2}$	$v_{\varsigma=0.3}$	$v_{\varsigma=0.4}$	$v_{\varsigma=0.5}$	v_{opt}
		PSNR and SSIM and GMSD and BEI and CT					
5		32.95	32.71	32.39	32.12	31.96	32.48
		0.961	0.965	0.961	0.957	0.955	0.960
		0.165	0.160	0.169	0.173	0.178	0.171
		10.63	10.25	10.04	9.92	9.98	10.09
		0.741	0.949	1.166	1.567	1.911	1.481
10		32.40	31.81	31.30	31.07	30.92	32.23
		0.957	0.955	0.949	0.941	0.937	0.956
		0.169	0.179	0.190	0.194	0.200	0.177
		10.29	10.04	10.14	10.15	10.20	10.21
		0.741	0.922	1.233	1.637	1.961	0.926
15		31.63	30.87	30.46	30.27	30.12	31.81
		0.948	0.937	0.926	0.917	0.912	0.949
		0.180	0.202	0.213	0.220	0.223	0.185
		10.32	10.25	10.32	10.38	10.39	10.25
		0.727	0.933	1.154	1.550	2.063	0.798
20		30.97	30.08	29.83	29.66	29.61	31.48
		0.936	0.916	0.898	0.887	0.879	0.941
		0.197	0.219	0.236	0.244	0.247	0.191
		10.45	10.34	10.39	10.43	10.43	10.50
		0.720	0.914	1.203	1.476	2.348	0.739
30		30.03	29.34	29.04	28.94	28.89	30.75
		0.901	0.862	0.832	0.816	0.803	0.920
		0.227	0.257	0.266	0.272	0.275	0.204
		10.59	10.52	10.54	10.55	10.67	10.62
		0.901	0.947	1.237	1.500	1.996	0.690
40		29.38	28.77	28.57	28.50	28.45	30.14
		0.856	0.795	0.756	0.734	0.717	0.899
		0.252	0.277	0.286	0.290	0.291	0.221
		10.79	10.57	10.71	10.68	10.67	10.72
		0.736	0.791	1.169	1.534	2.047	0.678

In addition, in order to comprehensively evaluate the impact of different iteration stop conditions on the performance of reconstructed images, this paper combined the above three indicators to form a composite index PSB (PSB = PSNR × SSIM/BEI) for evaluating the quality of reconstructed images. The relationship between PSB and Noise-std of the reconstructed image under different iterations was researched in this article, so as to explore the relationship between PSB and TSR. Figure 7 shows the corresponding relationship between the PSB, Noise-std, and TSR under the above six different iteration stop conditions of Lena. It can be seen from Figure 7a that compared with the other five sparsity-based (ς) reconstruction algorithms, the v_{opt}-based error analysis reconstruction algorithm generally has relatively good performance under different noise backgrounds. Similarly, Figure 7b shows that the v_{opt}-based error analysis reconstruction algorithm has advantages over other algorithms at different total sampling rates.

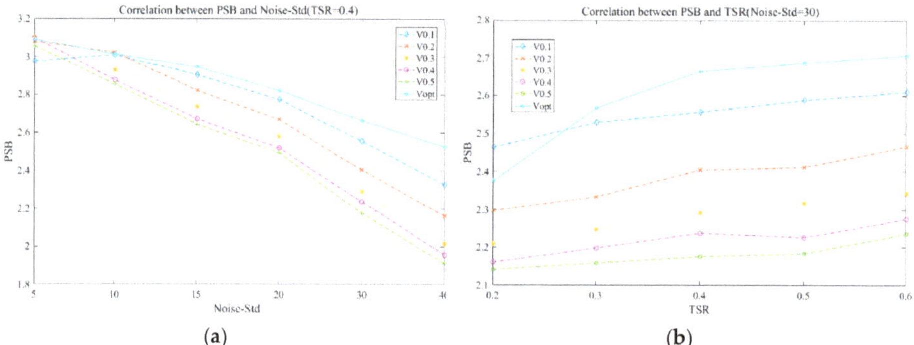

(a)   (b)

Figure 7. The correlation between the PSB, Noise-std, and TSR under the six different iteration stop conditions of Lena: (**a**) PSB changes with Noise-std, (**b**) PSB changes with TSR.

Furthermore, the differential threshold (γ)-based reconstruction algorithm and the v_{opt}-based error analysis reconstruction algorithm were compared in this article. Two standard test images and two real-nowadays images are adopted for the comparative experiment at the condition of Noise-std = 20 and TSR = 0.5. Experimental results show that the v_{opt}-based error analysis reconstruction algorithm has a significant advantage over the γ-based reconstruction algorithm in both PSNR and PSGBC (another composite index: PSGBC = PSNR × SSIM/GMSD/BEI/CT), which can be seen from Table 4, although there is a slight loss in BEI. Figure 8 shows the reconstruction images of these four images with differential threshold (γ) and error analysis (v_{opt}) as the iterative stop condition.

Table 4. The performance indexes of test images under different iterative stop condition.

Stop Condition	Images	Lena	Baboon	Flowers	Oriental Gate	Index
$\gamma = 300$		32.17	29.84	31.40	33.05	PSNR
		0.9562	0.8703	0.9744	0.9698	SSIM
		0.1661	0.2095	0.1812	0.1636	GMSD
		10.16	10.72	9.866	8.845	BEI
		0.8310	0.855 0	0.844 0	1.050	CT
		21.93	13.52	20.28	21.09	PSGBC
$\gamma = 1$		31.99	29.75	31.50	32.66	PSNR
		0.9574	0.8731	0.9768	0.9693	SSIM
		0.1638	0.1738	0.1571	0.1459	GMSD
		10.01	10.58	9.583	8.745	BEI
		2.905	2.704	2.823	3.578	CT
		6.429	5.224	7.240	6.934	PSGBC
v_{opt}		**32.75**	**30.03**	**31.78**	**33.15**	PSNR
		0.9603	0.8729	**0.9771**	0.9637	SSIM
		0.1471	0.1956	**0.1538**	0.1471	GMSD
		9.898	10.68	9.693	9.025	BEI
		0.7630	0.8900	0.8500	**0.8740**	CT
		28.31	**14.10**	**24.50**	**27.53**	PSGBC

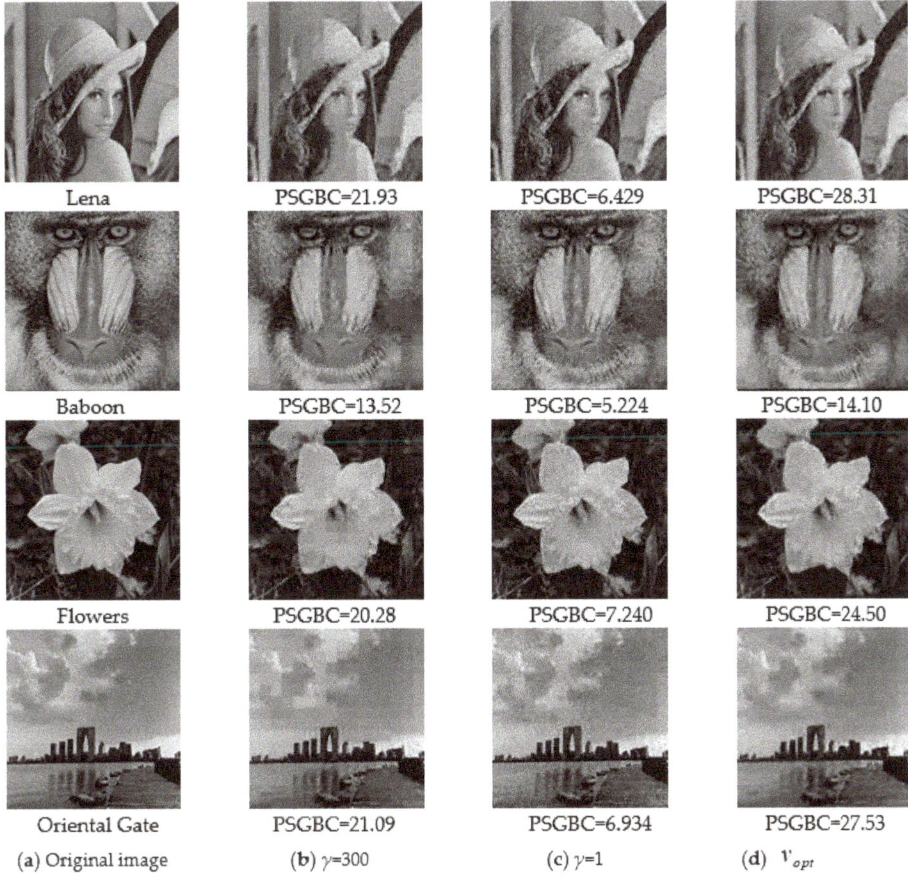

Lena	PSGBC=21.93	PSGBC=6.429	PSGBC=28.31
Baboon	PSGBC=13.52	PSGBC=5.224	PSGBC=14.10
Flowers	PSGBC=20.28	PSGBC=7.240	PSGBC=24.50
Oriental Gate	PSGBC=21.09	PSGBC=6.934	PSGBC=27.53
(a) Original image	(b) γ=300	(c) γ=1	(d) v_{opt}

Figure 8. Iterative reconstruction images based on γ and v_{opt} at the condition of Noise-std = 20 and TSR = 0.5.

5.2.2. Impact of Noise-Std and TSR on v_{opt}

Since v_{opt} is important to the proposed algorithm in this paper, it is necessary to analyze its influencing factors. According to Equation (44), v_{opt} is mainly determined by the measurement dimension of the signal and the added noise intensity under the BIC condition. In this section, the test image Lena is divided into 256 sub-images, and the relationship between the optimal iterative recovery stop condition (v_{opt}^{i}) of each sub-image, the TSR and the Noise-std is analyzed, and the experimental results are recorded in Figure 9. It can be seen from Figure 9a that the correlation between v_{opt} and TSR is small, but it can be seen from Figure 9b that v_{opt} has a strong correlation with Noise-std, that is, the larger the Noise-std, the smaller the v_{opt}.

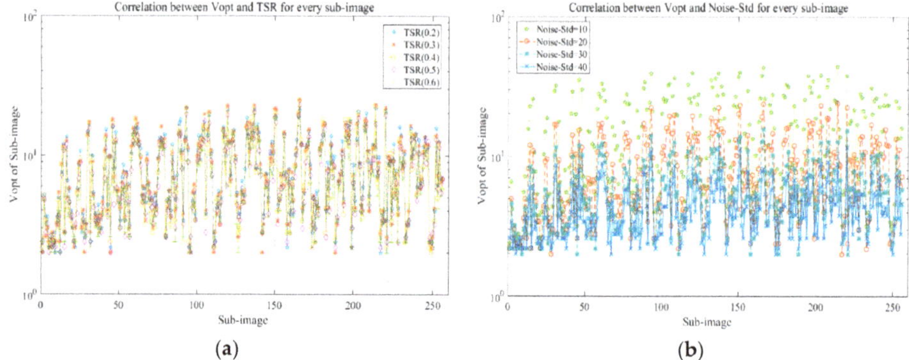

Figure 9. Correlation between v_{opt}, TSR, and Noise-std of sub-images: (**a**) Noise-std = 20, (**b**) TSR = 0.4.

5.3. Application and Comparison Experiment of FE-ABCS Algorithm in Image Compression

5.3.1. Application of FE-ABCS Algorithm in Image Compression

Although the FE-ABCS algorithm belongs to the CS theory which is mainly used for reconstruction of sparse images at low sampling rates, the algorithm can also be used for image compression after modification. The purpose of conventional image compression algorithms (such as JPEG, JPEG2000, TIFF, and PNG) is to reduce the amount of data and maintain a certain image quality through quantization and encoding. Therefore, the quantization and encoding module are added to the FE-ABCS in Figure 2b to form a new algorithm for image compression, which is shown in Figure 10 and named FE-ABCS-QC.

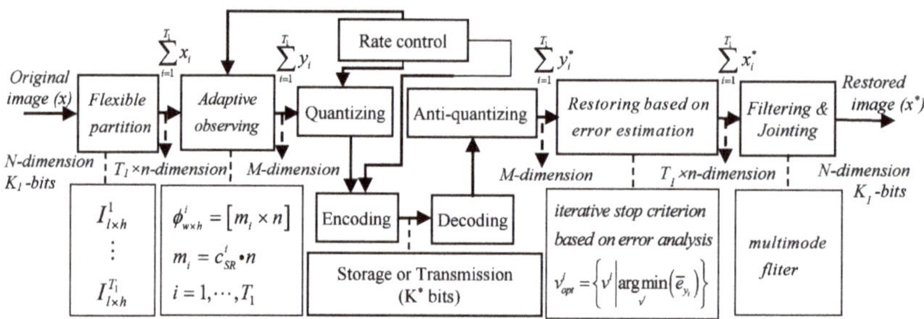

Figure 10. The workflow of the FE-ABCS-QC algorithm.

In order to demonstrate the difference between the proposed algorithm and the traditional image compression, without loss of generality, the JPEG2000 algorithm is selected as the comparison algorithm and is shown in Figure 11. Comparing Figures 10 and 11, it is found that the modules of FDWT and IDWT in the JPEG2000 algorithm are replaced by the observing module and the restoring module in the proposal respectively, and the dimensions of the input and output signals are both different in the observing and restoring module ($M < T_1 \times n = N$), that is different from the modules of FDWT and IDWT in which dimensions of the input and output signals are the same (both $T_1 \times n = N$). These differences make the proposed algorithm have a larger compression ratio (CR) and smaller bits per pixel (bpp) than JPEG2000 under the same quantization and encoding conditions.

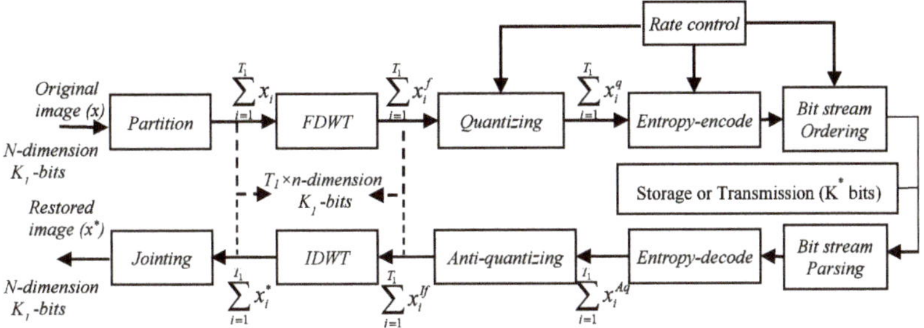

Figure 11. The workflow of the JPEG2000 algorithm.

5.3.2. Comparison Experiment between the Proposed Algorithm and the JPEG2000 Algorithm

In general, the evaluation of image compression algorithms is performed by rate-distortion performance. For the comparing of the FE-ABCS-QC and JPEG2000 algorithms, the indicators of PSNR, SSIM, and GMSD are adopted in this section. In addition, the definition of Rate (bpp) in the above two algorithms is as follows:

$$Rate = \frac{K^*}{N} \tag{56}$$

where, K^* is the number of bits in the code stream after encoding, and N is the number of pixels in the original image.

In order to compare the performance of the above two algorithms, multiple standard images are tested, and Table 5 records the complete experimental data for the two algorithms at various rates when using Lena and Monarch as test images. At the same time, the relationship between PSNR (used as the main distortion evaluation index) and the Rate of the two test images is illustrated in Figure 12. Based on the objective data of Table 5 and Figure 12, it can be seen that, compared with the JPEG2000 algorithm, the advantage of the FE-ABCS-QC algorithm becomes stronger with the increase of the rate, that is, at the small rate, the JPEG2000 algorithm is superior to the FE-ABCS-QC algorithm, while at medium and slightly larger rates, the JPEG2000 algorithm is not as good as the FE-ABCS-QC algorithm.

Table 5. The comparison results of different test-images under the various conditions (bits per pixel (bpp)) based on the JPEG2000 algorithm and the FE-ABCS-QC algorithm.

Test Image	Method	JPEG2000 (PSNR/SSIM/GMSD)	FE-ABCS-QC (PSNR/SSIM/GMSD)	ΔP/ΔS/ΔG
Lena	bpp = 0.0625	**31.64/0.9387/0.1842**	30.58/0.7341/0.2478	−1.06/−0.2046/0.0636
	bpp = 0.125	**33.38/0.9697/0.1399**	32.79/0.9413/0.1702	−0.59/−0.0284/0.0303
	bpp = 0.2	**34.59/0.9807/0.1161**	34.20/0.9710/0.1339	−0.39/−0.0097/0.0178
	bpp = 0.25	35.25/0.9850/0.0996	**37.80/0.9932/0.0612**	2.55/0.0082/−0.0384
	bpp = 0.3	35.73/0.9875/0.0917	**38.28/0.9941/0.0554**	2.55/0.0066/−0.0363
Monarch	bpp = 0.0625	**30.56/0.8184/0.2335**	27.52/0.3615/0.2726	−3.04/−0.4569/0.0391
	bpp = 0.125	**31.31/0.9074/0.1881**	29.12/0.6388/0.2568	−2.19/−0.2686/0.0687
	bpp = 0.2	32.49/0.9466/0.1554	**32.91/0.9473/0.1507**	0.42/0.0007/−0.0047
	bpp = 0.25	32.81/0.9572/0.1476	**35.77/0.9886/0.0682**	2.96/0.0314/−0.0794
	bpp = 0.3	33.40/0.9679/0.1305	**36.17/0.9896/0.0664**	2.77/0.0217/−0.0641

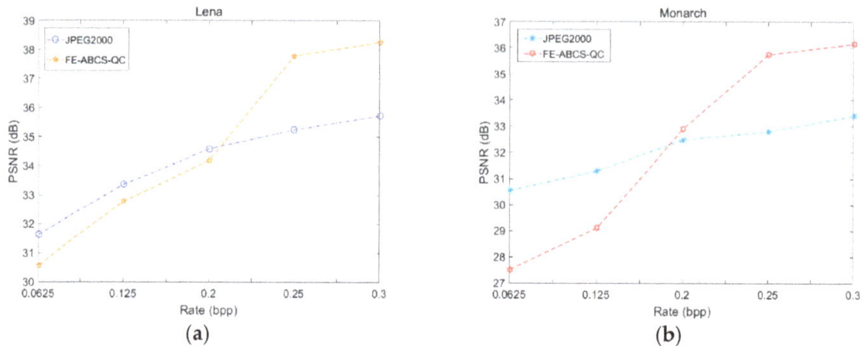

Figure 12. Rate-distortion performance for JPEG2000 and FE-ABCS-QC: (**a**) Lena, (**b**) Monarch.

Furthermore, the experiment results are recorded in the form of images in addition to the objective data comparison. Figure 13 shows the two algorithms' comparison of the compressed image restoration effects in the case of bpp = 0.25 when using Bikes as the test image. Comparing (**b**) and (**c**) of Figure 13, the image generated by the FE-ABCS-QC algorithm is slightly better than the one of the JPEG2000 algorithm, either from the perception of objective data or subjective sense.

(**a**) Original (**b**) JPEG2000 (**c**) FE-ABCS-QC

Figure 13. The two algorithms' comparison of test image Bikes at the condition of bpp = 0.25: (**a**) original image, (**b**) JPEG2000 image (PSNR = 29.80, SSIM = 0.9069, GMSD = 0.1964), (**c**) image by the FE-ABCS-QC algorithm (PSNR = 30.50, SSIM = 0.9366, GMSD = 0.1574).

Finally, the following conclusions could be gained by observing experimental data and theoretical analysis.

- Small Rate (bpp): the reason why the performance of the FE-ABCS-QC algorithm is worse than the JPEG2000 algorithm at this condition is that the small value of M which changes with Rate causes the observing process to fail to cover the overall information of the image.
- Medium or slightly larger Rate (bpp): the explanation for the phenomenon that the performance of the FE-ABCS-QC algorithm is better than the JPEG2000 algorithm in this situation is that the appropriate M can ensure the complete acquisition of image information and can also provide a certain image compression ratio to generate a better basis for quantization and encoding.
- Large Rate (bpp): this case of the FE-ABCS-QC algorithm is not considered because the algorithm belongs to the CS algorithm and requires M << N itself.

6. Conclusions

Based on the traditional block-compression sensing theory model, an improved algorithm (FE-ABCS) was proposed in this paper, and its overall workflow and key points were specified.

Compared with the traditional BCS algorithm, firstly, a flexible partition was adopted in order to improve the rationality of partitioning in the proposed algorithm, secondly the synthetic feature was used to provide a more reasonable adaptive sampling basis for each sub-image block, and finally error analysis was added in the iterative reconstruction process to achieve minimum error between the reconstructed signal and the original signal in the noisy background. The experimental results show that the proposed algorithm can improve the image quality in both noiseless and noisy backgrounds, especially in the improvement of a reconstructed image's composite index under a noisy background, and will be beneficial to the practical application of the BCS algorithm, and the application of the FE-ABCS algorithm in image compression.

Author Contributions: Conceptualization, methodology, software, validation and writing, Y.Z.; data curation and visualization, Q.S. and Y.Z.; formal analysis, W.L. and Y.Z. supervision, project administration and funding acquisition, W.L.

Funding: This work was supported by the National Natural Science Foundation of China (61471191).

References

1. Donoho, D.L. Compressed sensing. *IEEE Trans. Inf. Theory* **2006**, *52*, 1289–1306. [CrossRef]
2. Candès, E.J.; Wakin, M.B. An introduction to compressive sampling. *IEEE Signal Process. Mag.* **2008**, *25*, 21–30. [CrossRef]
3. Shi, G.; Liu, D.; Gao, D. Advances in theory and application of compressed sensing. *Acta Electron. Sin.* **2009**, *37*, 1070–1081.
4. Sun, Y.; Xiao, L.; Wei, Z. Representations of images by a multi-component Gabor perception dictionary. *Acta Electron. Sin.* **2008**, *34*, 1379–1387.
5. Xu, J.; Zhang, Z. Self-adaptive image sparse representation algorithm based on clustering and its application. *Acta Photonica Sin.* **2011**, *40*, 316–320.
6. Wang, G.; Niu, M.; Gao, J.; Fu, F. Deterministic constructions of compressed sensing matrices based on affine singular linear space over finite fields. *IEICE Trans. Fundam. Electron. Commun. Comput. Sci.* **2018**, *101*, 1957–1963. [CrossRef]
7. Li, S.; Wei, D. A survey on compressive sensing. *Acta Autom. Sin.* **2009**, *35*, 1369–1377. [CrossRef]
8. Palangi, H.; Ward, R.; Deng, L. Distributed compressive sensing: A deep learning approach. *IEEE Trans. Signal Process.* **2016**, *64*, 4504–4518. [CrossRef]
9. Chen, C.; He, L.; Li, H.; Huang, J. Fast iteratively reweighted least squares algorithms for analysis-based sparse reconstruction. *Med. Image Anal.* **2018**, *49*, 141–152. [CrossRef]
10. Gan, L. Block compressed sensing of natural images. In Proceedings of the 15th International Conference on Digital Signal Processing, Cardiff, UK, 1–4 July 2007; pp. 403–406.
11. Unde, A.S.; Deepthi, P.P. Fast BCS-FOCUSS and DBCS-FOCUSS with augmented Lagrangian and minimum residual methods. *J. Vis. Commun. Image Represent.* **2018**, *52*, 92–100. [CrossRef]
12. Kim, S.; Yun, U.; Jang, J.; Seo, G.; Kang, J.; Lee, H.N.; Lee, M. Reduced computational complexity orthogonal matching pursuit using a novel partitioned inversion technique for compressive sensing. *Electronics* **2018**, *7*, 206. [CrossRef]
13. Qi, R.; Yang, D.; Zhang, Y.; Li, H. On recovery of block sparse signals via block generalized orthogonal matching pursuit. *Signal Process.* **2018**, *153*, 34–46. [CrossRef]
14. Figueiredo, M.A.T.; Nowak, R.D.; Wright, S.J. Gradient projection for sparse reconstruction: Application to compressed sensing and other inverse problems. *IEEE J. Sel. Areas Commun.* **2007**, *1*, 586–597. [CrossRef]
15. Lotfi, M.; Vidyasagar, M. A fast noniterative algorithm for compressive sensing using binary measurement matrices. *IEEE Trans. Signal Process.* **2018**, *66*, 4079–4089. [CrossRef]
16. Yang, J.; Zhang, Y. Alternating direction algorithms for l_1 problems in compressive sensing. *SIAM J. Sci. Comput.* **2011**, *33*, 250–278. [CrossRef]
17. Yin, H.; Liu, Z.; Chai, Y.; Jiao, X. Survey of compressed sensing. *Control Decis.* **2013**, *28*, 1441–1445.
18. Dinh, K.Q.; Jeon, B. Iterative weighted recovery for block-based compressive sensing of image/video at a low subrate. *IEEE Trans. Circ. Syst. Video Technol.* **2017**, *27*, 2294–2308. [CrossRef]

19. Liu, L.; Xie, Z.; Yang, C. A novel iterative thresholding algorithm based on plug-and-play priors for compressive sampling. *Future Internet* **2017**, *9*, 24.

20. Wang, Y.; Wang, J.; Xu, Z. Restricted p-isometry properties of nonconvex block-sparse compressed sensing. *Signal Process.* **2014**, *104*, 1188–1196. [CrossRef]

21. Mahdi, S.; Tohid, Y.R.; Mohammad, A.T.; Amir, R.; Azam, K. Block sparse signal recovery in compressed sensing: Optimum active block selection and within-block sparsity order estimation. *Circuits Syst. Signal Process.* **2018**, *37*, 1649–1668.

22. Wang, R.; Jiao, L.; Liu, F.; Yang, S. Block-based adaptive compressed sensing of image using texture information. *Acta Electron. Sin.* **2013**, *41*, 1506–1514.

23. Amit, S.U.; Deepthi, P.P. Block compressive sensing: Individual and joint reconstruction of correlated images. *J. Vis. Commun. Image Represent.* **2017**, *44*, 187–197.

24. Liu, Q.; Wei, Q.; Miao, X.J. Blocked image compression and reconstruction algorithm based on compressed sensing. *Sci. Sin.* **2014**, *44*, 1036–1047.

25. Wang, H.L.; Wang, S.; Liu, W.Y. An overview of compressed sensing implementation and application. *J. Detect. Control* **2014**, *36*, 53–61.

26. Xiao, D.; Xin, C.; Zhang, T.; Zhu, H.; Li, X. Saliency texture structure descriptor and its application in pedestrian detection. *J. Softw.* **2014**, *25*, 675–689.

27. Haralick, R.M.; Shanmugam, K.; Dinstein, I.H. Texture features for image classification. *IEEE Trans. Syst. Man Cybern.* **1973**, *3*, 610–621. [CrossRef]

28. Cao, Y.; Bai, S.; Cao, M. Image compression sampling based on adaptive block compressed sensing. *J. Image Graph.* **2016**, *21*, 416–424.

29. Shen, J. Weber's law and weberized TV restoration. *Phys. D Nonlinear Phenom.* **2003**, *175*, 241–251. [CrossRef]

30. Li, R.; Cheng, Y.; Li, L.; Chang, L. An adaptive blocking compression sensing for image compression. *J. Zhejiang Univ. Technol.* **2018**, *46*, 392–395.

31. Liu, H.; Wang, C.; Chen, Y. FBG spectral compression and reconstruction method based on segmented adaptive sampling compressed sensing. *Chin. J. Lasers* **2018**, *45*, 279–286.

32. Li, R.; Gan, Z.; Zhu, X. Smoothed projected Landweber image compressed sensing reconstruction using hard thresholding based on principal components analysis. *J. Image Graph.* **2013**, *18*, 504–514.

33. Gershgorin, S.; Donoho, D.L. Ueber die Abgrenzung der Eigenwerte einer Matrix. *Izv. Akad. Nauk. SSSR Ser. Math.* **1931**, *1*, 749–754.

34. Beheshti, S.; Dahleh, M.A. Noisy data and impulse response estimation. *IEEE Trans. Signal Process.* **2010**, *58*, 510–521. [CrossRef]

35. Beheshti, S.; Dahleh, M.A. A new information-theoretic approach to signal denoising and best basis selection. *IEEE Trans. Signal Process.* **2005**, *53*, 3613–3624. [CrossRef]

36. Bottcher, A. Orthogonal symmetric Toeplitz matrices. *Complex Anal. Oper. Theory* **2008**, *2*, 285–298. [CrossRef]

37. Duan, G.; Hu, W.; Wang, J. Research on the natural image super-resolution reconstruction algorithm based on compressive perception theory and deep learning model. *Neurocomputing* **2016**, *208*, 117–126. [CrossRef]

Article

Wiener–Granger Causality Theory Supported by a Genetic Algorithm to Characterize Natural Scenery

César Benavides-Álvarez [1,*,†], Juan Villegas-Cortez [2,†], Graciela Román-Alonso [1,†] and Carlos Avilés-Cruz [2,†]

[1] Electrical Engineering Department, Autonomous Metropolitan University Iztapalapa, Av. San Rafael Atlixco 186, Leyes de Reforma 1ra Secc, Mexico City 09340, Mexico
[2] Electronics Department, Autonomous Metropolitan University Azcapotzalco, Av. San Pablo 180, Col. Reynosa, C.P., Mexico City 02200, Mexico
* Correspondence: cesarba@xanum.uam.mx; Tel.: +52-553-114-9579
† These authors contributed equally to this work.

Received: 9 May 2019; Accepted: 21 June 2019; Published: 26 June 2019

Abstract: Image recognition and classification have been widely used for research in computer vision systems. This paper aims to implement a new strategy called Wiener-Granger Causality theory for classifying natural scenery images. This strategy is based on self-content images extracted using a Content-Based Image Retrieval (CBIR) methodology (to obtain different texture features); later, a Genetic Algorithm (GA) is implemented to select the most relevant natural elements from the images which share similar causality patterns. The proposed method is comprised of a sequential feature extraction stage, a time series conformation task, a causality estimation phase, causality feature selection throughout the GA implementation (using the classification process into the fitness function). A classification stage was implemented and 700 images of natural scenery were used for validating the results. Tested in the distribution system implementation, the technical efficiency of the developed system is 100% and 96% for resubstitution and cross-validation methodologies, respectively. This proposal could help with recognizing natural scenarios in the navigation of an autonomous car or possibly a drone, being an important element in the safety of autonomous vehicles navigation.

Keywords: classification; content–based image retrieval; genetic algorithms; image retrieval; image classification; Wiener-Granger causality

---

1. Introduction

One of the challenges researchers face today is developing an artificial authentication system that has acquisition and processing capabilities similar to those possessed by humans [1]. Artificial vision is defined as the capacity of a machine to see the world that surrounds it in a 3-Dimensional form starting from a group of 2-Dimensional images [2]. Since there is no effective algorithm that can fully recognize any object one can imagine in the entire environment, computer vision is considered an open problem. A computer vision system is composed of different stages that work together for solving a particular problem [3].

Automatic image recognition is among the problems that might be solved using computer vision systems. Researchers are eager to develop these systems and different techniques have been implemented for their improvement, such as machine learning, pattern recognition and evolutionary algorithms.

One of the tasks of an automated image recognition system is to successfully classify and identify natural scenery images (It is said that a scene is natural if the image has no intervention or alteration by human hands). Currently, thousands of images are generated via different kinds of sources on a daily basis and the constant increase of the Internet has influenced human life.

More than half of the information on the Internet is images, 85% of which were taken with mobile devices with a final estimation of 5 trillion images reported so far [4].

In order to use this information efficiently, an image recovery system based on Content-Based Image Retrieval (CBIR) is necessary. It will help users to find relevant images based on their self-content features or those which are "seen" to e related to them, from our visual perception, even when there is no previous knowledge of the database, such as manual labeling of the images.

Our previous work successfully applied the CBIR technique to the face recognition problem [5,6]. The multiple textures, objects in unknown positions and their different compositions in natural scenery images challenge the proposals that combine different techniques for obtaining a better performance of natural scenery image classification. In this work, we use CBIR feature extraction as an input of a texture causality engine to characterize 5 scenery types, manually defining a base dictionary conformed by 4 textures. In future work, conforming this dictionary is planned to be dynamical, considering more base textures and scenery types to improve classification performance.

In this work, an image retrieval system of natural scenery images is developed by applying the Wiener-Granger Causality (WGC) theory [7] as a tool for analyzing images throughout self-content information. The causal relationships between local textures contained in an image were identified, leading to characterization of a descriptive pattern of a set of scenes inside an image dataset. The selection of causality relationships was carried out using genetic algorithm (GA) implementation as an evolutionary process.

The major stages involved in the developed system are the following (See Figure 1):

1. Scenery reading: First, images are read from the data set and then a change of space color format is applied from Red-Green-Blue RGB to Hue-Saturation-Intensity HSI.
2. Feature extraction: The statistical CBIR feature extraction is generated within a neighborhood in a grid.
3. Time series conformation: The texture features are organized as a time series for each image.
4. Causality analysis: The WGC analysis is applied to calculate the causal relationship matrix among different textures.
5. Genetic Algorithm (GA) implementation: GA is executed to find the characterization of causality relationships that perform better for natural element retrieval of the images that have similar causality texture patterns for a particular scene.

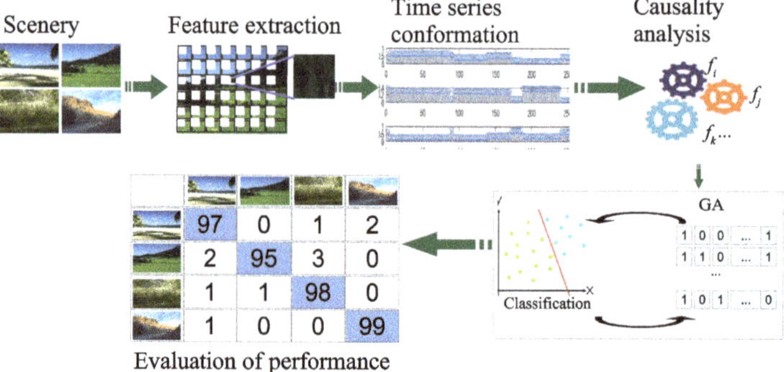

Figure 1. Proposed general methodology applied for image recognition.

The paper proposes a causality analysis of the natural scenery classes based on a pre-established texture dictionary and the WGC analysis from the CBIR methodology [5,8] in order to provide a whole dataset characterization.

This approach aims to improve the optimization process of evolutionary algorithms. In this case, since the GA [9] shows a simple and fast implementation, it was employed to select the relationships of the local-texture statistical features handled as time series.

Finally, an improvement in the classification accuracy obtained by our proposed strategy is reported, getting 100% on re-substitution and up to 96% for cross-validation methodologies. This approach was implemented using the computer power of a 19-processor cluster and the MPI parallel programming tool.

The current methodology was probed with two databases of natural scenery:

- Vogel and Shiele (V_S) [10], with 700 Images classified as: 144 coast, 103 forest, 179 mountain, 131 prairie, 111 river/lake and 32 sky/cloud.
- Oliva and Torralba (O_T) [11], with 1472 Images classified as: 360 coast, 328 forest, 374 mountain and 410 prairie.

Visualizing the future implementation of an autonomous system of recognition of natural scenes mounted on a car—which will be managed by our proposal as the autonomous system [12–14]—recognizing natural scenarios in the navigation of an autonomous car or possibly a drone, with a 100% certainty, this proposed system will be an important element in the safety of autonomous vehicles.

The rest of the paper is organized as follows: Section 2 presents the state of the art of the problem of image analysis from the CBIR criterion and the WGC theory used in our project, as well as the theoretical support of the WGC model to be applied; in Section 3, the proposed methodology for applying the WGC theory in the natural scenery image characterization is presented; in Section 4 our GA implementation approach to optimize the selection of texture causality relationships is explained; the parallel implementation of our proposal to get good efficiency when processing a large number of images is provided in Section 5; finally, the results and conclusions are presented in Sections 6 and 8, respectively.

2. State of The Art

The problem of image classification and recognition has been studied with different approaches for supporting visual search for different purposes.

Several techniques have been applied successfully to the face recognition problem [6,15–18]. The solutions are favored by controlling the way in which the images are obtained by determining the amount of light, the orientation, the distance, and so forth, in order to obtain ideal face images. In addition, the points to be identified on a face image are well known. The multiple textures, objects in unknown positions and their different compositions make it quite difficult to recognize and identify natural scenery in an image or group of images.

One of the most recent solutions for the classification of natural scenery is the use of the deep learning technique [19], which consists of a set of neural networks connected with each other in successive layers, where each layer network performs a convolution operation on the information of the previous layer, as we can see in Reference [20]. This methodology has the disadvantage of requiring high-end computational resources (memory and CPU) for the training task, unlike the CBIR technique which can be implemented in systems with few resources.

When using CBIR for scenery image classification, significant descriptors are determined considering the image self texture attributes to have an important and effective recovery. In this system, a user presents an image query and the system returns similar images from the database. In Figure 2, the general diagram of a CBIR-based classification system of natural scenery images is shown.

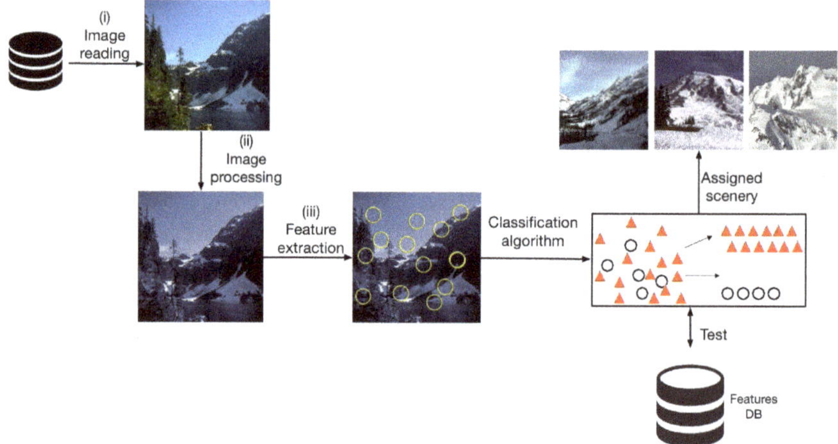

Figure 2. Classical methodology of image classification.

One of the first papers that uses the CBIR methodology for natural scenery classification is that by J. Vogel [21]; this work defines a regular 10×10 grid on the image; from each grid coordinate an analysis window is opened. Local information is extracted from a window texture and compared with a base texture dictionary; then, the author defines a classification system for natural scenery. A point to be improved is the definition of the base texture dictionary that is set manually, including only typical textures perceived by the researchers. This approach obtains up to 75% average for the cross validation classification test.

Unlike a grid, in Reference [8], random points are thrown on the image and around each point a window is opened; from each window, statistical texture local information is extracted to be grouped, conforming dynamically to a base texture dictionary. The testing is performed considering the generated dictionary obtaining an 85% classification average of natural scenery data bases.

In Reference [22], the CBIR approach is presented to classify natural scenery images through the composition of relevant features in relation to the texture, like in Reference [23], the shape and distribution of the luminosity.

CBIR, being an unsupervised learning technique, still has some disadvantages, since the information extracted is only treated as a histogram that represents the composition of textures in a scenery. This way of characterizing scenery has not been able to obtain more than 85% classification, that is why new proposals that use hybrid methodologies to give CBIR greater robustness arise [24]

In References [24,25], the authors combine the CBIR information with certain semantic content introducing high-level concept objects, trying to link content-based images to objects extracted inside them. This work obtained a percentage of natural scenery classification not greater than that of References [21] and [8].

In this work, a hybrid method of three components is presented. The basic component is CBIR, which generates the information regarding the local texture features of the image. Unlike performing only statistical management of the obtained features (using histograms), in this proposal the second component is responsible for applying a causality technique based on the Wiener Granger causality theory to identify the causal relationships that exist within the basic textures of a type of scenery. Since the causality component generates different configurations of causal relationships, the third component consists of a GA that allows the selection of the configuration that obtains the best classification percentage for each scenery.

Evolutionary Computational Vision (ECV) as a research area is currently growing in artificial intelligence through two areas of work—computational vision and evolutionary computation. Beginning from a practical point of view, ECV seeks to design the software and hardware solutions necessary to solve hard computer vision problems [1]. Bio-inspired computation within computational vision contains a set of techniques that are frequently applied to hard optimization problems. Its chief objective is to generate solutions formulated in a synthetic way and the artificial evolutionary process based on the evolutionary theory developed by Charles Darwin is the one frequently applied in Reference [9].

2.1. Theoretical Fundamentals Of Wgc

The causal inference paradigm has been used in different fields of science, for example, in neurology the WGC theory [26] is used to examine areas of the brain and the causal relationships among them. WGC analysis was carried out using sensors [27,28], and, lately in MRI images [29–31], the WGC theory is being used for the study of causal relationships among areas of the brain. Other science fields where WGC theory has been applied is video processing for indexing and retrieval [32]. Video processing for massive people and vehicle identification [33–35] and complex scenery analysis [36]. In this proposal, for the first time, WGC theory is applied to a natural elements and natural scenes retrieval.

In this section, the theoretical framework of the WGC is established. For simplicity and in order to avoid extending mathematically, the theory is presented only for three random processes, being extendable to $n-$processes. In our approach, a random process corresponds to a signal reading associated to one type of texture within a natural scenery; so, for the present analysis, each texture reading corresponds to one stochastic process represented by Ti, being i the i-th texture which has a stochastic behavior disposed into a scenery.

2.2. Stochastic Autoregressive Model

We assume that each texture can be represented by an autoregressive model into time series. In the current analysis, we will only carry out with three signals, $\{T1, T2,$ and $T3\}$, being easily extendable to n signals/textures. Let $T1$, $T2$, and $T3$ be three stochastic processes, individually and jointly stationary. Each stationary process can be represented by an autoregressive model in the following way:

$$T1(t) = \sum_{k=1}^{\infty} C_{T1}^1(k)T1(t-k) + \eta_{T1}^1, \quad \text{with} \quad \Sigma_{T1}^1 = \text{var}(\eta_{T1}^1), \tag{1}$$

$$T2(t) = \sum_{k=1}^{\infty} C_{T2}^1(k)T2(t-k) + \eta_{T2}^1, \quad \text{with} \quad \Sigma_{T2}^1 = \text{var}(\eta_{T2}^1), \tag{2}$$

$$T3(t) = \sum_{k=1}^{\infty} C_{T3}^1(k)T3(t-k) + \eta_{T3}^1, \quad \text{with} \quad \Sigma_{T3}^1 = \text{var}(\eta_{T3}^1), \tag{3}$$

being η_{T1}^1, η_{T2}^1 and η_{T3}^1 random Gaussian noise with zero mean and unit standard deviation; $C_{T1}^1(k)$, $C_{T2}^1(k)$ and $C_{T3}^1(k)$ are the coefficients of the regression model for textures $T1$, $T2$ and $T3$, respectively.

The joint autoregressive model for the three textures is defined by the equations:

$$T1(t) = \sum_{k=1}^{\infty} C_{T1}^{1,1}(k)T1(t-k) + \sum_{k=1}^{\infty} C_{T2}^{1,2}(k)T2(t-k) + \sum_{k=1}^{\infty} C_{T3}^{1,3}(k)T3(t-k) + \eta_{T1}^2,$$

$$\text{with} \ \Sigma_{T1}^2 = \text{var}(\eta_{T1}^2) \tag{4}$$

$$T2(t) = \sum_{k=1}^{\infty} C_{T1}^{2,1}(k)T1(t-k) + \sum_{k=1}^{\infty} C_{T2}^{2,2}(k)T2(t-k) + \sum_{k=1}^{\infty} C_{T3}^{2,3}(k)T3(t-k) + \eta_{T2}^2,$$

$$\text{with } \sum_{T2}^2 = \text{var}(\eta_{T2}^2) \quad (5)$$

$$T3(t) = \sum_{k=1}^{\infty} C_{T1}^{3,1}(k)T1(t-k) + \sum_{k=1}^{\infty} C_{T2}^{3,2}(k)T2(t-k) + \sum_{k=1}^{\infty} C_{T3}^{3,3}(k)T3(t-k) + \eta_{T3}^2,$$

$$\text{with } \Sigma_{T3}^2 = \text{var}(\eta_{T3}^2) \quad (6)$$

where Σ_{T1}^2, Σ_{T2}^2 and Σ_{T3}^2 are the variance of the residual terms η_{T1}^2, η_{T2}^2 and η_{T3}^2, respectively. On the other hand, the terms $C_{Tl}^{i,j}(k)$ $\forall i,j,l \in [1,2,3]$, are the regression coefficients for textures $T1(t)$, $T2(t)$ and $T3(t)$, respectively.

Now let us analyze the variances/covariances of the residual terms η_{Ti}^2 by means of the following Σ matrix form Equation (7):

$$\Sigma = \begin{pmatrix} \Sigma_{T1}^2 & Y_{1,2} & Y_{1,3} \\ Y_{2,1} & \Sigma_{T2}^2 & Y_{2,3} \\ Y_{3,1} & Y_{3,2} & \Sigma_{T3}^2 \end{pmatrix} \quad (7)$$

where $Y_{1,2}$ is the covariance between η_{T1}^2 and η_{T2}^2 (defined as $Y_{1,2} = cov(\eta_{T1}^2, \eta_{T2}^2)$); $Y_{1,3}$ is the covariance between η_{T1}^2 and η_{T3}^2 (defined as $Y_{1,3} = cov(\eta_{T1}^2, \eta_{T3}^2)$), and so on.

Based on the earlier conditions and using the concept of statistical independence between two random processes at the same time (in pairs), causality can be defined in time. An example of the causality between $T1$ and $T2$ is as in the following expression:

$$F_{T2,T1} = \ln \left[\frac{\Sigma_{T1}^1 \times \Sigma_{T2}^1}{\Sigma_{T1^2} \times \Sigma_{T2}^2} \right] \quad (8)$$

The Equation (8) is commonly known as the causality in the time domain. From this equation, if the random processes $T1(t)$ and $T2(t)$ are statistically independent, then $F_{T1,T2} = 0$; otherwise there will be causality from one to another.

In the Equation (1), Σ_{T1}^1 measures the precision of the autoregressive model to predict $T1(t)$, established on the past samples.

Then again, Σ_{T1}^2 in the expression (4) measures the precision to predict $T1(t)$ based on the previous values of $T1(t)$, $T2(t)$ and $T3(t)$ at the same time. Returning to the case of taking only 2 textures at the same time $T1(t)$ and $T2(t)$ and according to References [37] and [7], if $\Sigma_{T2}^2 < \Sigma_{T1}^1$ then it is said that $T2(t)$ has a causal influence on $T1(t)$. The causality is defined by the following equation:

$$F_{T2 \to T1} = \ln \left[\frac{\Sigma_{T1}^1}{\Sigma_{T1}^2} \right] \quad (9)$$

It is relatively easy to see that if $F_{T2 \to T1} = 0$ then there is no causal influence from $T2(t)$ towards $T1(t)$, at any other values, the result will be otherwise. On the other hand, the causal influence of $T1(t)$ towards $T2(t)$ is established using the following equation:

$$F_{T1 \to T2} = \ln \left[\frac{\Sigma_{T2}^1}{\Sigma_{T2}^2} \right] \quad (10)$$

3. Methodology

In the current section, we describe the methodology developed for the WGC technique with a GA support applied to natural scenery.

For the use of the CBIR, there are different determining factors that must be taken into account while extracting the information from the images, such as luminosity, orientation, scale, homogeneity, and so forth. The main characteristic in our proposed patterns is texture, such that we try to create a base dictionary to later create the time series from the reading of the images and their comparison with the dictionary, with which the theory of WGC was applied.

For the development of the dictionary, a set of k textures are manually selected on the images to be studied, which we will call *reference textures*. The k generated textures represent parts of objects such as the sky, clouds, grass, rock, and so forth, trying to make a manual segmentation of the scenery as shown in Figure 3. In Section 6, the $k = 4$ textures test is shown for 6 scenery-classes.

Once the set of the k reference textures has been obtained, the values in the HSI color space of each of them are examined to create a range of maximum and minimum values which represent them, these values help us to define the thresholds of comparison for the test textures of a query image.

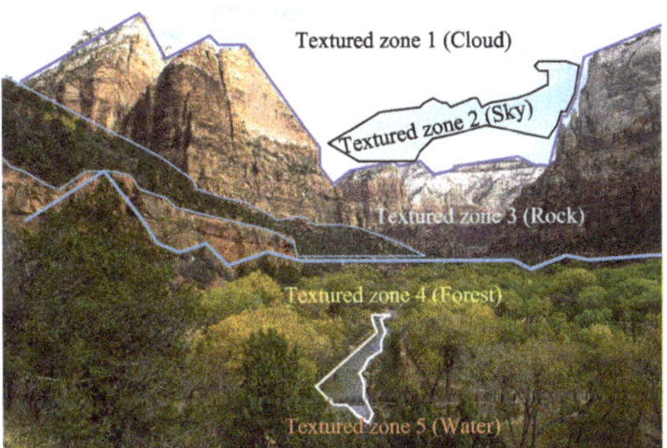

Figure 3. Example of segmentation texture zone in a natural scene.

The proposed methodology for the identification and classification of scenery by WGC is shown in Figure 4. The blocks of the architecture are described below.

1. Natural Scenery Database (NSDB). Represents the set of images to be analyzed, it contains the images of the natural scenery.
2. Reading the images. Is responsible for obtaining the images from the database, which will be processed in (Red-Green-Blue) RGB color format.
3. Pre-processing. Pre-processing of the images, erasing the noise, to be used in the next step.
4. Change to the (Hue-Saturation-Intensity) HSI color space. The RGB color space does not give us the necessary information for the feature extraction, therefore we pass them/it to the HSI color space, which gives us the information related to the texture.
5. Feature extraction. This block consists of three important stages:

   - Grid image. The work done in Reference [21] is taken as a reference, a regular grid of 10×10 windows is considered for the CBIR texture analysis; in our proposal, we use a grid of $r \times c$ windows, which has the property of $r \neq c$, where $c :=$*number of windows in horizontal (columns)*, and $r :=$*number of windows in vertical (lines)*.

- Neighborhood construction. In each of the resulting frames of the grid, the size of the neighborhood $p \times p$ pixels is extracted, starting from the top left corner of each window, as shown in Figure 5, such that $p < r$ and $p < c$.
- CBIR feature extraction. The image is read from the neighborhoods in the following way: It starts in the top left corner of the image and it moves following a descending vertical order through the neighborhoods, processing each of them. Once it reaches the last line, it moves one step to the right neighborhood and goes up to the first line; when the first line is attended again, it moves to the right column within the neighborhood and goes down again (like a snake moving), this reading is repeated for the entire image until the last neighborhood is reached, as shown in Figure 6. Each neighborhood section creates a pattern of size 1×3, that is, one feature per channel (HSI) of the image. After the feature extraction of all the neighborhoods was read in the established order, a matrix M_s^i with size $w \times 3$ is created, where $w = r * c$ is the number of neighborhoods analyzed for each i-th image of the class C_s.

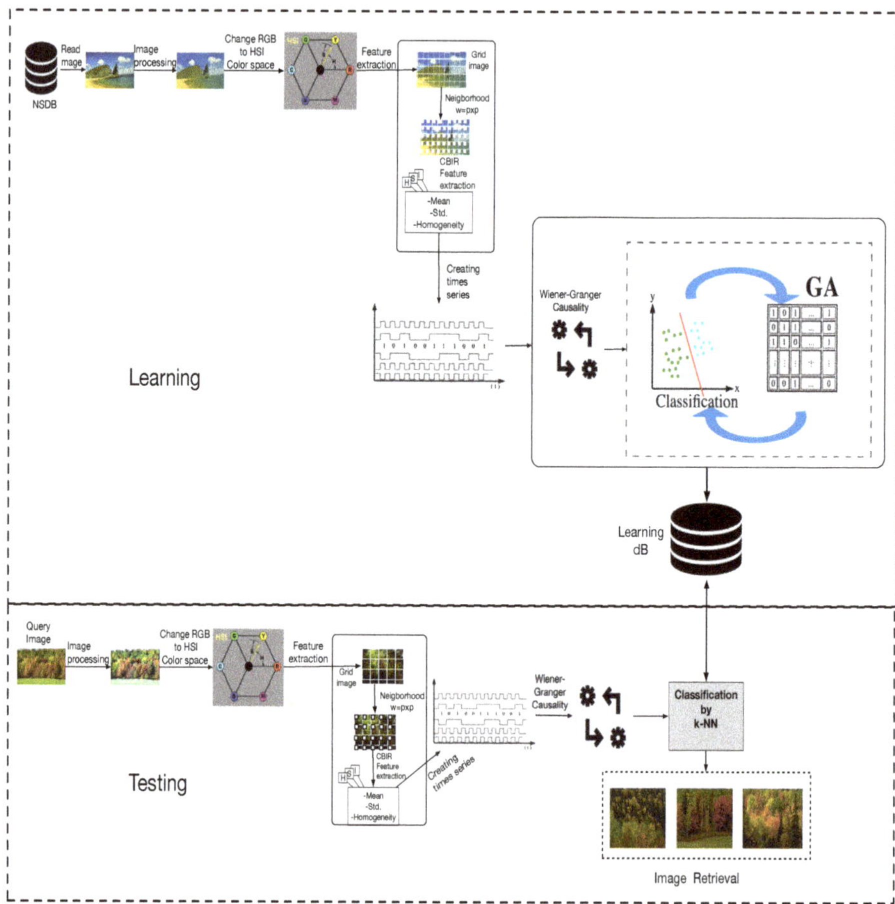

Figure 4. Learning and testing architecture for the classification system.

6. Generation of time series.For each M_s^i of the previous step, each matrix entry is compared to the $k-$textures of the dictionary to construct a discrete signal as a time series T_s^i, defined as a matrix of size $k \times w$.

    In comparison, the value 1 is assigned if the feature neighborhood approaches the dictionary texture and 0 if not, according to the threshold values which characterize each texture as they were previously presented. After processing the entries of all M_s^i, the set of signals for each scenery is stored in the FM_s, the time series matrix corresponding to a class s that contains Img_s images.

7. Wiener-Granger Causality analysis. Each FM_s matrix created in the previous step was carried to the WGC analysis to obtain the causal relationships, contained among each one of the base textures. A matrix of causality relationships, η_s, related to the training images was generated, as shown in Figure 7; therein, darker colors represent stronger relationships and these can be depicted through a state diagram where continuous lines represent only the stronger ones. The analysis of causality was computed with the causality toolbox MVGC [38], which was invoked as an external system call.

Once the causality analysis has been made for each of the Cs scenery, we get a causality relationships matrix η_s of size $k \times k$, with the total of the causal relationships F_{T_i,T_j} from the texture $T_i \to T_j$ (as given in Equation (11)), such that if a value of $F_{T_i,T_j} = 0$ means that there is no causal relationship of the texture $i \to j$, and in the measure that the value increases with respect to other η_s values, we say that the causal relationship is significant with respect to others.

$$\eta_s = \begin{bmatrix} F_{T_1,T_1} & F_{T_1,T_2} & \cdots & F_{T_1,T_k} \\ F_{T_2,T_1} & F_{T_2,T_2} & \cdots & F_{T_2,T_k} \\ \vdots & \vdots & \vdots & \vdots \\ F_{T_k,T_1} & F_{T_k,T_2} & \cdots & F_{T_k,T_k} \end{bmatrix} \tag{11}$$

The causality matrices η_s are normalized according to the total sum of their values, being $N_s = \sum_{i,j=1}^{k} F_{Ti,Tj}$, such that η_s^N is the normalized matrix of the $s-$th scenario, for $s = 1, \ldots, C_s$, with C_s: the number of scenery types considered, as given in the Equation (12). From this resulting matrix the values of the main diagonal are not taken into account because these values do not generate force in the causality relationship; as observed in the theory, there is no causal relationship between the same variables.

At the end, for C_s classes or scenery, the *total concentration of the matrices*, Γ, is defined as given in the Equation (13).

$$\eta_s^N = \begin{bmatrix} F_{T_1,T_1} & F_{T_1,T_2} & \cdots & F_{T_1,T_k} \\ F_{T_2,T_1} & F_{T_2,T_2} & \cdots & F_{T_2,T_k} \\ \vdots & \vdots & \vdots & \vdots \\ F_{T_k,T_1} & F_{T_k,T_2} & \cdots & F_{T_k,T_k} \end{bmatrix} * \frac{1}{N_s} \tag{12}$$

$$\Gamma = \cup_{l=1}^{Cs} \eta_l^N = \{\eta_1^N, \eta_2^N \ldots \eta_{Cs}^N\} \tag{13}$$

The Γ matrices as entries serve as a descriptive pattern for each scenery or class contained in the database.

8. Selection of causal relationships by means of Genetic Algorithm. To look for the causal relationships among different variables that are more important or relevant, for each of the scenes, this can be accomplished in a simple way by eliminating the relationships that have a numerical value less than a previously established threshold.

However, one disadvantage of this method is the establishment of the threshold to be used, because there is no *a priori* knowledge of the optimal value; in addition, the complexity increases when the number of textures increases in the dictionary, along with the number of classes and images to be examined. Other drawback of this solution is that some of the weak relationships could also be important in order to characterize a scenery. So there is a need to implement an automatic selection which discriminates the relevant relationships as a combinatorial optimization process. Genetic algorithms (GA) have been used successfully in several computer vision problems together with the digital image processing [39] and classification [1,9,40–42]. In this work the GA is also the right solution for the required optimization.

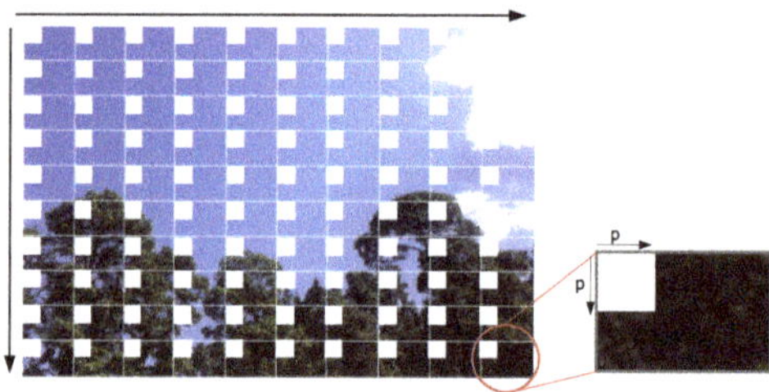

Figure 5. A 10×10 grid partition image example, every grid has a window of 10×10 pixels size.

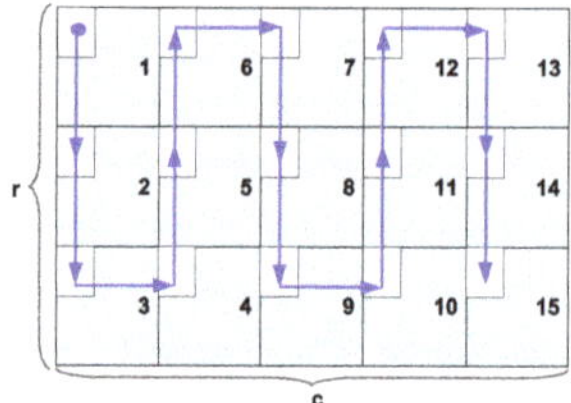

Figure 6. Reading image example among the grid neighborhood of the image.

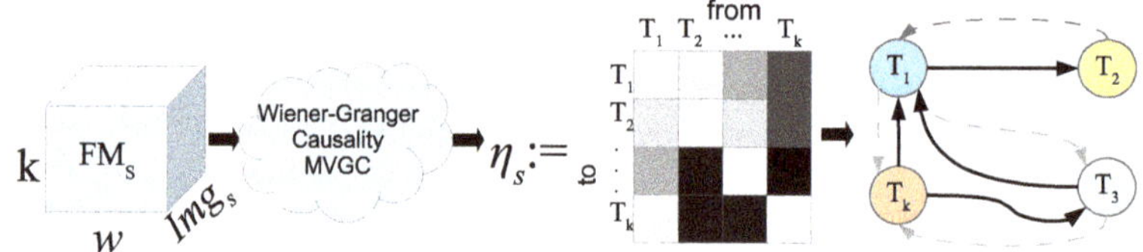

Figure 7. Generation of a texture-based causality relationship matrix, η_s, using the WGC analysis.

4. Genetic Algorithm Proposal

Looking for the analysis of the Γ matrices generated by the WGC to find the significant causality relationships for one scenery, we propose each matrix to be treated with a GA implementation. In this section we provide the GA proposal in detail.

In this approach, each matrix $\eta_s^N \in \Gamma$ is expressed using vector representation, see Figure 8 parts *(a)* and *(b)*; this is achieved only by concatenating the rows of the matrix η_s^N, then the entries of the diagonals are eliminated as shown in Figure 8 part *(c)*. In Figure 8, part *(d)*, a reallocation of the values after the previous elimination is adjusted. This provides a vector of continuous index having the size of each vector $1 \times (k^2 - k)$ for each s–th row, one per scenery. Following this process, finally, the matrix τ is created, which contains the linear conformation of each matrix η_s^N, with $s = \{1, 2, \ldots, Cs\}$ in different rows, as shown in Figure 8 part *(e)*.

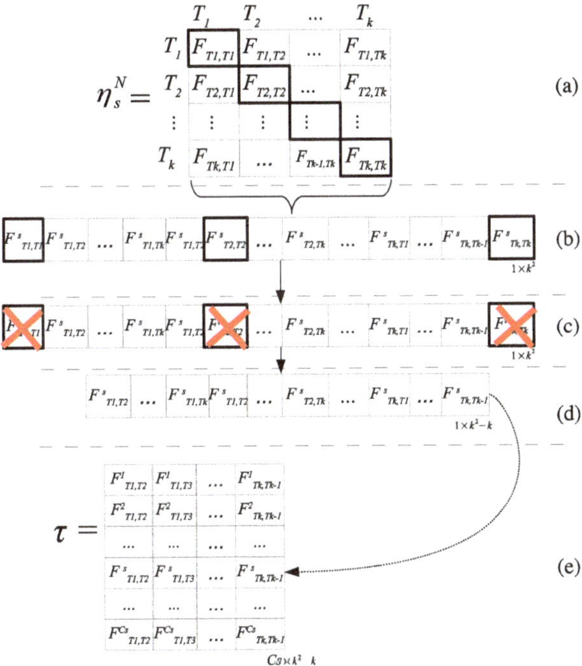

Figure 8. The τ matrix generation process, for every $s-$th row $\in \tau$.

4.1. Individual Codification

An individual binary representation for one scenery, $\tau[i]$, is conformed in order to create a filter type array of size $1 \times (k^2 - k)$ of zeros and ones, such that if an input or causal relationship is selected in that array, the value 1 is used, and 0 if not. So we have Cs rows, one row per scenery, it is intended that each row of the filter matrix could be different from the other lines, with the purpose of characterizing each type of scenery in a unique way.

It is then necessary to apply an automatic process to determine which values of the matrix τ are relevant features to distinguish the causal relationships of each scenery, and based on this result, it selects which values are going to be removed for the preset number of textures. With the selection of the most relevant causal values, it is sought to have a classification by means of a distance classifier, towards the matrix τ for each one of the query images.

4.2. Fitness Function

The fitness evaluation of each individual is generated in several parts. First, the Equation (14) is applied to the individual G_x, representing a texture relationships selection for the $s-$scenery in question, using the matrix τ in Figure 9.

$$\rho_s^{G_x} = \frac{\prod\limits_{l=1}^{k^2-k} G_x(l) * \tau_{s,l}}{\sum\limits_{m=1}^{Cs} \prod\limits_{l=1}^{k^2-k} G_x(l) * \tau_{m,l}}, \text{ such that } G_x(l) \neq 0 \tag{14}$$

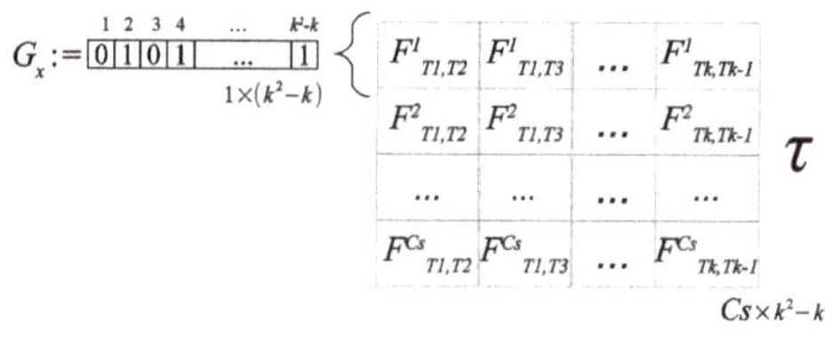

$$G_x := \underbrace{\boxed{0\,|\,1\,|\,0\,|\,1\,|\quad\ldots\quad|\,1}}_{1\times(k^2-k)} \left\{ \begin{array}{ccccc} F^1_{T1,T2} & F^1_{T1,T3} & \ldots & F^1_{Tk,Tk-1} \\ F^2_{T1,T2} & F^2_{T1,T3} & \ldots & F^2_{Tk,Tk-1} \\ \ldots & \ldots & \ldots & \ldots \\ F^{Cs}_{T1,T2} & F^{Cs}_{T1,T3} & \ldots & F^{Cs}_{Tk,Tk-1} \end{array} \right\} \tau \qquad Cs\times k^2-k$$

Figure 9. The G_x genome construction.

There, $\prod G_x(l) * \tau_{s,l}$ refers to the product of the τ entries located at $s-$scenery (row s) and column l, specifying a causal relationship, accomplishing $G_x(l)$ is a valid non zero entry of the genome. Thus $\rho_s^{G_x}$ is the total probability for the individual G_x applied to all scenery.

Based on these data, by means of the probability theory, the individual G_x is required to meet the condition: $\rho_s^{G_x} > \rho_j^{G_x}$, such that $s \in \{1, 2, \ldots, Cs\}$, $s \neq j$, and $1 \leq j \leq Cs$.

That is, Cs probabilities corresponding to each scenery evaluating the individual G_x are obtained with the calculation of $\rho_j^{G_x}$. Equation (15) gives the first step for the optimization process, considering the maximum probability related to the casual relationships which best characterized the $s-$scenery versus the others.

$$\overset{G_x}{\underset{s}{V}} = \begin{cases} \rho_s^{G_x} & \text{if} \quad \rho_s^{G_x} = Max\{\rho_j^{G_x}\}_{j=1,\ldots,Cs} \\ 0 & \text{if} \quad \rho_s^{G_x} \neq Max\{\rho_j^{G_x}\}_{j=1,\ldots,Cs} \end{cases} \tag{15}$$

Then, the fitness function, $f_s(G_x)$, is determined as Equation (16).

$$f_s(G_x) = \begin{cases} CP_s & \text{if} \quad \overset{G_x}{\underset{s}{V}} > 0 \\ 0 & \text{if} \quad \overset{G_x}{\underset{s}{V}} = 0 \end{cases} \tag{16}$$

To this end, the images contained in the $s-$scenery are consulted, using the re-substitution test. Each image query gives the scenery which belongs to filling the information of a confusion matrix that is used to calculate the percentage of classification. The image consult query process is described in the following paragraph. Later, in Section 4.3 the *global fitness* is taken into account for the population evolution in the GA loop process.

4.2.1. Creating a Query from a Single Image

In order to classify an s-scenery image considering the relationships specified in a G_x individual, a related causal relationship matrix needs to be constructed.

The first step consists of creating a set of M synthetic images, $L_1, L_2, \ldots, L_M$, from a single L image is performed. This is produced by means of manipulating the first reading of the image, making a circular shift of d positions for each new synthetic image, in order to create several samples of the same image as shown in Figure 10. In this way, the respective query matrix of size $|k \times w \times M|$ ($k :=$ number of textures in the dictionary, $w :=$ number of neighborhoods, and $M :=$ number of synthetic images) is generated to feed the WGC analysis process and to obtain the resulting normalized causal relationship matrix η_L^N of size $|k \times k|$. These steps are carried out by Equations (11) and (12).

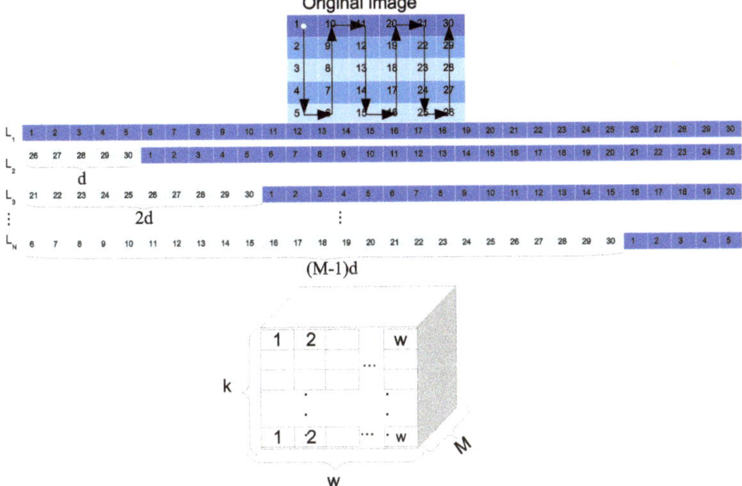

Figure 10. Representation of a query image construction into N samples.

Then the manipulation of η_L^N is performed as shown in the stage presented in Figure 8 to obtain the linear representation of the matrix. The last query step consists of applying the k-NN classifier (with $k := 1$) to determine which τ scenery (line) has the closest relationship to the linear relationship representation of image L, considering only the relationship indicated with the G_x non-zero values.

4.3. GA Implementation

A genetic algorithm is applied for each τ line to automatically select the most representative causal relationship of each scenery. Figure 11a shows the general algorithm flowchart of this approach.

An initial population, PG, of $sizeP$ individuals is randomly generated, where $sizeP$ is an odd number and each individual is of size $k^2 - k$, the size in columns of the matrix τ.

Then the PG individuals are evaluated with the fitness function, Equation (16), for a particular s−scenery, considering the total set of images that conform it, as shown in Figure 11b. The individual's fitness is stored inside a fitness array $\{\bullet\}$, as in Equation (17). The $\{\bullet\}$ array is consequently ordered, from highest to lowest, to find the best individual with the highest fitness.

$$\{\bullet\} = \{f_1(G_x), f_2(G_x), \ldots, f_{sizeP}(G_x)\}, \tag{17}$$

such that $f_p(G_x) \geq 0$ for $1 \leq p \leq sizeP$. For this proposal, size population $sizeP = 21$, the genome length is 12, and the number of iterations was $maxGen = 100$ generations.

To generate the new population, $(sizeP - 1)/2$ triplets of random numbers are generated, e.g., {1,5,1} or {2,4,0}, where the first two numbers are the selected individual numbers that generate the new individuals, the third element of the triplet is one of the two possible operations to be executed; either crossover "1" or mutation "0".

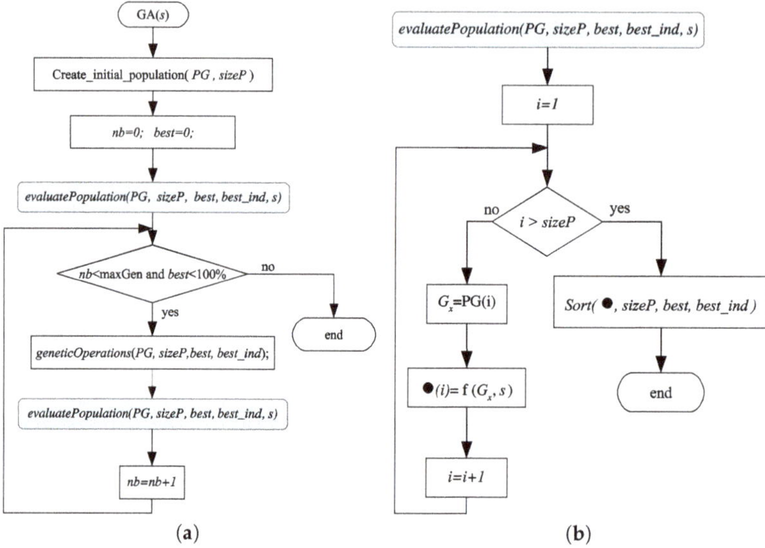

(a)   (b)

Figure 11. Flowchart showing the general GA algorithm implementation. (a) Implementation of the proposed General flowchart GA, and (b) GA implementation for a particular s-scenery.

For derivatives of each triplet, we generate two new individuals if it is by crossover operator, or if it is by mutation operator the two selected individuals are altered separately, in order to have two new elements for the new generation, 1 mutated individual from one single individual.

The crossover operator is applied at one uniform random point of the two participating chromosomes, and the mutation operator is performed over 10% of the elements of a chromosome, as shown in Figure 12.

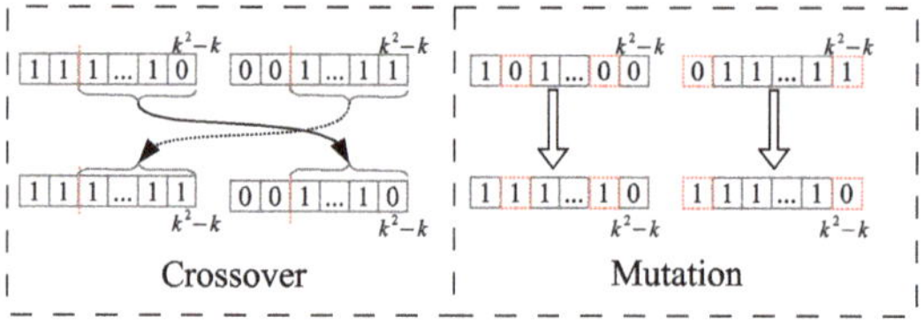

Figure 12. Genetic operators application.

The genetic operations of mutation and crossover are applied to 30% and 70% of the population respectively, favoring the selection of the highest fitness individuals to be reproduced. The individual with the best fitness passes to the next generation applying elitism. In this way, the population will evolve towards a selection of relevant causal relationships to be able to characterize each scenery.

The end of the GA or stop criteria is given when reaching the 100% classification percentage or a number of generations is attained.

After the GA is applied Cs times, the individuals that contain the most relevant relationships for each scenery are found. Then, the τ matrix is updated and its entries are replaced by a zero value whenever the corresponding individual entries have a zero and they keep their value in other case.

5. Parallel Approach

In this section, a parallel algorithm to speed-up the performance of the proposed WGC methodology is presented. The parallel approach works on a distributed memory architecture using MPI library; that is, there is a set of processes without shared memory, and these processes work in parallel, and the communication goes through message exchange to determine the relevant causal relationships of all the scenery. Each process can access the NSDB to extract and work with the corresponding set of images. The algorithm complexity in this proposal is given for the Equation (18).

$$O(N_{class} \times k \times CIP \times r \times c \times nImg \times t_{comp} \times WGC_{tb}) \tag{18}$$

where, $N_{class} :=$ number of classes, $k :=$ number of textures, $CIP :=$ constant for every image processing, $r :=$ number of rows in the grid, $c :=$ number of cols in the grid, $nImg :=$ total number of images, $t_{comp} :=$ comparison time against the base textures, $WGC_{tb} :=$ causality analysis time. (e.g., for an image of size $640x480$ pixels, $r = (640/20)$, $c = (480/30)$, and p is the size of the neighborhood, such that 10×10 implies $p = 10$) That means, if the number of rows r, cols c in the grid increases, then the number of images $nImag$ in NSDB increases, and the number of base textures k in the dictionary increases, and the number of classes N_{class} increases, thus the computational cost increases. In this way it is necessary to conceive a parallel architecture to solve this problem in a large number of images related to Big Data problems.

The Algorithm 1 shows the procedure which is executed simultaneously by each process and the general process of this parallel proposal is depicted in Figure 13.

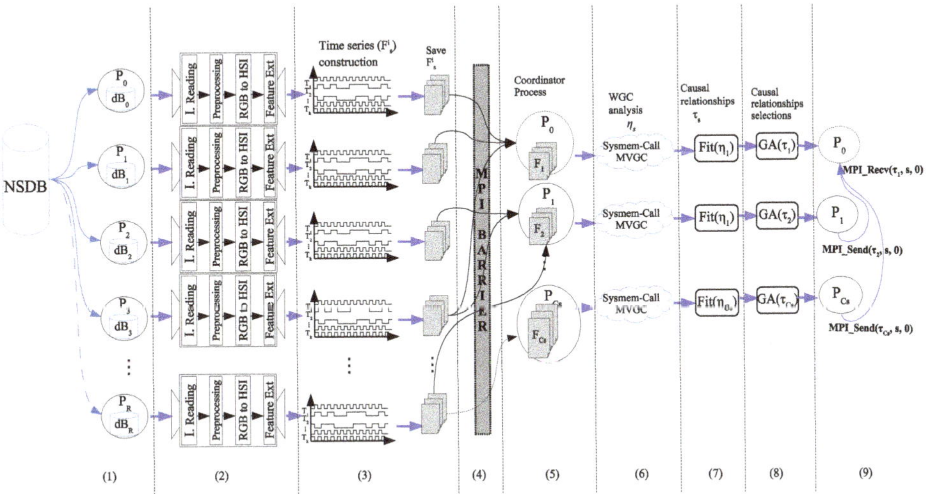

Figure 13. The proposed parallel algorithm structure.

At the beginning (line 2 of Algorithm 1, Figure 13, tag (1)), each process determines the amount set of images to be read (*ImgBlock*), taking into consideration the total number of images (*Total_IMGs*), the total number of processes (*Total_procs*) and the process identifier (*rank*). A single process can work with images belonging to different scenery (e.g., *Total_IMGs* = 700, *Total_procs* = 70, *imgBlock* = 700/10 = 10 for the *NSDB*).

Algorithm 1 Parallel algorithm for the causality matrix construction.

```
 1:  procedure CAUSALITY MATRIX CONSTRUCTION(rank)
 2:      initialization(ImgBlock,Total_IMGs,Total_procs,rank);
 3:      for every i in ImgBlock do
 4:          Img_RGB_read(image, s, i);
 5:          Img_preprocessing(image);
 6:          RGB_to_HSI(image);
 7:          M_s^i = Feature_extraction(image);
 8:          F_s^i = time_series_construction(M_s^i, Texture_Dictionary);
 9:          Save_TimeSeries(F_s^i);
10:      end for
11:      Barrier_synchronization();
12:      if rank in {Scenery coordinator ranks} then
13:          F_s = Load_all_time_series(s);
14:          η_s = System_call(MVGC(F_s));
15:          τ_s = Fitting(η_s);
16:          Genetic_Algorithm(τ_s);
17:          Send(τ_s, s, General_coordinator rank);
18:      end if
19:      if (rank == General_coordinator rank) then
20:          for (every id in {Scenery coordinator ranks}) do
21:              Recv(τ_s, s, id);
22:              τ(s) = τ_s;
23:          end for
24:      end if
25:  end procedure
```

Each process works simultaneously with the section of the NSDB, *ImgBlock*, which was assigned to it, performing the following steps (lines 3–10). The reading of the i-image in RGB space is the first action to be executed, next up the scenery, s, such that i-image $\in s$, is also obtained (line 4). Then the image preprocessing and the conversion from RGB to HSI domains are carried out (lines 5 and 6, respectively). In line 7, the statistical features are calculated, including the construction of the image grid and neighborhoods, then the CBIR features per each neighborhood generates the M_s^i matrix; Figure 13, tag (2), represents the execution of lines 4 to 7 of Algorithm 1. Then, M_s^i and the texture dictionary are used to construct the respective time series, F_s^i, that is stored on file (lines 8,9 of the algorithm, Figure 13, tag (3)).

Up to this point, all processes work independently; however, in order to ensure that every process has fully accomplished its task, a parallel barrier synchronization (line 11 of the algorithm, Figure 13, tag (4)) should be introduced before continuing with the next step. Here (line 12), only the processes identified as *scenery coordinators* (one process per scenery) continue with the construction of the corresponding F_s matrix (line 13, Figure 13, tag (5)), by loading the respective set of F_s^i matrices (one per scenery image), previously generated. Then, a system call is performed (line 14, Figure 13, tag (6)) to run the MVGC toolbox and obtain the causality relationship matrix, η_s, from the WGC analysis.

The $Fit(\eta_s)$ function in line 15 (Figure 13, tag (7)) is in charge of normalization and vector representation of the causality relationship matrix, η_s. The respective τ_s is thus generated, corresponding to the s-th row (scenery) of τ matrix. Line 16 of Algorithm 1 (Figure 13, tag (8)) shows the GA call that is executed by each one of the Cs scenery coordinator processes, with Cs being the number of scenery. After identifying the most relevant causal relationships by means of the GA, τ_s is updated and sent to the general coordinator (line 17) through a message.

Finally, in lines 19–24, the general coordinator process receives, by means of several messages, the results generated by the scenery coordinator processes (Figure 13, tag (9)). When all message receptions are achieved, the matrix τ is successfully constructed.

6. Experimental Results

The proposal evaluation was generated using the computer power of a 19-processor dual core cluster. Each processor is an Intel©Xeon©CPU E5-2670 v3 2.30 GHz, and 74 GB RAM.

Four image textures $k = 4$ where selected to conform the base dictionary, as shown in Table 1. For each texture the generated values were obtained manually within the images of the database, a set of 20 texture samples were taken from a set of 5 images per class, from each texture the average was extracted in the layer H plus twice the standard deviation, with this the maximum and minimum threshold values for each texture were generated.

Table 1. HSI ranges of the base texture dictionary.

Texture	H-max	H-min	S-max	S-min	I-max	I-min
Cloud	180	0	25	0	255	61
Sky	113	93	255	25	61	255
Rock	24	6	255	20	190	30
Forest	102	28	255	10	229	3

The *NSDB* used for the evaluation consists of the following data:

- Vogel and Shiele (V_S) [10], including 6 scenery with 700 images classified as: 144 coast, 103 forest, 179 mountain, 131 prairie, 111 river/lake, and 32 sky/cloud.
- Oliva and Torralba (O_T) [11], including 4 scenery with 1472 Images classified as: 360 coast, 328 forest, 374 mountain and 410 prairie.

The images were adapted so that some typical classification challenges were considered. The whole set of images was tested in a normal state and introducing Gaussian noise (GN), salt and pepper noise (S&P) of 1%, 3%, and 10% levels respectively, as shown in Figure 14. A rotation transformation was also introduced on each image considering 0°, 45°, 90°, 135°, and 180°, as shown in Figure 15. An image consult query was performed following the same procedure described in Section 4.2.1.

Figure 14. Example of images with Gaussian, salt, and pepper noise of 1%, 3%, and 10%, respectively.

The results in this section are organized as follows. The image classification performance obtained when applying the WGC theory is first presented. Then the execution times of the proposed parallel methodology are shown.

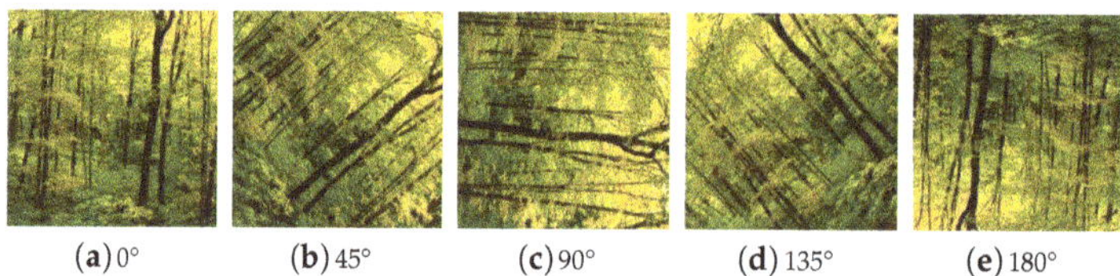

(**a**) 0°   (**b**) 45°   (**c**) 90°   (**d**) 135°   (**e**) 180°

Figure 15. Rotation degrees applied to the images set.

6.1. Classification Results

To show that the proposed GA implementation was a good solution to select some relevant texture relationships describing a scenery, we compared our proposal (GA version) to the manual strategy (Manual version) introduced in Reference [36]; under the manual strategy only the highest relationship values were selected, establishing a specific threshold. In both versions, the methodology presented in Section 3 for the construction of τ matrix, was executed. Table 2 shows the resulting τ values.

Because there were no *a priori* criteria to determine a threshold value, in the Manual version 25% of the less significant causal relationships per scenery were deleted. Table 3 shows the updated τ matrix after the manual selection.

Table 2. The obtained τ matrix values.

Scene/F	$F_{T1,T2}$	$F_{T1,T3}$	$F_{T1,T4}$	$F_{T2,T1}$	$F_{T2,T3}$	$F_{T2,T4}$	$F_{T3,T1}$	$F_{T3,T2}$	$F_{T3,T4}$	$F_{T4,T1}$	$F_{T2,T4}$	$F_{T3,T4}$
Forest	0.370	0.289	0.0051	0.0269	0.037	0.0013	0.108	0.075	0.0849	0.00332	1.611×10^{-5}	0.00062
Sky & cloud	0.0441	0.0481	0.0321	0.0269	0.343	0.197	0.0044	0.0385	0.211	0.011	0.0070	0.0371
Coast	0.0099	0.233	0.188	0.021	0.0363	0.0503	0.0542	0.0014	0.0646	0.1558	0.0772	0.109
Mountain	0.085	0.405	0.0132	0.0497	0.0799	0.0698	0.1115	0.0102	0.0162	0.1076	0.0392	0.0133
Prader	0.2401	0.1143	0.2400	0.0045	0.0140	0.0268	0.2151	0.1146	0.0006	0.0161	0.0140	0.0002
River	0.1619	0.4053	0.1112	0.0061	0.0046	0.0035	0.0377	0.0322	0.1794	0.0066	0.0084	0.0432

Table 3. The τ matrix resulting from the manual selection of the highest significant causal relationships.

Scene/F	$F_{T1,T2}$	$F_{T1,T3}$	$F_{T1,T4}$	$F_{T2,T1}$	$F_{T2,T3}$	$F_{T2,T4}$	$F_{T3,T1}$	$F_{T3,T2}$	$F_{T3,T4}$	$F_{T4,T1}$	$F_{T2,T4}$	$F_{T3,T4}$
Forest	0.370	0.289	0.0051	0.0269	0.037	0	0.108	0.075	0.0849	0.00332	0	0
Sky & cloud	0.0441	0.0481	0.0321	0.0269	0.343	0.197	0	0.0385	0.211	0	0	0.0371
Coast	0	0.233	0.188	0	0.0363	0.0503	0.0542	0	0.0646	0.1558	0.0772	0.109
Mountain	0.085	0.405	0	0.0497	0.0799	0.0698	0.1115	0	0	0.1076	0.0392	0
Prader	0.2401	0.1143	0.2400	0	0.0140	0.0268	0.2151	0.1146	0	0.0161	0.0140	0
River	0.1619	0.4053	0.1112	0	0	0	0.0377	0.0322	0.1794	0	0	0.0432

When executing the GA version for the selection of τ matrix relationships, a larger space of solutions was explored trying to look for the causal relationships that best represent one scenery. The obtained individuals are presented in Table 4, and the updated τ matrix is shown in Table 5.

Both, GA and manual versions where tested using 300 images, 50 per scenery. The manual version only obtained obtaining a 12.53% general classification percentage. The confusion matrix showing the image association per scenery can be seen in Table 6; we observe that most of the images were associated to the coast scenery, and as a result the manual selection test gave a poor classification percentage.

Table 4. The best individuals resulting from the evaluation of the GA per scenery.

Scene/F	1	2	3	4	5	6	7	8	9	10	11	12
Forest	1	0	0	0	0	0	0	0	1	0	0	0
Sky	0	1	1	0	1	1	1	1	0	0	0	1
Coast	0	0	1	1	0	0	0	1	0	0	1	1
Mountain	1	0	0	1	0	0	1	1	0	0	0	1
Prader	1	0	1	0	1	0	1	1	1	0	1	0
River	0	0	1	1	0	0	1	1	0	0	0	1

Table 5. Final τ values, applying the better individuals of the GA.

Scene/F	$F_{T1,T2}$	$F_{T1,T3}$	$F_{T1,T4}$	$F_{T2,T1}$	$F_{T2,T3}$	$F_{T2,T4}$	$F_{T3,T1}$	$F_{T3,T2}$	$F_{T3,T4}$	$F_{T4,T1}$	$F_{T2,T4}$	$F_{T3,T4}$
Forest	0.370	0	0	0	0	0	0	0	0.0849	0	0	0
Sky&cloud	0	0.0481	0.0321	0	0.343	0.197	0.0044	0.0385	0	0	0	0.0371
Coast	0	0	0.188	0.021	0	0	0	0.0014	0	0	0.0772	0.109
Mountain	0.085	0	0	0.0497	0	0	0.1115	0.0102	0	0	0	0.0133
Prader	0.2401	0	0.2400	0	0.0140	0	0.2151	0.1146	0.0006	0	0.0140	0
River	0	0	0.1112	0.0061	0	0	0.0377	0.0322	0	0	0	0.0432

Table 6. Confusion matrix for a test with 50 images per scenery, using a manual selection of causal relationships.

$Scene_i/Scene_j$	Forest	Sky	Coast	Mount	Prad	Riv
Forest	0	0	31	19	0	0
Sky	1	0	39	10	0	0
Coast	3	0	36	11	0	0
Mount	2	0	47	1	0	0
Prad	5	0	40	5	0	0
Riv	6	0	32	11	1	0

With the information in the Table 5, the most representative relations of each natural scenery are generated as a visual representation, the graphs representing the intensity of the causal relations between the $k = 4$ base textures of the dictionary. These graphs will show how textures are related within the corresponding scenery, obtaining the pattern which represents each of them, as it can be appreciated in Figure 16.

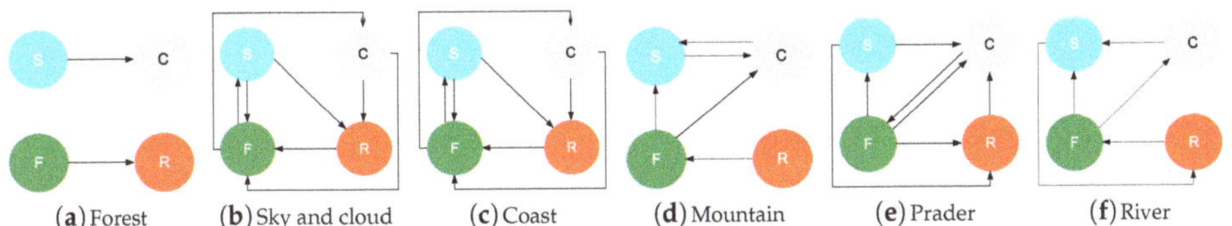

Figure 16. Evolutionary texture causal relationship graphs resulted for each scenery.

Given these first results, it can be observed that not necessarily the relationships with higher values were the best ones to be selected.

To measure the technical efficiency of our proposal using the GA, the *Recall* (managed as classification percentage), *Precision*, *Accuracy* and *F1 Score* were estimated from the confusion matrices of every test.

The classification results of Figure 17 show that that rotating the images by $45°$, $90°$ and $315°$, the classification performance decays significantly. Also, the noise (GN and S&P) significantly alters the classification which is expected in natural scenery images since the texture is a representative of the type of image, and the alterations with noise on it degenerate into another possible meaning. In normal conditions, avoiding noise and rotations, the classification performance reaches 100%.

Additionally, Figure 18 shows the estimations of the precision (Figure 18a), recall (Figure 18b), accuracy (Figure 18c), and F1 Score (Figure 18d) averages for the classes contained in the NSDB. In general, the classification of ideal images (0°) without rotations and noise obtains 100% classification, However, when rotations and noise are added this percentage decreases, particularly the sky class is the most affected.

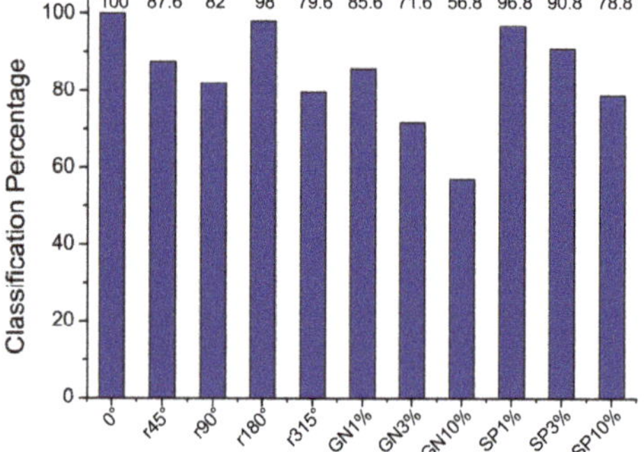

Figure 17. Classification results of our proposal using the GA, considering different noise type and rotati

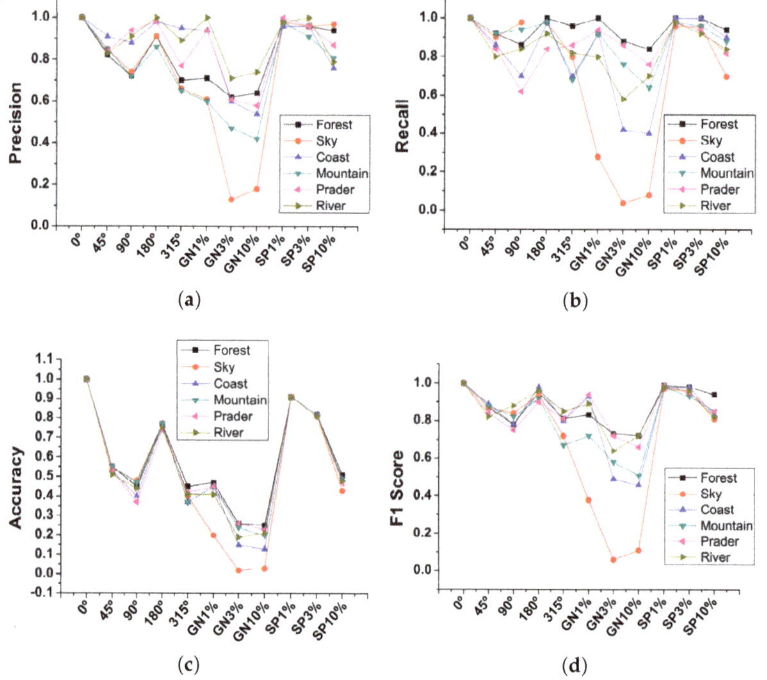

Figure 18. Technical efficiency measures for the best GA individuals for each scenery: (**a**) precision measure, (**b**) recall measure, (**c**) accuracy measure and (**d**) F1 score measure.

The average result for the GA evaluation is depicted in Figure 19. Figure 19a shows the fitness evolution within a run with 100 generations. Figure 19b shows that, while in some classes the highest fitness is achieved in the first iteration, in the other ones 100 generations are not enough to achieve the best fitness. Figure 19c shows the best fitness obtained through 200 runs considering 100 generations per run and population size set to 21 individuals; all fitness converges near the expected value of 100% classification.

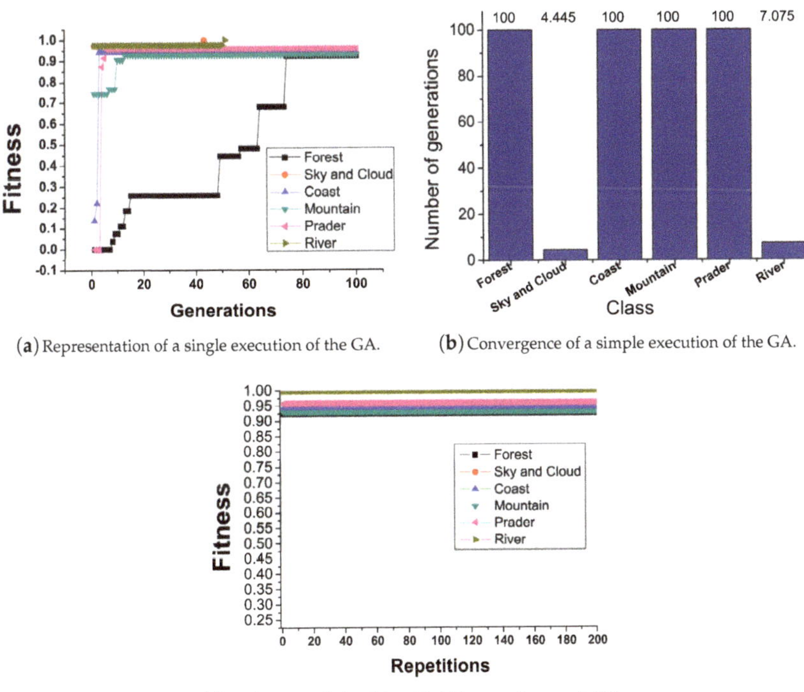

(**a**) Representation of a single execution of the GA.

(**b**) Convergence of a simple execution of the GA.

(**c**) Evaluation of the GA with 200 executions and 100 generations per execution.

Figure 19. Performance of the GA for the natural scenery contained in the NSDB.

6.2. Parallel Methodology Performance

The execution time taken by the proposed parallel causality methodology applied to the identification and classification of natural scenery, for a total of 700 images, 6 different scenery, and varying the number of processes, are shown in Figure 20. These values were the average time taken for 200 executions. We can observe that the execution time decreased rapidly while increasing the number of processes, getting the best execution time when 125 concurrent processes were defined. With this configuration each process worked with 5 or 6 images, favoring the internal scheduling that used more efficiently the computer resources so that the sequential version execution time decreased by 88.9%.

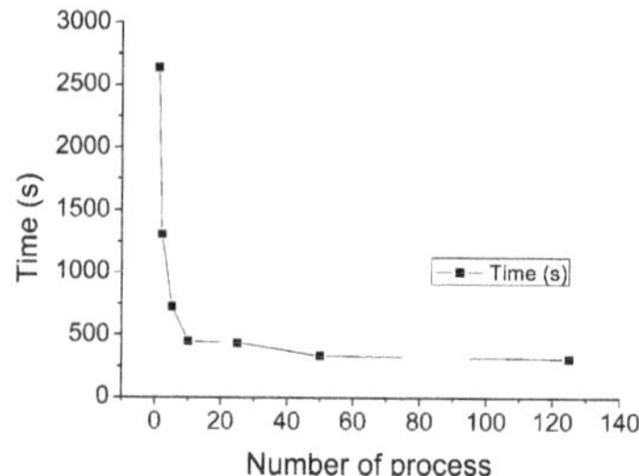

Figure 20. Performance of the parallel methodology algorithm.

7. Discussion

To find the texture causality relationships for characterizing a natural scenery, we found the necessity of GA implementation. With this solution, the automatic discrimination process for selecting the causal relationships that are important or relevant for the classification of the proposed scenery was successfully achieved. Compared to some of the articles in the literature, the possibility of our approach to perform the selection in an automatic way allows the study of the scenery classification problem considering more parameters. In this way, a larger number of scenery or base textures for future implementations of several purposes of texture classification could be taken into account considering the efficient methodology and evolutionary algorithm proposed in this work. This proposal could help in recognizing natural scenarios in the navigation of an autonomous car or possibly a drone, being an important element in the safety of autonomous vehicles navigation.

As we can see in Figure 17, the classification percentage obtained by means of the selected features for the evolutionary process surpasses those obtained by the manual selection version. This result is important because the representative causal relationships of the scenery are selected in such a way that they numerically escape the manual perspective; that is to say, in the manual selection strategy a non-significant threshold value is specified, such that any value lower than the threshold is set to zero, but evolutionary strategy turns out that some of these causal relationships are relevant to classify the scenery, marking differences with another similar scenery. From the causality theory applied to the image reading sequences, we are trying to infer the order of appearance of the textures typified in the base dictionary seeking to represent them as temporal visual reading that we see as a type of natural scenery.

8. Conclusions

In this paper a novel proposal for the use of the Wiener-Granger causality theory supported by a genetic algorithm was presented, along with the CBIR self-content analysis, applied for the identification and classification of 6 natural scenery: coast, forest, mountain, prairie, river/lake, and sky/cloud. Considering the new formulation it was possible to find a set of descriptors from the causality matrices in order to represent a scenery class, from a base set of reference textures, proposing a characterization of images based on the continuous appearance of textures within them; the base dictionary in this approach included the textures: Cloud, Sky, Rock, and Forest. Unlike others approaches, our methodology deals with the rotation and image noise considerations, and the results show excellent classification percentages.

Under this approach we have 100% image classification for the whole dataset, and the methodology provided the next good classification rate for 180° rotation, and the sensitivity for intermediate rotation levels (45°, 90°), and had good results for the salt and pepper image noise.

In relation to the proposal performance, the design of a parallel computing algorithm was developed. A reduction in execution times was achieved using a 19-processor dual-core cluster server, and the MPI tool, reaching an 88.9% decrease of the sequential version execution time when 125 processes were launched.

Future work includes the study of performance of this proposal using other parallel architectures; e.g., the GPU technology could perform efficiently for the image feature extraction stage, as well as the implementation of other evolutionary algorithms, such as Genetic Programming in order to analyze all together the image textures looking to characterize the whole scenery and its associations with paradigm of visual comprehension.

Author Contributions: Writing–review and editing, C.B.-A.; investigation, C.B.-A., C.A.-C. and J.V.-C.; resources, J.V.-C. and C.A.-C.; writing–original draft preparation, C.B-A; validation, G.R.-A., C.A.-C. and J.V.-C.; conceptualization G.R.-A, C.B.-A., C.A.-C.; formal analysis, C.A.-C., G.R.-A., J.V.-C.; methodology, C.B.-A., J.V.-C. and G.R.-A; C.B.-A. supervised the overall research work. All authors contributed to the discussion and conclusion of this research.

Funding: This research received no external funding.

Acknowledgments: Cesar Benavides thanks the CONACyT for the scholarship support. This work has been supported by Fundación Carolina, Spain, under the scholarship program 2016–2017. This work has been done under project EL006-18, granted by the Metropolitan Autonomous University, Unidad Azcapotzalco, Mexico.

Conflicts of Interest: The authors declare that there is no conflict of interests regarding the publication of this paper.

References

1. Gustavo, O. *Evolutionary Computer Vision: The First Footprints*, 1st ed.; Natural Computing Series; Springer: Berlin/Heidelberg, Germany, 2016.
2. Nalwa, V.S. *A Guided Tour of Computer Vision. Volume 1 of TA1632*; Addison Wesley: Boston, MA, USA, 1993.
3. Gonzalez, R.C.; Woods, R.E. *Digital Image Processing*; Addison-Wesley Longman Publishing Co., Inc.: Boston, MA, USA, 1992.
4. Tyagi, V. *Content-Based Image Retrieval: Ideas, Influences, and Current Trends*, 1st ed.; Springer: Singapore, 2017.
5. Benavides, C.; Villegas-Cortez, J.; Roman, G.; Aviles-Cruz, C. Reconocimiento de rostros a partir de la propia imagen usando técnica cbir. In Proceedings of the X Congreso Español sobre Metaheurísticas, Algoritmos Evolutivos y Bioinspirados (MAEB 2015), Merida Extremadura, Spain, 4–6 February 2015; pp. 733–740.
6. Benavides, C.; Villegas-Cortez, J.; Roman, G.; Aviles-Cruz, C. Face classification by local texture analysis through cbir and surf points. *IEEE Latin Am. Trans.* **2016**, *14*, 2418–2424. [CrossRef]
7. Granger, C.W. Investigating causal relations by econometric models and cross-spectral methods. *Econometrica* **1969**, *37*, 424–438. [CrossRef]
8. Serrano-Talamantes, J.F.; Aviles-Cruz, C.; Villegas-Cortez, J.; Sossa-Azuela, J.H. Self organizing natural scene image retrieval. *Expert Syst. Appl. Int. J.* **2012**, *40*, 2398–2409. [CrossRef]
9. Deb, K.; Kalyanmoy, D. *Multi-Objective Optimization Using Evolutionary Algorithms*; John Wiley & Sons, Inc.: New York, NY, USA, 2001.
10. Vogel, J.; Schiele, B. Performance evaluation and optimization for content-based image retrieval. *Pattern Recognit.* **2006**, *39*, 897–909. [CrossRef]
11. Oliva, A.; Torralba, A. Modeling the shape of the scene: A holistic representation of the spatial envelope. *Int. J. Comput. Vis.* **2001**, *42*, 145–175. [CrossRef]
12. Begum, R.; Halse, S.V. The smart car parking system using gsm and labview. *J. Comput. Math. Sci.* **2018**, *9*, 135–142. [CrossRef]
13. Blaifi, S.; Moulahoum, S.; Colak, I.; Merrouche, W. Monitoring and enhanced dynamic modeling of battery by genetic algorithm using labview applied in photovoltaic system. *Electr. Eng.* **2017**, *100*, 1–18. [CrossRef]
14. Alam, A.; Jaffery, Z. A Vision-Based System for Traffic Light Detection: SIGMA 2018, Volume 1, pages 333–343. 01 2019. *IEEE Latin Am. Trans.* **2016**, *14*, 2418–2424.

15. Baltrušaitis, T.; Robinson, P.; Morency, L. Openface: An open source facial behavior analysis toolkit. In Proceedings of the 2016 IEEE Winter Conference on Applications of Computer Vision (WACV), Lake Placid, NY, USA, 7–10 March 2016; pp. 1–10.

16. Sultana, M.G. *A Content Based Feature Combination Method for Face Recognition*; Springer: Heidelberg, Germany, 2013; Volume 226, pp. 197–206.

17. Madhavi, D.; Patnaik, R. *Genetic Algorithm-Based Optimized Gabor Filters for Content-Based Image Retrieval*; Springer: Singapore, 2018; pp. 157–164.

18. Desai, R.; Sonawane, B. *Gist, Hog, and Dwt-Based Content-Based Image Retrieval for Facial Images*; Springer: Singapore, 2017; Volume 468, pp. 297–307.

19. Gao, J.; Yang, J.; Zhang, J.; Li, M. Natural scene recognition based on convolutional neural networks and deep boltzmannn machines. In Proceedings of the 2015 IEEE International Conference on Mechatronics and Automation (ICMA), Beijing, China, 2–5 August 2015; pp. 2369–2374.

20. Meng, F.; Wang, X.; Shao, F.; Wang, D.; Hua, X. Energy-efficient gabor kernels in neural networks with genetic algorithm training method. *Electronics* **2019**, *8*, 105. [CrossRef]

21. Vogel, J.; Schiele, B. Semantic modeling of natural scenes for content-based image retrieval. *Int. J. Comput. Vis.* **2007**, *72*, 133–157. [CrossRef]

22. Traina, A.J.M.; Balan, A.G.R.; Bortolotti, L.M.; Traina, C. Content-based image retrieval using approximate shape of objects. In Proceedings of the 17th IEEE Symposium on Computer-Based Medical Systems, Bethesda, MD, USA, 25 June 2004; pp. 91–96.

23. Dabbiru, L.; Aanstoos, J.V.; Ball, J.E.; Younan, N.H. Screening mississippi river levees using texture-based and polarimetric-based features from synthetic aperture radar data. *Electronics* **2017**, *6*, 29. [CrossRef]

24. Liu, Y.; Zhang, D.; Lu, G.; Ma, W.Y. A survey of content-based image retrieval with high-level semantics. *Pattern Recognit.* **2007**, *40*, 262–282. [CrossRef]

25. Zeng, P.; Li, Z.; Zhang, C. Scene Classification Using Spatial and Color Features. In Proceedings of the 8th International Conference on Intelligent Information Processing (IIP), Hangzhou, China, 17–20 October 2014; Volume AICT-432 of Intelligent Information Processing VII; Shi, Z., Wu, Z., Leake, D., Sattler, U., Eds.; Springer: Berlin/Heidelberg, Germany, 2014; pp. 259–268.

26. Bressler, S.L.; Seth, A.K. Wiener-granger causality: A well established methodology. *NeuroImage* **2011**, *58*, 323–329. [CrossRef] [PubMed]

27. Matias, F.S.; Gollo, L.L.; Carelli, P.V.; Bressler, S.L.; Copelli, M.; Mirasso, C.R. Modeling positive granger causality and negative phase lag between cortical areas. *NeuroImage* **2014**, *99*, 411–418. [CrossRef] [PubMed]

28. Mannino, M.; Bressler, S.L. Foundational perspectives on causality in large-scale brain networks. *Phys. Life Rev.* **2015**, *15*, 107–123. [CrossRef] [PubMed]

29. Zhang, H.; Li, X. Effective connectivity of facial expression network by using granger causality analysis. *Parallel Process. Images Optim. Med. Imaging Process.* **2013**, *8920*, 89200K.

30. Friston, K. Causal modelling and brain connectivity in functional magnetic resonance imaging. *PLoS Biol.* **2009**, *7*, e1000033. [CrossRef] [PubMed]

31. Kim, E.; Kim, D.S.; Ahmad, F.; Park, H. Pattern-based granger causality mapping in fmri. *Brain Connect.* **2013**, *3*, 569–577. [CrossRef] [PubMed]

32. Fablet, R.; Bouthemy, P.; Pérez, P. Nonparametric motion characterization using causal probabilistic models for video indexing and retrieval. *IEEE Trans. Image Process* **2002**, *11*, 393–407. [CrossRef] [PubMed]

33. Kular, D.; Ribeiro, E. *Analyzing Activities in Videos Using Latent Dirichlet Allocation and Granger Causality*; Springer International Publishing: Cham, Switzerland, 2015; pp. 647–656.

34. Prabhakar, K.; Oh, S.; Wang, P.; Abowd, G.D.; Rehg, J.M. Temporal causality for the analysis of visual events. In Proceedings of the 2010 IEEE Conference on Computer Vision and Pattern Recognition (CVPR), San Francisco, CA, USA, 13–18 June 2010; pp. 1967–1974.

35. Zhang, C.; Yang, X.; Lin, W.; Zhu, J. Recognizing human group behaviors with multi-group causalities. In Proceedings of the Proceedings of the The 2012 IEEE/WIC/ACM International Joint Conferences on Web Intelligence and Intelligent Agent Technology—Volume 03, WI-IAT '12, Macau, China, 4–7 December 2012; pp. 44–48.
36. Fan, Y.; Yang, H.; Zheng, S.; Su, H.; Wu, S. Video sensor-based complex scene analysis with granger causality. *Sensors* **2013**, *13*, 13685–13707. [CrossRef] [PubMed]
37. Winner, N. *The Theory of Prediction*; McGraw-Hill: New York, NY, USA, 1958; Chapter 8.
38. Barnett, L.; Seth, A.K. The {MVGC} multivariate granger causality toolbox: A new approach to granger-causal inference. *J. Neurosci. Methods* **2014**, *223*, 50–68. [CrossRef] [PubMed]
39. Nag, S. Vector quantization using the improved differential evolution algorithm for image compression. *Genet. Program. Evolv. Mach.* **2019**, *20*, 187–212. [CrossRef]
40. Shirali, A.; Kordestani, J.K.; Meybodi, M.R. Self-adaptive multi-population genetic algorithms for dynamic resource allocation in shared hosting platforms. *Genet. Program. Evol. Mach.* **2018**, *19*, 505–534. [CrossRef]
41. Karpov, P.; Squillero, G.; Tonda, A. Valis: An evolutionary classification algorithm. *Genet. Program. Evol. Mach.* **2018**, *19*, 453–471. [CrossRef]
42. Martínez, Y.; Trujillo, L.; Legrand, P.; Galvan-Lopez, E. Prediction of expected performance for a genetic programming classifier. *Genet. Program. Evol. Mach.* **2016**, *17*, 409–449. [CrossRef]

Article

Development Design of Wrist-Mounted Dive Computer for Marine Leisure Activities

Jeongho Lee [1] and Dongsan Jun [2,*]

1 Department of Convergence IT Engineering, Kyungnam University, Changwon 51767, Korea; ljhscorpion@hanmail.net
2 Department of Information and Communication Engineering, Kyungnam University, Changwon 51767, Korea
* Correspondence: dsjun9643@kyungnam.ac.kr

Received: 29 February 2020; Accepted: 27 April 2020; Published: 28 April 2020

Abstract: Divers conventionally use underwater notepad or flash to communicate each other in the water. For safe marine leisure activities, touchscreen based intuitive means of communications such as drawing and writing are needed to be integrated into the conventional dive computers. In this paper, we propose a wrist-mounted dive computer, so called DiverPAD, for underwater drawing and writing. For the framework design of proposed DiverPAD, firmware, communication protocol, user interface (UI), and underwater touchscreen functions are designed and integrated on DiverPAD. As a key feature, we deployed an electrical insulator based capacitive touchscreen which enables divers to perform underwater drawing and writing for clear and immediate information delivery in the water.

Keywords: wrist-mounted DiverPAD; electrical insulator; capacitive touchscreen; marine leisure activities

---

1. Introduction

Recent developments in image and video based technologies have enabled new services in the field of multimedia and recognition areas [1]. Those emerging services have been developed along with touchscreen based image processing and communication technologies. As a kind of emerging multimedia services, the necessity of underwater communication-device is increasing [2–4]. As industrial development of modern society and economic affluence have led to improvement in people's living standards, there are increasing demands for various marine leisure activities that will improve the quality of life.

Since underwater activities are not free and brain activity is reduced by breathing on air tanks, those limitations make it difficult to quickly respond to the surrounding risk factors (lack of air, dangerous marine life, and rapid algae). Therefore, a new type of dive computer is developing for marine leisure personnel to communicate underwater information [5–7].

In order to communicate clear and immediate information delivery through touchscreen based user friendly interface in the water, we implemented a new underwater communication-device that can be applicable to both professionals and publics for conducting marine leisure activities. The proposed underwater communication-device can support image/text data processing, and powerful visibility on the touchscreen. While conventional dive computers support several functionalities in terms of depth of water, water temperature, time, ascent/descent excessive speed alarm, and compass, the proposed communication-device, so called DiverPAD, can provide divers with touchscreen based memo functions (drawing and writing) as well as conventional functionalities.

The remainder of this paper is organized as follows: the previous works for recent dive computers are introduced in Section 2. The design and implementation of the proposed DiverPAD are described in Section 3. Finally, field test and conclusions are presented in Sections 4 and 5, respectively.

2. Previous Works

Because scuba diving, as a kind of typical marine leisure activities, has a limitation by the breathing in the water, most divers have been using the dive computer to observe underwater information [8,9]. In addition, divers can obtain underwater information such as diver's location from tablet screen within a waterproof case [10]. As depicted in Figure 1, typical dive computers are manufactured to be worn on a human wrist and they provide a variety of underwater information about dive time, remaining air, water dive depth, and temperature. Since the information directly affects to people's lives in an emergency, the accuracy and reliability of the information obtained by the dive computer is important factor to the divers.

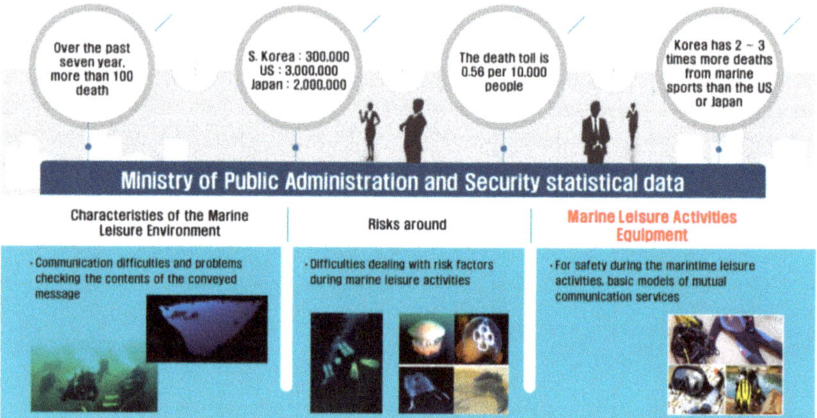

Figure 1. Statistical data in terms of marine safety accidents.

Even though conventional diver computers provide the basic information of underwater environment on display panel as shown in Figure 2, these cannot provide efficient communication methods among divers in the water. In addition, communication methods by using hand signals or writing board have limitations in transmitting and receiving accurate information in the water. In order to communicate underwater information about emergency situations in the water, new underwater device for supporting correct communication is needed to exchange accurate underwater information in the type of touchscreen based texts, symbols, and pictures [11].

In the case of conventional touch panels, when a point on the display screen is pressed or touched with a finger, a process to recognize the location corresponding to the pressed point works in three types, which are capacitive type, resistive film type, and infrared/ultrasonic type. Capacitive type is generally used in the conventional touch panels and it forms a constant capacitive layer on an insulating layer.

Because the weak electrical signal could not be properly detected in the water, when a finger touches a pad that is a transparent electrode on a substrate of the capacitive layer, touch signal is generated with its position. This is a difference between the existing touch panel method which detect the static signal and the proposed method to detect the weak electrical signal from the finger.

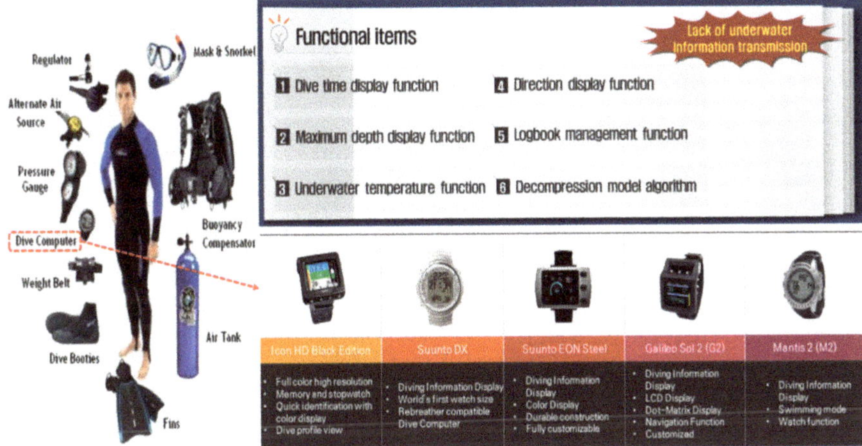

Figure 2. Functional items of conventional dive computers.

In this paper, DiverPAD was implemented with the capacitive method to enhance the correctness of electrical signal during marine leisure activities as shown in Figure 3.

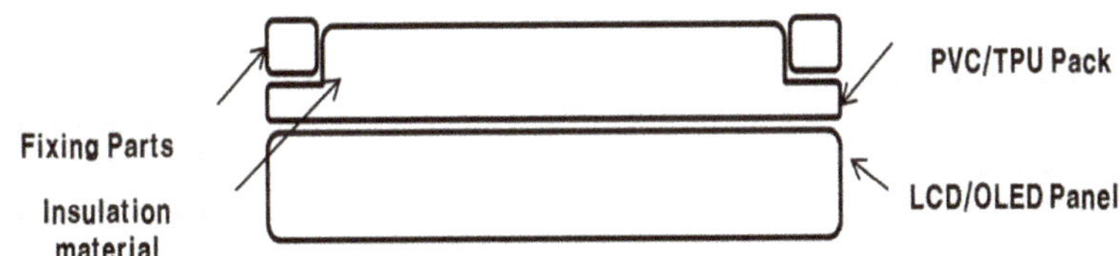

Figure 3. Touch panel of DiverPAD.

The PVC (Poly vinyl chloride)/TPU (Thermo Plastic Polyurethane) of Figure 3 is a material that is less corrosive and resistant to chemicals and it has the characteristics of generating static electricity. Since it has the characteristics of plastic, it can protect the touch panel from sea water and it is possible to transfer the user's touch input to the LCD (Liquid Crystal Display)/OLED (Organic Light Emitting Diode) panel by inducing static electricity. Inside the PVC/TPU, an insulating material, so called glucerine, was filled to allow smooth touch input.

3. Proposed Wrist-Mounted Dive Computer for Underwater Drawing and Writing First Bullet

The proposed DiverPAD are composed of applications processor (AP), battery charger, power management, display, sensors, and Bluetooth based connectivity modules as depicted in Figure 4. While DiverPAD can provide the functions of conventional dive computers, its main feature enables to provide the function of underwater drawing. Therefore, DiverPAD was newly modified on four major modules for supporting the underwater drawing, which are firmware, protocol, user interface, and underwater touch screen.

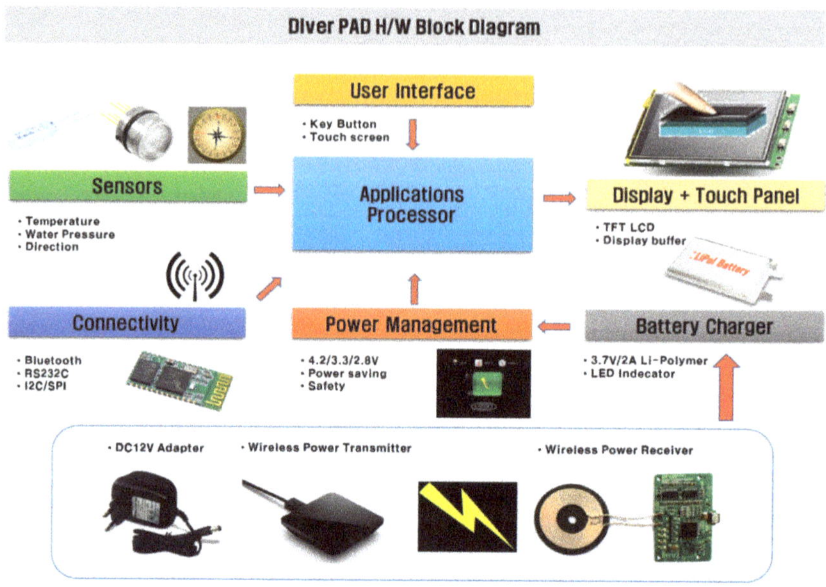

Figure 4. Block diagram of DiverPAD.

In addition, we set DiverPAD's target values in terms of supply voltage, operating time, operating temperature range, waterproof depth, charging time, and battery capacity to ensure drawing capabilities as well as existing conventional functions as described in Table 1.

Table 1. Target values of DiverPAD.

Item	Target Value	Description
Supply voltage	2.75~3.25 V	Portable terminals is optimized as 5 V
Operation time	1 h/3 h	Charging time about 1 h, use time about 3 h
Operating temperature	−20~70 °C	5 °C low in summer ~10 °C high in winter
Waterproof depth	50 m	Water pressure is 1 atmospheric per 10 m, up to 50 m
Charging time	1 h	As a wireless charging technology (2000 mA capacity: 1 h)
Battery capacity	2 mA more	Charging wirelessly (within 1 h)

3.1. Firmware

There are three main operations in terms of firmware in DiverPAD as followings: First, it prescribes the specific module code that makes up the hardware. Second, it controls the user interface (UI) for terminal device operation. Lastly, it stores user data (sensor, text, and image data).

As depicted in Figure 5, DiverPAD's firmware is installed into the main controller and it is combined with power status information, sensor detection, screen configuration, mobile linkage, and environment setting to maintain function-based calling relationship. After the initializations of each module is performed, DiverPAD operates specific functions like drawing according to "DiverPADTask" as presented in Algorithm 1, which is obtained from the touch panel by pressing the touch pen. As the insulator came into the reservoir space, upper and lower layers contact. Following this, the touch panel unit recognizes the point of contact and performs the drawing and writing process. When the user releases the touch, the restoring force raises the upper layer, as a result the insulator flows back from the reservoir space to the touch space and returns to its original state.

Algorithm 1 DiverPAD firmware logic

Input: positive integer button, positive integer user
Output: Action

1:	**Initialize MCU()**.	// Initialize the device
2:	**PowerInit()**.	// Power supply initialization
3:	**PowerEnableAll()**.	// Power supply
4:	**TimerInit()**.	// Set Timer
5:	**LEDInit()**.	// Initialize LED
6:	**MenuInit()**.	// Menu Initial screen
7:	**While** (1) **do**	
8:	Initial_TFT_LCD().	// Initialize LCD
9:	Setting Load().	// Loading the setting value
10:	GUI_Load()	// Load GUI
11:	StateInit	// Initialize State
12:		
13:	TFT_clear_screen().	// Reset touch function
14:	GUI_system_info().	// GUI information based sensor
15:		
16:	**If** (MenuTak)	// Wait for menu
17:	**Diver**	// Function execution
18:	**Else**	
19:	PowerDisa	// Off the power
20:	**end while**	
21:	**Return** action.	

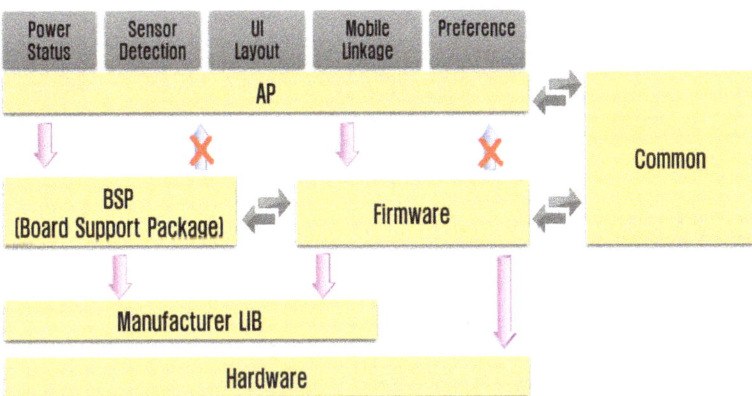

Figure 5. Firmware design of the proposed DiverPAD.

In addition, DiverPAD has eight supplementary LEDs as a means of emergency indications which are a real-time clock for dive time, a booster circuit, a switch module for various menu operations, a boot loader for firmware update, and a watchdog timer to prevent malfunction. Depending on the level of training of underwater activities, when the depth and time limit of each individual is exceeded, the LED module provides a notification function to the diver as presented in Algorithm 2. After the LED module was implemented in the form of firmware, we conducted the field tests of LED recognitions in both restricted and open waters. It can be seen that the LED brightness is high within 10 meters as shown in Figure 6.

Algorithm 2 dive qualification method

Input: positive integer User, integer depth, positive integer count
positive integer clock, char cert
Output: LED alarm 1: **if** (count < 5) then
2:          DiverRight <- beginner.
3:          **if** (depth > 18 m) then
4:               LED = Warning (RED).
5:          **Else**
6:               LED = Warning (Green)
7: **Else**
8:          if count is positive, then DiverRight = Openwater; Advenced; Rescue; Master;
9: **Return** DiverRight.
10:
11: **switch** (clock)
12:          **case** (9 < clock < 18)
13:               LED = Warning (Green)
14:          **if** (clock > 18)
15:               **if** (user = cert)
16:                    LED = Warning (Green)
17:          **Else**
18:               LED = Warning (RED)
19: **Return** LED alarm.

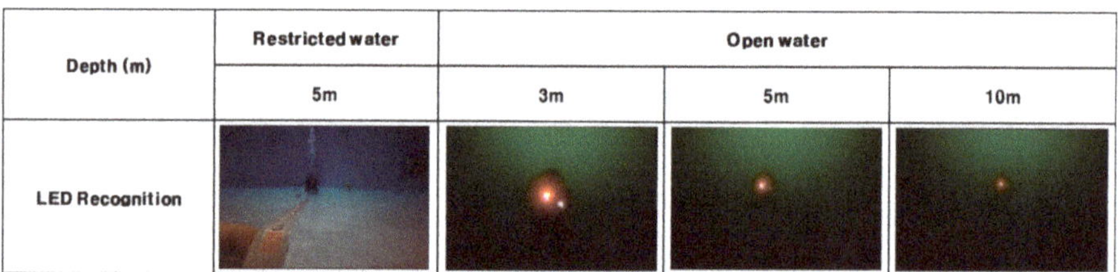

Depth (m)	Restricted water	Open water		
	5m	3m	5m	10m
LED Recognition				

Figure 6. Field tests of LED recognition.

3.2. Communication Protocol

DiverPAD allows diver to store text or image data in the water. In order to communicate those data from DiverPAD to handheld devices (e.g., smartphone), we present a communication protocol for DiverPAD, which is shown in Figure 7. Since the communication function is not smooth in water, an optimization method for fast and accurate information transmission was implemented based on the existing communication standards.

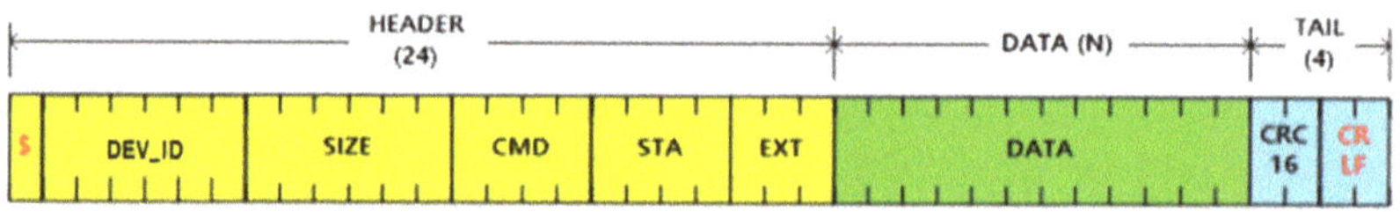

Figure 7. Communication protocol of DiverPAD.

As shown in Table 2, some fields of the header and tail are defined as constant values in the data packet of the DiverPAD and are designed to transmit text and image data in addition to the existing sensor data.

Table 2. Structure of DiverPAD data packet.

No	Division	Field Name	Description
1		START	Start of data packet
2		DEV_ID	ID of device
3	Header	SIZE	Length of total data
4		CMD	Command value
5		STA	Status value
6		EXT	Extended value: Reserved
7	Data	DATA	Data (Sensor values, text, and image)
8	Tail	CRC	CRC16 value (Error Check)
9		END	End of Data CR (=0x0D), LF (=0x0A)

Based on the presented data packet structure, the bidirectional communication between the DiverPAD and a handheld device is performed, and both the sensor information generated in the water and the information about the diver's marine leisure activity are transmitted into a handheld device. In particular, as shown in Table 3, CMD (Command) values in the packet structure are newly proposed in this paper to implement user message transmission about log data request, environment information setting, and data transmission of the DiverPAD (sensor information, text, image), so as to overcome the limitation of DiverPAD's memory capacity.

Table 3. Description of Command values.

No	CMD Value	Description
1	0000	Send custom phrase message
2	0001	Request log data information
3	0002	Set preference information
4	0003	Requests all data

3.3. Design of User Interface (UI)

As shown in Figure 8a, we show user environment, hardware environment, content management, and technical constraints by applying the UI design process. After wearing the DiverPAD, as a way to recognize the risk situation between marine leisure personnel, the depth and time were set according to the qualification situation, the number of times, and the curriculum. In case of increasing and descending, the update cycle for perceiving the risk situation was set in 1 s increments. The averages of previous unit time (10 s) section and the current unit time (10 s) section were analyzed.

As shown in Figure 8b, if the difference between the maximum and minimum depth for 1 min is within 3 m, the average depth value for the 1 min section is set as the current depth, and if it is within +2~−2 m from the current depth, the swim is displayed. In addition, when a difference in water depth of +2~−2 m or more occurs in the current depth, it is displayed increasing or descending, and when a difference of +10~−10 m or more occurs, a risk situation display screen is designed and implemented.

The above pattern analysis of marine leisure activity is divided into education, emotion, and emergency modes. First, in the education mode, the educational contents of divers are stored based on the water depth and displayed according to the rise and fall. In the emotion mode, the user displays self-written contents using the terminal device buttons. The emergency mode is activated if the control system determines the situation to be an emergency, sending emergency signals to the display. The module was implemented to identify any emergencies by comparing the values using an acceleration sensor, timer, and others. The above emergency situation information (maximum depth, dive time, water temperature, rise alert and visual, nap time, date, time, frequency, and others) is compared with the values of the sensor values and database thresholds.

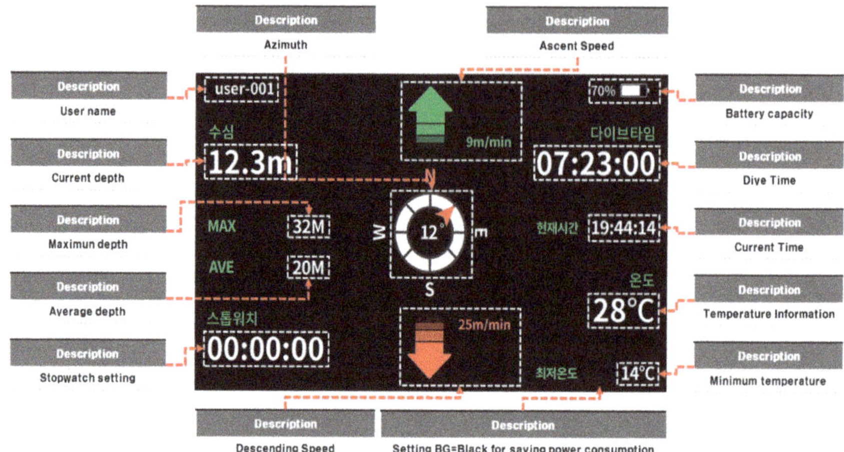

(**a**) Description of DiverPAD display

(**b**) Examples of DiverPAD display

Figure 8. User interface of DiverPAD.

3.4. Electrical Insulator Based Capacitive Touchscreen

DiverPAD adopted novel underwater touch function to conveniently communicate various information among divers in the water. The touch function of DiverPAD was designed as a means of the accurate and immediate communication tools in the process of data sending and receiving. The module to operate underwater touch function consists of touch panel unit and touch input controller. In general, touch function makes it possible to detect the force of touch pen pressing on the surface of LCD or OLED panel.

In this paper, underwater touch function of DiverPAD was implemented to adaptively adjust the force of touch pen pressing according to depth of water. DiverPAD allows a user to freely express intention by inputting through a touch input underwater. Thus, a touch input diver pad comprising a module is configured so that a sensor measurement module having a temperature sensor for sensing underwater temperature and a depth sensor for detecting water depth, as well as touch input and output modules including a touch panel section displaying data by the sensor unit, a battery module for supplying power and control was obtained.

The pressure protection module is made of tempered glass and it is made as thin as possible to lighten the weight and sharpen the design defined as shown in Figure 9.

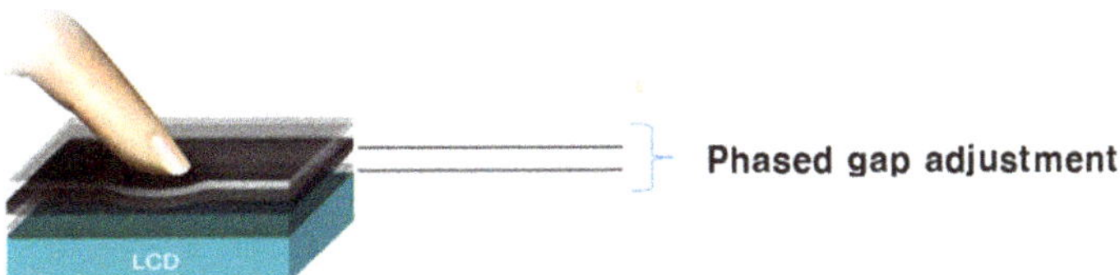

Figure 9. Phased gap adjustment corresponding to the different depth of water.

At the end of the touch input and output modules, a sealing module was installed to prevent the inflow of water and an insulated touch module was attached to the top due to unpredictable water flow and strong pressure of surrounding environment. In addition, the insulation touch module was configured to include an insulating layer between the upper layer and the lower layer as a pressure protection section. Accordingly, when there is no user's touch in the water, the upper layer and the lower layer are separated by the insulating layer so that no touch occurs.

The insulating layer is configured to serve to prevent malfunction of the touch panel unit due to water pressure in water. Based on this, the upper part of the DiverPAD terminal device was divided into an insulation touch module, and a space division divided into a touch space and a reservoir space, so that an empty chamber was formed between the space division and the insulation touch module. In this state, when the user presses and touches the touch space, the insulator is pushed to the reservoir space, and the upper layer and the lower layer come into contact.

The touch panel module senses the point of contact, and thus touch sensing is performed. In addition, when the user releases the touch, the upper layer rises due to the restoring force, and the insulator flows back from the reservoir space to the touch space and returns to the original state.

A method of attaching an adhesive film to improve durability against external pressure was used between the thin film display module and the touch panel module and between the touch panel module and the pressure protection module. The pressure protection module includes a pair of tempered glass, and is configured to withstand strong water pressure by configuring a pressure regulating gas to be injected between the pair of tempered glass. The weight can be reduced by excluding thick tempered glass to prepare for high pressure, and it is possible to prevent fogging inside due to low water temperature due to the external environment and the role of an intermediate buffer layer inside the body.

Based on this, it is possible to conveniently and freely input the contents of the person's intention during underwater leisure activities and to facilitate communication with the other party.

4. Experimental Result

4.1. Test Conditions

To verify the performance in terms of drawing, we defined underwater test conditions. Because it is not possible to measure the data of the same underwater environment due to the flow of water, it is difficult to compare the performance of DiverPAD with those of conventional dive computers.

After two divers wear both DiverPAD and conventional diver computer on their wrist, they checked various underwater information at each of five stages in a range of 0 m to 50 m of underwater and this experiment was repeated four times as shown in Figure 10. We confirmed that there is no difference in the measurement between the two dive devices as demonstrated in Figure 11.

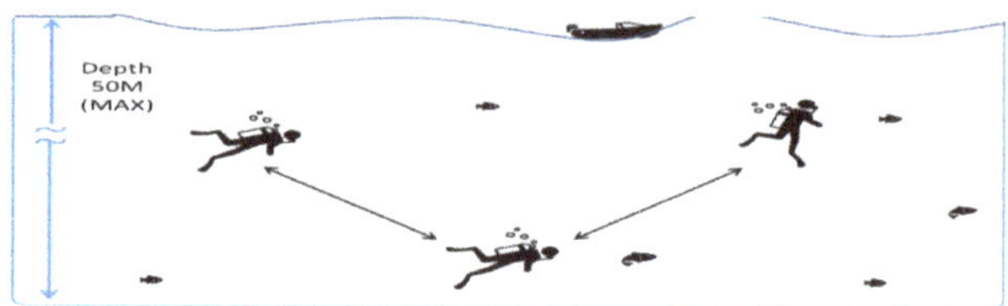

Figure 10. Test environment of DiverPAD.

		0 – 10	10 – 20	20 – 30
Depth (m)		8.5m(Dive Com) 8.6m(DiverPAD)	20.7m(Dive Com) 20.9m(DiverPAD)	28.3m(Dive Com) 28.1m(DiverPAD)
		※ Difference in height of wearing or wave movement		
Error Rate	Depth of water	±0.1(m)	±0.2(m)	±0.2(m)
	Dive time	±0:00:00(sec)	±0:00:00(sec)	±0:00:00(sec)
	Temperature	±0.1(°C)	±0.2(°C)	±0.2(°C)
	Ascent speed	±0.1(m)	±0.1(m)	±0.1(m)
	Descending speed	±0.2(m)	±0.2(m)	±0.5(m)
	Azimuth	±0.1(°)	±0.1(°)	±0.2(°)

Figure 11. Test conditions of DiverPAD.

In addition to data sensing, drawing and writing functions of DiverPAD were confirmed in the water as shown in Figure 12. Although the performance of the device could vary depending on the water pressure, inputs for drawing and writing were successfully processed in a range of 0 m to 50 m of underwater.

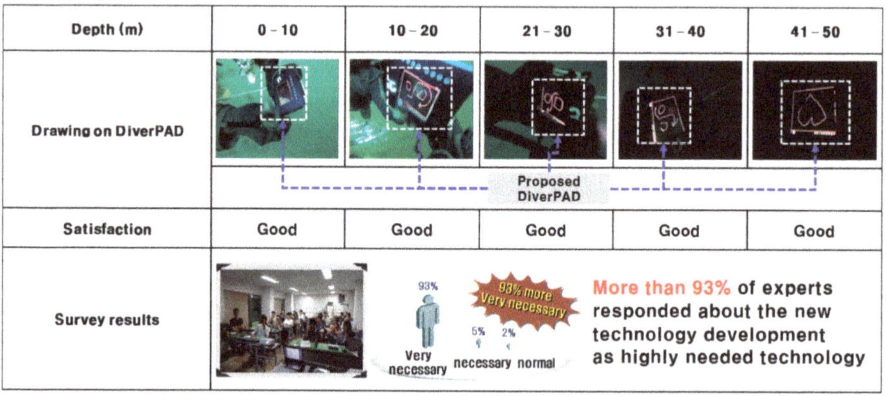

Depth (m)	0 – 10	10 – 20	21 – 30	31 – 40	41 – 50
Drawing on DiverPAD			Proposed DiverPAD		
Satisfaction	Good	Good	Good	Good	Good
Survey results					More than 93% of experts responded about the new technology development as highly needed technology

※ Number of test and experts : 7 experiments by 4 experts for three days

Figure 12. Drawing tests on DiverPAD.

4.2. Prototype Implementation and Field Test

Based on DiverPAD's prototype as shown in Figure 13, a pre-production sample test was conducted to examine the problems that may occur in the development process. In addition, an optimization method was derived by reviewing the product prototype, and conducting product assessment by repeatedly comparing the test results. Figure 14 shows that field tests were conducted on DiverPAD prototype to make the display and button operating parts waterproof through arranging insulated touch panel.

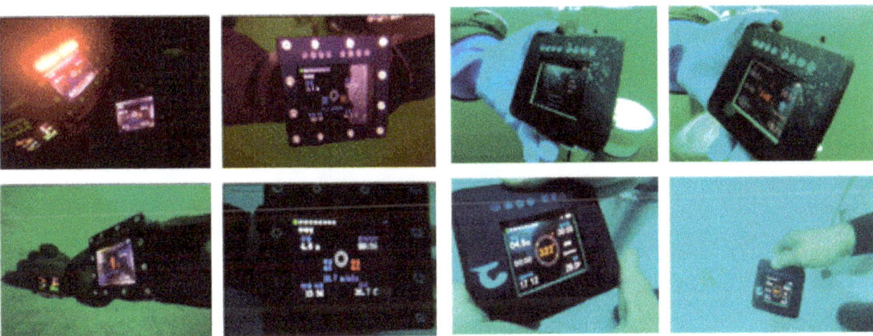

Figure 13. Prototype of DiverPAD.

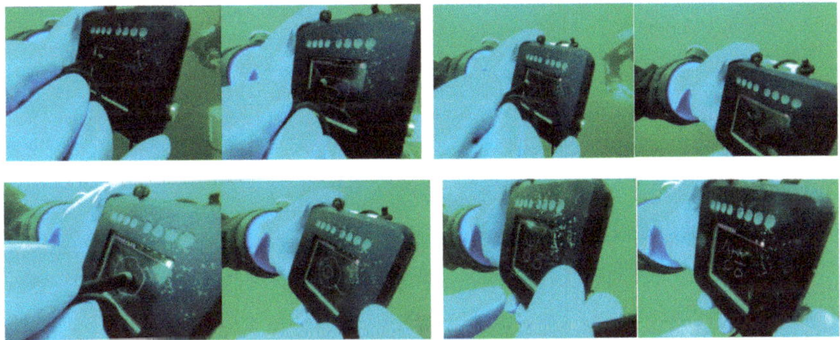

Figure 14. Field test of DiverPAD.

In the field tests, information such as temperature and distance depending on water depth can be checked on the touch screen of the DiverPAD. Measured values of temperature and depth on the DiverPAD can be analyzed for errors through Wavelet technique. In the fabrication of the DiverPAD, the distinctive features are an electrical insulator based capacitive touchscreen device that enables drawing and writing in the water, and modules for control and communication functions. The DiverPAD is manufactured using sensor devices that are the same with conventionally used diver computers. Therefore, it is less possibility that there is an error in the measurement value depending on the water depth compared to the existing diver computer. In addition, when comparing the two devices (i.e., DiverPAD in water, it was confirmed that there was no difference in measured values.

5. Conclusions

In this paper, we proposed a new product that is anticipated to provide a strong foundation for future technological advancements in marine environment communication, and to be used widely in marine leisure activities, including scuba diving. Figure 15 shows that the proposed DiverPAD will result in making marine leisure activities safer, more diverse, and systematic, especially with the rapidly growing number of divers every year. It is expected that the marine leisure industry will expand its accessibility.

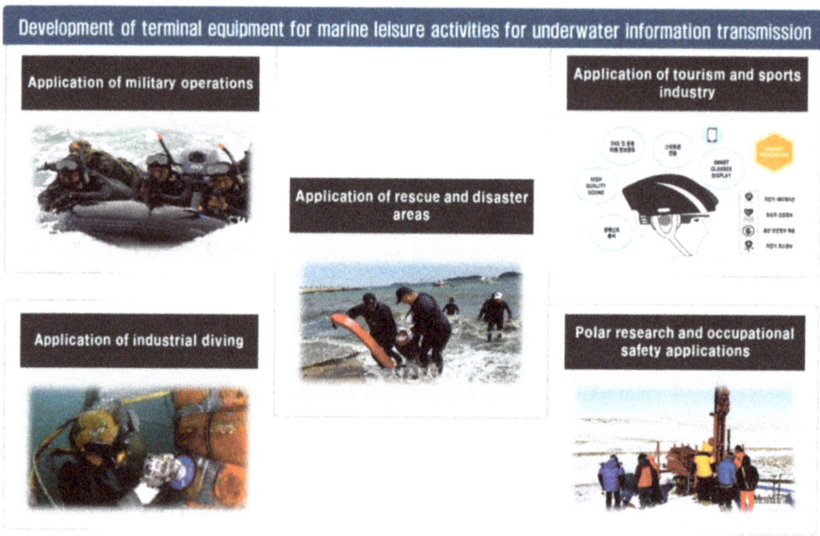

Figure 15. Application areas of DiverPAD.

Author Contributions: Conceptualization, J.L.; methodology, J.L.; software, J.L.; validation, D.J.; formal analysis, J.L.; writing—original draft preparation, J.L.; writing—review and editing, D.J.; supervision, D.J. All authors have read and agreed to the published version of the manuscript.

Funding: This research received no external funding.

Conflicts of Interest: The authors declare no conflict of interest.

References

1. Lee, J.H.; Kim, S.J.; Jun, D.S. Analysis of Machine Learning Engine Structure for Road Traffic Hazard Awareness. *Inst. Electron. Inf. Eng.* **2019**, *8*, 54–56.
2. Azzopardi, E.; Sayer, M. A Review of the Technical Specifications of 47 Models of Diving Decompression Computer. *Int. J. Soc. Underwater* **2009**, *29*, 63–72. [CrossRef]
3. Collins, K.J.; Baldock, B. Use of Diving Computers in Brittlestar Surveys. *Int. J. Soc. Underwater* **2007**, *27*, 115–118. [CrossRef]
4. Lucrezi, S.; Egi, S.M.; Pieri, F.; Burman, T.; Ozyigit, D.; Cialoni, G.; Thomas, A.; Marroni, M.; Saayman, M. Safety Priorities and Underestimations in Recreational Scuba Diving Operations: A European Study Supporting the Implementation of New Risk Management Programmes. *Front. Psychol.* **2018**, *9*, 383. [CrossRef] [PubMed]
5. Egi, S.M.; Cousteau, P.Y.; Pieri, M.; Cerrano, C. Designing a Diving Protocol for Thermocline Identification Using Dive Computers in Marine Citizen Science. *Appl. Sci.* **2018**, *8*, 2315. [CrossRef]
6. Eun, S.J.; Kim, J.Y.; Lee, S.H. Development of Customized Diving Computer Based on Wearable Sensor for Marine Safety. *IEEE Access* **2019**, *7*, 17951–17957. [CrossRef]
7. Kuch, B.; Koss, B.; Buttazzo, G.; Sieber, A. Underwater navigation and communication: A novel GPS/GSM diving computer. In Proceedings of the 35th Annual Scientific Meeting of the European Underwater and Baromedical Society (EUBS 2009), Aberdeen, UK, 25–28 August 2009.

8.	Kuch, B.; Buttazzo, G.; Azzopardi, E.; Sayer, M.D. GPS diving computer for underwater tracking and mapping. *Int. J. Soc. Underwater* **2012**, *189*, 189–194. [CrossRef]

9.	Pucciarelli, G. Wavelet Analysis in Volcanology: The Case of Phlegrean Fields. *J. Environ. Sci. Eng.* **2017**, *6*, 300–307.

10.	Bruno, F.; Barbieri, L.; Muzzupappa, M.; Tusa, S. Enhancing Learning and Access to Underwater Cultural Heritage Through Digital Technologies: The Case Study of the "Cala Minnola" Shipwreck Site. *Digit. Appl. Archaeol. Cult. Herit.* **2019**, *13*, e00103. [CrossRef]

11.	Bühlmann, A.A. *Tauchmedizin*; Springer: Berlin, Germany, 1995.

Article

An Efficient Separable Reversible Data Hiding Using Paillier Cryptosystem for Preserving Privacy in Cloud Domain

Ahmad Neyaz Khan [1,*], Ming Yu Fan [1,*], Muhammad Irshad Nazeer [2], Raheel Ahmed Memon [2], Asad Malik [3] and Mohammed Aslam Husain [4]

[1]   School of Computer Science and Engineering, UESTC, Chengdu 611731, China
[2]   Department of Computer Science, Sukkur IBA University, Sukkur 65200, Pakistan;
      irshad.nazeer@iba-suk.edu.pk (M.I.N.); raheelmemon@iba-suk.edu.pk (R.A.M.)
[3]   School of Information Science & Technology, SWJTU, Chengdu 614200, China; asad@my.swjtu.edu.cn
[4]   EED, REC, Ambedkarnagar 224122, India; mahusain87@gmail.com
*   Correspondence: ahmadnk500@gmail.com (A.N.K.); ff98@163.com (M.Y.F.); Tel.: +86-182-8024-9324 (A.N.K.)

Received: 23 April 2019; Accepted: 3 June 2019; Published: 17 June 2019

Abstract: Reversible data hiding in encrypted image (RDHEI) is advantageous to scenarios where complete recovery of the original cover image and additional data are required. In some of the existing RDHEI schemes, the image pre-processing step involved is an overhead for the resource-constrained devices on the sender's side. In this paper, an efficient separable reversible data hiding scheme over a homomorphically encrypted image that assures privacy preservation of the contents in the cloud environment is proposed. This proposed scheme comprises three stakeholders: content-owner, data hider, and receiver. Initially, the content-owner encrypts the original image and sends the encrypted image to the data hider. The data hider embeds the encrypted additional data into the encrypted image and then sends the marked encrypted image to the receiver. On the receiver's side, both additional data and the original image are extracted in a separable manner, i.e., additional data and the original image are extracted independently and completely from the marked encrypted image. The present scheme uses public key cryptography and facilitates the encryption of the original image on the content-owner side, without any pre-processing step involved. In addition, our experiment used distinct images to demonstrate the image-independency and the obtained results show high embedding rate where the peak signal noise ratio (PSNR) is $+\infty$ dB for the directly decrypted image. Finally, a comparison is drawn, which shows that the proposed scheme is an optimized approach for resource-constrained devices as it omits the image pre-processing step.

Keywords: reversible data hiding (RDH); image processing; cloud computing; public key cryptography (PKC); security

---

1. Introduction

Data hiding is one of the techniques used for securing data, apart from encrypting data. In data encryption technique, the original data are converted into a non-interpretable form so that adversary cannot extract any useful information. In data hiding, additional data are embedded into the carrier cover media (text, audio, video and image) in such a way that they remain concealed and can be extracted from the cover media later. Cover media is the original media that is used to carry additional data.

However, it is notable that, in data hiding, when data are extracted from the cover media, some form of distortion remains in the recovered cover media. In some scenarios (e.g., medical and satellite imagery), distortion in the cover image is inadmissible. That is, the image to be recovered on the receiver's side needs to be lossless. To cope with this problem, several reversible data hiding (RDH) techniques have been proposed. RDH is a technique to manipulate pixel bits of the cover image to create some space for embedding the additional data into the cover image, where both the additional data and the original cover image can be recovered completely. This is done while maintaining the perceptible quality of the carrier media. RDH schemes can be broadly classified into three categories: difference expansion [1], lossless compression [2] and histogram shifting [3].

With the growth of cloud-based applications [4,5], data outsourcing is one of the fields where the users are dependent on cloud for processing and storage. Security concern of the data owners using cloud for the cover media is addressed using encryption for the cover image. For the sake of management (using timestamp, tagging, image source information, etc.) of the encrypted media, the data hider embeds some additional data into the encrypted image. To achieve security and reversibility, the technique of RDH is used in the encrypted domain.

Reversible data hiding in encrypted images (RDHEI) is a method where additional data are embedded by the data hider into the encrypted cover image (obtained from the content-owner) to obtain marked encrypted image. This marked encrypted image is sent to the receiver, where recovery of both additional data and the cover image from the marked encrypted image is made losslessly.

Many RDHEI schemes based on the symmetric key have been proposed until now. In symmetric key cryptography, there is only one secret-key for both encryption and decryption, and key management is needed to share the secret key between the sender and the receiver. Some of the RDH schemes based on symmetric key cryptography are discussed below.

In 2008, Puech et al. [6] proposed the first RDHEI scheme where the Advanced Encryption Standard (AES) is used for encryption. Here, the encrypted image is divided into non-overlapping blocks of n-pixels each, and each block is responsible to carry one bit of additional data. The local standard deviation of the marked encrypted image is used to retrieve additional data on the receiver's side. In 2011, Zhang [7] successfully recovered an image similar to the original image with a secret key, which is encrypted using a stream cipher. With the help of data hiding key and spatial co-relation in the image, additional data and original image are recovered losslessly. Embedding is done using a two-step block division of the encrypted original image, and using least significant bit (LSB) flipping to identify the type of embedded bit (0 or 1). Hong et al. [8] improved Zhang's scheme [7] by exploiting the correlations in neighboring border pixels, which were not taken into consideration by Zhang [7].

At the receiver side, additional data in RDHEI are broadly recovered in two ways, namely non-separable and separable techniques, which are respectively depicted in Figure 1a,b. To extract the additional data from the marked encrypted image using the aforementioned schemes, the receiver must have the data-hiding key and the secret key. As the additional data can only be extracted after image decryption, this type of schemes falls under the category of non-separable RDHEI, as depicted in Figure 1a, where the additional data cannot be extracted without decrypting the marked encrypted image.

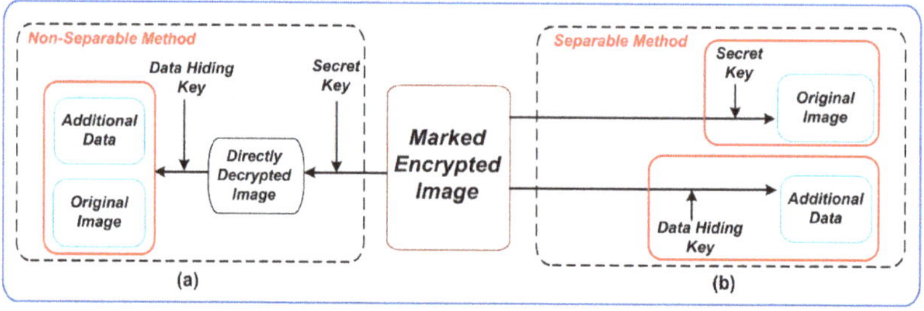

Figure 1. (**a**) Non-separable; and (**b**) separable methods in reversible data hiding.

Figure 1b depicts the separable method used in RDHEI, where the receiver can extract the additional data independent of image decryption, with the use of data hiding key only. The original image can be recovered by only using the secret key.

Zhang [9] proposed the first separable RDHEI method in 2012. RDHEI methods are fit for the cloud applications, where the data hider using the cloud can embed additional data for the sake of management. In addition, the receiver can extract additional data without knowing the content of the original image. The privacy of the cover image is still preserved as the additional data can be extracted without image decryption. Yin et al. [10] proposed a separable RDHEI scheme by breaking the cover image into non-overlapping blocks and the additional data are embedded into the blocks using block smoothness order and peak points.

Generally, at the content-owner's side, creating space for embedding data is done mainly in two ways: Vacating room after encryption (VRAE) (Figure 2a) and vacating room before encryption (VRBE) (Figure 2b). In VRAE, the space to embed additional data by the data hider is created after the image encryption. However, in VRBE, the space to embed additional data by the data hider is created before the image encryption. The framework followed by the above schemes is VRAE.

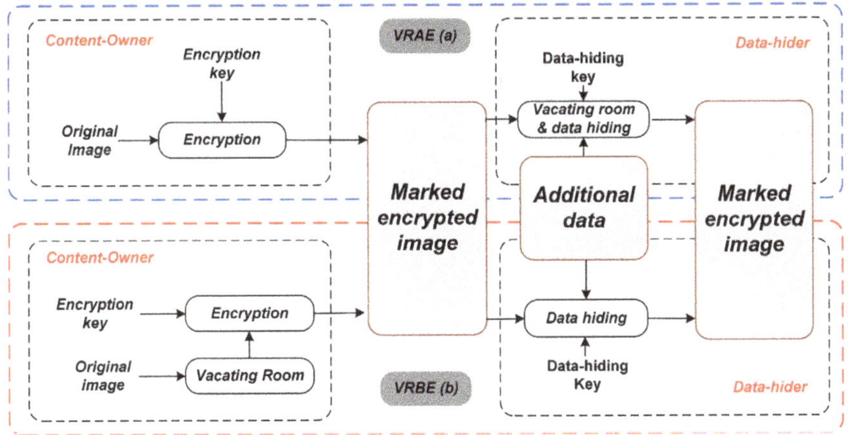

Figure 2. (**a**) Vacating room after encryption (VRAE); and (**b**) vacating room before encryption (VRBE).

After the image encryption, entropy rises, which allows less space for payload (additional data). To tackle this problem, Ma et al. [11] proposed the first VRBE scheme by preprocessing the cover image, using traditional RDH scheme. Zhang et al. [12] preprocessed the cover image by estimating some pixels in the cover image. Then, they embedded additional data by using histogram shifting for estimated prediction errors.

Qian et al. [13] significantly enhanced image quality and embedding rate using histogram modification, based on n-nary histogram intensities. However, the image histogram leakage reduced image security. Zhang et al. [14] embedded compressed additional data into an encrypted image using low-density parity check code. Zheng et al. [15] compressed the pixel LSBs, in the chaotic-encrypted image, using Hamming distance to embed data. Cao et al. [16] used patch level sparse representation to embed data, using the sparse coding technique. Self-embedding of leading residual errors and embedding over-complete dictionary (created by using sparse coefficients to represent cover image) into the encrypted image is used in this technique.

Recently, public key cryptography is used for efficient key management over cloud [17,18], specifically in RDHEI [19–24]. In asymmetric key (or public key) cryptography, the key for encryption (public-key) and the key for decryption (private-key) are different. In the context of RDHEI using public-key, Chen et al. [19] proposed the first signal based reversible data hiding, for multiple signals and data hiders, using Paillier cryptosystem [25]. Shiu et al. [20] improved the work of Chen et al. [19], having a drawback of inherent overflow, by grouping 64-pixels of the encrypted image followed by compression. Li et al. [21] used difference histogram shifting based on the additive homomorphic property to embed additional data. Wei et al. [22] enhanced the work of Li et al. [21], by taking the cross-shaped division mask instead of block-based division mask to use all the embedding opportunities. Zhang et al. [23] proposed reversible and lossless RDHEI in public-key, with better performance in terms of PSNR of the directly decrypted image compared to some of the previous schemes.

Tai et al. [24] proposed RDHEI scheme based on Paillier's cryptosystem. Here, the image is preprocessed before encryption by dividing each pixel into two parts, called encrypted units: EU^1 and EU^2. The order of the two encrypted parts of a pixel, namely EU^1 and EU^2, is exploited to embed the bits of additional data in the data hiding phase. If the additional bit to be embedded is 1 and $EU^1 < EU^2$, then swap EU^1 and EU^2. If the additional bit to be embedded is 0 and $EU^1 > EU^2$, then swap EU^1 and EU^2. On the receiver side, the additive homomorphic property of Paillier's cryptosystem is used to recover the original image losslessly. Again, the order of the two encrypted parts, EU^1 and EU^2, is used to extract the bits of additional data from the encrypted image. In [24], the image is preprocessed before encryption and this makes the content-owner to send double the size of image to the data hider. This issue is the one addressed in our proposed scheme.

In this paper, an efficient separable reversible data hiding scheme for encrypted images is proposed. It enjoys the benefits of using Paillier cryptosystem [25], which provides privacy preserving advantage over the cloud. The usage of our scheme befits cloud domain. Thus, a real-life application scenario of the proposed scheme over cloud is vividly explained in Section 3.6. Paillier's cryptosystem is used to encrypt the original image with the public key. When the encrypted image is received by the data hider, the data-hiding key and the public key is used to embed additional data into the encrypted image to obtain marked encrypted image. The additional data and the original image are recovered losslessly at the receiver's side using respective data hiding and private keys, independent of each other. This is done in a separable manner, which means for embedded additional data extraction, encrypted image is not required to be decrypted and vice versa.

After the brief discussion of previous schemes, we move on to a brief introduction to Paillier's public key cryptosystem in Section 2. In Section 3, the proposed scheme is discussed with an example. Experimental results and discussions are covered in Section 4. Finally, a conclusion is drawn in Section 5.

2. Paillier Cryptosystem

Paillier cryptosystem [25] is one of the most widely used public key cryptosystems, based on homomorphic properties along with probabilistic properties. Our scheme uses Paillier cryptosystem [25] for the encryption, decryption and its homomorphic properties have been exploited for data embedding. Homomorphic implies that arithmetic operations will be preserved from plaintext space to ciphertext space. It has given a strong base for secure computing on cloud, as it deals with privacy-preserving concerns of data owners. This is due to its property of semantic security, i.e., the same plaintext gives different ciphertexts, which obviously can be recovered using the private key. It means one cannot distinguish between the different ciphertexts generated from the same plaintext.

2.1. Key Generation

Two large prime numbers p and q of equal length are randomly chosen, satisfying $gcd\ (pq, (p-1) \times (q-1)) = 1$. Subsequently, the message sender calculates N and λ using $N = pq$ and $\lambda = lcm(p-1, q-1)$. Again, some integer g following $g \in \mathbb{Z}_N^*$ is randomly selected, such that it satisfies $gcd(L(g^\lambda mod N^2), N) = 1$, where $L(x) = (x-1)/N$. As a result, we get the public key (N, g) and private key (λ).

2.2. Encryption

Let m be the given message to be encrypted, where integer $m \in \mathbb{Z}_N$. Select an integer $r \in \mathbb{Z}_N^*$ randomly. The corresponding ciphertext c can be obtained using:

$$c = E[m, r] = g^m \times r^N mod N^2 \tag{1}$$

where $E[\cdot]$ represent encryption function having the property of Paillier cryptosystem, and the ciphertext c lies in the set $\mathbb{Z}^*_{N^2}$.

2.3. Decryption

Using the private key (λ) in decryption function $D[\cdot]$, the user can decrypt the ciphertext c to get original message m, using:

$$m = D[c] = \frac{L(c^\lambda mod N^2)}{L(g^\lambda mod N^2)} mod N. \tag{2}$$

2.4. Homomorphic Property

The encryption function is additive homomorphic, i.e., the multiplication of two ciphertexts will decrypt to the sum of their corresponding plaintexts. For two plaintexts $m_1, m_2 \in \mathbb{Z}_N$ and randomly selected integers $r_1, r_2 \in \mathbb{Z}^*_N$, the corresponding ciphertexts $c_1, c_2 \in \mathbb{Z}^*_{N^2}$ can be calculated as $c_1 = E[m_1, r_1] = g^{m_1} \times r_1^N mod N^2$ and $c_2 = E[m_2, r_2] = g^{m_2} \times r_2^N mod N^2$. In addition, c_1 and c_2 satisfy Equations (3) and (4):

$$c_1 \times c_2 = g^{m_1+m_2} \times (r_1 \times r_2)^n mod N^2, \tag{3}$$

$$D[g^{m_1+m_2} \times (r_1 \times r_2)^N mod N^2] = m_1 + m_2 mod N^2. \tag{4}$$

Semantic security is assured with the homomorphic property of the Paillier cryptosystem, as shown for a message $m \in \mathbb{Z}_N$ in Equation (5), where $r_1, r_2 \in \mathbb{Z}^*_N$, $c_1 = E[m, r_1]$ and $c_1 = E[m, r_2]$.

$$c_1 \neq c_2 \tag{5}$$

3. Proposed Scheme

The idea of our scheme is based on the separable method, as illustrated in Figure 3. The notations used in the proposed scheme are denoted in Table 1. It comprises of three stakeholders: the content-owner, the data-hider, and the receiver.

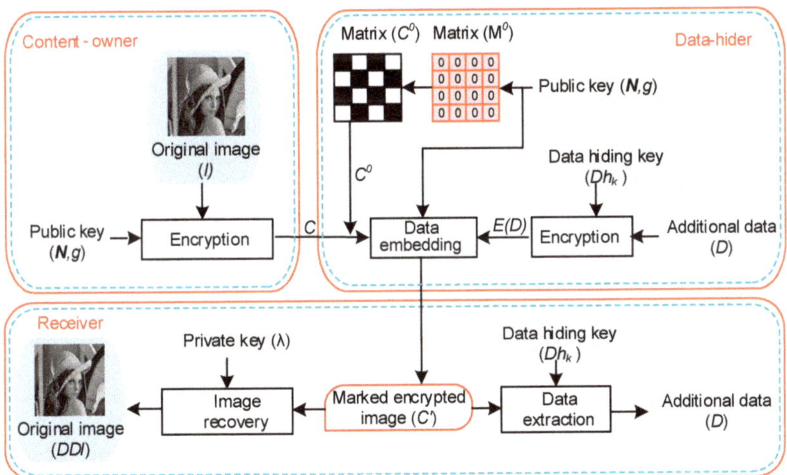

Figure 3. Working of the proposed scheme.

Table 1. Key notations used in the proposed scheme.

Notations	Description
(N, g)	A public key for encryption
(Dh_k)	A data hiding key for hiding and recovery of additional data
λ	A private key possessed by the receiver for image recovery
I	An original image of size $L \times B$
k	Index for each pixel where $1 \le k \le L \times B$
I_k	kth pixel of the original image I
C_k	An encrypted value of I_k i.e., kth encrypted pixel of the encrypted image C
$E[\cdot]$	An encryption function
$D[\cdot]$	A decryption function
r_k	A randomly selected integer for each I_k such that $r_k \in \mathbb{Z}_N^*$
C	An encrypted image generated from all C_k achieved by pixel by pixel encryption
D	Additional data of size $L \times B$ bits to be embedded
$E(D)$	Encrypted additional data using (Dh_k)
M^0	A zero matrix of size $L \times B$ where all the elements are zero
C^0	A matrix resulting from encryption of matrix M^0
C_k^0	kth encrypted value of "0" from the matrix C^0
PU	A padded unit
PU_k	kth padded unit consisting of pair (C_k^0, C_k) where $1 \le k \le L \times B$
C'	A marked encrypted image *(MEI)* constituted from all the padded units *(PUs)*
DDI	A directly decrypted image
DDI_k	kth pixel of directly decrypted image *(DDI)*

Data-hider uses (Dh_k) to encrypt the additional data (D) and uses (N, g) to embed the encrypted additional data $E(D)$ into the encrypted image. At the receiver's end, additional data are recovered from the marked encrypted image C' using (Dh_k), and the original image I is completely recovered in a separable manner, using (λ) owned by the receiver.

3.1. Image Encryption

In this step, the content-owner scans each pixel of the original image I in the order from left to right and top to bottom, as shown in Figure 4 (for an image of size 5×5), to get the index k for each pixel. The size of I is $L \times B$, where $1 \le k \le L \times B$. This order is used throughout our proposed scheme to get the value at index k.

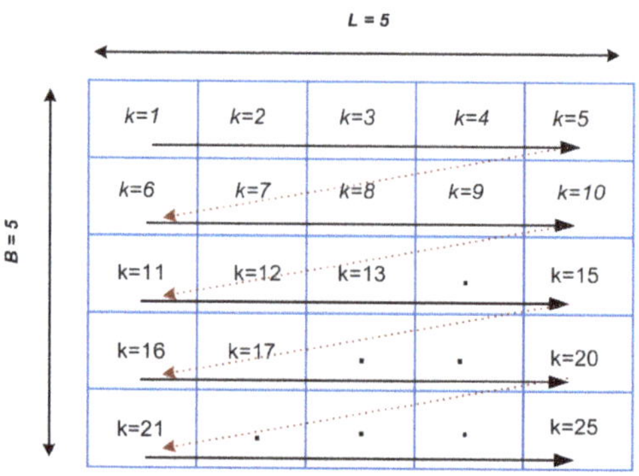

Figure 4. Scanning order to get index k for an image of size 5×5.

After scanning, the content-owner encrypts each I_k taken from original image I. Public key and Equation (1) are used to encrypt I_k to get the corresponding encrypted value C_k as follows:

$$C_k = E[I_k, r_k] = (g)^{I_k} \times (r_k)^N \ mod \ \mathbf{N}^2 \tag{6}$$

where $E[\cdot]$ is the encryption function and r_k is randomly selected integer for each I_k such that $r_k \in \mathbb{Z}_N^*$. Step-wise encryption of the original image is as follows.

Step 1. Scan I for each pixel I_k in order from left to right and top to bottom (Figure 4).
Step 2. Encrypt each I_k using the public key (N, g) as in Equation (6) to get the encrypted pixel C_k.
Step 3. After encrypting all the k pixels, the encrypted image generated is C.

3.2. Data Embedding

In this step, initially, the data hider uses the data hiding key (Dh_k), to encrypt the additional data (D) to obtain $E(D)$. Here (also refer to Section 3.5), it is assumed that the number of bits in $E(D)$ is equal to the number of pixels in the encrypted image C. Before embedding, the data hider generates a separate zero matrix M^0 of size $L \times B$, where all the elements are zero. The data hider uses Equation (1) and the public key to encrypt M^0, to obtain the encrypted matrix C^0. Thus, after encryption the size of matrix C^0, is the same as the size of the encrypted image C. It can be noted that each element of the matrix C^0 is distinct but of the same size, by the property of semantic security (Equation (5)) inherent in Pallier cryptosystem.

In the next step, each bit of $E(D)$ is embedded into the corresponding C_k of the encrypted image C. That is, each encrypted pixel C_k is responsible for carrying one bit of $E(D)$. For embedding, the data hider has to pad kth encrypted value of "0" (i.e., C_k^0) from the matrix C^0, with the corresponding kth encrypted pixel (i.e., C_k) of C. This padding results into a matrix of $L \times B$ padded units (PUs). As each PU is composed of two encrypted values (C_k, C_k^0), $L \times B$ padding units make $2 \times L \times B$ encrypted values. This makes the size of the PU matrix equal to $2 \times L \times B$, which is double the size of the C i.e., $L \times B$. All the PUs constitute the encrypted image with additional data, i.e., the marked encrypted image C'.

Padding Procedure

In padding, four cases arise as depicted in Figure 5: two cases, if the bit of $E(D)$ is "0", and two cases, if the bit of $E(D)$ is "1". For embedding, we compare values C_k and C_k^0.

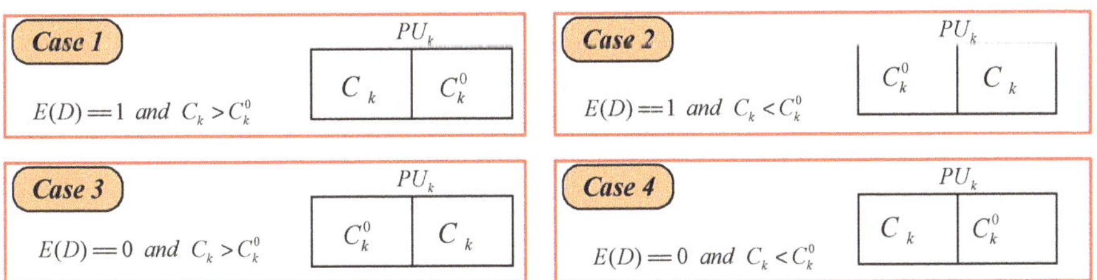

Figure 5. Four cases for padded unit (PU) construction.

When the bit of $E(D)$ is equal to 1, the padding is done such that the bigger value (between C_k and C_k^0) is at the first position in PU_k. Thus, if $C_k > C_k^0$, then C_k is placed first in the PU_k, as in Figure 5 (Case 1). If $C_k < C_k^0$, then C_k^0 is placed first in the PU_k, as in Figure 5 (Case 2).

When the bit of $E(D)$ is equal to 0, the padding is done in such a way that the bigger value (between C_k and C_k^0) will be at the second position in the PU_k. If $C_k > C_k^0$, then C_k is placed second in the PU_k, as in Figure 5 (Case 3). If $C_k < C_k^0$, then C_k^0 is placed second in the PU_k, as in Figure 5 (Case 4).

The step by step procedure for data embedding into the encrypted image can be understood from the following algorithm:

Let us assume C_k to be the encrypted image for the pixel k where $k \in [1, (L \times B)]$ and C_k^0 be the encrypted value of "0" for pixel k.

Step 1. With the help of data hiding key (Dh_k), bits of additional data (D) are encrypted in order to generate bits of encrypted additional data $E(D)$.

Step 2. If the bit of $E(D)$ to be embedded is 1:

If $C_k > C_k^0$ in the selected PU_k, then we rearrange the order in PU_k by appending C_k^0 after C_k. (That is, the bigger value is first and the smaller value is second in the PU_k (Figure 5, Case 1).)

Otherwise, we append C_k after C_k^0. (That is, the bigger value is first and the smaller value is second in the PU_k (Figure 5, Case 2).)

Step 3. If the bit of $E(D)$ to be embedded is 0:

If $C_k > C_k^0$ in the selected PU_k, then we rearrange the order in PU_k by appending C_k after C_k^0. (That is, the smaller value is first and the bigger value is second in the PU_k (Figure 5, Case 3).)

Otherwise, we append C_k^0 after C_k. (That is, the smaller value is first and the bigger value is second in the PU_k (Figure 5, Case 4).)

Step 4. After the encrypted image C has been embedded with the $E(D)$, the marked encrypted image C' is obtained with all the PU_k.

3.3. Data Extraction

After the receipt of the marked encrypted image C' on the receiver side, the embedded $E(D)$ is extracted using the data-hiding key (Dh_k). The step by step procedure for data extraction from C' is as follows:

Step 1. Scan the marked encrypted image C' in the same manner as used in the encryption and embedding phase, i.e., left to right and top to bottom (Figure 4).

Step 2. For each of the selected PU_k, Steps 3 and 4 are performed.

Step 3. If the first value of the pair (C_k^0, C_k) in the selected PU_k is bigger than the second value, then the embedded bit of $E(D)$ is "1".

In this case, "1" will be extracted.

Step 4. If the first value of the pair (C_k^0, C_k) in the selected PU_k is smaller than the second value, then the embedded bit of $E(D)$ is "0".

In this case, "0" will be extracted.

Step 5. After extracting all the bits, the encrypted additional data $E(D)$ is constituted.

Data hiding key (Dh_k) is used to regenerate the original additional data (D).

3.4. Image Recovery

In this step, if the receiver wants to recover the original image I, he must own the private key (λ). After receiving the marked encrypted image C', the receiver applies homomorphic multiplication on each pair of (C_k^0, C_k) in PU_k of C', to get the corresponding C_k, and then, decrypts each C_k to get kth pixel of the directly decrypted image DDI_k with private key (λ) using:

$$DDI_k = D[C_k] = \frac{L((C_k)^{\lambda} mod N^2)}{L(g^{\lambda} mod N^2)} mod N, \tag{7}$$

where $D[\cdot]$ is the decryption function, λ is the private key and DDI is the directly decrypted image. In our scheme, DDI results in the completely recovered original image I, i.e., no post-processing on DDI is further required. Step-wise procedure used to recover the original image using the private key (λ) is as follows:

Step 1. Scan the marked encrypted image C' in the same manner as used in the encryption and embedding phase, i.e., left to right and top to bottom (Figure 4).

Step 2. For each selected PU_k.

Step 3. Apply homomorphic multiplication ($\times$) to each pair (C_k^0, C_k) in PU_k, to obtain C_k, such that $C_k = C_k^0 \times C_k$.

(The order of C_k^0 and C_k does not affect the result.)

Step 4. Each C_k is decrypted using the private key (λ) to give the corresponding DDI_k.

Step 5. DDI is obtained constituting all the DDI_k.

It can be noted that, in Step 3., the original values (unencrypted values) of C_k^0 and C_k are "0" and I_k respectively. When homomorphic multiplication is applied to (C_k^0, C_k), it means internally zero is added to the value I_k. Thus, we get encrypted value C_k (i.e., $C_k = (C_k^0 \times C_k)$ = encrypted value of $(0 + I_k)$).

3.5. Exemplifying Our Proposed Scheme

Figure 6 shows the working of the proposed scheme at the data-hider side. It is supported by the following example: our example shows the working for the 1st pixel value $I_1 = 65$, of the original image. Let the public key = (1763, 94) and the private key = (840). Using the public key and the encryption function $E[\cdot]$, we get $E(65) = C_1 = (184, 481)$ and $E(0) = C_1^0 = (304, 186)$. Let the bit to be embedded be 1, i.e., $E(D) = 1$. For embedding, we compare C_1 and C_1^0. According to Step 2 of Section 3.2 (Figure 5, Case 2), when $C_1^0 > C_1$, we put C_1 after C_1^0 in PU_1. Thus, here, $PU_1 = (304, 186; 184, 481)$, i.e., the bigger value is first and the smaller value is second. The receiver extracts the first bit of $E(D)$, by reading the order in PU_1. In $(304, 186; 184, 481)$, the first value is bigger than the second value, so the embedded bit of $E(D)$ is 1. However, this bit will be further decrypted using data-hiding key (Dh_k) to get D. Furthermore, the receiver having the private key (840) gets the first pixel value DDI_1, for the directly decrypted image DDI, from the $PU_1 = (304, 186; 184, 481)$ by using homomorphic multiplication. This is done as: $\lambda(PU_1) = \lambda((C_1^0 \times C_1) mod N^2) = \lambda((304, 186 \times 184, 481) mod 1763^2) = \lambda(0 + 65) = 65$. Thus, the pixel value DDI_1 is same as the original pixel value $I_1 = 65$.

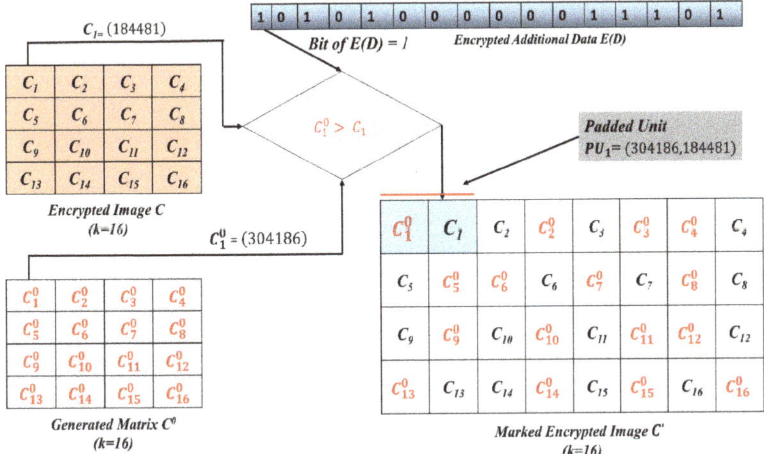

Figure 6. Data embedding process in our proposed scheme after getting an encrypted image C of size (4×4) from the content-owner. C_1^0 and C_1 are the values for index $k = 1$ of C^0 and C, respectively.

3.6. Proposed Scheme in Cloud Domain

There are a number of resource constrained devices based on cloud services; to show the process flow, involved hardware and software services we take an example of closed circuit television (CCTV) cameras installed on roads for monitoring the traffic against any traffic law violation. As shown in Figure 7, the workflow of the given scenario can be divided into following three phases:

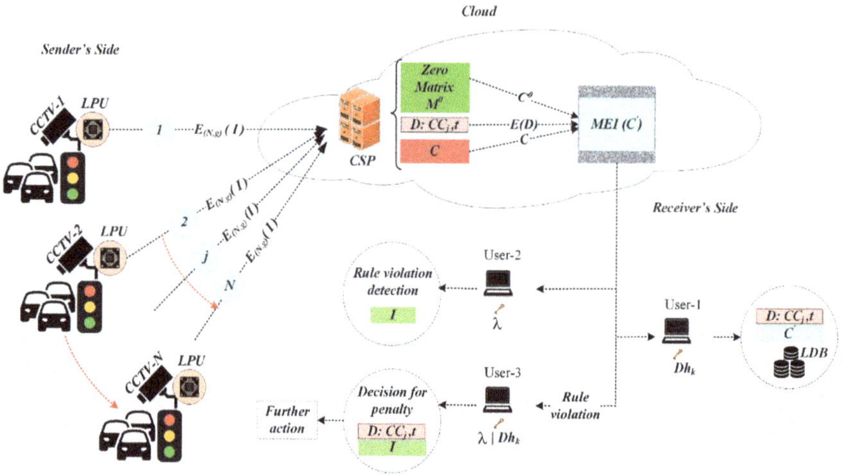

Figure 7. Of the proposed scheme in the cloud scenario.

(1) Image capturing and encryption at the sender's end

(2) Generating MEI over the cloud

(3) Authorized complete recovery of the D and I

(1) Image capturing and encryption: As shown in Figure 7, at the sender's end, there can be a number of CCTVs (1, 2, 3, ..., j, ..., N), all of which are programmed to capture the images with a fixed time interval. To obtain the encrypted image C, the local processing unit (LPU) encrypts the image (I) captured from jth CCTV. This is done using public key (N, g) and the encryption algorithm $E[\cdot]$. Once the image is encrypted, it is sent to the cloud service provider (CSP) acting as the data hider, for marking.

(2) Generating marked encrypted image (MEI): The data hider embeds the additional data D, such as the time (t) of the captured image or CCTV camera-id (CC_j). To make D secret, it is encrypted with the data hiding key (Dh_k) to obtain $E(D)$. A zero matrix M^0 equal to the size of the original image (I), is encrypted with public key (N, g), to obtain C^0. C^0 is used, to embed $E(D)$ into the encrypted image C, to obtain C' as *MEI*.

(3) Authorized access and recovery: There are three type of end users with different access policies. The first category of users hold the data-hiding key (Dh_k) These users are authorized to access additional data (e.g., the camera-id CC_j, capturing time t, etc.) from the MEI. This type of data can be used to categorize and store the MEI in a local data base (LDB). The second type of users hold private key (λ) for retrieving original image. This type of authorized users exploit the original image for detecting any traffic rule violation such as wrongful crossing, incorrect overtake, etc. The third type of users hold both data hiding (Dh_k) and private (λ) keys. This enables users to access both additional data and the original image. Thus, for any rule violation, the image can be crosschecked with the corresponding additional data (time t, camera-id CC_j, location, etc.) for validity. Reasonable penalties can be applied to the subjects flouting the rules.

4. Experimental Results

To evaluate our proposed scheme, we used miscellaneous dataset [26] of gray-scale images, each having size 512×512. For simulation purpose, MATLAB 2015 b was used. We measured our results on the basis of embedding rate in terms of bit per pixel (bpp), visual quality in terms of peak signal noise ratio (PSNR) and structural similarity index matrix (SSIM). The embedding rate (ER) can be calculated using Equation (8) as follows:

$$Embeding\ Rate(ER) = \frac{Number\ of\ total\ embedded\ bits}{Number\ of\ total\ pixels\ of\ the\ image}. \tag{8}$$

To calculate PSNR, Equation (9) is used

$$PSNR = 10 \times \log_{10} \frac{255^2}{MSE}(dB), \tag{9}$$

where mean square error (MSE) is calculated as follows:

$$MSE = \frac{1}{L \times B} \sum_{i=0}^{L-1} \sum_{j=0}^{B-1} (I(i,j) - R(i,j))^2. \tag{10}$$

Structural similarity index matrix (SSIM) is a method to measure the structural similarity between the recovered and reference images, where the reference image is the original image. It is used as a measure of perceived degradation in structural information of the recovered image with respect to the original image. The range of SSIM is $[-1, 1]$, where 1 shows that the images are identical. For original image I and the recovered image R, SSIM is calculated as follows

$$SSIM(I, R) = \frac{(2\mu_I\mu_R + c_1)(2\sigma_{IR} + c_2)}{(\mu_I^2 + \mu_R^2 + c_1)(\sigma_I^2 + \sigma_R^2 + c_2)}. \tag{11}$$

where μ_I, μ_R are the mean of images I and R, respectively, and σ_I^2, σ_R^2 are the variance of images I and R, respectively. σ_{IR} is the covariance of I and R, respectively. Here, $c_1 = (k_1 L)^2, c_2 = (k_2 L)^2$, where L is the dynamic range of values of pixel and $k_1 = 0.01$, $k_2 = 0.03$ by default.

4.1. Results Showing Independence of the Proposed Scheme for Different Images

From the selected database [26] we experimented on four standard gray-scale images, as depicted in Figure 8a–d (Lena, Baboon, Boat and Airplane, respectively) with size 512×512 each. The respective images when decrypted directly from the marked encrypted image are depicted in Figure 8e–h. The embedding rate is 1 for all the four images, using Equation (8).

In the proposed scheme, for each pixel, one padded unit (PU) is constructed and each PU is responsible to carry 1-bit of additional data. This results in embedding rate of 1-bit per pixel, i.e., 1 bpp as each pixel is responsible to carry one bit of additional data. PSNR for directly decrypted images, calculated using Equation (9), was found to be $+\infty$ dB for all four images. It implies that the recovery of all the four original images was complete. This also shows that no post-processing on the directly decrypted image was needed to completely recover the original image. Using Equation (11), SSIM value for the four images was 1. This means that the structural similarity index of the recovered images against the original images was the same. Values for PSNR and SSIM support that the perceptual quality of the directly decrypted image with respect to the original image was the same. Table 2 shows the results for the metrics embedding rate, PSNR, and SSIM for all four images. It is inferred that the embedding rate, PSNR and SSIM are independent of the texture of chosen images, as these metric values remained constant for the four distinct images. This implies that, for a complex image, the embedding rate will be the same as that for a smooth image for our proposed scheme.

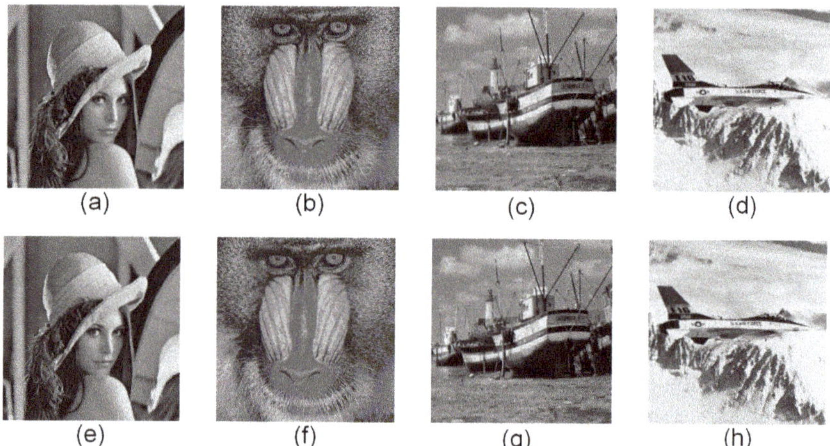

Figure 8. (**a–d**) The four original standard gray-scale images of size 512×512; and (**e–h**) the directly decrypted images (DDI) to check embedding rate, peak signal noise ratio (PSNR), structural similarity index matrix (SSIM) for different images in our scheme.

Table 2. Embedding rate, peak signal noise ratio (PSNR), structural similarity index matrix (SSIM) for four distinct standard test images (Figure 8a–d) in the proposed scheme.

Test Images	Embedding Rate (bpp)	PSNR	SSIM
Lena, Baboon, Boat and Airplane	1.0	$+\infty$	1

4.2. Comparative Analysis with Other Standard Schemes in RDHEI

To compare the proposed scheme with other schemes [9–12,16,24], we used the test image Lena (Figure 9a) from the database [26]. Figure 10 shows the comparison on the basis of PSNR of the directly decrypted image and their corresponding embedding rate.

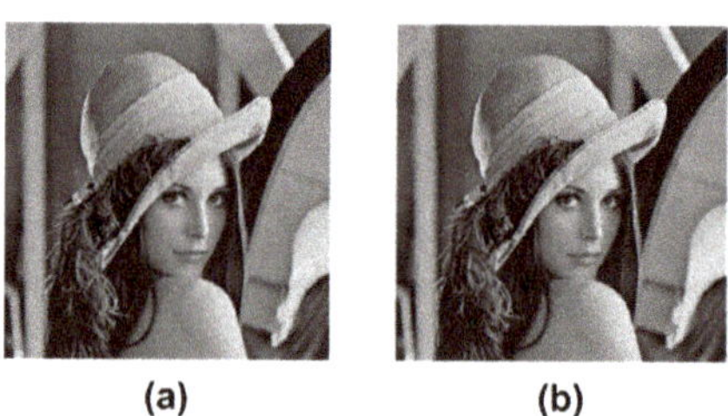

Figure 9. The test image Lena: (**a**) Original; and (**b**) directly decrypted image.

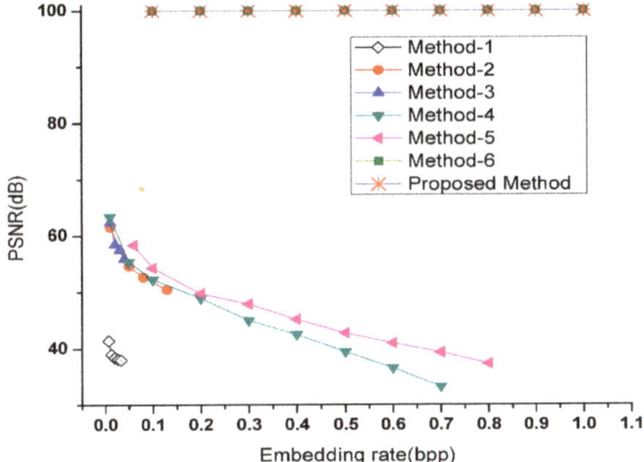

Figure 10. Performance comparison on the test image Lena for compared schemes Method-1 [9], Method-2 [10], Method-3 [12], Method-4 [11], Method-5 [16], Method-6 [24].

For Methods-1–5 [9–12,16], PSNR decreased with the increase in embedding rate. For Method-6 [24], PSNR was $+\infty$ dB and was independent of the embedding rate with maximum embedding rate of 1 bpp. It is notable that the maximum PSNR for any of the highest embedding rate for the directly decrypted image in Methods-1–5 [9–12,16] was less than 55.34 dB, showing that recovery of the directly decrypted image is incomplete without post-processing. However, in Method-6 [24] and the proposed scheme, the directly decrypted image is the same as the original image and no post-processing is required.

In the proposed scheme, the PSNR for the directly decrypted image (Figure 9b) was independent of the embedding rate and the maximum embedding rate is 1 bpp. The image quality of other compared schemes (except Method-6 [24]) was significantly less as compared to our proposed scheme when PSNR for the directly decrypted image was taken into account.

Table 3 shows a property-wise comparison of different schemes [9–12,16,24]. The maximum embedding rate for Zhang's scheme [9] was 0.033 bpp with PSNR = 38.0 dB, which decreased with the increase in the embedding rate. The maximum embedding rate for Zhang et al.'s scheme [12] was 0.04 bpp with PSNR = 55.34 dB. For Yin et al.'s scheme [10], maximum embedding rate is 0.1294 bpp with PSNR = 50.51 dB.

Table 3. Scheme-wise property comparison.

Schemes	Image Pre-Processing	Encryption	Receiver	Maximum Embedding Rate (bpp)	PSNR (dB) of Directly Image	Data Expansion
Zhang [9]	No	Stream cipher	Separable	0.033	38.0	No
Zhang et al. [12]	Yes	Stream cipher	Separable	0.04	55.34	No
Yin et al. [10]	No	Stream cipher	Separable	0.1294	50.51	No
Ma et al. [11]	Yes	Stream cipher	Separable	0.7	33.273	No
Cao et al. [16]	Yes	Stream cipher	Separable	0.8	37.375	No
Tai et al. [24]	Yes	Public key	Separable	1.0	$+\infty$	Yes
Proposed	No	Public key	Separable	1.0	$+\infty$	Yes

The maximum embedding rate for schemes of Ma et al. [11], Cao et al. [16] and Tai et al. [24] are 0.7 bpp, 0.8 bpp and 1.0 bpp with PSNR equal to 33.273 dB, 37.375 dB and $+\infty$ dB, respectively. For our proposed scheme, the maximum embedding rate was 1 bpp with PSNR = $+\infty$ dB. Here, PSNR is $+\infty$ dB was irrespective of the embedding rate.

The schemes compared in Table 3 achieved complete recovery of image Lena (Figure 9a) at the receiver's side after post-processing. All schemes including the proposed scheme are separable. A stream cipher is used for encryption in the schemes in [9–12,16], while public key cryptography is used for encryption in the scheme in [24] and the proposed scheme. The public-key cryptosystem used in our scheme is responsible for the inevitable data expansion caused due to homomorphic properties unlike in the schemes in [9–12,16]. However, with this disadvantage comes an advantage that the property of homomorphic addition inherent in the Paillier cryptosystem used in our scheme makes it suitable for privacy-preserving environment needed for cloud computing.

Sharing the similarity of using Paillier public-key cryptosystem with Tai's scheme [24], our scheme is also a separable reversible data hiding scheme, where the data hiding key is used by the receiver to extract the additional data from the marked encrypted image. Although image pre-processing is not done in the schemes in [9,10], the quality of the directly decrypted image is lesser as compared to our scheme. In the scheme in [24], the image preprocessing step needed for data hiding is an unnecessary overhead for the content-owner. This overhead is gracefully transferred to the data hider's side using the benefits of cloud in our scheme, which enhances the efficiency in the resource-constrained environment on the content-owner's end because the size of the pre-processed image (in the scheme in [24]) to be encrypted and sent is reduced to half in our scheme (see Table 4).

Table 4. Comparative analysis of our scheme in terms of bit-size for image ($L \times B$) with Tai et al. [24].

Schemes	Size (in bits)		
	Content-Owner	Data-Hider	Receiver
Tai et al. [24]	$2 \times L \times B \times (\lfloor \log_2 N^2 \rfloor + 1)$	$2 \times L \times B \times (\lfloor \log_2 N^2 \rfloor + 1)$	$2 \times L \times B \times (\lfloor \log_2 N^2 \rfloor + 1)$
Proposed	$1 \times L \times B(\lfloor \log_2 N^2 \rfloor + 1)$	$2 \times L \times B \times (\lfloor \log_2 N^2 \rfloor + 1)$	$2 \times L \times B \times (\lfloor \log_2 N^2 \rfloor + 1)$

In Table 4, a test image of size $L \times B$ is taken for comparison between our proposed scheme and the scheme in [24]. It is inferred that the proposed scheme has an edge over the scheme in [24] as we reduced the image preprocessing step on the sender's side, which is an additional step in the scheme in [24]. The size of the encrypted image is $L \times B \times (\lfloor \log_2 N^2 \rfloor + 1)$ bits in our proposed method. For the same image, the size of the encrypted image in the scheme in [24] is $2 \times L \times B \times (\lfloor \log_2 N^2 \rfloor + 1)$ bits. This is because the scheme in [24] involves a preprocessing step of dividing the original image into two parts and then encrypting it on the content-owner side. This can be also be seen as a reduction in the encryption cost to half on the content-owner side because, in the scheme in [24], the original image pixel is divided into two parts, each having the size of the original pixel. Thus, for encryption of a single image of size $L \times B$, the scheme in [24] has to encrypt $L \times B$ two times, i.e., for the two preprocessed parts of the same image. In the case of our scheme, for encrypting the image of size $L \times B$, encryption has to be done only once. It is a boon for those resource constrained devices on the sender side which entrust cloud for storage and processing. On the cloud side, the processing power is extremely large, thus attaching an equal payload to embed the additional data and its transmission is a trivial task.

5. Conclusions

In this work, an efficient separable reversible data hiding framework in encrypted images is proposed where Paillier cryptosystem is adequately used to preserve the privacy of the content in the cloud environment. Here, data extraction is accomplished in a separable manner, where the embedded additional data can be extracted using data hiding key and an independent complete image recovery is achieved using the private key.

In addition, the cost for encryption is reduced on the content owner's side (using limited processing and memory) with respect to Tai et al.'s scheme [24] by gracefully transferring the pre-processing step to the data hider side using cloud. Moreover, this proposed scheme exploits vast storage and memory resources available on the cloud. The proposed scheme was well explained with a real life application over cloud. Future works may include improving the efficiency of our scheme using new techniques in RDHEI.

Author Contributions: Conceptualization, A.N.K.; Data curation, A.N.K., M.I.N., A.M. and M.A.H.; Formal analysis, A.N.K. and A.M.; Funding acquisition, M.A.H.; Investigation, R.A.M.; Methodology, A.N.K.; Supervision, M.Y.F.; Validation, A.N.K.; Writing—original draft, A.N.K.; Writing—review & editing, A.N.K.

Funding: This research was funded by TEQIP-III of REC Ambedkar Nagar, grant number-TEQIP 3-RECABN and the APC was funded by TEQIP-III of REC Ambedkar Nagar.

Acknowledgments: The support of TEQIP-III of REC Ambedkar Nagar for this work is highly acknowledged.

Conflicts of Interest: The authors declare no conflict of interests.

References

1. Tian, J. Reversible data embedding using a difference expansion. *IEEE Trans. Circuits Syst. Video Technol.* **2003**, *13*, 890–896. [CrossRef]
2. Celik, M.; Sharma, G.; Tekalp, A.M.; Saber, E. Lossless generalized-LSB data embedding. *IEEE Trans. Image Process.* **2005**, *14*, 253–266. [CrossRef] [PubMed]
3. Ni, Z.; Shi, Y.Q.; Ansari, N.; Su, W. Reversible data hiding. *IEEE Trans. Circuits Syst. Video Technol.* **2006**, *16*, 354–362.
4. Malik, A.; Wang, H.; Wu, H.; Abdullahi, S.M. Reversible Data Hiding with Multiple Data for Multiple Users in an Encrypted Image. *Int. J. Digit. Crime Forensics* **2019**, *11*, 46–61. [CrossRef]
5. Shi, Y.-Q.; Li, X.; Zhang, X.; Ma, B.; Wu, H. Reversible Data Hiding: Advances in the Past Two Decades. *IEEE Access* **2016**, *4*, 1. [CrossRef]
6. Puech, W.; Chaumont, M.; Strauss, O. A reversible data hiding method for encrypted images. *Proc. SPIE* **2008**, *6819*, 68191E.
7. Zhang, X. Reversible Data Hiding in Encrypted Image. *IEEE Signal Process. Lett.* **2011**, *18*, 255–258. [CrossRef]
8. Hong, W.; Chen, T.-S.; Wu, H.-Y. An Improved Reversible Data Hiding in Encrypted Images Using Side Match. *IEEE Signal Process. Lett.* **2012**, *19*, 199–202. [CrossRef]
9. Zhang, X. Separable Reversible Data Hiding in Encrypted Image. *IEEE Trans. Inf. Forensics Secur.* **2012**, *7*, 826–832. [CrossRef]
10. Yin, Z.; Luo, B.; Hong, W. Separable and Error-Free Reversible Data Hiding in Encrypted Image with High Payload. *Sci. World J.* **2014**, *2014*, 1–8. [CrossRef]
11. Ma, K.; Zhang, W.; Zhao, X.; Yu, N.; Li, F. Reversible Data Hiding in Encrypted Images by Reserving Room Before Encryption. *IEEE Trans. Inf. Forensics Secur.* **2013**, *8*, 553–562. [CrossRef]
12. Zhang, W.; Ma, K.; Yu, N. Reversibility improved data hiding in encrypted images. *Signal Process.* **2014**, *94*, 118–127. [CrossRef]
13. Qian, Z.; Han, X.; Zhang, X. Separable Reversible Data hiding in Encrypted Images by n-nary Histogram Modification. In Proceedings of the 3rd International Conference on Multimedia Technology (ICMT 2013), Guangzhou, China, 29 November–1 December 2013; pp. 201–204.
14. Zhang, X.; Qian, Z.; Feng, G.; Ren, Y. Efficient reversible data hiding in encrypted images. *J. Vis. Commun. Image Represent.* **2014**, *25*, 322–328. [CrossRef]
15. Zheng, S.; Li, D.; Hu, D.; Ye, D.; Wang, L.; Wang, J. Lossless data hiding algorithm for encrypted images with high capacity. *Multimed. Tools Appl.* **2016**, *75*, 13765–13778. [CrossRef]
16. Cao, X.; Du, L.; Wei, X.; Meng, D.; Guo, X. High Capacity Reversible Data Hiding in Encrypted Images by Patch-Level Sparse Representation. *IEEE Trans. Cybern.* **2016**, *46*, 1. [CrossRef] [PubMed]
17. Kuribayashi, M.; Tanaka, H. Fingerprinting protocol for images based on additive homomorphic property. *IEEE Trans. Image Process.* **2005**, *14*, 2129–2139. [CrossRef] [PubMed]
18. Wu, H.-T.; Cheung, Y.-M.; Huang, J. Reversible data hiding in Paillier cryptosystem. *J. Vis. Commun. Image Represent.* **2016**, *40*, 765–771. [CrossRef]

19. Chen, Y.-C.; Shiu, C.-W.; Horng, G. Encrypted signal-based reversible data hiding with public key cryptosystem. *J. Vis. Commun. Image Represent.* **2014**, *25*, 1164–1170. [CrossRef]

20. Shiu, C.-W.; Chen, Y.-C.; Hong, W. Encrypted image-based reversible data hiding with public key cryptography from difference expansion. *Signal Process. Image Commun.* **2015**, *39*, 226–233. [CrossRef]

21. Li, M.; Xiao, D.; Zhang, Y.; Nan, H. Reversible data hiding in encrypted images using cross division and additive homomorphism. *Signal Process. Image Commun.* **2015**, *39*, 234–248. [CrossRef]

22. Liu, W.-L.; Leng, H.-S.; Huang, C.-K.; Chen, D.-C. A Block-Based Division Reversible Data Hiding Method in Encrypted Images. *Symmetry* **2017**, *9*, 308. [CrossRef]

23. Zhang, X.; Long, J.; Wang, Z.; Cheng, H.; Wang, J. Lossless and Reversible Data Hiding in Encrypted Images with Public Key Cryptography. *IEEE Trans. Circuits Syst. Video Technol.* **2016**, *26*, 1. [CrossRef]

24. Tai, W.-L.; Chang, Y.-F. Separable Reversible Data Hiding in Encrypted Signals with Public Key Cryptography. *Symmetry* **2018**, *10*, 23. [CrossRef]

25. Paillier, P. Public-Key Cryptosystems Based on Composite Degree Residuosity Classes. In Proceedings of the Advances in Cryptology—EUROCRYPT '99, Prague, Czech Republic, 2–6 May 1999; pp. 223–238.

26. CVG-UGR—Image Database. Available online: http://decsai.ugr.es/cvg/dbimagenes/g512.php (accessed on 21 April 2019).

 *electronics*

Article

Wavelet-Integrated Deep Networks for Single Image Super-Resolution

Faisal Sahito [1], Pan Zhiwen [1,*], Junaid Ahmed [2] and Raheel Ahmed Memon [3]

[1]   National Mobile Communications Research Laboratory, Southeast University, Nanjing 210096, China;
     Faisal@seu.edu.cn
[2]   Department of Electrical Engineering, Sukkur IBA University, Sukkur 65200, Pakistan;
     j.bhatti@iba-suk.edu.pk
[3]   Department of Computer Science, Sukkur IBA University, Sukkur 65200, Pakistan;
     raheelmemon@iba-suk.edu.pk
*   Correspondence: pzw@seu.edu.cn

Received: 25 April 2019; Accepted: 14 May 2019; Published: 17 May 2019

Abstract: We propose a scale-invariant deep neural network model based on wavelets for single image super-resolution (SISR). The wavelet approximation images and their corresponding wavelet sub-bands across all predefined scale factors are combined to form a big training data set. Then, mappings are determined between the wavelet sub-band images and their corresponding approximation images. Finally, the gradient clipping process is used to boost the training speed of the algorithm. Furthermore, stationary wavelet transform (SWT) is used instead of a discrete wavelet transform (DWT), due to its up-scaling property. In this way, we can preserve more information about the images. In the proposed model, the high-resolution image is recovered with detailed features, due to redundancy (across the scale) property of wavelets. Experimental results show that the proposed model outperforms state-of-the algorithms in terms of peak signal-to-noise ratio (PSNR) and structural similarity index measure (SSIM).

Keywords: wavelet analysis; deep learning; super-resolution; deep neural architecture; pattern mining; multi-scale analysis

---

1. Introduction

Single image super-resolution (SISR) is generally posed as an inverse problem in the image processing field. Here, the task is to recover the original high-resolution (HR) image from a single observation of the low-resolution (LR) image. This method is generally used in applications where the HR images are of importance, such as brain image enhancement [1], biometric image enhancement [2], face image enhancement [3], and standard-definition television (SDTV) and high definition television (HDTV) applications [4]. The problem of SISR is considered a highly ill-posed problem, because the number of unknown variables from an HR image is much higher compared to the known ones from an LR image.

In the literature for SISR, a number of algorithms have been proposed for the solution of this problem. They can be categorized as including an interpolation algorithm [5], edge-based algorithm [6], and example-based algorithms [7–9]. The interpolation and edge-based algorithms provide reasonable results. However; their performance severely degrades with the increase in an up-scale factor. Recently, the neural network-based algorithms have captured the eye of researchers for the task of SISR [10–12]. The main reasons can be the huge capacity of the neural network models and end-to-end learning, which helps researchers to get rid of the features used in the previous approaches.

However, the algorithms proposed so far are unable to achieve better performance for higher scale-ups. The proposed algorithm is a wavelet domain-based algorithm inspired by the category of the

SISR algorithms in the wavelet domain [13–17]. Most of these algorithms give state-of-the-art results. However, their computational cost is quite high. With the advances in deep-learning algorithms, the task of computational cost is much reduced with acceptable quality.

Authors in [16], proposed a wavelet domain-based deep learning algorithm with three layers, inspired by the super-resolution convolution neural network (SRCNN) [8] and using a discrete wavelet transform (DWT), and achieved good results. However, the authors fail to capture the full potential of deep learning and wavelets. In this paper, we propose a wavelet domain-based algorithm for the task of SISR. We incorporate the merits of neural network-based end-to-end learning and large model capacity [18], along with the properties of the wavelet domain, such as sparsity, redundancy, and directionality [19,20]. We propose the use of stationary wavelet transform (SWT) for the wavelet domain analysis and synthesis, owing to its up-sampling property over the DWT down-sampling. By doing so, we want to preserve more contextual information about the images. Moreover, we propose the use of deep neural network architecture in the wavelet domain.

More specifically, we train our network between the wavelet approximation images and their corresponding wavelet sub-band images for the task of SISR. By experimental analysis, we show that the proposed deep-network architecture in the wavelet domain can improve performance for the task of SISR with a reasonable computational cost. The proposed algorithm is compared with recent and state-of-the-art algorithms in terms of peak signal-to-noise ratio (PSNR) and structural similarity index measure (SSIM) over the publicly available data sets of "Set5", "Set14", "BSD100", and "Urban100" for different scale factors.

The rest of the paper is organized as follow. Section 2 describes the details about related work. Section 3 describes the details about the proposed method. Section 4 gives an experimental discussion about the properties of the proposed model. Section 5 given the discussion about the experiments and comparative analysis, and Section 6 concludes the paper.

2. Related Work

The proposed algorithm falls into the category of wavelet domain-based SISR algorithms. Authors in [13] proposed a dictionary learning-based algorithm in the wavelet domain. The proposed algorithm learns compact dictionaries for the task of SISR. A similar approach utilizing dictionary learning is proposed in [14], utilizing the DWT. Authors in [15] proposed coupled dictionary learning in the wavelet domain, utilizing the properties of the wavelets with the coupled dictionary learning approach. Another algorithm that utilizes the dual-tree complex wavelet transform (DT-CWT), along with the coupled dictionary and mapping learning for the task of SISR, is proposed in [17]. Authors in [16] utilize the convolution neural networks in the wavelet domain using the DWT, and propose an efficient model for the task of SISR.

In the wavelet-based SISR approaches [13–16], the main point to note is that they assume the LR image as the level-1 approximation image of the wavelet decomposition. Here, to recover the HR image, the task is to estimate the wavelet sub-band images representing this approximation image, and finally doing one-level inverse wavelet transform. By doing so, authors induce sparsity and directionality along with compactness in the algorithms, which helps boosts the performance of the algorithms as well as improve their convergence speed.

Dong et al. [8] exploited a fully convolution neural network (CNN). In this method, they proposed a three-layer network where complex non-linear mappings are learned between the HR and LR image patches. Authors in [18] propose deep network architecture for the task of SISR. Instead of using the HR and LR images for training, they utilized the residual images, and to boost the convergence of their algorithm, they utilized adjustable gradient clipping. Authors in [8] further propose the sped-up version of the super-resolution convolution neural network (SRCNN) algorithm, called a fast super-resolution convolution neural network (FSRCNN) [21] algorithm. They achieve this by learning the mappings between the HR and LR images without interpolations, along with shrinking the mappings in the feature learning step. Also, the authors decrease the size of filters and increase the number of layers. Authors in [22] propose a deep residual learning network with batch normalizations for the task of SISR, called a deep-network convolution neural network (DnCNN) algorithm. Authors in [23] propose an information distillation network (IDN) algorithm for the task of SISR. They propose a compact network that utilizes the mixing of features and compression to infer more information for the SISR problem. Authors in [24] propose a super-resolution with multiple degradations (SRMD) algorithm for the problem of SISR. They propose the deep network model for SR, utilizing the degradation maps achieved using the dimensionality reduction of principle component analysis (PCA) and then stretching. By doing so, they learned a single network model for multiple scale-ups.

There are several applications related to single image super-resolution, pattern recognition, neural networks, etc., which can be applied in our human's daily life as well as in human biology. In [25,26], authors have applied different algorithms of neural networks that focus on magnetic resonance imaging (MRI), while in [27–29], authors have applied different algorithms of neural networks that focus on human motion and character control. Likewise, our proposed work can be applied in different applications: brain image enhancement, face image enhancement, and SDTV and HDTV applications. The proposed model can be effectively extended to other image processing and pattern recognition applications.

3. Proposed Method

We propose a deep neural network model based on wavelets and gradient clipping for SISR. The wavelet domain-based algorithm was chosen because of the unique properties of the wavelets: they exploit multi-scale modeling, and wavelet sub-bands are significantly sparse. Moreover, instead of DWT, we propose the use of SWT. DWT is a down-sampling process and SWT is an up-sampling process, so the size of the wavelet approximation and sub-bands remains the same, while preserving all the essential properties of the wavelets.

The DWT and SWT decompositions are shown in Figure 1. Further, the wavelet domain-based algorithms consider the LR image as the wavelet approximation image of the corresponding HR image. The task is to estimate its detailed coefficients, as done in [30–33].

$$A_q(m,n) = \sum_{l=j}^{M} \sum_{j=1}^{N} h_{l=1}^{1} h_{j=1}^{2} A_q(l,j), \tag{1}$$

$$H_q(m,n) = \sum_{l=j}^{M} \sum_{j=1}^{N} h_{l=1}^{1} h_{j=1}^{2} A_q(l,j), \tag{2}$$

$$V_q(m,n) = \sum_{l=j}^{M} \sum_{j=1}^{N} g_{l=1}^{1} h_{j=1}^{2} A_q(l,j), \tag{3}$$

$$D_q(m,n) = \sum_{l=j}^{M} \sum_{j=1}^{N} g_{l=1}^{1} g_{j=1}^{2} A_q(l,j), \tag{4}$$

where h_m^1, h_n^2, g_m^1, and g_n^2 are the wavelet analysis filters for the SWT. $A_{q-1}(m,n)$, $H_{q-1}(m,n)$, $V_{q-1}(m,n)$, and $D_{q-1}(m,n)$ are the wavelet approximation image, horizontal sub-band image, vertical sub-band image, and diagonal sub-band image, respectively. The practical decomposition is shown in Figure 2. In the experimental analysis, we have chosen the sym29 wavelet filters, following the convention from [13,15,17]. The wavelet synthesis equation can be given as

$$A_{q+2}(m,n) = \sum_{l=j}^{M} \sum_{j=1}^{N} \tilde{h}^1_{l-m}\tilde{h}^2_{j-n}\tilde{\tilde{A}}_q(l,j)A_{q+2}(m,n)$$
$$+ \sum_{l=j}^{M} \sum_{j=1}^{N} \tilde{h}^1_{l-m}\tilde{g}^2_{j-n}\tilde{H}_q(l,j) \sum_{l=j}^{M} \sum_{j=1}^{N} \tilde{g}^1_{l-m}\tilde{h}^2_{j-n}\tilde{V}_q(l,j) \qquad (5)$$
$$+ \sum_{l=1}^{M} \sum_{J=1}^{N} \tilde{g}^1_{l-m}\tilde{h}^2_{j-n}\tilde{D}_q(l,j).$$

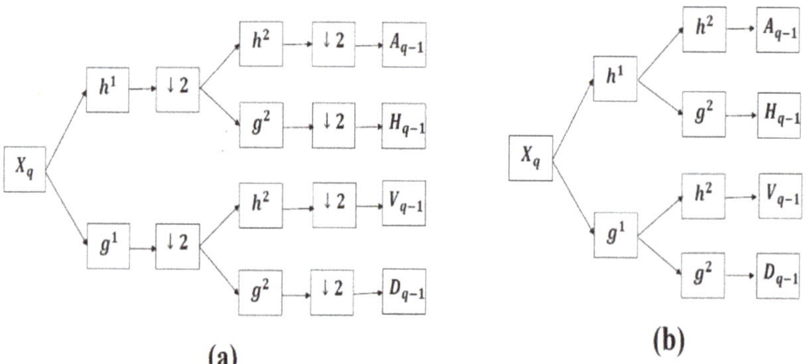

Figure 1. (a) Discrete wavelet transform (DWT) decomposition and (b) stationary wavelet transform (SWT) decomposition.

Figure 2. Wavelet decomposition. (a) Original image; (b) from left to right and top to bottom: approximation, horizontal, vertical, and diagonal images.

After getting the desired unknown wavelet coefficients, one-level inverse wavelet transform is required to get the desired HR output. Figure 2 shows the wavelet decomposition at level one of the hibiscus image. It can be seen from the image that a strong dependency is present between the wavelet coefficients at the given level and its sub-bands.

There have been several attempts to handle the problem of dimensionality reduction. In [34], authors propose a local linear embedded (LLE) approach that computes low-dimensional, neighborhood-preserving embeddings of high-dimensional inputs. The LLE approach maps its inputs into a single global coordinate system of lower dimensionality, and its optimizations do not involve local minima. LLE is able to learn the global structure of nonlinear manifolds, such as those generated by images of faces or documents of text. In [35], the authors describe an approach that combines the classical techniques of dimensionality reduction, such as principal component analysis (PCA) and multidimensional scaling (MDS) features. This approach is capable of discovering the nonlinear degrees of freedom that underlie complex natural observations, such as human handwriting or images of a face under different viewing conditions. In [36], authors have compared PCA, kernel principal component analysis (KPCA), and independent component analysis (ICA) to a support vector machine (SVM) for feature extraction. Furthermore, the authors described that the KPCA method is best among three for feature extraction. In [37], authors have proposed a geometrically motivated algorithm for representing the high-dimensional data, which provides a computational approach to dimensionality reduction compared to previous classical methods like PCA and MDS. The algorithm proposed learns a single network model for multiple scale-ups. However, the proposed algorithm utilizes the wavelet domain decomposition before the training of the network, and the wavelet sub-band images are used as the input the training. As can be seen from Figure 2, which shows the wavelet decomposition of a single image, the wavelet sub-band images are significantly sparse, and represent the directional fine features of the images. Further implying the dimensionality results will result in the loss of such directional fine features.

However, in spite of the sparsity property of the wavelets, the assumption of independence of wavelet coefficients at consecutive levels is somewhat limited for the task of SISR. This assumption fails to take into account the intra-scale dependency of the wavelet coefficients that capture the useful structures from the given images.

We make use of this dependency on the task of SISR. The proposed algorithm is different from the previous neural network- and wavelet domain-based methods in the following aspects.

- We use the SWT wavelet decomposition of the image and estimate the wavelet coefficients;
- We propose the deep network architecture similar to very deep super-resolution (VDSR) algorithm [18], but we train the network on the wavelet domain images instead of residual images—whereas, the authors of [16] utilize the DWT with a three-layer neural network inspired by SRCNN [8];
- We take a step further and design the deep network with 20 layers in the wavelet domain. The proposed wavelet-integrated deep-network (WIDN) model for super resolution estimates the sparse output, thus improving its reconstruction accuracy and training efficiency.

For the WIDN, the deep-network architecture is inspired by the Simonyan and Zisserman [38]. The network configuration can be found in Figure 3.

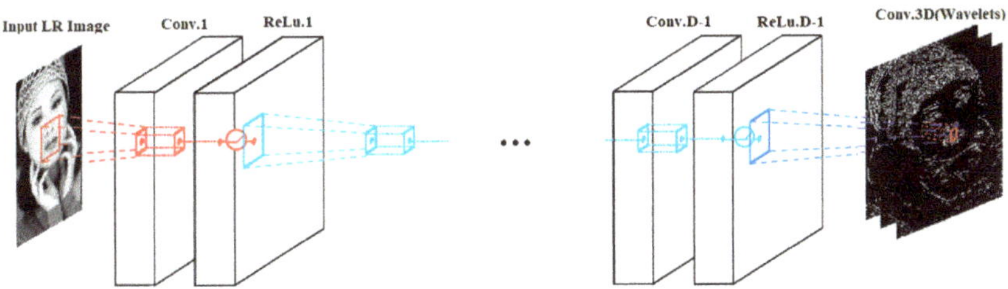

Figure 3. The wavelet deep network configuration.

In our network model, we utilize D layers. All the layers in our network are the same except the first and the last. In our network, the first layer has a total of 64 filters. The size of each filter is $1 \times 3 \times 3 \times 64$. These filters operate at a 3×3 spatial size on 64 channels. These channels are also called feature maps. The first layer is used for the LR input image, and the last layer reconstructs the output image. As the last layer is used for the output image reconstruction, it has three filters, each of size $3 \times 3 \times 3 \times 64$. Our network is trained between the input LR image and its corresponding wavelet coefficients. Thus, given an input LR image, the network can predict the corresponding wavelet coefficients for HR image reconstruction. Modeling the image details in the wavelet domain has certain usefulness for the task of SISR [13–15]. The proposed model shows that by using wavelet details, the performance of SISR is highly improved. One of the problems pertaining to the deep convolution networks is that size of the output feature maps get reduced after each layer as the convolution operation is performed.

The problem is maintaining the same output size after each convolution operation is performed. Some authors suggest the use of surrounding pixels can give information about the center pixel [8]. This is quite handy when it comes to the problem of SISR. However, for the boundary of the image this can fail; cropping may be utilized to solve this problem. To alleviate the problem of size reduction and boundary condition, we employed zero paddings before the convolution operation. We find that by doing so, the size of the features remains constant, and the boundary condition problem is also solved. Once the three wavelet sub-bands are predicted, we add back the LR input image and do one-level wavelet reconstruction to get the HR image estimate.

Data preprocessing is a very important step to make features invariant to input scale and reduce dimensionality in the machine learning process (a restricted Boltzmann machine, or RBM), which is likely to be used for preprocess the input data. In [39], the authors note that the RBM is an undirected graphical model with hidden variables and visible variables along with a feature learning approach, which is used to train an RBM model separately for audio and video. After learning the RBM, the posteriors of the hidden variables given the visible variables can be used as a new representation of the data. This model is used for multimodal learning as well as for pre-training the deep networks. In [40], the authors present the sparse feature representation method based on unsupervised feature learning. By using the RBM graphical model, which consists of visible nodes and hidden nodes, the visible nodes represent input vectors, while hidden nodes are feature-learned by training the RBM. This method helps to pre-process the data. In [41], the authors present a method in which a number of motion features computed by a character's hand is considered. The motion features are preprocessed using restricted Boltzmann machines (RBMs). RBM pre-processing performs a transformation of the feature space based on an unsupervised learning step. In our proposed model, we have utilized the data augmentation technique for pre-processing the data, inspired by VDSR [18] and FSRCNN [21] algorithms. However, implementing the RBMs will definitely be considered as a future task of our approach.

3.1. Training

For the training of our model, we require a set of HR images. As we train our model between the wavelet approximation image and its corresponding sub-band coefficient images, we do a one-level wavelet decomposition on the HR images from the training data set. The wavelets have a very unique property of redundancy across the scale.

Given the wavelet approximation image at a certain scale and its coefficients, one can perfectly reconstruct the preceding approximation image. Thus, the wavelet coefficient contains all the information about the preceding approximation image. We utilize this property of the wavelet and learn the mappings between the wavelet approximation image and its corresponding coefficients for the task of SISR. Let X denotes the level1 wavelet LR image and Y denote the detail sub-band images. The task is to learn the relationship between the LR approximation image and its corresponding same-level wavelet sub-band images (horizontal, vertical, and diagonal).

In the algorithm SRCNN [8], one problem is that the network has to preserve the information about input details as the output is obtained, using these learned features alone, and the input image is not utilized and discarded. If the network is deep, having many weight layers, this corresponds to an end-to-end learning problem, which requires a huge memory.

Due to this reason, the problem of the vanishing/exploding gradient [42] arises and needs to be solved. We can solve this problem by wavelet coefficient learning. As we assume the dependency between the wavelet LR approximation image and its corresponding same-level detailed coefficients, we define the loss function as

$$L(\Theta) = \frac{1}{k} \sum_{i=1}^{k} (\sum_{b=1}^{3} \|f(X_{i},\Theta)^{b} - y_{i}^{b}\|), \tag{6}$$

where k is the number of training samples, X is the tensor containing the LR approximation images, and Y is the tensor containing the wavelet sub-band images (horizontal, vertical, and diagonal). T represents the network parameters, and b represents the sub-band index. For the training, we use the gradient descent-based algorithm from [43]. This algorithm works on the mini-batch of images and utilizes the back-propagation approach to optimize the objective function. In our model, we set the momentum parameter to be 0.9, with the regularizing penalty on the weight decay as 0.0001. Now, to boost the speed on training, one can use a high learning rate. However, if a high learning rate is utilized, the problem of vanishing/exploding gradients [42] becomes evident. To solve this, we utilize the adjustable gradient clipping.

Gradient Clipping

Gradient clipping is generally used for training the recurrent neural networks [38]. However, it is seldom used in the CNN training. There are many ways in which gradients can be clipped. One of them can be to clip them in a pre-defined range $(-\theta, \theta)$. In the process of clipping, the gradient lies in a specific range. If the stochastic gradient descent (SGD) algorithm is used for training, we multiply the gradient with the learning rate for step size adjustment. If we want our network to train much faster, we need a high rate of learning; to achieve this value, the gradient θ must be high.

However, high gradient values will cause the exploding gradients problem. We can avoid this problem by using a smaller learning rate. However, if the learning rate is made smaller, the effective gradient approaches zero, and the training may take a lot of time. For this purpose, we propose to clip the gradients to $\left[-\frac{\theta}{\gamma}, \frac{\theta}{\gamma}\right]$, where γ is the learning rate. By doing so, we observe that the convergence of our network becomes faster. It is worth mentioning here that our network converges within 3 h, just like in [44], while the SRCNN [16] takes several days to train. Despite the fact that the deep models proposed nowadays have greater performance capability, if we want to change the scale-up the parameter, the network is trained for that scale again, and hence for each scale, we need a different training model.

Considering the fact that the scale factor is used often and is important, we need to find a way across this problem. To tackle this problem, we propose to train a multi-scale model. By doing so, we can utilize the parameters and features from all scales jointly. To do so, we combine all the approximation images and their corresponding wavelet sub-bands across all predefined scale factors, and form a big data set of training images.

4. Properties of the Proposed Model

Here we discuss the properties of the proposed model. First, we say that the large depth networks can give good performance for the task of SISR. Very deep networks make use of the contextual information of an image, and can model complex functions with many non-linear layers. We experimentally validate our claim. Second, we argue that the proposed network gives a significant boost in performance, with an approximately similar convergence speed to VDSR.

4.1. Deep Network

Convolution neural networks make use of the spatial–local correlation property. They enforce the connecting patterns between the neurons of adjacent layers in the network model. In other words, for the case of hidden units, the output from the layer m − 1 is an input to the layer m in the network model. By doing so, a receptive field is formed that is spatially contagious. In this network model, the corresponding hidden unit in the network only corresponds to the receptive field, and is invariant to the changes outside its receptive field. Due to this fact, the filters learned can efficiently represent the local spatial patterns in the vicinity of the receptive field.

However, if we stack a number of such layers to form a network model, the output ends up being global—i.e., it corresponds to bigger pixel space. The other way around, a filter having large spatial support can be broken into a number of filters with smaller spatial support. Here we use 3×3 size filters to learn the wavelet domain mappings. The filter size is kept the same for all corresponding layers. This means that the receptive field for the layer has the 3×3 filter size. For the corresponding proceeding layer, this size is increased by a factor of two. The depth of the receptive field in our model has the size of $(2D + 1) \times (2D + 1)$. For the task of SISR, if one has more contextual information about the high-frequency components, it can be used to infer and generate a high-quality image. In this paradigm of neural networks, a bigger receptive field can serve the purpose of extracting more contextual information. As the problem of super-resolution is highly ill-posed, using more contextual information is bound to give better results.

Another advantage of using deep networks is that they can model non-linearity very well. In our proposed network architecture, we utilize 19 ReLUs, which allows our network to model highly complex non-linear functions. We experimentally evaluated the performance of deep networks by calculating the network's PSNR as depth values increased from 5 to 20, only counting the weight layers and excluding the non-linearity layers. The results are shown in Figure 4. In most cases, the performance increases as depth increases.

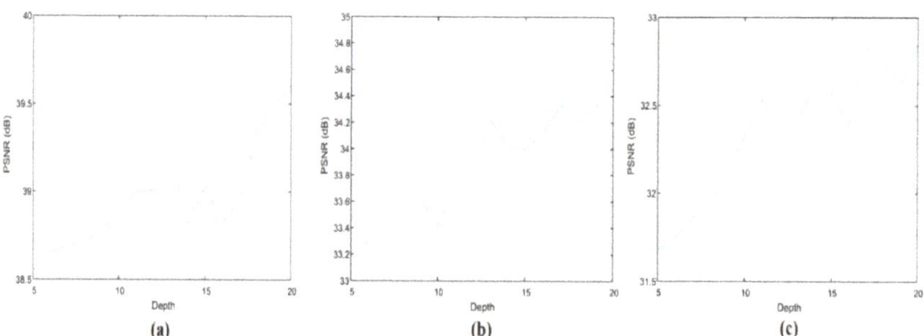

Figure 4. Depth performance of the network on dataset Set5: (**a**) at scale 2, (**b**) at scale 3, and (**c**) at scale 4.

There are a number of different techniques in machine learning to solve computational problems. Some of them we discuss here and compare with our proposed WIDN. In [45], authors have proposed a recurrent neural network (acRNN), which synthesizes highly complex human motion variations of arbitrary styles, like dance or martial arts, without asking from the database. In [46], the authors have proposed dilated convolutional neural network for capturing temporal dependencies in the context of driver maneuver anticipation. In [47], authors have proposed CNN for speech recognition within the framework of a hybrid NNHMM model. Hidden Markov models (HMMs) are used in state-of-the-art automatic speech recognition (ASR) to model the sequential structure of speech signals, where each HMM state uses a Gaussian mixture model (GMM) to model a short-time spectral representation of the speech signal. In [48], authors have briefly explained in detail the number of graphical models that can be used to express speech recognition systems. The main idea of the proposed work is the wavelet domain-based deep-network algorithm. In our proposed model, we use the wavelet sub-band images as the input to the network, and learn a single model for multiple degradations. One can try such an implementation with other DNN-based algorithms, but the first one needs to investigate whether the DNN will be compatible with the wavelet sub-band images or itself. One also has to account for the sparsity and directionality of the wavelet sub-band images. We have proposed the DNN model of the

VDSR [18] algorithms, as it utilizes the residual images obtained by subtracting the LR from HR images for the training of the network. The wavelet sub-band images possess quite similar properties as the residual images for the task of SISR. Experimental analysis validated our assumption, and comparative analysis proved the efficacy of the proposed model.

4.2. Wavelet Learning

In this work, we propose a network structure that learns wavelet sub-band images. We now study this modification to the VDSR approach. First, we show that for approximately similar convergence, the network gives better performance. We use a depth of 20(weight layers) and the scale parameter is 2. Performance curves for various learning rates are shown in Figure 5. All use the same learning rate scheduling. It can be seen that the proposed algorithm gives superior performance.

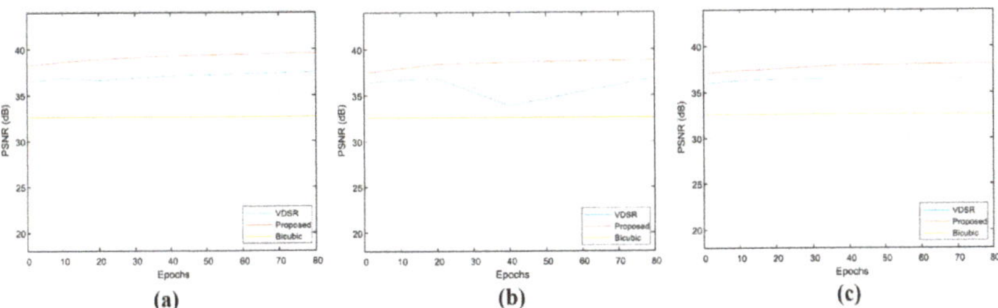

Figure 5. Performance comparison with VDSR and bicubic algorithms based on learning rates: (**a**) 0.1, (**b**) 0.01, and (**c**) 0.001.

5. Experiments and Results

Here we give the details about the experiments and results. Data preparation in our case is similar to SRCNN [8], with a minute difference. In our model, the patch size of the input image is made the same as the receptive field of the network. We do not utilize the overlap condition while extracting the patches to form a mini-batch. A single mini-batch in our model has a total of 64 sub-images. Also, the sub-images corresponding to the difference scales can also be combined to form a mini-batch. We implement our model using the publicly available MatConvNet package [44]. For the training data set, we used the 291 images with augmentation (rotations), as done in [21].

For the test data sets, we used the most commonly used data sets of "Set5", "Set14", "Urban100", and "BSD100", as used in previous works [18,21,23,24]. The depth of our network model is 20. The batch size used is 64. The momentum used is 0.9 with the decay rate of 0.0001. The network was trained for 80 epochs, and initially, the learning rate was set to 0.1; after every 20 iterations, we decreased it by a factor of 10. The training of our model normally takes about 3 h using the GPU Titan Z. However, if we use a small training set like that in [49], we can increase the speed of learning. Table 1 shows the average PSNR values of the proposed algorithm with increasing numbers of epochs and on different learning rates. It can be seen from the Table 1 that the proposed algorithm provides good results by employing the deep neural network architecture in the wavelet domain.

Table 1. Performance table (peak signal-to-noise ratio, or PSNR) for the proposed and VDSR [18] networks ("Set5" dataset, ×2).

(a) 0.1 rate of learning			
Epoch	**VDSR** [18]	**Proposed**	**Difference**
10	36.90	38.66	1.76
20	36.64	38.95	2.31
40	37.12	39.32	2.20
80	37.53	39.61	2.08
(b) 0.01 rate of learning			
Epoch	**VDSR** [18]	**Proposed**	**Difference**
10	36.82	37.98	1.16
20	36.90	38.42	1.52
40	36.98	38.63	1.65
80	37.06	38.86	1.8
(c) 0.001 rate of learning			
Epoch	**VDSR** [18]	**Proposed**	**Difference**
10	36.42	37.41	0.99
20	36.58	37.58	1
40	36.69	38.01	1.32
80	36.79	38.13	1.34

The visual results are shown in Figures 6–11. Figures 6 and 7 show the comparative results for the scale-up parameter of 2. Almost all the algorithms perform better. However, the proposed wavelet domain-based algorithm provides more sharp edges and textures. Figures 8 and 9 show the comparative results from the BSD100 test set images for the scale-up parameter of 3.

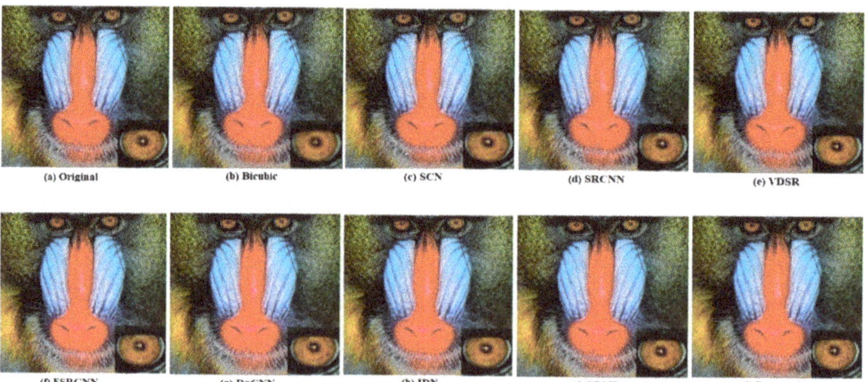

Figure 6. Visual comparison for a baboon image at the scale-up factor of 2.

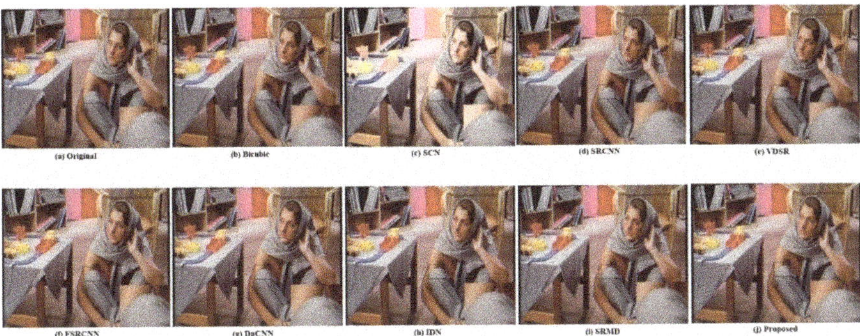

Figure 7. Visual comparison for the Barbara image at the scale-up factor of 2.

Figure 8. Visual comparison for the tiger image at the scale-up factor of 3.

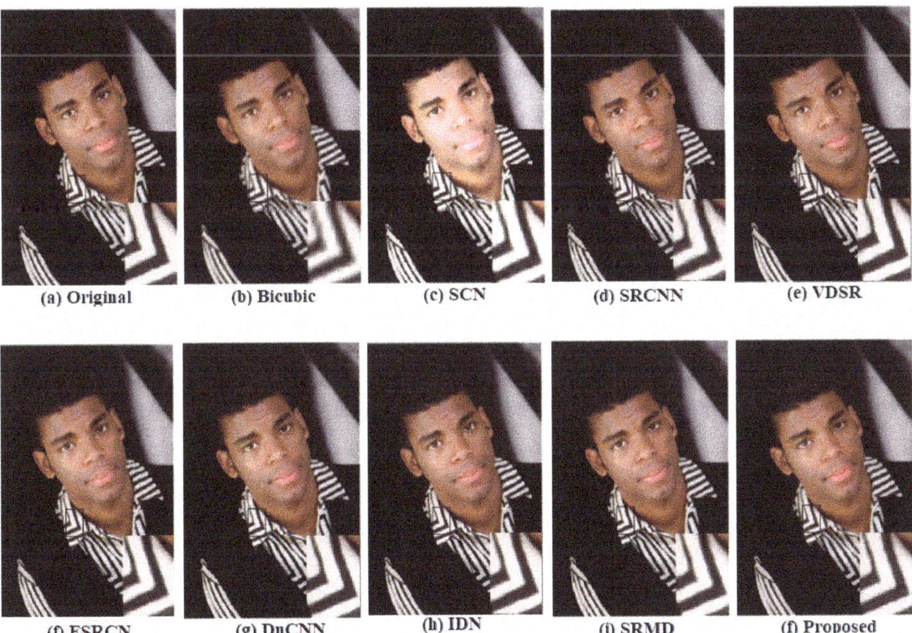

Figure 9. Visual comparison for the man image at the scale-up factor of 3.

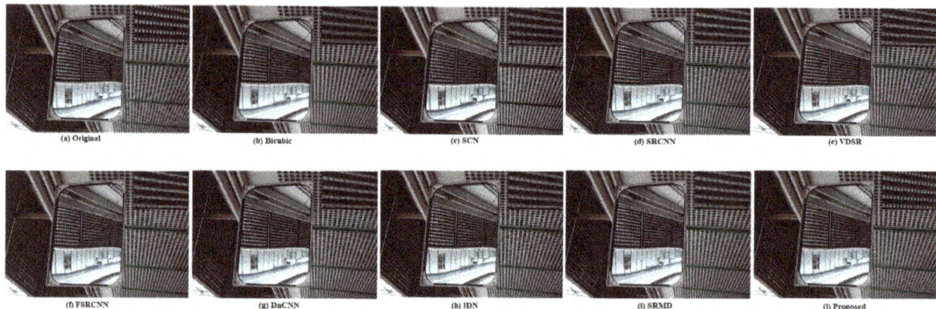

Figure 10. Visual comparison for the Urban04 image at the scale-up factor of 4.

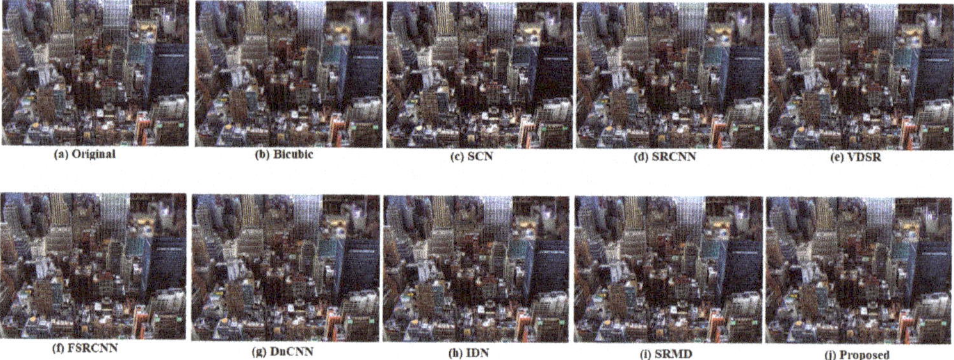

Figure 11. Visual comparison for the Urban73 image at the scale-up factor of 4.

Here the algorithms under comparison fail to provide good results; however, the proposed algorithm provides better results. Figures 10 and 11 are taken from a more challenging image data set of Urban100. Here, the scale-up parameter used is 4. Looking at Figures 10 and 11, the proposed algorithm is able to recover the sharp edges and texture where other algorithms fail.

The quantitative analysis based on PSNR and SSIM is shown in Table 2. The algorithms under comparison include the bicubic technique, SRCNN algorithm [8], SCN algorithm [11], VDSR algorithm [18], FSRCNN algorithm [21], DnCNN algorithm [22], IDN algorithm [23], and SRMD algorithm [24]. In the comparative analysis, the trained models used for these algorithms are provided by the authors. The proposed algorithm gives better results than the algorithms under comparison.

Table 2. Comparative results based on PSNR (left) and the structural similarity index measure (SSIM) (right).

Data Set	Scale	Bicubic	SRCNN [8]	SCN [11]	VDSR [18]	DnCNN [22]	FSRCNN [21]	SRMD [24]	IDN [23]	Proposed
Set 5	2	33.64/0.929	36.35/0.953	36.52/0.953	37.56/0.959	37.58/0.959	36.99/0.955	37.53/0.959	37.83/0.960	**39.60/0.983**
	3	30.39/0.866	32.74/0.908	32.60/0.907	33.67/0.922	33.75/0.922	33.15/0.913	33.86/0.923	34.11/0.952	**34.48/0.943**
	4	28.42/0.810	30.48/0.863	30.39/0.862	31.35/0.885	31.40/0.884	30.71/0.865	31.59/0.887	31.82/0.890	**32.85/0.929**
Set 14	2	30.22/0.868	32.42/0.906	32.42/0.904	33.02/0.913	33.03/0.912	32.73/0.909	33.12/0.914	33.30/0.915	**34.44/0.980**
	3	27.53/0.774	29.27/0.821	29.24/0.819	29.77/0.832	29.81/0.832	29.53/0.826	29.84/0.833	29.99/0.835	**30.95/0.931**
	4	25.99/0.702	27.48/0.751	27.48/0.751	27.99/0.766	28.04/0.767	27.70/0.756	28.15/0.772	28.25/0.773	**29.75/0.909**
BSD100	2	29.55/0.843	31.34/0.887	31.24/0.884	31.89/0.896	31.90/0.896	31.51/0.891	31.90/0.896	32.08/0.898	**33.52/0.979**
	3	27.20/0.738	28.40/0.784	29.32/0.782	28.82/0.798	28.85/0.798	28.52/0.790	28.87/0.799	28.95/0.801	**29.99/0.928**
	4	25.96/0.667	26.90/0.710	26.87/0.710	27.28/0.726	27.29/0.725	26.97/0.714	27.34/0.728	27.41/0.730	**28.10/0.910**
Urban 100	2	26.66/0.841	29.53/0.897	29.50/0.896	30.76/0.914	30.74/0.913	29.87/0.901	30.89/0.916	31.29/0.920	**32.48/0.952**
	3	24.46/0.737	26.25/0.801	26.21/0.801	27.13/0.828	27.15/0.827	26.42/0.807	27.27/0.833	27.42/0.846	**28.68/0.941**
	4	23.14/0.653	24.52/0.722	24.52/0.725	27.17/0.753	25.20/0.752	24.67/0.727	25.34/0.761	25.41/0.763	**26.41/0.903**

6. Conclusions

A scale-invariant, wavelet-integrated deep-network model is proposed for the task of SISR. To improve the training speed of the algorithm, the adjustable gradient clipping is used. Useful properties of the convolution neural networks, such as large model capacity, end-to-end learning, and high performance, are exploited in the wavelet domain for the task of SISR. The up-sampling SWT is proposed instead of the down-sampling DWT, to avoid the data loss. Experimental analysis is carried out to validate the efficacy of the proposed model. Quantitative results based on the PSNR and SSIM indicate that the proposed algorithm performs better in comparison with the recent state-of-the-art algorithm. Visual results also validate the quantitative ones. The proposed algorithm can be extended and modified for other super-resolution applications, such as face and brain image enhancement. Also, the proposed algorithm can be tested with other wavelet transforms, such as dual-tree complex wavelet transforms (DT-CWT).

Author Contributions: Conceptualization, F.S.; Data curation, J.A.; Formal analysis, R.A.M.; Funding acquisition, P.Z.; Investigation, P.Z.; Methodology, F.S.; Project administration, P.Z.; Resources, P.Z.; Software, F.S.; Supervision, P.Z.; Validation, J.A.; Visualization, R.A.M.; Writing—original draft, F.S.; Writing—review & editing, J.A.

Funding: This work is partially supported by national major project under Grants 2017ZX03001002-004 and 333 Program of Jiangsu under Grants [BRA2017366].

Conflicts of Interest: The authors declare no conflict of interest.

References

1. Zhang, J.; Zhang, L.; Xiang, L.; Shao, Y.; Wu, G.; Zhou, X.; Shen, D.; Wang, Q. Brain atlas fusion from high-thickness diagnostic magnetic resonance images by learning-based super-resolution. *Pattern Recognit.* **2017**, *63*, 531–541. [CrossRef]

2. Nguyen, K.; Fookes, C.; Sridharan, S.; Tistarelli, M.; Nixon, M. Super-resolution for biometrics: A comprehensive survey. *Pattern Recognit.* **2018**, *78*, 23–42. [CrossRef]

3. Chen, X.; Zhang, Z.; Wang, B.; Hu, G.; Hancock, E.R. Recovering variations in facial albedo from low-resolution images. *Pattern Recognit.* **2018**, *74*, 373–384. [CrossRef]

4. Park, S.C.; Park, M.K.; Kang, M.G. Super-resolution image reconstruction: A technical overview. *IEEE Signal Process. Mag.* **2003**, *20*, 21–36. [CrossRef]

5. Morse, B.S.; Schwartzwald, D. Image magnification using level-set reconstruction. In Proceedings of the IEEE Computer Society Conference on Computer Vision and Pattern Recognition (CVPR), Kauai, HI, USA, 8–14 December 2001.

6. Fattal, R. Image upsampling via imposed edge statistics. *ACM Trans. Graph.* **2007**, *26*, 95. [CrossRef]

7. Timofte, R.; Smet, V.D.; Gool, L.V. A$^+$: Adjusted anchored neighborhood regression for fast super-resolution. In Proceedings of the Asian Conference on Computer Vision, Singapore, 1–5 November 2014.

8. Dong, C.; Loy, C.C.; He, K.; Tang, X. Image super-resolution using deep convolutional networks. *IEEE Trans. Pattern Anal. Mach. Intell.* **2016**, *38*, 295–307. [CrossRef]

9. Huang, J.B.; Singh, A.; Ahuja, N. Single image super-resolution from transformed self-exemplars. In Proceedings of the IEEE Conference on Computer Vision and Pattern Recognition (CVPR), Boston, MA, USA, 7–12 June 2015.

10. Cui, Z.; Chang, H.; Shan, S.; Zhong, B.; Chen, X. Deep network cascade for image super-resolution. In Proceedings of the European Conference on Computer Vision (ECCV), Zurich, Switzerland, 6–12 September 2014.

11. Wang, Z.; Liu, D.; Yang, J.; Han, W.; Huang, T. Deep networks for image super-resolution with sparse prior. In Proceedings of the IEEE International Conference on Computer Vision (ICCV), Santiago, Chile, 7–13 December 2015.

12. Wang, L.; Huang, Z.; Gong, Y.; Pan, C. Ensemble-based deep networks for image super-resolution. *Pattern Recognit.* **2017**, *68*, 191–198. [CrossRef]

13. Nazzal, M.; Ozkaramanli, H. Wavelet domain dictionary learning-based single image super-resolution. *Signal Image Video Process.* **2015**, *9*, 1491–1501. [CrossRef]

14. Ayas, S.; Ekinci, M. Single image super-resolution based on sparse representation using discrete wavelet transform. *Multimed. Tools Appl.* **2018**, *77*, 16685–16698. [CrossRef]
15. Ahmed, J.; Waqas, M.; Ali, S.; Memon, R.A.; Klette, R. Coupled dictionary learning in wavelet domain for Single-Image Super-Resolution. *Signal Image Video Process.* **2018**, *12*, 453–461. [CrossRef]
16. Kumar, N.; Verma, R.; Sethi, A. Convolutional neural networks for wavelet domain super-resolution. *Pattern Recognit. Lett.* **2017**, *90*, 65–71. [CrossRef]
17. Ahmed, J.; Gao, B.; Tian, G.Y. Wavelet domain based directional dictionaries for single image super-resolution. In Proceedings of the IEEE International Conference on Imaging Systems and Techniques (IST), Beijing, China, 18–20 October 2017.
18. Kim, J.; Kwon, L.J.; Mu, L.K. Accurate image super-resolution using very deep convolutional networks. In Proceedings of the IEEE Conference on Computer Vision and Pattern Recognition (CVPR), Las Vegas, NV, USA, 26 June–1 July 2016.
19. Matsuyama, E.; Tsai, D.Y.; Lee, Y.; Tsurumaki, M.; Takahashi, N.; Watanabe, H.; Chen, H.-M. A modified undecimated discrete wavelet transform based approach to mammographic image denoising. *J. Digit. Imaging* **2013**, *26*, 748–758. [CrossRef] [PubMed]
20. Chen, Y.; Cao, Z. Change detection of multispectral remote-sensing images using stationary wavelet transforms and integrated active contours. *Int. J. Remote Sens.* **2013**, *34*, 8817–8837. [CrossRef]
21. Wang, Y.; Xie, L.; Qiao, S.; Zhang, Y.; Zhang, W.; Yuille, A.L. Multi-scale spatially-asymmetric recalibration for image classification. In Proceedings of the European Conference on Computer Vision (ECCV), Munich, Germany, 8–14 September 2018.
22. Zhang, K.; Zuo, W.; Chen, Y.; Meng, D.; Zhang, L. Beyond a gaussian denoiser: Residual learning of deep cnn for image denoising. *IEEE Trans. Image Process.* **2017**, *26*, 3142–3155. [CrossRef]
23. Hui, Z.; Wang, X.; Gao, X. Fast and accurate single image super-resolution via information distillation network. In Proceedings of the IEEE Conference on Computer Vision and Pattern Recognition, Salt Lake City, UT, USA, 18–23 June 2018.
24. Zhang, K.; Zuo, W.; Zhang, L. Learning a single convolutional super-resolution network for multiple degradations. In Proceedings of the IEEE Conference on Computer Vision and Pattern Recognition, Salt Lake City, UT, USA, 18–23 June 2018.
25. Rachmadi, M.F.; Valdés-Hernández, M.D.C.; Agan, M.L.F.; Di Perri, C.; Komura, T. Alzheimer's Disease Neuro imaging Initiative. Segmentation of white matter hyper intensities using convolutional neural networks with global spatial information in routine clinical brain MRI with none or mild vascular pathology. *Comput. Med. Imaging Graph.* **2018**, *66*, 28–43. [CrossRef]
26. Suk, H.I.; Wee, C.Y.; Lee, S.W.; Shen, D. State-space model with deep learning for functional dynamics estimation in resting-state fMRI. *Neuroimage* **2016**, *129*, 292–307. [CrossRef]
27. Mousas, C.; Newbury, P.; Anagnostopoulos, C.N. Evaluating the covariance matrix constraints for data-driven statistical human motion reconstruction. In Proceedings of the 30th Spring Conference on Computer Graphics, Smolenice, Slovakia, 28–30 May 2014.
28. Mousas, C.; Newbury, P.; Anagnostopoulos, C.N. Data-driven motion reconstruction using local regression models. In Proceedings of the IFIP International Conference on Artificial Intelligence Applications and Innovations, Rhodes, Greece, 19–21 September 2014.
29. Holden, D.; Komura, T.; Saito, J. Phase functioned neural networks for character control. *ACM Trans. Graph.* **2017**, *36*, 42. [CrossRef]
30. Kim, S.S.; Eom, I.K.; Kim, Y.S. Image interpolation based on the statistical relationship between wavelet subbands. In Proceedings of the IEEE International Conference on Multimedia and Expo, Beijing, China, 2–5 July 2007.
31. Kinebuchi, K.; Muresan, D.D.; Parks, T.W. Image interpolation using wavelet-based hidden Markov trees. In Proceedings of the IEEE International Conference on Acoustics, Speech, and Signal Processing, Salt Lake City, UT, USA, 7–11 May 2001.
32. Chan, R.H.; Chan, T.F.; Shen, L.; Shen, Z. Wavelet algorithms for high-resolution image reconstruction. *SIAM J. Sci. Comput.* **2003**, *24*, 1408–1432. [CrossRef]
33. Tian, J.; Ma, L.; Yu, W. Ant colony optimization for wavelet-based image interpolation using a three-component exponential mixture model. *Expert Syst. Appl.* **2011**, *38*, 12514–12520. [CrossRef]

34. Roweis, S.T.; Saul, L.K. Nonlinear dimensionality reduction bylocally linear embedding. *Science* **2000**, *290*, 2323–2326. [CrossRef]

35. Tenenbaum, J.B.; De Silva, V.; Langford, J.C. A globalgeometric framework for nonlinear dimensionality reduction. *Science* **2000**, *290*, 2319–2323. [CrossRef] [PubMed]

36. Cao, L.J.; Chua, K.S.; Chong, W.K.; Lee, H.P.; Gu, Q.M. A comparasion of PCA, KCPA and ICA for dimensionelity reduction in support vector machine. *Neurocomputing* **2003**, *55*, 321–336. [CrossRef]

37. Belkin, M.; Niyogi, P. Laplacian eigenmaps for dimensionality reduction and data representation. *Neural Comput.* **2003**, *15*, 1373–1396. [CrossRef]

38. Pascanu, R.; Mikolov, T.; Bengio, Y. On the difficulty of training recurrent neural networks. In Proceedings of the International Conference on Machine Learning (ICML), Atlanta, GA, USA, 16–21 June 2013.

39. Ngaim, J.; Khosla, A.; Kim, M.; Nam, J.; Lee, H.; Ng, A.Y. Multimodal deep learning. In Proceedings of the 28th International Conference on Machine Learning (ICML), Bellevue, WA, USA, 28 June–2 July 2011.

40. Nam, J.; Herrera, J.; Slaney, M.; Smith, J.O. Learning sparse feature representations for Music Annotation and Retrieval. In Proceedings of the 13th International Society for Music Information Retrieval Conference, Porto, Portugal, 8–12 October 2012.

41. Mousas, C.; Anangnostopoulos, C.N. Learning motion features for example based finger motion estimation for virtual characters. *3D Res.* **2017**, *8*, 136. [CrossRef]

42. Bengio, Y.; Simard, P.; Frasconi, P. Learning long-term dependencies with gradient descent is difficult. *IEEE Trans. Neural Netw.* **1994**, *5*, 157–166. [CrossRef] [PubMed]

43. LeCun, Y.; Bottou, L.; Bengio, Y.; Haffner, P. Gradient-based learning applied to document recognition. *Proc. IEEE* **1998**, *86*, 2278–2324. [CrossRef]

44. Vedaldi, A.; Lenc, K. MatConvNet: Convolutional Neural Networks for MATLAB. In Proceedings of the 23rd ACM International Conference on Multimedia, Brisbane, Australia, 26–30 October 2015.

45. Zimo, L.; Zhou, Y.; Xiao, S.; He, C.; Huang, Z.; Li, H. Auto-conditioned LSTM recurrent network for extended complex human motion synthesis. *arXiv* **2017**, arXiv:1707.05363.

46. Rekabdar, B.; Mousas, C. Dilated convolutional neural network for predicting driver's activity. In Proceedings of the 21st International Conference on Intelligent Transportation System (ITSC), Maui, Hawaii, USA, 4–7 November 2018.

47. Abdel-Hamid, O.; Mohamed, A.R.; Jiang, H.; Penn, G. Applying Convolutional neural networks concepts to hybrid NN-HMM model for speech recognition. In Proceedings of the IEEE International Conference on Acoustics, Speech and Signal Processing (ICAS), Kyoto, Japan, 25–30 March 2012.

48. Bilmes, J.A.; Bartels, C. Graphical model architectures for speech recognition. *IEEE Signal Process. Mag.* **2005**, *22*, 89–100. [CrossRef]

49. Yang, J.; Wright, J.; Huang, T.S.; Ma, Y. Image super-resolution via sparse representation. *IEEE Trans. Image Process.* **2010**, *19*, 2861–2873. [CrossRef]

Article

Reversible Data Hiding Using Inter-Component Prediction in Multiview Video Plus Depth

Jin Young Lee [1,†], Cheonshik Kim [2,*,†] and Ching-Nung Yang [3,†]

1   School of Intelligent Mechatronics Engineering, Sejong University, Seoul 05006, Korea; jinyounglee@sejong.ac.kr
2   Department of Computer Engineering, Sejong University, Seoul 05006, Korea
3   Department of Computer Science and Information Engineering, National Dong Hwa University, Hualien 97401, Taiwan; cnyang@gms.ndhu.edu.tw
*   Correspondence: mipsan@paran.com
†   These authors contributed equally to this work.

Received: 5 April 2019; Accepted: 22 April 2019; Published: 9 May 2019

Abstract: With the advent of 3D video compression and Internet technology, 3D videos have been deployed worldwide. Data hiding is a part of watermarking technologies and has many capabilities. In this paper, we use 3D video as a cover medium for secret communication using a reversible data hiding (RDH) technology. RDH is advantageous, because the cover image can be completely recovered after extraction of the hidden data. Recently, Chung et al. introduced RDH for depth map using prediction-error expansion (PEE) and rhombus prediction for marking of 3D videos. The performance of Chung et al.'s method is efficient, but they did not find the way for developing pixel resources to maximize data capacity. In this paper, we will improve the performance of embedding capacity using PEE, inter-component prediction, and allowable pixel ranges. Inter-component prediction utilizes a strong correlation between the texture image and the depth map in MVD. Moreover, our proposed scheme provides an ability to control the quality of depth map by a simple formula. Experimental results demonstrate that the proposed method is more efficient than the existing RDH methods in terms of capacity.

Keywords: 3D; depth map; inter-component prediction; MVD; reversible data hiding; texture

---

1. Introduction

Data hiding (DH) [1] plays an important role in secret communication. For this purpose, secret information and metadata are embedded in the cover media, such as still image, video, audio, 3D video, and so on. The visual quality and capacity of the cover image are important criteria for DH schemes. In addition, reversible DH (RDH) techniques are developed to extract the embedded secret information and restore losslessly the cover image.

Up to date, various RDH algorithms have been proposed, e.g., difference expansion-based algorithms [2–6], histogram shifting [7–10], prediction-error expansion (PEE) [11–14] and integer-to-integer transform [15–17], etc.

The approaches of difference expansion (DE) show good performance in respect of high-capacity. The first introduction of this algorithm was by Tian, and the research has been extended by [3,4]. For data embedding, it has to make a room for a secret bit through a pixel extension, and inserts data therein. Alattar [3] improved the performance of Tian's work by generalizing a DE technique for all integer conversions. The method proposed by Sachnev et al. is that image pixels are separated into black and white squares of a chessboard with two identical sets diagonally connected. This prediction method, called rhombus, is superior to the existing prediction methods (e.g., median edge detector predictor (MED) and gradient-adjusted predictor (GAP)) by making the average value of adjacent neighbors of a specific pixel as a predictive value. Thereafter, various methods for improving prediction performance were introduced.

The histogram shifting (HS) technique is also known as a method having relatively a few distortion in a cover image. However, it requires a location map in RDHs to embed data and restore a cover image. Al-Qershi and Khoo proposed a 2-dimensional DE (2D-DE) scheme achieving about 1-BPP performance [6]. These histogram-based schemes may achieve good visual quality and adequate embedding capacity, but it has the drawback having to send a pair of peaks and zero points to the receiver.

PEE involves a process that obtains the prediction error (PE) from the neighborhood of a pixel and embeds information bits into the expanded errors. If the difference between an original pixel and a predicted pixel is large, the distortion of the cover image is greatly enlarged during the embedding process. In this case, by enhancing the data embedding capacity of the region of low frequency in the cover image, it may maintain the image quality and embedding capacity of the cover image. Compared to DE and HS-based methods, it is well known that PEE performs better. When we are considering the existing PEE methods, DH with order prediction has less distortion at low embedding rates.

Meanwhile, with the rapid development of multiview video technologies, viewers can experience more realistic 3D scenes with highly advanced multimedia systems, such as 3D television and free-viewpoint television. To overcome a limited bandwidth, the multiview video plus depth (MVD) is adopted as a 3D video format [18,19]. In MVD (see Figure 1), a texture image indicates intensities of an object, whereas a depth map represents a distance between an object and a camera as a grayscale image having values between 0 and 255. Because MVD enables the advanced video system to arbitrarily generate virtual views by using a depth image-based rendering (DIBR) method [20], a small number of view information can be transmitted.

Figure 1. MVD consisting of a texture image (**left**) and its corresponding depth map (**right**).

Until now, various watermarking technologies [21–25] have been introduced for marking 3D videos. Asikuzzaman and Pickering [21] proposed a digital watermarking approach that inserts a watermark into the DT-CWT coefficients. Pei and Wang [22] introduced a 3D watermarking technique based on the D-NOSE model which can detect the suitable region of the depth image for watermark embedding. Since view synthesis is very sensitive to variations in-depth values, this scheme focuses mainly on the synthesis error. Wang et al. [23] exploited scale-invariant feature transform (SIFT)-based feature points to synchronize a watermark but focused on only signal processing and omitted geometric attacks.

Based on MVD, Chung et al. [26] and Shi et al. [27] proposed RDHs for depth maps using a depth no-synthesis-error (D-NOSE) model [28] and PEE. Each pixel in the depth maps has an allowable pixel range and the pixel value is increased or decreased in an allowable range. However, it may be guaranteed that there are no errors in the synthesized image using D-NOSE. Taking advantage of this characteristic may be used to improve embedding capacity for RDH. Chung et al. first proposed a method based on PEE that could effectively hide data in depth maps of 3D images. The disadvantage of Chung et al.'s method is that it does not provide sufficient embedding capacity. Shi et al. proposed a way to use all of the acceptable range of pixels to solve the problem proposed by Chung et al.'s method. However, Shi et al. did not suggest a systematic way of adjusting the embedding capacity considering a quality of depth map.

In this paper, we first analyze the disadvantages of Chung et al.'s reversible data hiding algorithm and propose a PEE-based DH technique that completely uses the allowable range of each pixel using an inter-component prediction method. The performances of the proposed RDH are improved by using the correlation between the texture and the depth map, which is an advantage of the inter-component prediction. Also, we propose a method to control embedding rates and image quality systematically which may be applied to various RDH applications such as medical or military fields.

The remainder of this paper is organized as follows. Section 2 briefly discusses view synthesis, difference expansion, related RDH and watermarking methods. In Section 3, we introduce a reversible data hiding method based on D-NOSE model, PEE, and inter-component prediction. In Section 4, we compare and analyze the experimental result with conventional RDHs and our proposed RDH. Finally, this paper concludes in Section 5.

2. Related Works

In Section 2, we explain a 3D view synthesis principle, a difference expansion (DE) method, and Chung et al.'s [26] and Shi et al.'s [27] RDH methods based on 3D view synthesis. In addition, Zhu et al.'s [24], Wang et al.'s [23], and Asikuzzaman et al.'s [25] digital watermarking methods are also introduced.

2.1. View Synthesis

We can obtain a 3D version of the classical 2D videos with depth information via 3D view synthesis. Depth information plays a key role in synthesizing virtual views and the quality of synthesized views is critical in 3D video systems. In view synthesis, a pixel in the texture image is mapped to a new position in the virtual view by using the corresponding depth value. First, the disparity d in a pixel of depth map is obtained using the following equation.

$$d_i = \frac{f \cdot l}{255} \left(\frac{1}{z_{near}} - \frac{1}{z_{far}} \right) \times q_i. \tag{1}$$

where f and l denote the focal length and the baseline distance between two horizontally adjacent cameras, respectively, and z_{near} and z_{far} mean the nearest and farthest depth values, respectively. The pixel q indicates the i_{th} depth pixel value. After the calculation of a disparity d_i, it is rounded into an integer value. Then, the pixel (x', y') is filled by shifting the pixel (x, y) with d (see Figure 2).

$$\begin{pmatrix} x' \\ y' \end{pmatrix} = \begin{pmatrix} x \pm d \\ y \end{pmatrix} \tag{2}$$

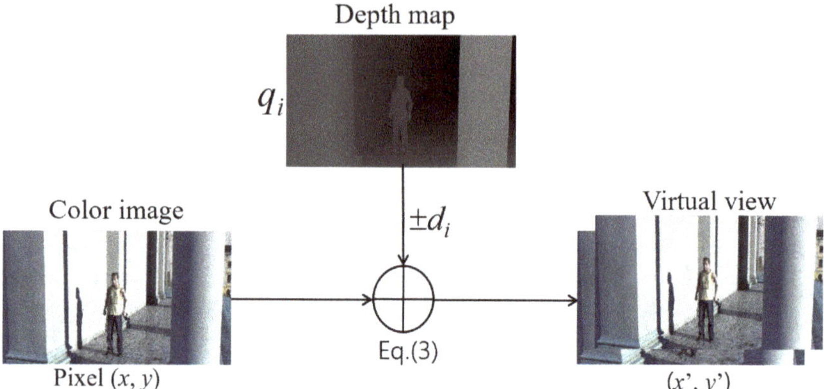

Figure 2. View synthsis based on disparity information.

Based on the D-NOSE model [28], the symbols L and U indicate the minimum and maximum depth values within the allowable distortion range. If the marked pixel q'_i is still in the range $([L, U])$ after hiding the data in q_i, the virtual view will not be distorted. The notation φ is a set collecting depth pixel value with the disparity $(d_i = n)$. N denotes the number of pixels (width $\times$ height) in the depth map.

For example, assuming that a disparity $n = 32$ and the minimum and maximum pixels belonging to n are $[L_q, U_q] = \{21...24\}$. It means that there are four pixels corresponding to disparity $d_i = 32$, i.e., the pixels are $21, \ldots, 24$. The way to figure out the allowable range for each pixel of the depth map is shown stepwise in the Algorithm 1.

$$\varphi_n = \{q_i \in (d_i = n)\}_1^N \tag{3}$$

Algorithm 1 DRange(q)

Input: input pixel q
Output: $[L, U]$

1: input pixel q
2: $d \leftarrow$ depth2disprity(q) // Equation (1)
3: **for** $i = q - 1$ To 0 Step -1 **do**
4:     **if** depth2disprity(i)! $= d$ **then** break
5:     **end if**
6: **end for**
7: $L \leftarrow i + 1$
8: **for** $i = q + 1$ To 255 Step 1 **do** //8-bit depth pixel
9:     **if** depth2disprity(i)! $= d$ **then** break
10:     **end if**
11: **end for**
12: $U \leftarrow i - 1$
13: **return** [L, U]

2.2. Difference Expansion

In this section, we describe the concept of RDH using pixel prediction and DE, where p and $\hat{p}$ denote an original pixel and a predicted pixel. A prediction error is determined by $e_{i,j} = p_{i,j} - \hat{p}_{i,j}$. If $e_{i,j} < T$ and no overflow and underflow on each pixel, the secret bits b may be embedded into the pixel

p as $p'_{i,j} = p_{i,j} + e_{i,j} + b$. If $|e_{i,j}| \geq T$, it is not appropriate to embed secret bits, because the carrier pixel p' may have higher prediction error than the other embedded pixels. This pixel is modified as follows:

$$p'_{i,j} = \begin{cases} p_{i,j} + T, & \text{if } (e_{i,j} \geq T) \\ p_{i,j} - (T-1), & \text{if } (e_{i,j} \leq -T) \end{cases}$$

The location information of the underflow or overflow is recorded in the location map and is used for decoding. If the predictive values are the same before and after data hiding, the reversibility of the watermarking is guaranteed. It should restore the shifted values T before obtaining the predictive value. Then, it is possible to obtain the predictive values correctly.

2.3. Chung et al.'s Reversible Data Hiding

The D-NOSE model can guarantee zero synthesis distortion by determining an allowable distortion range for each depth pixel. The previous two existing methods [26,27] use the rhombus prediction to obtain the prediction pixel $\hat{p} = \left\lceil \frac{p_1 + p_2 + p_3 + p_4}{4} \right\rceil$, and the prediction error $e = p - \hat{p}$ means the difference between the original pixel p and the predicted depth pixel $\hat{p}$. Here, p_1, p_2, p_3, and p_4 denote the four adjacent pixels, where they are placed on the top, left, bottom, and right sides of p, respectively. The average value is rounded up in the calculation. If the number of hidden bits m for a pixel p is $\lfloor log_2(U - \hat{p} + 1) \rfloor$ when $e = 0$ and $m = \lfloor log_2(p_i - L + 1) \rfloor - 1$ when $e = -1$. Otherwise, $m = 0$.

The marked pixel p' is obtained from $p' = \hat{p} + b$ when $e = 0$ and $p' = \hat{p} - b - 1$ when $e = -1$. Here, b means a binary number of m bits, and the allowable range is $[L, U]$. On the receiving side, the secret bits b can be simply obtained by the expression $b = (e' \bmod 2^m)$, where $e' = p' - \hat{p}$.

The location map in Chung et al.'s method is a simple way, i.e., mark "1" if there is no hidden data at the position and "0" otherwise. If the surface of the image is the same color, most of the location map may be zeros since most of prediction errors may be $e = 0$. For this reason, Chung et al. used arithmetic coding to compress the location map and then hide the location map in the front or back end of the depth map. Chung et al.'s method achieves the purpose of RDH for 3D synthesis images. Unfortunately, it does not embed a sufficient amount of data.

There is a vulnerability in Chung et al.'s method. For example, if a pixel p and an error e is 85 and 0, respectively, and acceptable range of the pixel p is [83, 87], then it is allowed to hide only 1-bit in the pixel p, because of $m = \lfloor log_2(87 - 85 + 1) \rfloor = 1$. If a pixel is $p = 87$ and $e = 0$, then it does not allow to embed bits. That is because a room to hide bits is determined by the position of the predicted pixel in an allowable range of a pixel. Thus, there is no room to hide a bit when the pixel is $p = 87$. Another vulnerability is that the quality of the depth map is not considered sufficiently, because the quality only depends on the feature of the depth map.

2.4. Shi et al.'s Reversible Data Hiding

Shi et al. [27] proposed a RDH based on D-NOSE model, where the method embed information into double layer of depth maps. Here, the prediction method for PEE obtains $\hat{p}$ by the method of rhombus prediction. In order to use of the allowable range fully, it is embed the data into the prediction-error ($e = p - \hat{p}$) value 0, and the pixel is expanded toward either the maximum or the minimum values. In the allowable range $[l_n, u_n]$ of the disparity $d_i = n$ in the pixel, the number of bits for the pixel q_i is $m_i = log_2(u_n^* - l_n^* + 1)$ where $\hat{q}_i \leq u_n^* \leq u_n$ and $l_n \leq l_n^* \leq \hat{q}_i$ when $e_i = 0$. Otherwise, it is $m_i = 0$. The marked pixel q'_i may be expressed as $q'_i = l_n^* + b$, where $b = \{0, 1, 2, \ldots, 2^m - 1\}$. For example, if a pixel p and an error e is 85 and 0, respectively, and acceptable range of the pixel p is [83, 87], then it is allowed to hide $m = log_2 5$.

2.5. Zhu et al.'s Digital Watermarking

Zhu et al. [24] propose a watermarking method for a new viewpoint video frame generated by DIBR (Depth Image-Based Rendering). To preserve the watermark information during the generation of the viewpoint video frame, the blocks in foreground object of original video frame is selected to embed the watermark because pixels in this kind of object are more likely to be preserved in the warping. Here, for watermarking, DCT transformation of the blocks in foreground object is firstly done. Then, after embedding the watermark into the DCT, IDCT should be done before the DIBR.

2.6. Wang et al.'s Digital Watermarking

Wang et al. [23] propose a novel watermarking method for DIBR 3D images by using SIFT to select the area where watermarking should be embedded. Then, the watermark information is embedded into the DCT coefficients of the selected area by using Spread spectrum technology. The SIFT is used to select suitable areas in which watermarking should be embedded by applying $n \times n$ 2D-DCT to the areas selected. Next, spread spectrum technique and orthogonal spread spectrum code are applied to embed the watermark. In order to extract watermarks from images, we can compute the correlation between DCT coefficients of every area and the spread spectrum code to estimate the embedded message.

2.7. Asikuzzaman et al.'s Digital Watermarking

Asikuzzaman et al. [25] proposed a blind video watermarking algorithm in which a watermark is embedded into two chrominance channels using a double tree complex wavelet transform. The chrominance channel has a watermark and preserves the original video quality and the double tree composite wavelet transform ensures robustness against geometric attacks due to the shift invariant nature. The watermark is extracted from a single frame without the original frame. This approach is also robust to downscaling in arbitrary resolution, aspect ratio change, compression, and camcording.

3. Proposed Scheme

In this section, we introduce RDH based on the D-NOSE model by using PEE and inter-component prediction to improve the accuracy of predictive pixel in our proposed method. The accuracy of the prediction maximizes the performance of the proposed RDH based on the D-NOSE model using maximum allowable pixel range of each pixel. Moreover, our proposed scheme has the capability controlling the quality and embedding capacity of the depth map.

RDH methods based on the 2D texture images usually have a trade-off between the capacity and quality of the cover signal. However, since the depth map is used to synthesize virtual views as non-visual data, we may embed secret data into the depth map without degrading the quality of the virtual view synthesis. The D-NOSE model can guarantee zero synthesis distortion by determining an allowable distortion range for each depth pixel. Besides, since we apply the existing PEE technology to the 3D depth map and use the location map, our proposed method is also a perfect way to restore the original depth map.

3.1. Inter-Component Prediction

Here, we will introduce inter-component prediction to obtain a better prediction for high embedding capacity. Figure 3 illustrates the configuration of the depth pixel, and inter-component prediction based on the corresponding texture pixels. A depth map in Figure 3a is composed of marked '♦' and '◊'. The pixels in depth map are subdivided into two sets, $\Phi1(\in \{♦\ldots♦\})$ and $\Phi2(\in \{◊\ldots◊\})$. The inter-component prediction (in Figure 3b) selects adaptively the predicted direction for the depth pixel by comparing the corresponding texture pixel and its adjacent pixels.

When the depth pixel $q_{i,j}$ is predicted from neighbor pixels, $q_1, q_2, q_3,$ and q_4, as shown in Figure 3b, the corresponding texture pixel $t_{i,j}$ is compared with $t_1, t_2, t_3,$ and t_4 as follows

$$dirD = \min_{dirT}(|t_1 - t|, |t_2 - t|, |t_3 - t|, |t_4 - 4|), \tag{4}$$

where $dirT$ and $dirD$ denote the prediction direction of the texture image and the depth maps, respectively. For example, if the pixel having the minimum texture difference is t_1, q_{dirD} is set as q_1. The existing rhombus prediction method was an efficient prediction method for color and grayscale images. However, the rhombus method is less effective than the proposed method for depth map pixels. That is why we propose a way to predict the pixel of depth map by considering the texture image. Obviously, considering 3D synthesis, it seems that inter-component prediction is an excellent method.

The prediction is alternately performed for two sets, $\Phi 1$ and $\Phi 2$. That is, $\Phi 2$ is used to predict $\Phi 1$, and vice versa. When predicting $\Phi 2$, the prediction may not be accurate because there are hidden pixels in $\Phi 1$. To tackle this matter, we use the average of the unmarked pixel in $\Phi 1$. This is because the correlation between the marked depth pixels and the associated pixels is low, and the prediction is not accurate.

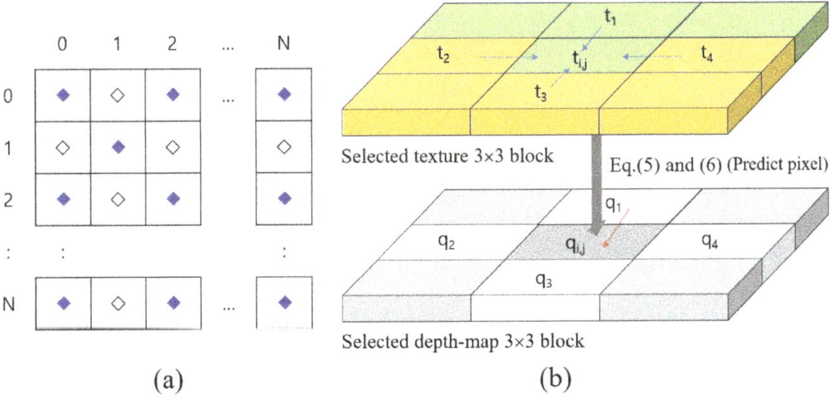

Figure 3. Diagram for configuration of the depth map, and inter-component prediction based on the corresponding texture pixels.

3.2. Embedding Algorithm

Here, we will examine in detail the data embedding procedure. To extract hidden bits and recover the original depth map, we have to record locations on whether each pixel contains hidden bits or not.

Step 1: Reads 3×3 block from the given depth map and assigns it to the variable B. Obtain the pixel $q_{i,j}$ and predicted pixel q_{dirD} from B using Equation (4).

Step 2: The prediction error e is calculated by subtracting the prediction pixel q_{dirD} from the original pixel $q_{i,j}$ as follow

$$e_{i,j} = q_{i,j} - q_{dirD}. \tag{5}$$

Step 3: If ($e_{i,j} = 0$), the number of bits m that can be embedded into the pixel $q_{i,j}$ is calculated using the allowable pixel range of the pixel $q_{i,j}$ (Equation (6)), where (L_q, U_q) is obtained from the Algorithm 1.

$$\begin{cases} m = \lfloor log_2(U_q - L_q + 1) \rfloor. \\ \text{if } (m < 1), \text{ goto Step 1} \end{cases} \tag{6}$$

Here, $(\lfloor x \rfloor)$ is the integer less than or equal to x.

Step 4: The binary secret bits η_m are embedded in $q_{i,j}$ using Equation (7). (Note: the function $b2d(\cdot)$ is to converts binary values to a decimal value). That is, the η_m is included in the expanded e' by DE.

$$q'_{i,j} = \left\{ \begin{array}{l} e'_{i,j} = 2^m \times e_{i,j} + b2d(\eta_m) \\ L_q + e'_{i,j} \end{array} \right. \tag{7}$$

Step 5: if $(e_{i,j} = 0)$, $LM_{i,j} = 0$, otherwise $LM_{i,j} = 1$. (Notes: location map $LM_{i,j} = 0$ means that hidden bits exists.)

$$LM_{i,j} = \left\{ \begin{array}{ll} 0, & \text{if } (e_{i,j} = 0) \\ 1, & \text{otherwise} \end{array} \right. \tag{8}$$

Step 6: Go to Step 1 until all pixels are processed.

After the embedding procedure is finished, all pixels including the hidden bits are still within the acceptable range. Therefore, 3D synthesis images have no distortion.

Example 1. *Given two blocks in Figure 4a, we demonstrate how to hide secret bits in the pixel $q_{i,j}$. First, we obtain the predictive pixel from the texture block and depth block. Applying Equation (4), it is observed that $t_1 = 51$ is the optimum pixel, so q_{dirD} becomes $q_{i-1,j} = 22$.*

After applying Equation (5), if $e_{i,j} = (q_{i,j} - q_{dirD}) = 22 - 22 = 0$, we may obtain disparity and allowable range through Algorithm 1. i.e., $[L_q, U_q] = (20, 23)$. The number of embedded bits in the pixel can be obtained via the Equation (6), i.e., $m = 2$. Thus, it takes 2-bits from the secret data and converts it into a decimal value. By applying the Equation (7), $q'_{i,j} = 20 + 2 = 22$. Finally, the pixel q' has secret bits $'10'_2$. In this case, $q_{i,j}$ is the same as $q'_{i,j}$, so there is no noise in the marked pixel $q'_{i,j}$.

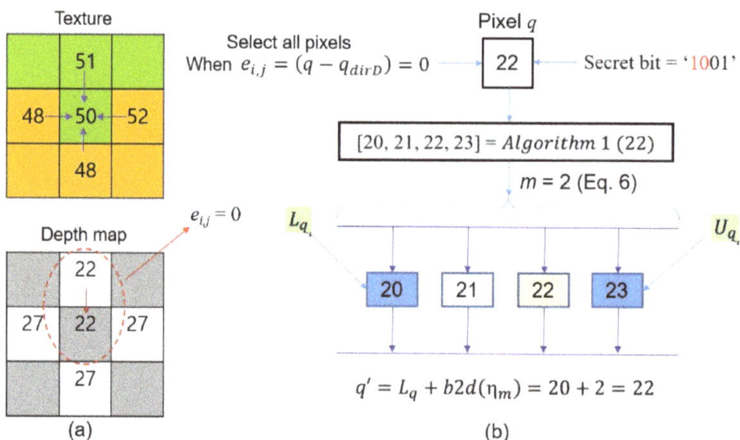

Figure 4. Example of the data embedding using the proposed method.

3.3. Extraction Algorithm

Suppose that a depth map containing secret data and a location map was delivered to the receiver side. We describe the procedure of extracting the hidden data and recover original pixels from the depth map. The detail (stepwise) is as follows.

Step 1: Reads 3×3 block from the given depth map and assigns it to the variable B. Obtains the pixel $q'_{i,j}$ and the predicted pixel q_{dirD} from the B, respectively.

Step 2: If the location map $LM_{i,j} = 0$, we obtain the size of hidden bits $m = \lfloor (U_q - L_q + 1) \rfloor$ from the allowable range of $q_{i,j}$.

Step 3: Obtain the error $e'_{i,j}$ from the pixel $q'_{i,j}$ including secret bits by using Equation (9). (Notes: $d2b(x, m)$ converts the decimal number x to a binary number of m bits.)

$$\eta_m = \begin{cases} e'_{i,j} = q'_{i,j} - L_q \\ b = d2b(e'_{i,j}, m) \end{cases} \tag{9}$$

Step 4: The original pixel $q_{i,j}$ is restored by replacing the pixel q' with q_{dirD}. Go to Step 1 until all pixels are processed.

Example 2. *First, Assuming that through the data embedding procedure (see Figure 4) in Example 1, the pixel $q'_{i,j}$ having the binary bits $'10'_2$ was transferred to the receiver. On the receiving side, some of the procedure for extracting the hidden bits in $q'_{i,j}$ are similar to the embedding procedure. The pixel q_{dirD} is determined using the neighboring pixels of the pixel $q'_{i,j}$ and the inter-component prediction method. In this case, the predicted pixel is $q_{dirD} = 22$. Since the location map $LM_{i,j} = 0$ at position (i,j) of $q'_{i,j}$, it can be seen that the data is hidden. Thus, the number of hidden bits from the allowable pixel range of the pixel $q'_{i,j}$ may be determined. That is, if we apply Equation (6) to the pixel $q'_{i,j}$, then we obtain the $m = \lfloor log_2 (U_q - L_q + 1) \rfloor = \lfloor log_2 (23 - 20 + 1) \rfloor = 2$. Next, the error bit $e'_{i,j} = 2$ obtains from $q'_{i,j} = 22$ and $L_q = 20$ by using Equation (9). The value hidden in the pixel $q_{i,j}$ is $e'_{i,j} = q'_{i,j} - L_q = 22 - 20 = 2$, which is converted to the binary number $'10'_2$. The marked pixel $q'_{i,j}$ is reconstructed to the original pixel $q_{i,j}$ by assigning the prediction value q_{dirD} to the position (i,j).*

3.4. Embedding/Extraction Procedure for Location Map

The location map (LM) is necessarily required for data extraction and depth map restoration, and thus, the location map should be transferred to the receiver side. There are many ways to transfer location map, but the most common way is to hide it into a cover image. At this step, it should minimize the location map, because the capacity of the secret data is reduced with the size of the location map. Thus, the compression of location map is a common procedure before embedding it into the cover image. There are several compression methods, but here we use arithmetic coding. The map size may be reduced by less than 10% by arithmetic coding, because the ratio of "0" in the map is more than 90%.

The location map LM compressed by arithmetic coding is assigned into variable δ. The compressed location data δ is embedded by the LSB replacement as follows, but the data are embedded considering the allowable pixel range.

$$q'_{i,j} = \begin{cases} \text{if } (U_q - L_q \geq 3) \ \{ \\ \quad q_{i,j} + 1, \quad \begin{cases} \text{if } (\delta_{i,j} = 0 \ \& \ q_{i,j} \% 2 = 1) \\ \text{if } (\delta_{i,j} = 1 \ \& \ q_{i,j} \% 2 = 0) \end{cases} \Big\} \equiv A \\ \quad q_{i,j} - 1, \quad \text{if } (q_{dirD} = U_q \ \& \ A) \\ \} \end{cases} \tag{10}$$

The compressed location data δ is embedded in front of the depth map. The data δ does not cause a synthesis error since it also adopted a D-NOSE model that hides the data within the allowable pixel range. The last position of the location information is sent on a separate channel. On the receiving side, the compressed location map can be extracted by the following Equation (11).

$$\delta_{i,j} = \begin{cases} \text{if } (U_q - L_q \geq 3) \ \{ \\ \quad 0, \quad \text{if } (q_{i,j} \% 2 = 0) \\ \quad 1, \quad \text{otherwise} \\ \} \end{cases} \tag{11}$$

3.5. Quanlity Control

In our proposed method, the quality of the depth map can be somewhat reduced if the allowable pixel range on all pixels is fully used. It can certainly be an advantage in terms of embedding capacity, but it is not desirable in terms of quality. Therefore, it is also important to find a balance between the two criteria. Adjustment of the allowable pixel range may be used to achieve such a purpose. Equations (12) and (13) can be used for managing depth map quality and embedding capacity through Equation (6). The control of depth map is achieved by using a limited allowable range $[L_q', U_q']$ instead of the $[L_q, U_q]$.

$$L_q' = \begin{cases} q_{dirD} - \sigma, & \text{if } (q_{dirD} - \sigma < L_q) \\ L_q, & \text{otherwise} \end{cases} \tag{12}$$

$$U_q' = \begin{cases} L_q' + 2^\sigma - 1, & \text{if } (L_q' + 2^\sigma - 1 > U_q) \\ U_q, & \text{otherwise} \end{cases} \tag{13}$$

Here, σ is an integer variable and the range of values is $\{\sigma \geq 1 \ \& \ \sigma \leq n\}$. That is, increasing the value of σ can increase the embedding capacity of RDH, while decreasing the value of σ may improve the quality of the depth map. We may increase the usefulness of the proposed method if we adjust the value of σ appropriately for the application.

4. Experimental Results

To better demonstrate the performance of our proposed scheme, we graphically show the results of experiments and analysis of 3D images with various features. All experiments are performed with eight 3D sized 1920×1088 (*or* 1024×768), "Poznan_Hall2","Poznan_Street", "Undo_Dancer", "GT_Fly", "Kendo", "Balloons", "Newspaper", and "Shark" (see Figure 5), which are often used for 3D video coding standards, such as 3D-AVC [19] and 3D HEVC [18], and the view synthesis reference software (VSRS) in JCT-3V [29]. Since the schemes such as Chung et al.'s, Shi et al.'s, and the proposed methods adopt the D-NOSE model, the synthesis using the original depth map is identical to that of the synthesis with the marked depth map.

In this paper, we use two criteria to evaluate the performance of the existing and our proposed schemes. The first criterion is the embedding rates (ERs) and the second is the peak signal-to-noise ratio (PSNR). The most well-known measurement method for objective evaluation of images is PSNR. That is, PSNR is the intensity of noise over the maximum intensity the signal can have. The MSE used in PSNR is an average difference in intensity between the marked depth map and the reference depth map. If the MSE value of the depth map is low, it is evaluated that the quality of the image is good. The MSE is calculated using a reference depth map p and the distorted depth map p' as follows.

$$MSE(p, p') = \frac{1}{N} \sum_{i=1}^{N} (p_i - p_i')^2 \tag{14}$$

The error value $\varepsilon = p_i - p_i'$ denotes the difference between the original depth map and the distorted depth map signal. The 255^2 means the allowable pixel intensity in Equation (15).

$$PSNR = 10 log_{10} \frac{255^2}{MSE} \tag{15}$$

Meanwhile, ER is a measurement of how much information is included in the marked depth map. That is, ER is the ratio of the embedded information contained in the marked depth map. In Equation (16), N is the total number of pixels and $||\eta||$ denotes the number of message bits.

$$\rho = \frac{||\eta||}{N} \tag{16}$$

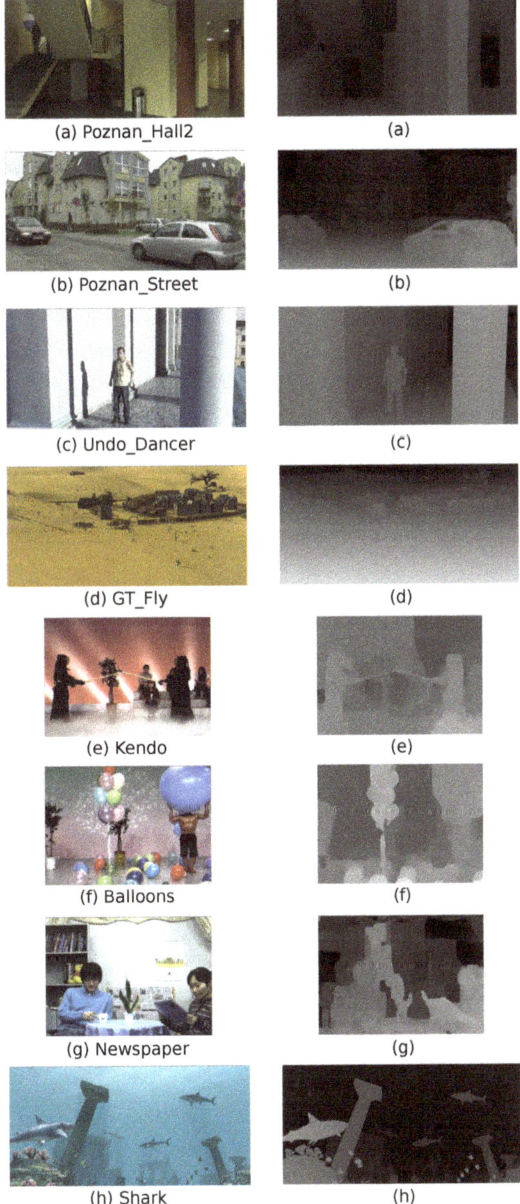

Figure 5. Test images; (**a–d,h**): 1920 × 1088, (**e–g**): 1024 × 768.

Figure 6 shows the comparison of data embedding rates using three methods (two existing methods and the proposed method) and eight depth maps: (a) Poznan_Hall2, (b) Poznan_Street, (c) Undo_Dancer, (d) GT_Fly, (e) Kendo, (f) Balloons, (g) Newspaper and (h) Shark; 1920 ×1088: (a)–(d) and (h), 1024 ×768: (e)–(g).

The ERs of Chung et al.'s method is much less than that of Shi et al.'s and our proposed methods. That is because Chung et al.'s method does not fully use the allowable range. On the other hand, Shi et al.'s and our proposed method may achieve better results by adopting an allowable pixel range. For example, we show that the ERs of "Poznan_Hall2" in Chung et al.'s method is very low. For the "GT_Fly", the embedding rates of our proposed method is 0.22 BPP higher than that of Shi et al.'s method. As shown in Figure 6, the coefficients of difference between these two methods show that the performance of our inter-component prediction is superior to the rhombus prediction. Therefore, our proposed method outperforms the existing two methods, including Chang et al.'s and Shi et al.'s method.

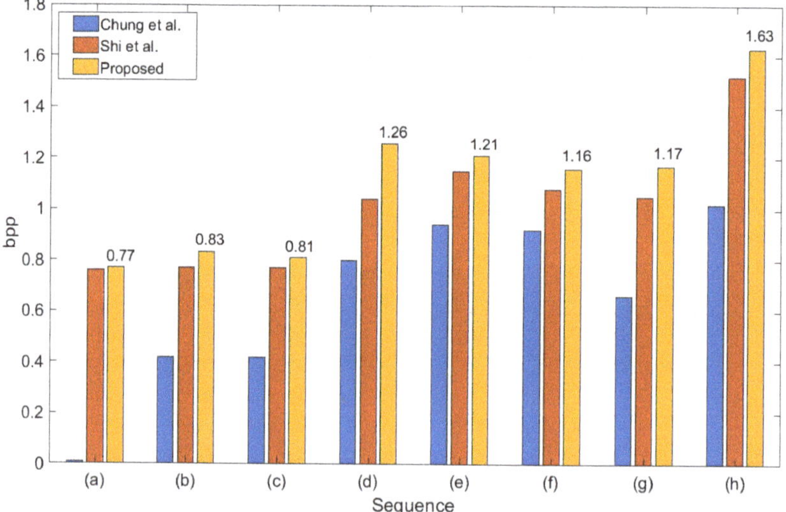

Figure 6. Performance comparison of three data hiding methods using eight depth maps.

In Table 1, we compare the performance of the proposed method and the existing methods by using PSNR with various BPPs and the depth map, "Shark". As the embedding rates increases, it appears that Chung et al.'s PSNR is sharply decreased. The reason is that Chung et al.'s method does not use the same method of quality control and the allowable depth ranges in the data embedding procedure. On the other hand, Shi et al.'s and our methods use the allowable pixel ranges and quality control, thus, they have good performance. In addition, our method shows slightly better performance than Shi et al.'s method because it uses an accurate prediction method.

Table 1. The comparision of the PSNR at the different BPP for the depth map, "Shark".

BPP	Chung et al.'s Method	Shi et al.'s Method	The Proposed Method
	PSNR (dB)	PSNR (dB)	PSNR (dB)
0.1	59.6257	61.1498	61.1504
0.2	54.2950	58.1325	58.1328
0.3	51.6756	56.3714	56.3744
0.4	49.6793	55.1210	55.1246
0.5	47.5926	54.1483	54.1521
0.6	46.6222	53.3583	53.3613
0.7	45.9716	52.6881	52.6907
0.8	45.5451	52.1081	52.1113

As shown in Table 2, we know that the maximum BPP of Chung et al.'s method is less than that of both Shi et al.'s and our proposed method. However, since the BPP is low, the PSNR is relatively higher than that of Shi et al.'s and our method.

Table 2. PSNR comparison for maximum BPP on each image using various methods.

Image	Chung et al.		Shi et al.		Our Scheme	
	BPP	**PSNR (dB)**	**BPP**	**PSNR (dB)**	**BPP**	**PSNR (dB)**
(a) Poznan_Hall2	0.0050	74.5704	0.7559	49.2263	0.7700	49.1044
(b) Poznan_Street	0.4222	53.9985	0.7713	49.3750	0.8339	48.9424
(c) Undo_Dancer	0.4215	53.6574	0.7693	49.0931	0.8094	48.8123
(d) GT_Fly	0.7956	48.6470	1.0363	44.5942	1.2603	43.5969
(e) Kendo	0.9393	51.4107	1.1485	45.6550	1.2093	45.4096
(f) Ballons	0.9179	51.5102	1.0813	45.5657	1.1633	45.2515
(g) Newspaper	0.6621	52.9301	1.0533	49.0001	1.1661	48.5315
(h) Shark	0.9608	45.2074	1.5194	41.0606	1.6274	40.7411

Therefore, if the BPP is the same for the three methods, the PSNRs of Shi et al.'s and our proposed method are better than that of Chang et al.'s method. In "Newspaper", there is the highest difference of PSNR between our proposed and the Shi et al.'s method, because it seems that the depth map is an image including a high-frequency property. Thus, it can be seen that the proposed method had high prediction competence on depth maps with high-frequency characteristics. Moreover, the average BPP of our proposed method is higher (by 0.09) than that of Shi et al.'s method, while our method is less than 0.39 dB compared to that of Shi et al.'s method in the aspect of PSNR. However, in this case, it is indistinguishable from the viewpoint of the usual human visual system. Therefore, it can be recognized that the proposed scheme improves somewhat regarding BPP.

In Figure 7, the control variable σ (Equations (12) and (13)) is applied to adjust the embedding capacity and quality of the depth map. Its principle is that the amount of embedding capacity increases in proportion to the value of the variable σ, while the quality of depth map decreases as σ increases. When the control variable $\sigma = 1$, the maximum embedding rate is 0.8 and the PSNR is 52 dB. When the variable $\sigma = 7$, BPP is measured from the lowest 0.1 to the maximum 1.6 and we obtain that the PSNR is from 48.5 dB to 41 dB. Under a strict communication environment, it may be useful to use the control variable for secret communication.

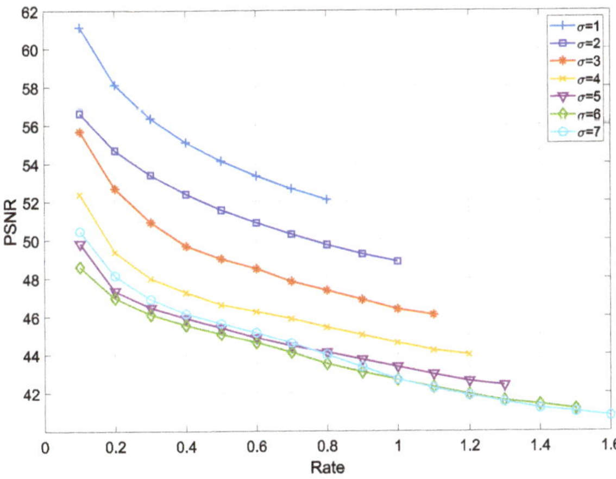

Figure 7. The relationship between BPP and PSNR according to control variable σ with depth map "Shark".

Table 3 shows PSNRs for the eight depth maps under various BPPs using three methods (two existing and the proposed methods) when $\sigma = 1$. In the table, we can see that depth maps (a), (e), and (f) may hide data up to 0.9 BPP.

Table 3. PSNR measurement for the eight depth maps when $\sigma = 1$.

BPP	PSNRs							
	(a)	(b)	(c)	(d)	(e)	(f)	(g)	(h)
0.1	61.1483	61.1333	61.1505	61.1533	61.1649	61.1649	61.1421	61.1504
0.2	58.1324	58.1279	58.1374	58.1310	58.1378	58.1378	58.1260	58.1328
0.3	56.3710	56.3680	56.3769	56.3762	56.3735	56.3735	56.3642	56.3744
0.4	55.1215	55.1214	55.1246	55.1220	55.1250	55.1250	55.1171	55.1246
0.5	54.1511	54.1503	54.1547	54.1511	54.1514	54.1514	54.1458	54.1521
0.6	53.3607	53.3599	53.3635	53.3602	53.3635	53.3635	53.3590	53.3613
0.7	52.6905	52.6890	52.6929	52.6900	52.6921	52.6921	52.6886	52.6907
0.8	52.1107	–	52.1138	52.1091	52.1116	52.1116	52.1085	52.1113
0.9	51.5992	–	–	–	51.6007	51.6007	–	–

The depth map of (a), (e), and (f) has a higher embedding capacity than the other depth maps; because the sum of pixels φ_n (Equation (3)) of these depth maps are high. In other words, there are a number of pixels having a wide range of allowable pixels. From this point of view, it can be seen that sum of pixels φ_n of the depth map (b) is low. It is proved that the proposed method maintains very high PSNR like the conventional DHs through various simulations.

Figure 8 is a visual comparison of the marked depth maps generated from the simulation results derived from three methods in "Poznan_Street". In Figure 8, the BPP for (c), (d), and (f) are 0.4222, 0.7713, and 0.8339, respectively, and the PSNRs of these are 53.9985 dB, 49.3750 dB, and 48.9424 dB, respectively. As mentioned above, the embedding rates of the proposed method are about twice as high as that of Chung et al.'s method, and also our proposed scheme provides a good depth map quality about 49 dB.

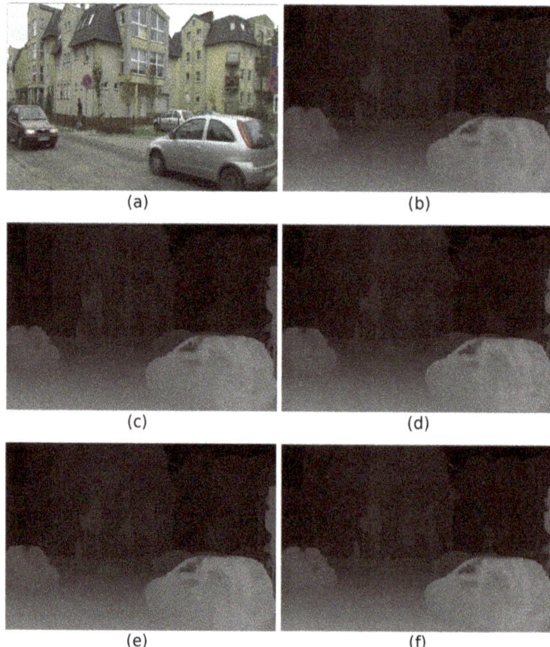

Figure 8. The relationship between BPP and PSNR according to control variable, σ in depth map on "Poznan_Street".

Figure 9 shows a visual representation of how much pixels are distorted during the data-embedding procedure for the depth map, "Shark". The two marked depth maps by Shi et al.'s and the proposed method include 0.4 BPP. We can easily observe the distortion of the original pixels through the comparison of the histogram made by Shi et al.'s and the proposed method. As shown in the histogram, it seems that the marked histograms are very similar to the original histogram. For this reason, the two marked depth maps show a very high image quality about 55 dB. In graylevel 106 and 107, we may recognize that there is a small difference between the the two methods. As a result, it is proved that our proposed method has fewer errors compared to Shi et al.'s method.

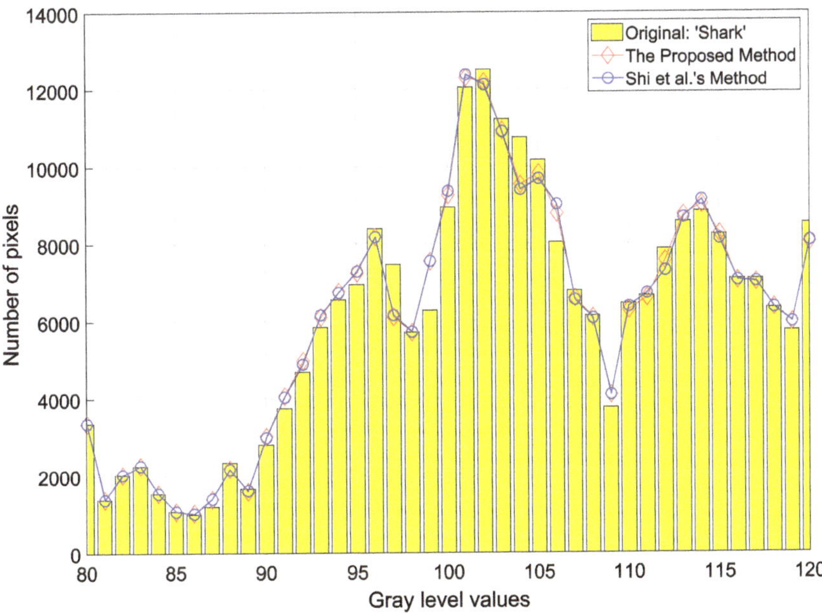

Figure 9. Comparison of histogram coefficients on depth map—"Shark" between Shi et al.'s method and the proposed method.

5. Conclusions

In this paper, we introduced a method to hide metadata in 3D videos using RDH technology, which is one of many watermarking technologies. The proposed RDH is the PEE method, which hides a large amount of data while minimizing the damage of cover image using LSB of depth map. The accuracy of the pixel prediction is very important to the performance of the PEE method. To improve the accuracy, we proposed an efficient MVD-based RDH using inter-component prediction that predicts depth pixels using MVD-related texture pixels. The newly introduced inter-component prediction may improve the performance of the RDH because the prediction precision is higher than the conventional diamond shape prediction. Especially, the prediction of the depth map in the texture image with high frequency characteristics showed excellent performance. Experimental results demonstrated that the proposed method achieves higher embedding capacity than the all the previous methods by improving the prediction accuracy.

Author Contributions: J.Y.L., C.K. conceived and designed the model for research and pre-processed and analyzed the data and the obtained inference. J.Y.L. simulated the design using Visual C ++. C.K. and C.-N.Y. wrote the paper. J.Y.L., C.K., C.-N.Y. checked and edited the manuscript. The final manuscript has been read and approved by all authors.

Funding: This research was supported in part by Ministry of Science and Technology (MOST), under Grant 107-2221-E-259-007. This research was supported by the Basic Science Research Program through the National Research Foundation of Korea (NRF) funded by (2015R1D1A1A01059253), and was supported under the framework of international cooperation program managed by NRF (2016K2A9A2A05005255). This work was supported by the National Research Foundation of Korea(NRF) grant funded by the Korea government(MSIT) (No. 2018R1C1B5086072).

Acknowledgments: The authors are grateful to the editors and the anonymous reviewers for providing us with insightful comments and suggestions throughout the revision process.

Conflicts of Interest: The authors declare no conflict of interest.

References

1. Moulin, P.; Koetter, R. Data-hiding codes. *Proc. IEEE* **2005**, *93*, 2083–2126. [CrossRef]
2. Tian, J. Reversible data embedding using a difference expansion. *IEEE Trans. Circuits Syst. Video Technol.* **2003**, *13*, 890–896. [CrossRef]
3. Alattar, A.M. Reversible watermark using the difference expansion of a generalized integer transform. *IEEE Trans. Image Process.* **2004**, *13*, 1147–1156. [CrossRef] [PubMed]
4. Kamstra, L.; Heijmans, H. Reversible data embedding into images using wavelet techniques and sorting. *IEEE Trans. Image Process.* **2005**, *14*, 2082–2090. [CrossRef] [PubMed]
5. Kim, H.J.; Sachnev, V.; Shi, Y.Q.; Nam, J.; Choo, H.G. A novel difference expansion transform for reversible data embedding. *IEEE Trans. Inf. Forensics Secur.* **2008**, *4*, 456–465.
6. Osamah, M.A.; Bee, E.K. Two-dimensional difference expansion (2D-DE) scheme with a characteristics-based threshold. *Signal Process.* **2013**, *93*, 154–162.
7. Tai, W.L.; Yeh, C.M.; Chang, C.C. Reversible data hiding based on histogram modification of pixel differences. *IEEE Trans. Circuits Syst. Video Technol.* **2009**, *19*, 906–910.
8. Ni, Z.; Shi, Y.Q.; Ansari, N.; Su, W. Reversible data hiding. *IEEE Trans. Circuits Syst. Video Technol.* **2006**, *16*, 354–362.
9. Lee, S.K.; Suh, Y.H.; Ho, Y.S. Reversible image authentication based on watermarking. In Proceedings of the IEEE ICME, Toronto, ON, Canada, 9–12 July 2006; pp. 1321–1324.
10. Hwang, J.; Kim, J.; Choi, J. A reversible watermarking based on histogram shifting. In Proceedings of the 5th IWDW, Jeju Island, Korea, 8–10 November 2006; pp. 348–361.
11. Sachnev, V.; Kim, H.J.; Nam, J.; Suresh, S.; Shi, Y.Q. Reversible watermarking algorithm using sorting and prediction. *IEEE Trans. Circuits Syst. Video Technol.* **2009**, *19*, 989–999. [CrossRef]
12. Hong, W.; Chen, T.S.; Shiu, C.W. Reversible data hiding for high quality images using modification of prediction errors. *J. Syst. Softw.* **2009**, *82*, 1833–1842. [CrossRef]
13. Hong, W. An efficient prediction-and-shifting embedding technique for high quality reversible data hiding. *EURASIP J. Adv. Signal Process.* **2010**, *2010*, 1–12. [CrossRef]
14. Wu, H.-T.; Huang, J. Reversible image watermarking on prediction errors by efficient histogram modification. *Signal Process.* **2012**, *92*, 3000–3009. [CrossRef]
15. Wang, X.; Li, X.; Yang, B.; Guo, Z. Efficient generalized integer transform for reversible watermarking. *IEEE Signal Process. Lett.* **2010**, *17*, 567–570. [CrossRef]
16. Peng, F.; Li, X.; Yang, B. Adaptive reversible data hiding scheme based on integer transform. *Signal Process.* **2012**, *92*, 54–62. [CrossRef]
17. Gui, X.; Li, X.; Yang, B. A novel integer transform for efficient reversible watermarking. In Proceedings of the ICPR, Tsukuba, Japan, 11–15 November 2012; pp. 947–950.
18. Sullivan, G.J.; Boyce, J.M.; Chen, Y.; Ohm, J.-R.; Segall, C.A.; Vetro, A. Standardized Extensions of High Efficiency Video Coding (HEVC). *IEEE J. Sel. Top. Signal Process.* **2013**, *7*, 1001–1016. [CrossRef]
19. Lee, J.Y.; Lin, J.-L.; Chen, Y.-W.; Chang, Y.-L.; Kovliga, I.; Fartukov, A.; Mishurosvskiy, M.; Wey, H.-C.; Huang, Y.-W.; Lei, S. Depth-Based Texture Coding in AVC-Compatible 3D Video Coding. *IEEE Trans. Circuit Syst. Video Technol.* **2015**, *25*, 1347–1360. [CrossRef]
20. Fehn, C. Depth-Image-Based Rendering (DIBR), Compression and Transmission for a New Approach on 3D-TV. *Proc. SPIE* **2004**, *5291*, 93–104.
21. Asikuzzaman, M.; Pickering, M.R. An Overview of Digital Video Watermarking. *IEEE Trans. Circuit Syst. Video Technol.* **2018**, *28*, 2131–2153. [CrossRef]

22. Pei, S.C.; Wang, Y.Y. Auxiliary metadata delivery in view synthesis using depth no synthesis error model. *IEEE Trans. Multimedia* **2015**, *17*, 128–133. [CrossRef]
23. Wang, S.; Cui, C.; Niu, X. Watermarking for DIBR 3D images based on SIFT feature points. *Measurement* **2014**, *48*, 54–62. [CrossRef]
24. Zhu, N.; Ding, G.; Wang, J. A Novel Digital Watermarking Method for New Viewpoint Video Based on Depth Map. In Proceedings of the International Conference on Intelligent Systems Design and Applications, Kaohsiung, Taiwan, 26–28 November 2008; pp. 1–5.
25. Asikuzzaman, M.; Alam, M.J.; Lambert, A.J.; Pickering, M.R. A Blind and Robust Video Watermarking Scheme Using Chrominance Embedding. In Proceedings of the 2014 International Conference on Digital Image Computing: Techniques and Applications (DICTA), Wollongong, NSW, Australia, 25–27 November 2008; pp. 1–6.
26. Chung, K.-L.; Yang, W.-J.; Yang, W.-N. Reversible data hiding fordepth maps using the depth no-synthesis-error model. *Inf. Sci.* **2014**, *269*, 159–175. [CrossRef]
27. Shi, X.; Ou, B.; Qin, Z. Tailoring reversible data hiding for 3D synthetic images. *Signal Process. Image Commun.* **2014**, *64*, 46–58. [CrossRef]
28. Zhao, Y.; Zhu, C.; Chen, Z.; Yu, L. Depth no-synthesis-error model for view synthesis in 3-D video. *IEEE Trans. Image Process.* **2011**, *20*, 2221–2228. [CrossRef] [PubMed]
29. Rusanovskyy, D.; Müller, K.; Vetro, A. Common Test Conditions of 3DV Core Experiments. Doc. JCT3V-E1100. 2013. Available online: https://phenix.int-evry.fr/jct2/doc_end_user/current_document.php?id=1344 (accessed on 25 February 2019).

MDPI
St. Alban-Anlage 66
4052 Basel
Switzerland
Tel. +41 61 683 77 34
Fax +41 61 302 89 18
www.mdpi.com

Electronics Editorial Office
E-mail: electronics@mdpi.com
www.mdpi.com/journal/electronics